Data Sketches

"Shirley and Nadieh are honest, entertaining, and insightful in their retrospectives, and their openness and humility make the practice less intimidating, inviting newcomers to get started."

– Mike Bostock, Creator of D3.js and Founder of Observable

"Written in an approachable, first-person angle, this book is a delightful behind-the-scenes look at the process for creating any sophisticated data visualization. They make it seem so easy you will want to start on your own project right away."

– Manuel Lima, Author, RSA Fellow and Founder of VisualComplexity.com

"The work of Nadieh Bremer and Shirley Wu is some of the most beautiful and exciting data visualization work being done today."

– Elijah Meeks, Author and Executive Director, Data Visualization Society

"The Data Sketches collaboration is a glorious tour de force: two people spur each other along a remarkable spiral of visualization creativity, and let the rest of us come along for the ride!"

– Tamara Munzner, Author and Professor, University of British Columbia

"Lay-chart readers cannot appreciate the expertise and thousands of decisions that go into a single visualization, but this wonderful behind-the-scenes peek reveals how there is never just one 'right' answer, but many possible answers — each of them beautiful, provocative, and shaped by the unique lens of its creator."

– Scott Murray, Author and Designer, O'Reilly media

AK Peters Visualization Series

This series aims to capture new developments and summarize what is known over the whole spectrum of visualization by publishing a broad range of textbooks, reference works, and handbooks. It will contain books from all subfields of visualization, including visual analytics, information visualization, and scientific visualization. The scope will largely follow the calls of the major conferences such as VIS: techniques, algorithms, theoretical foundations and models, quantitative and qualitative evaluation, design studies, and applications.

Series Editor: Tamara Munzner
University of British Columbia, Vancouver, Canada

Visualization Analysis and Design
Tamara Munzner

Information Theory Tools for Visualization
Min Chen, Miquel Feixas, Ivan Viola, Anton Bardera, Han-Wei Shen, Mateu Sbert

Data-Driven Storytelling
Nathalie Henry Riche, Christophe Hurter, Nicholas Diakopoulos, Sheelagh Carpendale

Interactive Visual Data Analysis
Christian Tominski, Heidrun Schumann

Visualizing with Text
Richard Brath

Data Sketches
Nadieh Bremer, Shirley Wu

For more information about this series please visit:
https://www.crcpress.com/AK-Peters-Visualization-Series/book-series/CRCVIS

CRC Press is an imprint of the
Taylor & Francis Group, an **informa** business

Data Sketches

NADIEH BREMER
SHIRLEY WU

First edition published 2021
by CRC Press
6000 Broken Sound Parkway NW, Suite 300, Boca Raton, FL 33487-2742

and by CRC Press
2 Park Square, Milton Park, Abingdon, Oxon, OX14 4RN

Library of Congress Cataloging-in-Publication Data
Names: Bremer, Nadieh, author. | Wu, Shirley, author.
Title: Data sketches / Nadieh Bremer, Shirley Wu.
Description: First edition. | Boca Raton : CRC Press, 2020. | Series: AK Peters visualization series | Includes bibliographical references and index.
Identifiers: LCCN 2020014301 | ISBN 9780367000127 (hbk) | ISBN 9780367000080 (pbk) | ISBN 9780429445019 (ebk)
Subjects: LCSH: Information visualization. | Charts, diagrams, etc.
Classification: LCC QA76.9.I52 B74 2020 | DDC 001.4/226--dc23
LC record available at https://lccn.loc.gov/2020014301

ISBN: 9780367000127 (hbk)
ISBN: 9780367000080 (pbk)
ISBN: 9780429445019 (ebk)

Cover by Alice Lee

Design by Praline

Edited by Stephanie Morillo and Tianna Mañón

To Ralph,
my first and forever love
Nadieh

To 妈 and 爸,
who gave me the freedom to dream.

And to Alex,
who gave me the courage to fearlessly
pursue those dreams.
Shirley

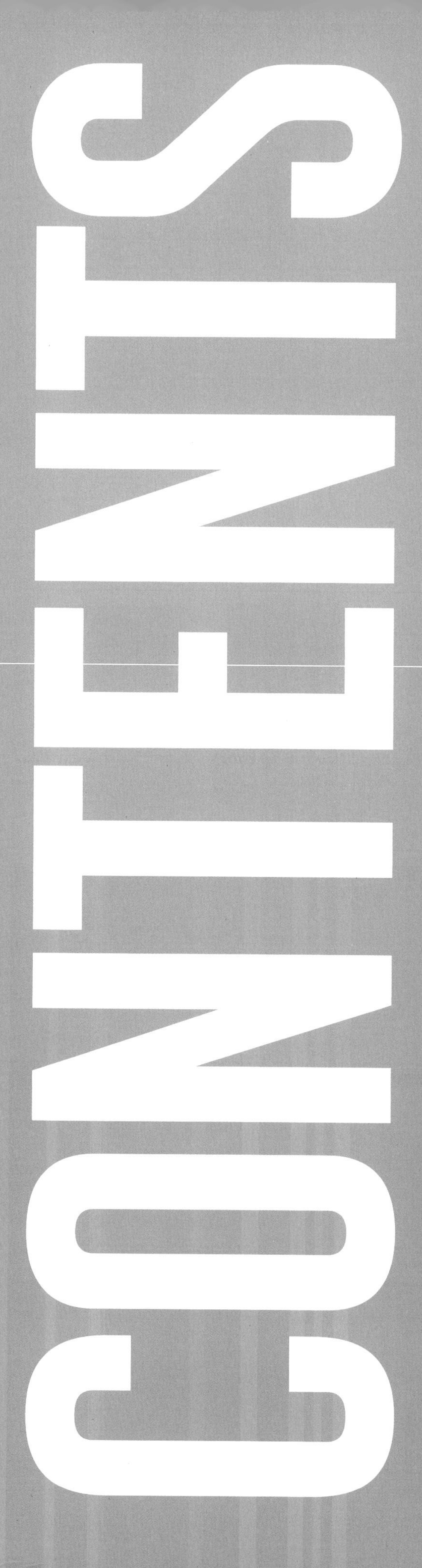

CONTENTS

Movies

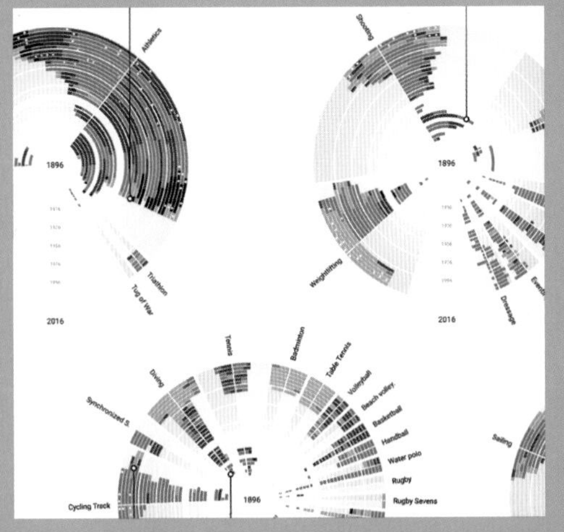

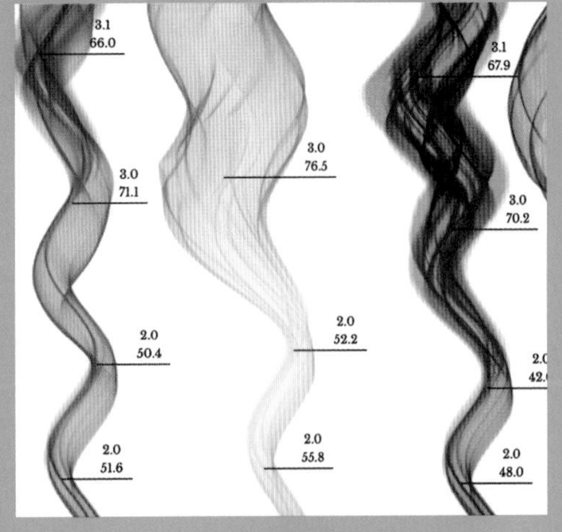

Olympics

Travel

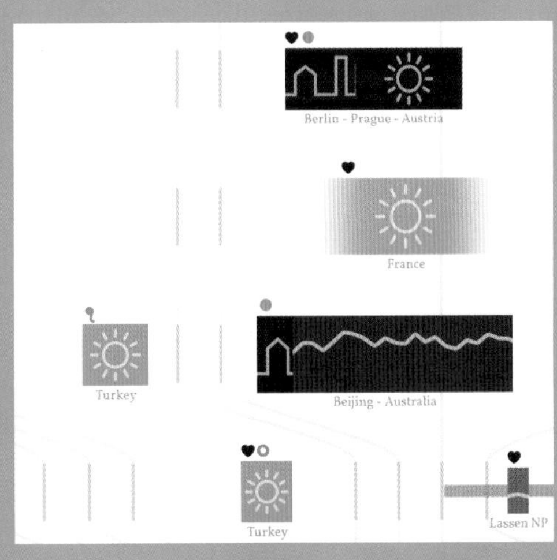

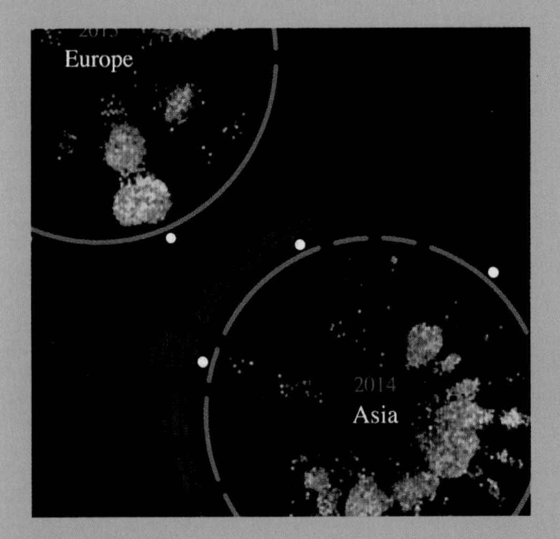

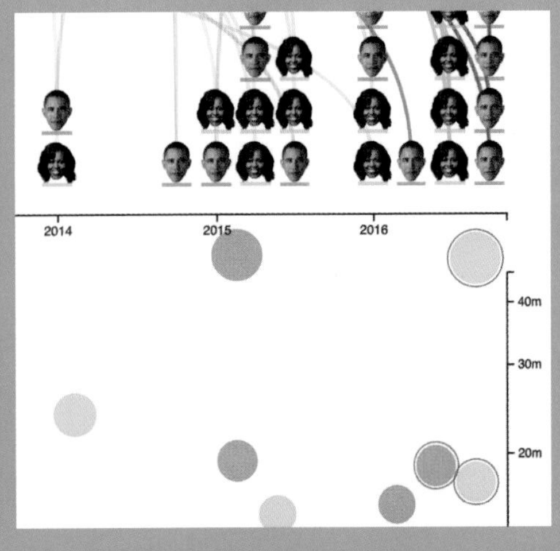

Presidents & Royals

Books

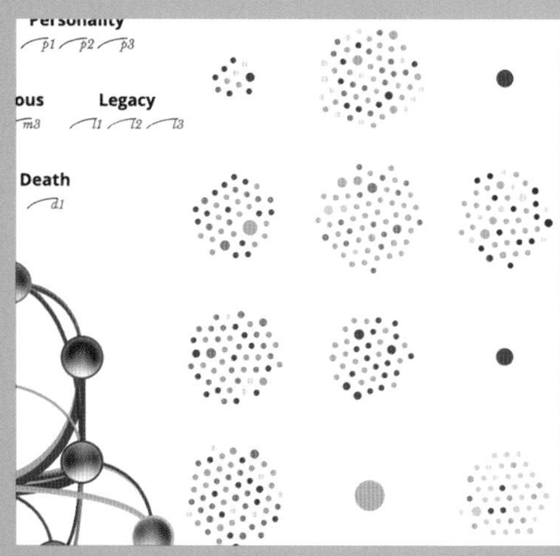

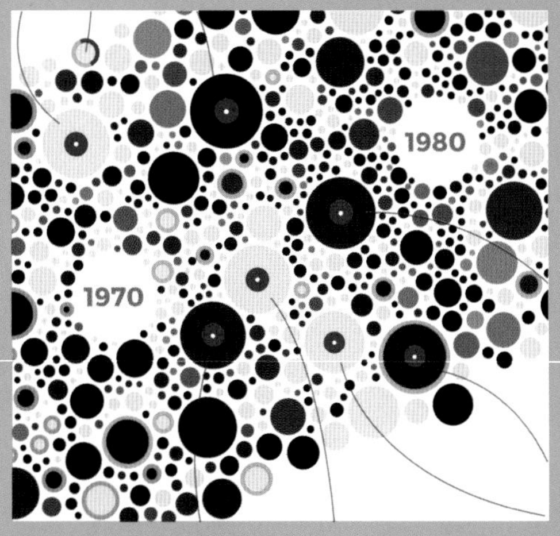

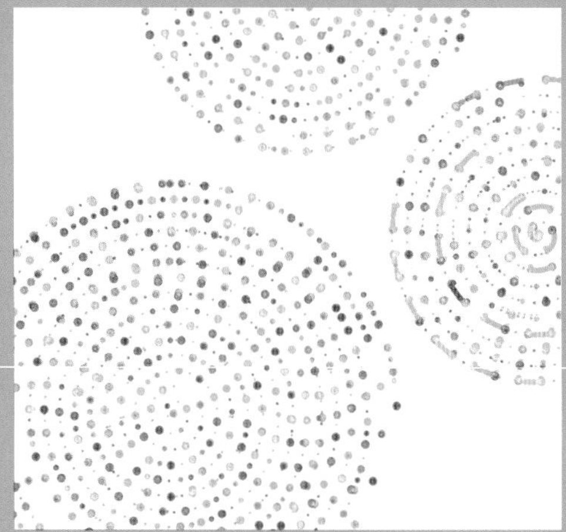

Music

Nostalgia

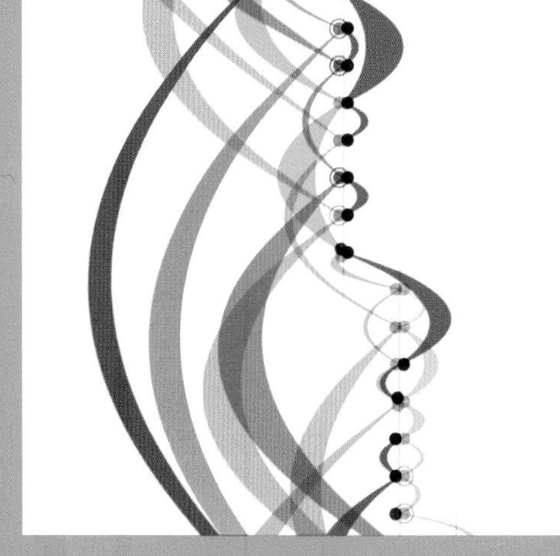

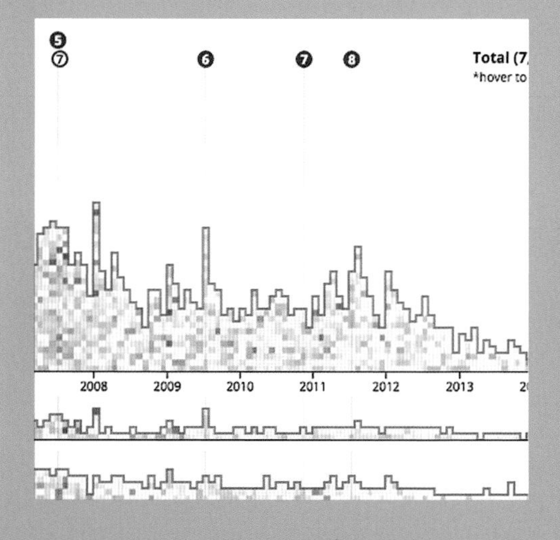

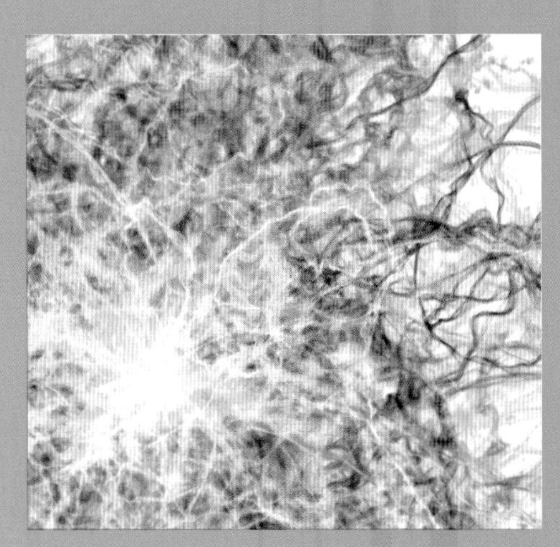

Nature

Culture

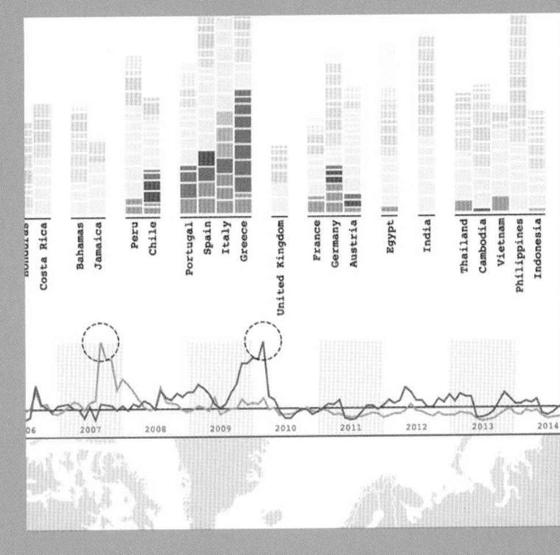

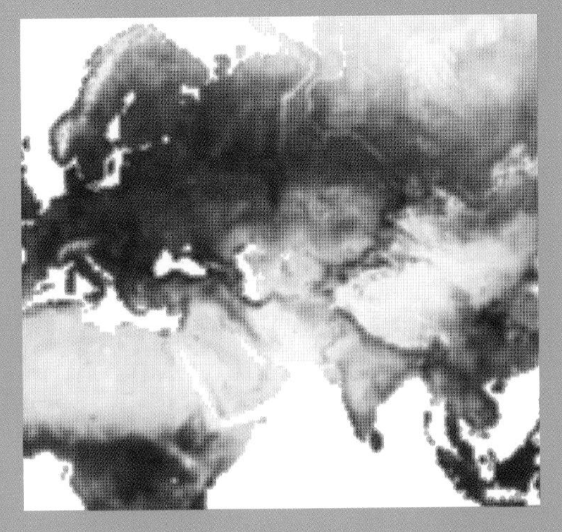

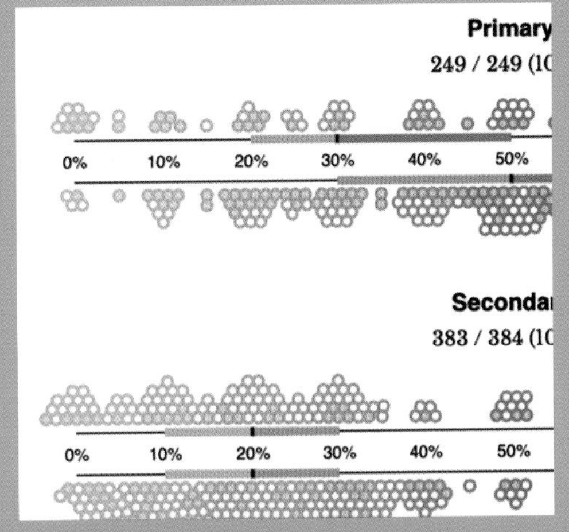

Community

Myths & Legends

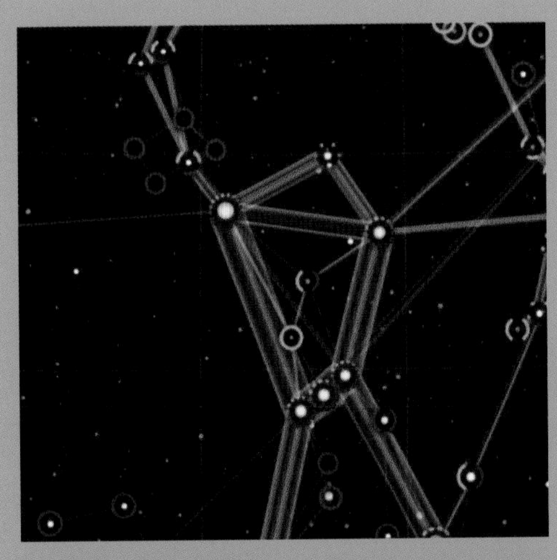

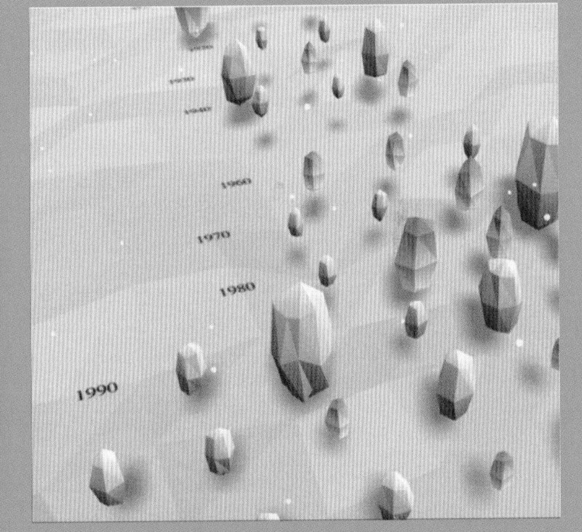

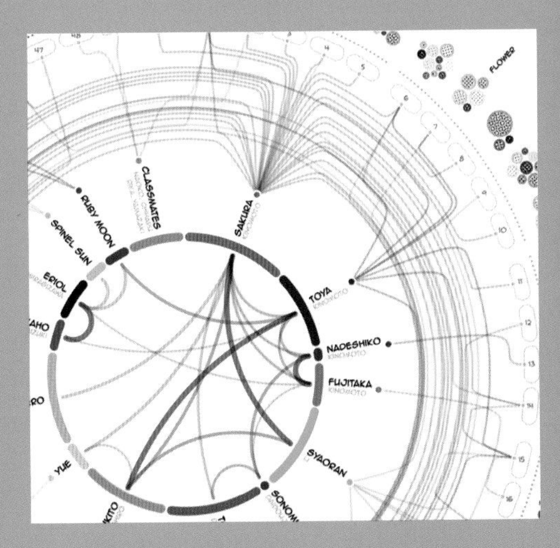

Fearless

ORTHODOXY & ECCENTRICITY

Alberto Cairo

Knight Chair at the University of Miami
and author of *How Charts Lie*

Orthodoxy and eccentricity are opposing but complementary forces in any field, and data visualization isn't an exception. Periods when the former prevails over the latter discourage whim, passion, and experimentation, and favor stability and continuity. When the opposite happens—when eccentrics take over—chaos and turmoil ensue, but progress becomes more rapid and invention more likely. This is a book where eccentricity abounds. That's a good thing.

The formalization and systematization of data visualization took decades and books by authors such as Jaques Bertin, John Tukey, William Cleveland, Naomi Robbins, Stephen Kosslyn, Leland Wilkinson, Tamara Munzner, and many others. It is thanks to them that we possess a common language to discuss what constitutes a well-designed chart or graph and principles that aid us when creating them. They deserve our gratitude.

What most of those authors have in common, though, is a background in statistics or the sciences, and I suspect that this has had an effect on the visual style favored for many years. Since the 1970s at least, data visualization has been governed by a vague consensus—an orthodoxy—that favors bare clarity over playfulness, simplicity over allegedly gratuitous adornments, supposed objectivity over individual expression.

As a consequence, generations of visualization designers grew up in an era of stern and pious sobriety that sadly degenerated sometimes into dismissive self-righteousness exemplified by popular slurs such as 'chart junk.'

It's time maybe not to abandon that orthodoxy outright—when the goal of a visualization is to conduct exploratory analysis, reveal insights, or inform decisions, prioritizing clarity and sticking to standard graphic forms, conventions, and practices is still good advice—but to acknowledge that other orthodoxies are possible and necessary. Visualization can be designed and experienced in various ways, by people of various backgrounds, and in various circumstances. That's why reflecting on the purpose of a visualization is paramount before we design it—or before we critique it.

Nadieh Bremer and Shirley Wu are wondrous eccentrics. Their splendid book is the product of a collaborative experimental project, *Data Sketches*, that might be one of the first exponents of an emerging visualization orthodoxy in which uniqueness is paramount and templates and conventions are seen with skepticism.

I discovered *Data Sketches* right after it was launched, back in 2016, and I was immediately enthralled, even if I couldn't understand many of its graphics. They are insanely complex and ornate, I thought, colorful, mysteriously organic in some cases, a departure from the strictures of classic graphs, charts, and maps. I felt that Nadieh and Shirley were not only pushing what was possible through technologies such as `D3.js`, but also wished to defy what was acceptable.

The book that you have in your hands reveals how Nadieh and Shirley think. This is useful. Visualization is, like written language, based on a body of symbols and a syntax that aids us in arranging those symbols to convey information. However, this system of symbols and syntax isn't rigid—again like written language—but flexible and in constant flux. That's why I've come to believe that visualization can't be taught as a set of rigid rules, but as a principled process of reasoning about how to make good decisions when it comes to what to show and how to show it.

This process ought to be informed by what we know about vision and cognitive science, rhetoric, graphic and interaction design, UX, visual arts, and many other fields. However, this knowledge shouldn't be a straitjacket. Rather, it's a foundation that opens up multiple possibilities, some more appropriate, some less, always depending on the purpose of each visualization and on its intended audience.

The education of visualization designers, whether it's formal or not, can't be based on memorizing rules, but on learning how to justify our own decisions based on ethics, aesthetics, and the incomplete but ever-expanding body of empirical evidence coming from academia. There are plenty of lengthy and detailed discussions in *Data Sketches* about how to balance out these considerations, and it's always useful to peek into the minds of great designers, if only to borrow ideas from them. Some of you will be persuaded by those discussions, and others will disagree and argue against them. That's fine. Conversation is what may help us determine whether certain novelties fail and should be discarded, or succeed and become convention. Today's eccentricity is tomorrow's orthodoxy.

Now go ahead: read, think, and discuss. And consider becoming a bit more of an eccentric.

OUR JOURNEYS INTO DATA VISUALIZATION

In the last decade, the amount of data collected has exploded, and many fields—including data visualization—have gained momentum to make sense of all this data. These data visualizations can take a wide variety of forms, such as dashboards, infographic posters, data art, and data-driven journalism. But what we enjoy the most is wielding data as a tool to explore our curiosities about the world around us. We love collecting a fun dataset, finding the insights and stories buried within it, and sharing that story in a beautiful, visual way that excites people. Our most successful projects have turned spreadsheets full of numbers into visualizations that entice people to dive in, explore, and learn all that it has to reveal.

To do this, we honed a wide variety of skills: data analysis, information design, coding, and storytelling. And because of this wide spectrum of skills (and the relative youth of the field as a whole), most data visualizers start from a variety of different backgrounds. We are no exceptions.

Nadieh's Start

In 2011, I graduated as an astronomer. I loved the topic and its gorgeous imagery, but knew that writing academic papers wasn't for me. Instead, I joined the new Analytics department of Deloitte Consulting as a data scientist. Apart from data analyses, I was creating tons of simple charts in PowerPoint, QlikView, and (mostly) R. In 2013, at a data science conference, I randomly joined an "Introduction to D3.js" workshop and my mind was blown by the possibilities! I didn't care that I had to learn JavaScript, CSS, and HTML; I was going to add D3.js to my repertoire! I still saw myself as a data scientist, though.

At another conference at the end of 2014, while waiting for Mike Freeman's talk, my eyes fell on his first slide. He called himself a Data Visualization Specialist. "Wait? What? That's a separate thing!?" It was like I was struck by lightning. I *immediately* knew that was where my true passion was, not the data analysis part. And from then on, I spent every moment of free time I had to become better in the visualization of data.

In December 2015, I joined Adyen as a full-time data visualization designer. However, after designing and creating dashboards for months, even in D3.js, I felt that I was still missing something; my creative side wasn't feeling fulfilled.

Shirley's Start

I loved math and art growing up. When I went to college, I studied business, but found much more enjoyment in the computer science classes I took. When I graduated in 2012, I started as a software engineer on a front-end team at Splunk, a big data company. There, I was introduced to D3.js, a JavaScript library for creating data visualizations on the web, and I immediately fell in love with being able to draw in the browser.

In 2013, I started frequenting the Bay Area D3 User Group, where I learned that D3.js was only a subset of a larger field called data visualization that people could specialize in. I loved that it was such a beautiful blend of art, math, and code—all of my favorite things.

For my second job, I joined an enterprise security start-up, specifically because they were looking for someone to specialize in creating data visualizations for their product. I grew a lot technically and loved all the data visualization aspects of my work, but not the industry I was in.

In 2016, I decided to take the leap into freelancing. I wanted to see the data-related challenges that companies faced and the problems they were trying to solve with data visualization; I wanted to find the industries I'd be the most excited to work in.

And this is where our stories converge, in early 2016...

Data Sketches, or "Shirley & Nadieh's Awesome Collaboration Marathon"

We met in April 2016 at OpenVis Conf in Boston and kept in touch when we got home (Shirley in San Francisco and Nadieh in Amsterdam). One June day, we were lamenting the fact that we'd had little free time to focus on personal data visualization projects and, as a result, hadn't completed many in the past year. But on that fateful June day, Shirley had just quit her full-time job and had plenty of free time to experiment and create. So she plucked up her courage and asked Nadieh a simple question: "Do you want to collaborate with me?" Nadieh excitedly agreed!

OpenVis Conf was an amazing annual data visualization conference in Boston that had a great blend of technical and design talks.

It's wild how dramatically this simple question and decision altered both of our careers.

In a series of rapid back-and-forth messages, we defined the structure of the project: 12 months, 12 topics, 12 projects each. We would gather our own data, create our own design, and code our visualizations from scratch, while also documenting the whole process. We encouraged each other to use these opportunities to try new approaches, explore new tools, and push the boundaries of what we could create. But most of all, we promised each other that we would have lots of fun.

We decided to call our project "Data Sketches," and went live on September 21st, 2016 with our first four visualizations on datasketch.es. We didn't think anyone beyond our friend group would care, and we certainly didn't expect the overwhelming response we got. But as it turned out, people really liked getting a behind-the-scenes look at our process, and we kept hearing how helpful and educational our write-ups were.

Our very first brainstorming document was titled "Shirley & Nadieh's Awesome Collaboration Marathon" before we eventually landed on "Data Sketches." Other names that we seriously considered included "Pencils&Code" and "Visual Wanderlust."

We first mused about a "Data Sketches" book not long after our launch, when we heard Alberto Cairo mention our project on a livestream. We were giddy with delight when we heard him say that if we ever created a book, he'd display it on his coffee table. We were enamored with the idea, but it felt like a far-away dream.

Nonetheless, we tried to talk to a few publishers. None of them seemed keen on the book we wanted to make: a beautiful coffee table book with large, indulgent images of our projects, side-by-side with our very technical process write-ups. We wanted it to be both aesthetically pleasing and educational, yet the publishers we talked to all seemed only to want one and not the other.

We were close to giving up when one summer day in 2017, Shirley reached out to Tamara Munzner for dinner during her visit to Vancouver. At the end of a long and delicious meal, Tamara asked if we had ever considered turning "Data Sketches" into a book. She wanted us to be part of her series of data visualization books, and we were more than thrilled! We had finally found an editor (and enthusiastic champion of our work) and a publisher willing to work with us to create our dream book.

Tamara gave us a renewed purpose in creating a "Data Sketches" book. She convinced us that because everything on the Internet eventually bit-rots, we needed to do it for archival purposes. We also knew that because we wanted to keep many of our online write-ups freely available, we therefore needed to make this book worth splurging on. We've tightened up our tangents, filled in gaps in explanation, and packed the book full of lessons we've learned along the way. This book is just as much about the 24 individual data visualizations we created as it is a celebration of the technical and personal growth we've gone through in this three-year journey. It is a snapshot in time to commemorate the immense impact "Data Sketches" has had on our lives, allowing us to quit our full-time jobs, launch thriving freelance careers, travel the world to talk about our work, and develop beautiful friendships.

We've added the time frame of each data visualization at the beginning of each chapter to give context.

Writing a book is hard. We knew this going in, and our original one-year project has since turned into three years. It's been a monumental three (four by the time of publishing) years, and we're so excited to hold *Data Sketches* in our hands, to flip through the pages, and see our work immortalized on paper. We want to thank you for all of your excitement and support, whether you've been with us from the very beginning, or have just picked up this book. Thank you for helping us get here.

We hope you enjoy this dream book of ours and that you have just as much fun reading it as we did working on our *Data Sketches*.

Shirley and Nadieh, on the left and right respectively, together in real life for a week in San Francisco during September 2017.

Nadieh Bremer is a graduated astronomer, turned data scientist, turned freelancing data visualization designer. She's worked for companies such as Google, UNESCO, Scientific American, and the New York Times. As 2017's "Best Individual" in the "Information Is Beautiful" Awards, she focuses on uniquely-crafted data visualizations that are both effective and visually appealing for print and online.

VisualCinnamon.com

Shirley Wu is an award-winning creative focused on data-driven art and visualizations. She has worked with clients such as Google, The Guardian, SFMOMA, and NBC Universal to develop custom, highly interactive data visualizations. She combines her love for art, math, and code into colorful, compelling narratives that push the boundaries of the web.

shirleywu.studio

HOW TO READ

Each chapter represents one of the 12 topics that we've visualized with *Data Sketches*, and each chapter consists of both of our individual write-ups.

Period during which we worked on this project

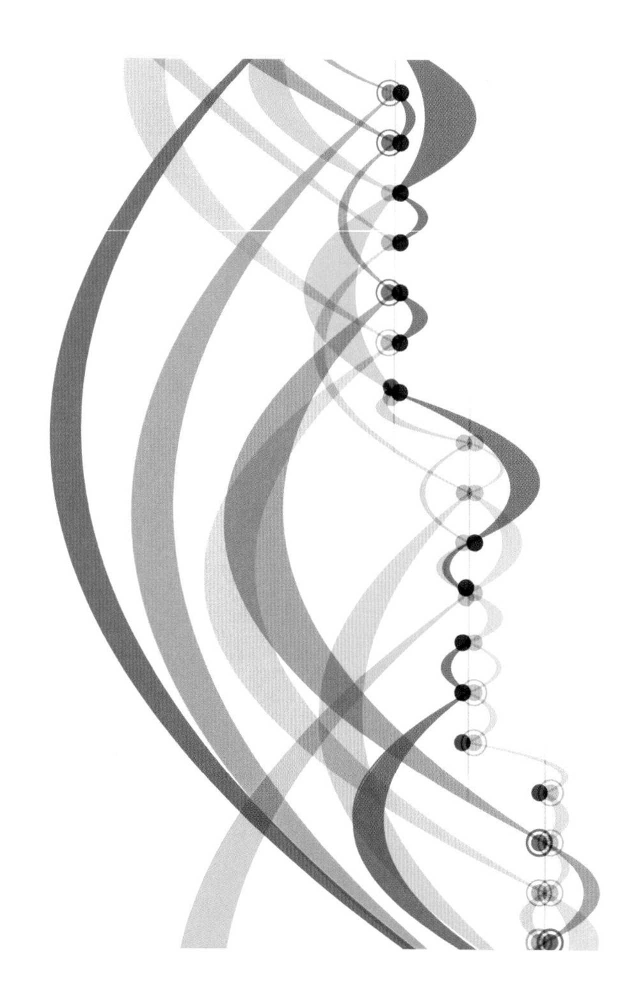

All the Fights in Dragon Ball Z

NADIEH

I had several choices for the topic of this project's "Nostalgia" theme—including video games, manga, and anime—and I didn't know which one would provide the most interesting dataset. I therefore scoured the web for all the topics I had in mind. I loved playing *The Legend of Zelda* on the Game Boy, especially "Minish Cap" and the "Oracle of Ages/Seasons" combo. There was something about the 2D bird's eye view of Game Boy's *Zelda* games that worked exceptionally well for me, whereas in 3D games I just kept falling off bridges and running into corners... Unfortunately I couldn't find any interesting data. I looked at the original *Cardcaptor Sakura* (CCS) manga but also didn't find enough data that I could use to create something elaborate. Neopets is where I first encountered HTML & CSS at 14, even though I didn't continue using any web languages until I saw the magic of `D3.js` some 12 years later. However, I found out my account was hacked, my pet stolen, and I wasn't in the mood to investigate that further at the time.

Eventually I turned my focus to the last subject I had in mind: *Dragon Ball Z* (DBZ). This anime was my very first introduction to the Japanese animation scene, airing on Cartoon Network in the Netherlands when I was about 13 or 14. Although I quickly turned solely to manga in general, I stayed loyal to the DBZ anime and loved watching it all the way to the Fusion Saga. (I eventually stopped watching DBZ because I was too annoyed by the characters Majin Buu and Gotenks ㅌ_ㅌ)

JANUARY 2017

219

NOSTALGIA

Each write-up has an introduction, **data, sketch, code,** and **reflections** (textual) section and ends with several pages showing the final visual. The data, sketch, and code titles refer to what we were doing at that point in the project, not literally what is shared in that section (i.e., we've avoided sharing actual code snippets in the code section).

The write-ups can be distinguished between Shirley and Nadieh through the base color. (Shirley's color is pink and Nadieh's color is green.)

Topic

Technology, referring to any kind of package or library, is displayed in this monotype font

Technical lessons have a blue background

Code is displayed with a light pink or green highlight

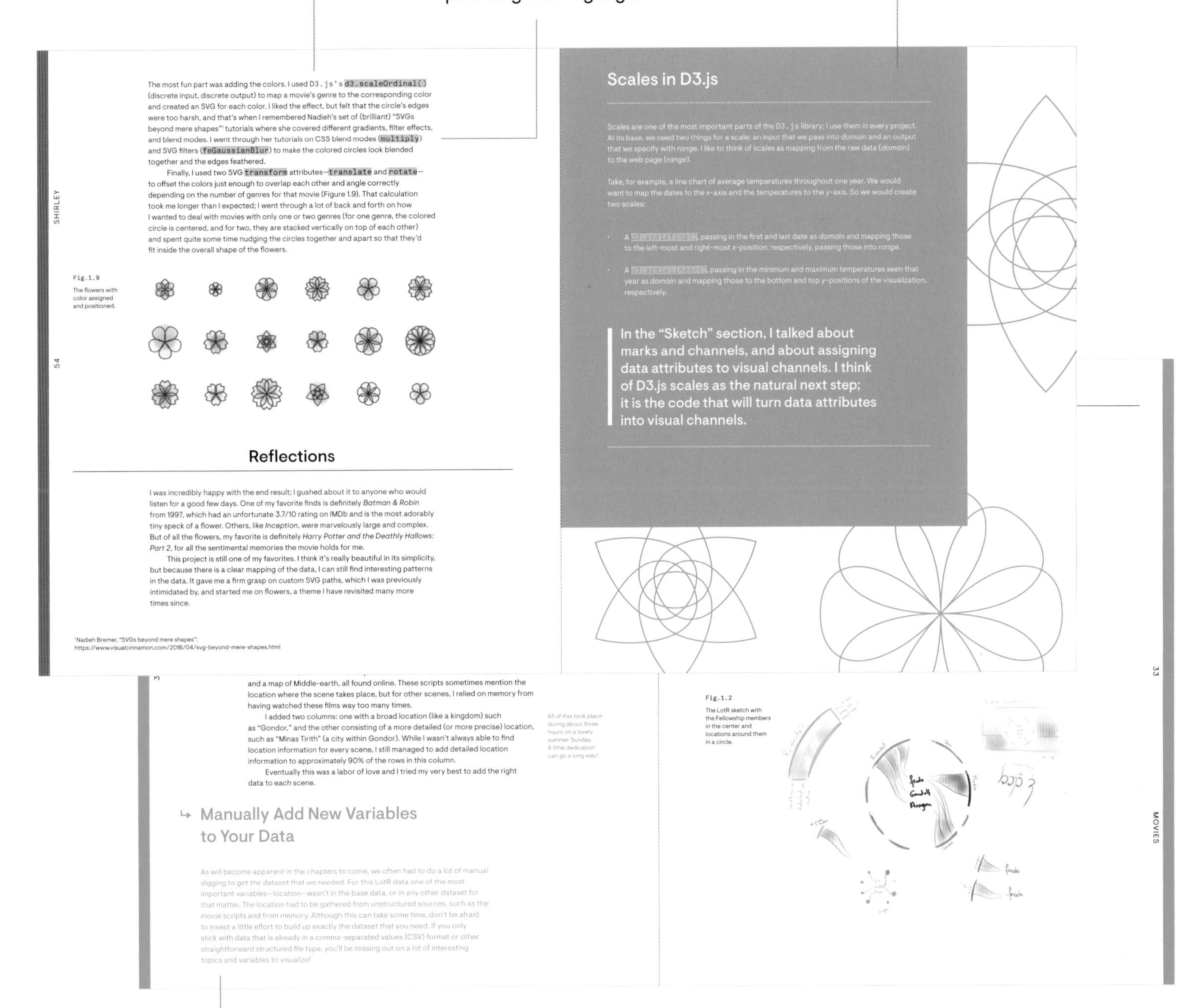

SHIRLEY

54

The most fun part was adding the colors. I used `D3.js`'s `d3.scaleOrdinal()` (discrete input, discrete output) to map a movie's genre to the corresponding color and created an SVG for each color. I liked the effect, but felt that the circle's edges were too harsh, and that's when I remembered Nadieh's set of (brilliant) "SVGs beyond mere shapes"[1] tutorials where she covered different gradients, filter effects, and blend modes. I went through her tutorials on CSS blend modes (`multiply`) and SVG filters (`feGaussianBlur`) to make the colored circles look blended together and the edges feathered.

Finally, I used two SVG `transform` attributes—`translate` and `rotate`—to offset the colors just enough to overlap each other and angle correctly depending on the number of genres for that movie (Figure 1.9). That calculation took me longer than I expected; I went through a lot of back and forth on how I wanted to deal with movies with only one or two genres (for one genre, the colored circle is centered, and for two, they are stacked vertically on top of each other) and spent quite some time nudging the circles together and apart so that they'd fit inside the overall shape of the flowers.

Fig.1.9

The flowers with color assigned and positioned.

Reflections

I was incredibly happy with the end result; I gushed about it to anyone who would listen for a good few days. One of my favorite finds is definitely *Batman & Robin* from 1997, which had an unfortunate 3.7/10 rating on IMDb and is the most adorably tiny speck of a flower. Others, like *Inception*, were marvelously large and complex. But of all the flowers, my favorite is definitely *Harry Potter and the Deathly Hallows: Part 2*, for all the sentimental memories the movie holds for me.

This project is still one of my favorites. I think it's really beautiful in its simplicity, but because there is a clear mapping of the data, I can still find interesting patterns in the data. It gave me a firm grasp on custom SVG paths, which I was previously intimidated by, and started me on flowers, a theme I have revisited many more times since.

[1]Nadieh Bremer, "SVGs beyond mere shapes": https://www.visualcinnamon.com/2016/04/svg-beyond-mere-shapes.html

Scales in D3.js

Scales are one of the most important parts of the `D3.js` library; I use them in every project. At its base, we need two things for a scale: an input that we pass into *domain* and an output that we specify with *range*. I like to think of scales as mapping from the raw data (*domain*) to the web page (*range*).

Take, for example, a line chart of average temperatures throughout one year. We would want to map the dates to the *x*-axis and the temperatures to the *y*-axis. So we would create two scales:

- A `d3.scaleTime()`, passing in the first and last date as *domain* and mapping those to the left-most and right-most *x*-position, respectively, passing those into *range*.
- A `d3.scaleLinear()`, passing in the minimum and maximum temperatures seen that year as *domain* and mapping those to the bottom and top *y*-positions of the visualization, respectively.

> In the "Sketch" section, I talked about marks and channels, and about assigning data attributes to visual channels. I think of D3.js scales as the natural next step; it is the code that will turn data attributes into visual channels.

and a map of Middle-earth, all found online. These scripts sometimes mention the location where the scene takes place, but for other scenes, I relied on memory from having watched these films way too many times.

I added two columns: one with a broad location (like a kingdom) such as "Gondor," and the other consisting of a more detailed (or more precise) location, such as "Minas Tirith" (a city within Gondor). While I wasn't always able to find location information for every scene, I still managed to add detailed location information to approximately 90% of the rows in this column.

All of this took place during about three hours on a lovely summer Sunday. A little dedication can go a long way!

Eventually this was a labor of love and I tried my very best to add the right data to each scene.

↳ Manually Add New Variables to Your Data

As will become apparent in the chapters to come, we often had to do a lot of manual digging to get the dataset that we needed. For this LotR data one of the most important variables—location—wasn't in the base data, or in any other dataset for that matter. The location had to be gathered from unstructured sources, such as the movie scripts and from memory. Although this can take some time, don't be afraid to invest a little effort to build up exactly the dataset that you need. If you only stick with data that is already in a comma-separated values (CSV) format or other straightforward structured file type, you'll be missing out on a lot of interesting topics and variables to visualize!

Fig.1.2

The LotR sketch with the Fellowship members in the center and locations around them in a circle.

33

MOVIES

General lessons are written in blue. Multiple chapters may refer to the same lesson, so all lessons are collected at the end of the book for quick reference.

THIS BOOK

TECH & TOOLS

To create our data visualizations, we relied on a diverse set of technologies and tools. Some were used by both of us for each project, others only for one specific visual. Here, you can read brief explanations about the technologies and tools that played an important role in our creation process.

Data

CSV & JSON

Comma-Separated Values (CSV) and JavaScript Object Notation (JSON) are two of the most common data formats on the web, and are frequently used in creating data visualizations. A CSV is a text file with one data point per row, uses a comma to separate each data point's values, and can be easily edited with Excel. JSON is also a text file but encodes data with attribute-value pairs instead and is especially useful for storing nested data and manipulating it with JavaScript.

Many European countries use semicolons instead of commas

We both prefer working with CSVs because it is easier to edit and results in smaller file sizes than JSON. We usually work with JSON only when we have a nested data structure, or the existing dataset is already in JSON.

Nadieh: All chapters
Shirley: All chapters

Microsoft Excel

Excel is a software program that allows users to organize, format, and perform calculations on data with formulas using a spreadsheet system.

Nadieh often uses Excel when a dataset is small, around 100 to 1,000 rows, and requires cleaning or editing that only needs to happen once and are faster to do manually instead of programmatically. She'll then load the cleaned dataset into R for further analysis. Shirley used to do her data gathering and cleaning in a text editor (manually adding commas and curly brackets) until she read about Nadieh using Excel. She now uses Excel often when doing manual data gathering and cleaning.

Nadieh: Chapters 1, 2, 3, 5, 7, 9, 11 & 12
Shirley: Chapters 11 & 12

R/RStudio

R is a programming language used for data wrangling, cleaning, preparation, and analysis. RStudio is the go-to GUI for working with R. R is only used by Nadieh, who learned this language while still a data scientist. This is one of the tools that she uses for each and every project in order to gather, prepare, and reshape the data. From there, she often creates some very simple plots to get a better grip on the values that are present within the data. Shirley has never worked in R.

Nadieh: All chapters

Node.js

Node.js is a JavaScript runtime environment that works outside of the browser and is used primarily for writing server-side code.

Shirley uses Node.js modules and various packages to scrape websites, make Twitter, Google, and Facebook APIs easier to interact with, and perform data processing. She tries to do as much data preparation as she can server-side with Node.js to ease the amount of calculations that need to happen client-side when a viewer visits. Nadieh didn't use Node.js during *Data Sketches* and only uses it in her professional work on larger client projects.

Shirley: Chapters 1, 3, 4, 7, 8, 11 & 12

Sketch

Pen & Paper

Just about as simple as it gets. There's little that beats getting your ideas out in the open than plain pen and paper. We often sketch out our initial thoughts on paper (or the "digital" paper of our tablets). Nadieh basically always has a small pocket notebook and pen within her reach, even when outside, so she can start a brainstorming session wherever she might be. Shirley used a sketchbook until she got an iPad Pro mid-way through *Data Sketches*.

Living in San Francisco, Shirley has had her backpack and sketchbooks stolen twice. It was so heartbreaking that she has moved everything to digital with cloud back-ups.

Nadieh: Chapters 2, 6, 7, 8, 9, 10, 11 & 12
Shirley: Chapters 1, 2, 3, 4, 5, 7, 9 & 10

iPad & Apple Pencil

A stand-in for the conventional "paper and pen," the introduction of high precision tablets, drawing tools, and apps makes it easy to take a "sketch" to the next level. We can undo an action, easily switch between colors and pen types and work with multiple layers, while still having the ease of drawing with our hands (as opposed to code).

Shirley likes the simplicity of the note taking app "Paper" by FiftyThree (now WeTransfer). She finds that its limited-but-curated set of drawing tools helps her focus on just getting the idea across, instead of obsessing and fine tuning every detail. Nadieh on the other hand always works with "Tayasui Sketches," which she finds to be the perfect fit with just the right amount of options and possibilities to be able to sketch the concept, while not getting lost in all the possible settings.

We both like different tools for the same reasons, which goes to show that what's important is to find a tool that works for you.

Nadieh: Chapters 1, 2, 3, 5, 9 & 11
Shirley: Chapters 8, 11 & 12

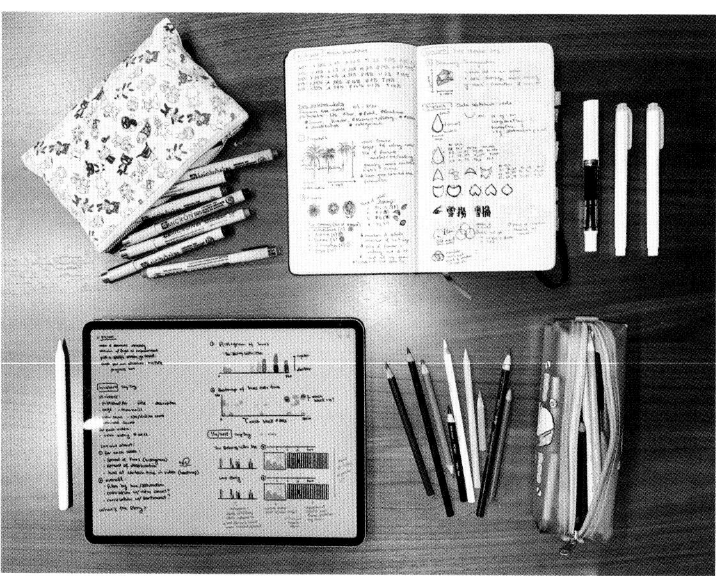

Nadieh's and Shirley's sketchbooks and iPads used to create all the sketches and doodles for *Data Sketches* on the left and right, respectively.

Adobe Illustrator

Adobe Illustrator is a proprietary software application meant for creating artwork based on vectors and is perfect for working with SVGs (with which many of our data visualizations are created).

Nadieh uses Illustrator to create visual elements that can be a pain to program, such as chart legends. She *always* uses Illustrator when the final visual will be a static image because it can be much simpler to add surrounding elements, such as annotations and titles or text, in Illustrator than to program them on the web. Shirley didn't use Illustrator for any *Data Sketches* projects but does use it in her other projects.

There are other vector-based drawing programs available, such as Affinity Designer, Sketch, and Inkscape..

Nadieh: Chapters 2, 3, 5, 6 & 12

Observable Notebook and Vega-Lite

Observable is a web-based notebook for exploring data and thinking with code, and it is conceptually similar to Python's Jupyter notebook. Shirley likes to use Observable with Vega-Lite, a JavaScript library for quickly creating commonly used charts for analysis. She likes its concise JSON syntax that allows her to create line charts, bar charts, scatterplots, heatmaps, etc., with just a few lines of code. Coupled with Observable, she's able to quickly and visually explore her dataset with them.

Shirley started to use Observable with Vega-Lite in her later projects, after trying a myriad of other data analysis approaches. It's the most effective approach she's found for herself thus far. Nadieh uses Observable (not with Vega-Lite) these days to search for code snippets and sometimes creates simple examples with interactive sliders to assess the visual result of changing values.

Before Observable was launched in 2018 Nadieh relied heavily on Bl.ock Builder to search for d3.js code examples.

Shirley: Chapters 10, 11 & 12

Local Development Server

To create data visualizations on the web, we need a browser to display the page and a server that sends back the files (such as datasets) that the webpage asks for. Most websites load files from giant server farms around the world, but when we're coding and testing our visualizations, we want to use our own computers as local servers so that the files can be loaded quickly. The simplest way to get started is by installing an Apache HTTP server on your local machine.

But Apache servers only provide the most basic of functionality, so a lot of people have built additional solutions that make it much easier to quickly develop and test websites. Our two favorite features are "hot reloading," which automatically refreshes the webpage every time a file is saved, and bundlers that compile multiple JavaScript files into one for faster loading.

Nadieh always creates both her prototypes and full projects on a locally running web server. Half of her screen is taken up by her preferred code editor—Visual Studio Code, which she's also programmed to take care of "hot reloading"—and the other half by Chrome with the DevTools open for debugging. Shirley does all of her development work locally as well and uses Atom for her text editor and different build systems (`create-react-app` for `React.js`, `vue-cli` for `Vue.js`, and `Parcel.js` for WebGL with `Vue.js`) for compiling and deploying to the web.

Nadieh: All chapters
Shirley: All but chapter 12

SVG

SVG is short for "Scalable Vector Graphics" and is a human-readable syntax (not using pixels) for defining shapes and paths. The major benefit is that an SVG image will look crisp and sharp at any size. Although not required, many of the visuals made with `D3.js` are built with SVG elements.

We use SVG when the visualization doesn't require rendering too many elements or performing expensive animations. Even if most of the visual is created with something else (such as canvas), there will still often be an SVG layer on top to keep annotations or other textual elements looking as crisp as possible.

Nadieh: All but chapter 8
Shirley: Chapters 1, 3, 4, 5, 7, 8, 9 & 10

HTML5 Canvas

Canvas is a web technology which can be used to "draw" graphics on a web page, on the fly, by using JavaScript. The result that you see in the browser is a (raster-based) PNG image. Its main benefit is that it performs much better than SVG. However, other things, such as interactions, are more difficult to program.

We both use canvas for larger datasets or more intricate animations when SVG just doesn't cut it.

A handy metaphor for Adobe users: SVG is like Illustrator, while canvas is like Photoshop.

Nadieh: Chapters 4 (redone in canvas months later), 8, 10, 11 & 12
Shirley: Chapters 2, 3, 5, 6, 7, 8 & 11

WebGL

WebGL is a JavaScript API for creating 2D and 3D interactive graphics for the web. The majority of SVG and canvas code runs on the Central Processing Unit (CPU), while WebGL code uses the Graphics Processing Unit (GPU). GPUs are much better at tasks such as physics simulations and image processing, which unlocks more sophisticated effects for WebGL.

We've always had WebGL on our list of technologies to experiment with and were both able to do so in later *Data Sketches* chapters. Nadieh tends to use it when the amount of data to show is too large even for canvas to handle smoothly, and Shirley uses `Three.js` (a JavaScript library built on top of WebGL to make coding in 3D easier) to create 3D data visualizations and explore more computationally expensive visual effects.

One of Shirley's long-time dreams is to programmatically generate watercolor paintings, which is possible with WebGL.

Nadieh: Chapter 10
Shirley: Chapter 11

D3.js

D3.js is a JavaScript library for creating data visualizations on the web, often using SVG or canvas. D3.js has many extremely useful functions to prepare data for a specific visual form, translate from data to the pixels on a screen, and to implement interactions and transitions commonly found in data visualizations.

It is our most consistently used technology; we might not always use D3.js to render the data, but it is always used in other ways, such as for data preparation and handling. In that way, we see D3.js as a large collection of tools and building blocks where each project uses a different combination, giving an amazing amount of creative freedom for each design we have in mind.

Nadieh: All chapters
Shirley: All but chapter 6

React & Vue

Both `React.js` and `Vue.js` are JavaScript libraries for building and managing complex UIs and Single Page Applications. These are great at keeping track of interactions in one part of the webpage that affect another area of the page —a common use case in modern web apps.

Shirley has always used one of these libraries with D3.js to manage complex data updates in her visualizations, including interactions that result in filtering, aggregating, or sorting the data and automatically updating visualizations to reflect the changes. Having given `Vue.js` only a half-hearted try once (and never even having touched `React.js`), Nadieh happily manages to get by without using either.

The classic example is in the first version of Facebook chat. When sending messages in the chat box, the friends list would also update to reflect the new messages.

Shirley: All but chapters 1 & 12

Greensock Animation Platform (GSAP)

Greensock Animation Platform (GSAP) is a JavaScript framework for managing animations on the web. It optimizes for performance across browsers and makes keeping track of complex animations (especially those that chain multiple animations that trigger at different start times) very easy with timelines.

Shirley really enjoys creating animated visual explainers with GSAP, and will often combine it with `Scrollama.js` to create scrollytelling pieces. During *Data Sketches*, Nadieh didn't know GSAP, but has since used it for a few client projects with many elaborate animations.

Shirley: Chapters 8 & 9

MO

VIES

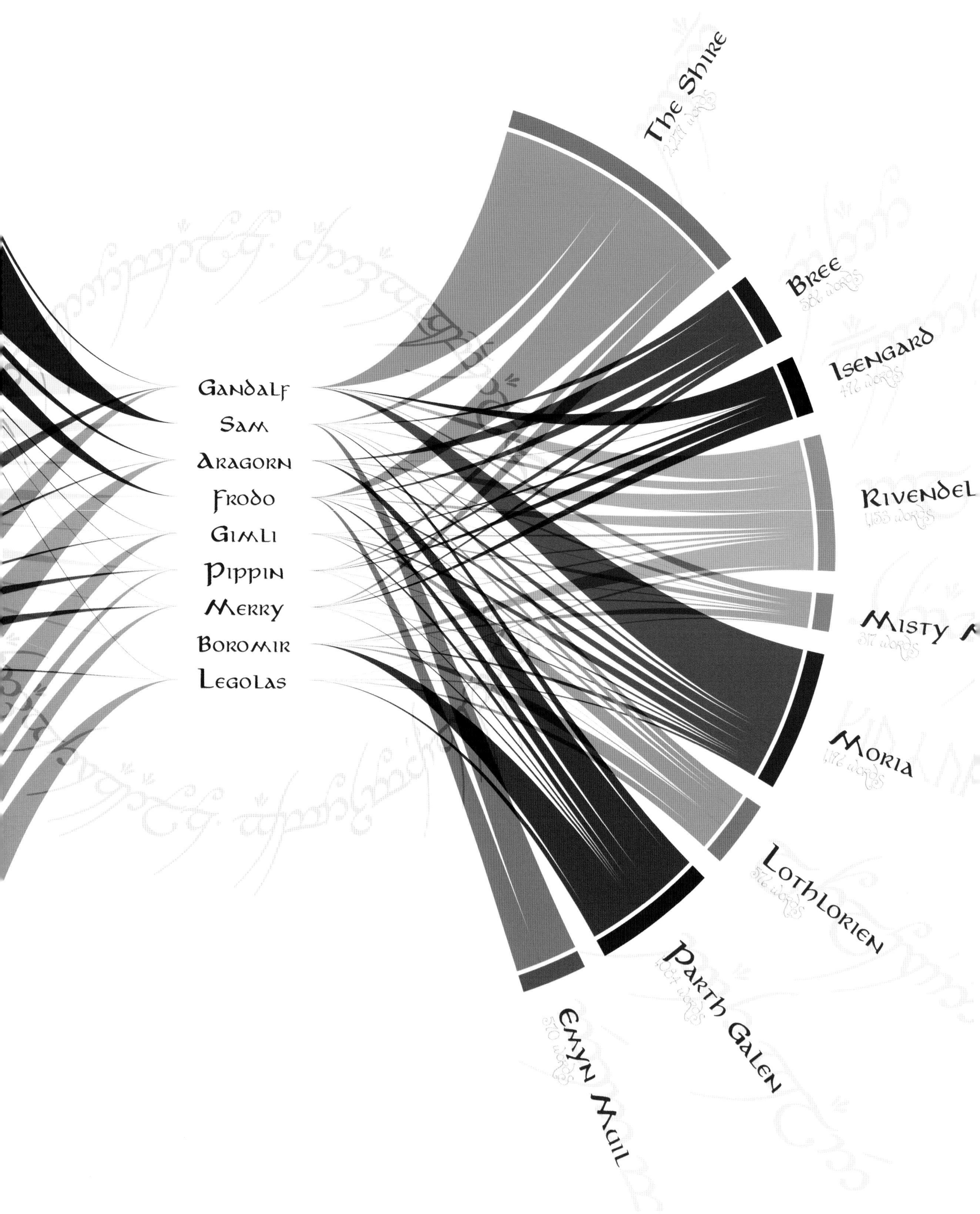

The Shire
2,279 words
Bree
582 words
Isengard
496 words
Rivendel
1,153 words
Misty M
317 words
Moria
1,196 words
Lothlorien
576 words
Parth Galen
1,064 words
Emyn Muil
510 words
Gandalf
Sam
Aragorn
Frodo
Gimli
Pippin
Merry
Boromir
Legolas

The Words of the Lord of the Rings

NADIEH

For our very first *Data Sketches* project we chose the topic "Movies," and we wanted to have some sort of personal connection to whatever dataset we'd end up visualizing. Starting with an open mind I went ahead and did a general search of movies to get a feeling of what might be out there.

I quickly came across budget information per movie and found my way to the OMDb API[1] and IMDb Datasets[2] where you can download huge files with lots of information on movies and series. Having access to such large databases seemed like a very good starting point, but I wanted to make the process more personal and relatable. So I decided to search for data on my favorite movie trilogy: *the Lord of the Rings* (LotR).

With the popularity of the movies, I was quite surprised that I couldn't find any structured datasets about them. Thankfully, after digging through more search results and using variations of the search query "Lord of the Rings dataset" on Google, I found a fascinating dataset in a GitHub repo[3] with the number of words spoken by each character in each scene, in all three extended(!) editions (see an excerpt in Figure 1.1). How *amazing* is that?! (ﾉ◕ヮ◕)ﾉ*:･ﾟ✧ I did a few manual checks, comparing the word count in the dataset to scripts available online, and they coincided pretty well. In this case I didn't need a perfect match, as I was more interested in the aggregated results.

[1] OMDb API, the Open Movie Database: http://www.omdbapi.com
[2] IMDb Datasets: https://www.imdb.com/interfaces/
[3] LotR by @jennybc: https://github.com/jennybc/lotr

Data

Fig.1.1

The original LotR dataset with the number of words spoken by each character in each scene of the three movies.

	Film	Chapter	Character	Race	Words
1	Film	Chapter	Character	Race	Words
2	The Fellowship Of The Ring	01: Prologue	Bilbo	Hobbit	4
3	The Fellowship Of The Ring	01: Prologue	Elrond	Elf	5
4	The Fellowship Of The Ring	01: Prologue	Galadriel	Elf	460
5	The Fellowship Of The Ring	02: Concerning Hobbits	Bilbo	Hobbit	214
6	The Fellowship Of The Ring	03: The Shire	Bilbo	Hobbit	70
7	The Fellowship Of The Ring	03: The Shire	Frodo	Hobbit	128
8	The Fellowship Of The Ring	03: The Shire	Gandalf	Wizard	197
9	The Fellowship Of The Ring	03: The Shire	Hobbit Kids	Hobbit	10
10	The Fellowship Of The Ring	03: The Shire	Hobbits	Hobbit	12
11	The Fellowship Of The Ring	04: Very Old Friends	Bilbo	Hobbit	339

After finding the dataset, my first thought was: "How many words did each member of the Fellowship speak at each place/scene/location?" Focusing on the nine members of the Fellowship seemed like a nice and structured way to filter the total dataset.

I'm still a bit sad that meant that I had to ignore a few other wonderful characters such as Galadriel and Saruman.

Although the original dataset was structured around scenes, it didn't contain information about the scenes' locations. I found the scenes to be a bit arbitrary; scenes, unlike locations, are more connected to the making of the movie versus the actual universe where the movie takes place. Therefore, I decided to manually add the location to each of the ±700 rows of data. ᕦ(ò_óˇ)ᕤ

To do this, I looked at the scripts of the extended and non-extended editions and a map of Middle-earth, all found online. These scripts sometimes mention the location where the scene takes place, but for other scenes, I relied on memory from having watched these films way too many times.

I added two columns: one with a broad location (like a kingdom) such as "Gondor," and the other consisting of a more detailed (or more precise) location, such as "Minas Tirith" (a city within Gondor). While I wasn't always able to find location information for every scene, I still managed to add detailed location information to approximately 90% of the rows in this column.

All of this took place during about three hours on a lovely summer Sunday. A little dedication can go a long way!

Eventually this was a labor of love and I tried my very best to add the right data to each scene.

↳ Manually Add New Variables to Your Data

As will become apparent in the chapters to come, we often had to do a lot of manual digging to get the dataset that we needed. For this LotR data one of the most important variables—location—wasn't in the base data, or in any other dataset for that matter. The location had to be gathered from unstructured sources, such as the movie scripts and from memory. Although this can take some time, don't be afraid to invest a little effort to build up exactly the dataset that you need. If you only stick with data that is already in a comma-separated values (CSV) format or other straightforward structured file type, you'll be missing out on a lot of interesting topics and variables to visualize!

Sketch

For my sketches, I used an iPad Pro 9.7" with an Apple Pencil that I had recently bought and was eager to try out. After having tried out a handful of apps that were recommended for drawing, I used the one that charmed me most: *Tayasui Sketches*. I found it to have the right level of options—not too little, as I felt with the app *Paper*, and not too many, such as the multitude of features in *Procreate*.

About a month before I started on this project I got an email from a potential client with a rough sketch of a chart that is known as a "chord diagram," but with extra circles in the center. A chord diagram is a type of chart that reveals flows or connections between a group of entities and can be used to show many different types of datasets, such as import/export flows between countries or how people switch between phone brands. It seemed very intriguing, and I have a fond memory of "hacking" the chord diagram for other purposes (such as turning it into a circular version of a more general flow chart to reveal how students went from their educational degree to the type of job they end up doing a year and a half after graduating). During a dataviz design brainstorm for this project, I remembered that sketch and thought something along those lines, with data also present in the center, would be a perfect way
to visualize my LotR data.

I decided to place the Fellowship characters in the center with the locations spread around them in a circle. Each character would be connected to the location where they spoke and the thickness of the chord/string represented the number of words spoken by that character at that location.

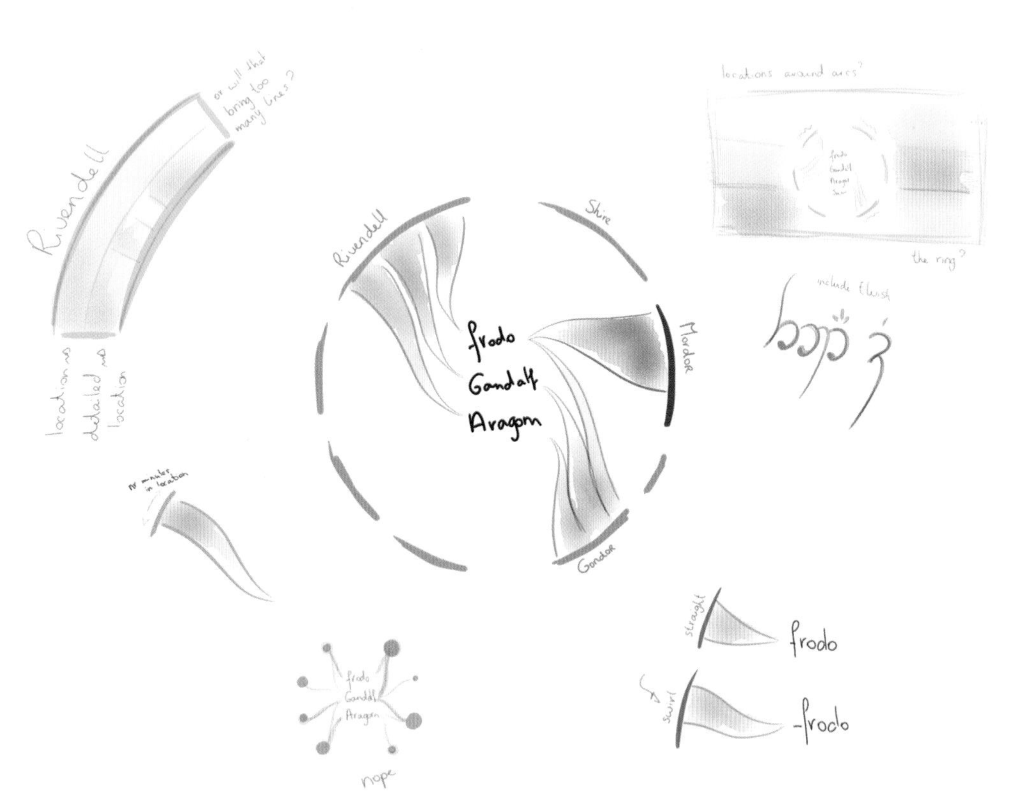

Fig.1.2

The LotR sketch with the Fellowship members in the center and locations around them in a circle.

Initially, I had the idea of placing the detailed locations within the general location arcs as a second layer (see the top left side of Figure 1.2).

Because it's good practice, I tried coming up with some other ideas. One sketch involved creating a timeline, placing circles that were sized to the number of words spoken on the approximate time that the scene took place.

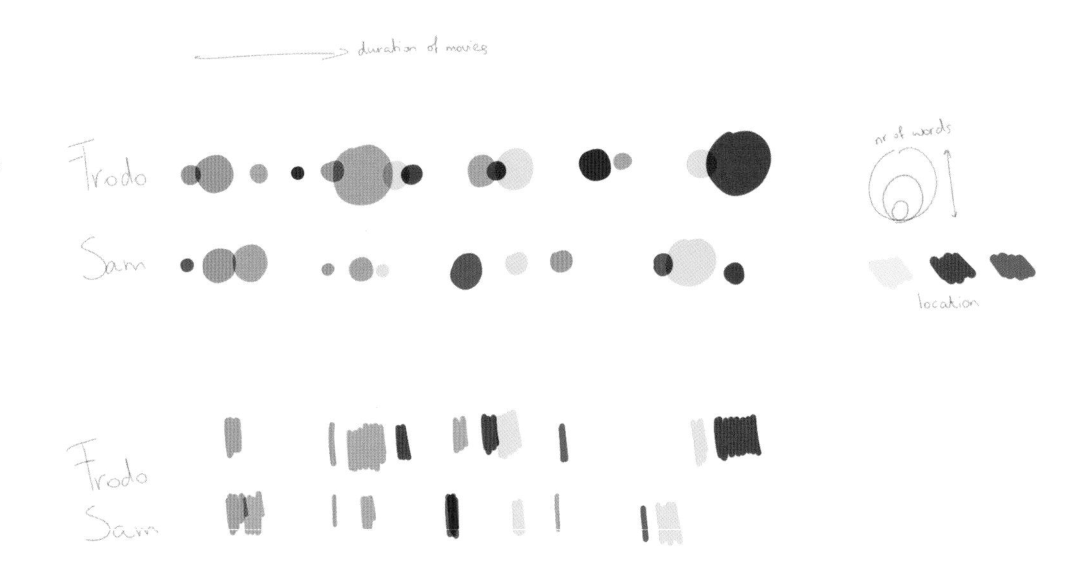

Fig.1.3

Another sketch, which places the number of words spoken on a timeline per character.

I also tried something more abstract where each location would be a spirograph, with the number of "petals" being the number of detailed locations, each sized to the total number of words spoken. But I definitely liked the chord-diagram-like concept the most.

Fig.1.4

A final idea that turned the words spoken into spirograph-like figures.

Sketching on the iPad was quite fun. It was very easy to combine techniques, move parts around, and undo things. And even though I don't have the same amount of control that I have with a regular pen on paper, it has been part of my "dataviz repertoire" since this project.

Code

As the design that I sketched wasn't a standard "chart," I had to figure out how to make it a reality. When this happens (which is often) I always try and find something that lies close to the idea that I've sketched and work from there. In this case, the starting point was already obvious: the chord diagram. I therefore started with the `D3.js` based code of a basic chord diagram.

For a short explanation of `D3.js`, please see "Technology & Tools" at the beginning of the book.

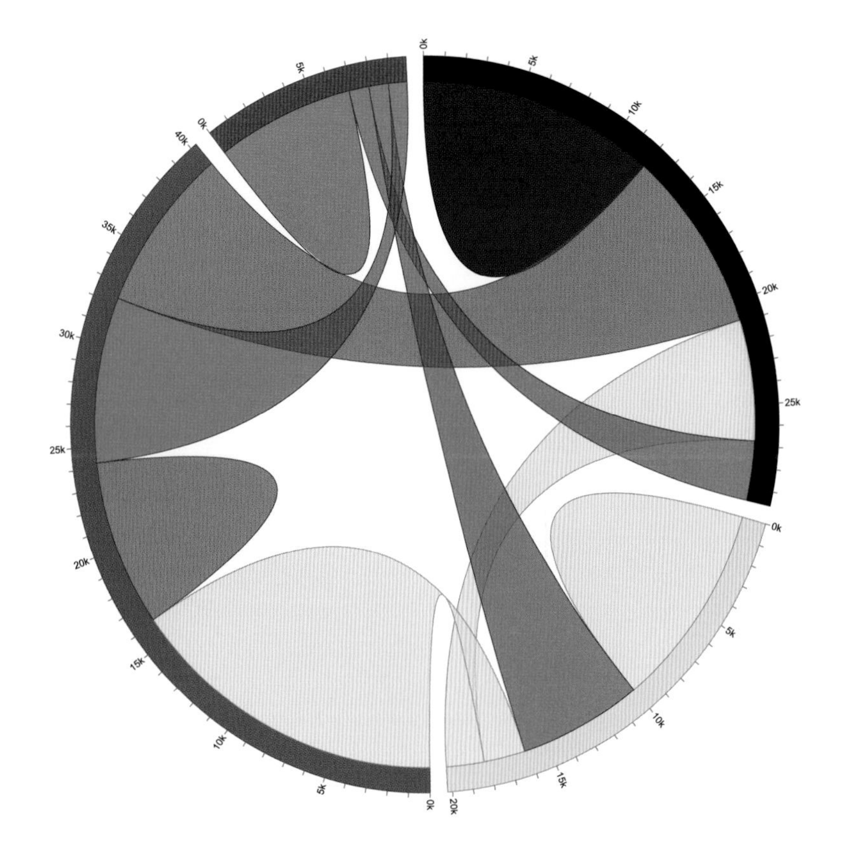

Fig.1.5

A basic chord diagram is a chart that's often tricky to understand, but it can display a wealth of information.

This approach helps me stay flexible and possibly change a design mid-way when some difficult step isn't working out or looking as intended. For this chord diagram, the most important step was to ensure all the central chords had one end in the center of the circle. If I could pull that off, then I figured I could manage the rest of the design as well. (If I couldn't, I'd have to think of a plan B!)

> I never plan out all of my steps; instead I try and focus on the most fundamental change/addition to be made at that point, see if I can make it work, and then think about the next step.

As making the chords move towards the center was far beyond the capabilities of the normal chord diagram, I knew that I had to dive straight into the source code, copy it, and wrangle it to what I wanted it to do. I did all kinds of tests to understand *exactly* what was happening in each line of `D3.js` chord diagram code. After a while, I felt comfortable enough to make more substantial changes and, in less time than I'd expected, I was looking at a new version with all the chords ending in the center!

The chord diagram code is part of `D3.js` itself.

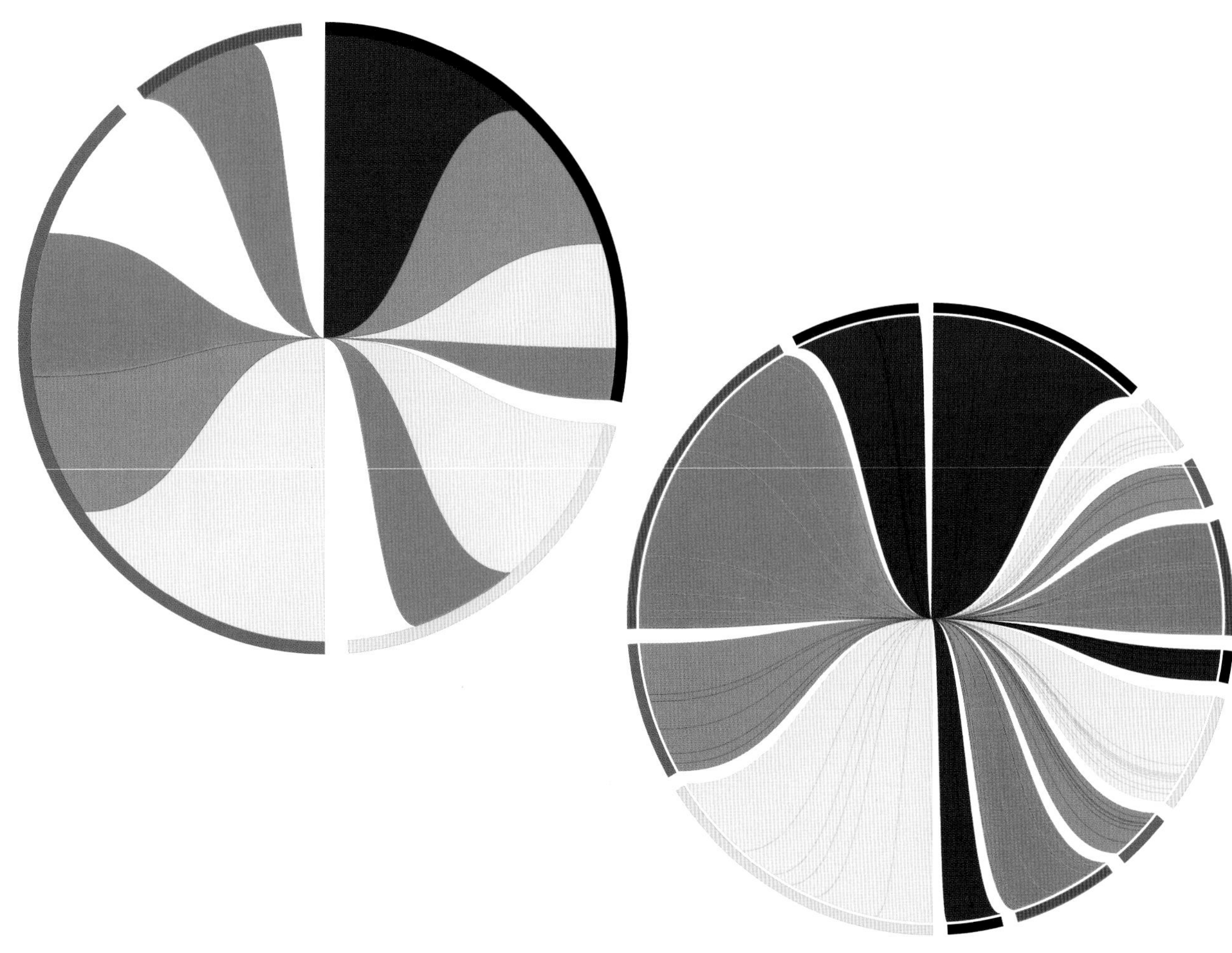

Fig.1.6

All the chords now flowed towards the center, which created some gaps around the edges.

Fig.1.7

The LotR data applied to the "tweaked" chord diagram.

After updating some of the formulas to calculate the sizes of the outer arcs to get rid of the gaps, the visual was ready to handle the actual LotR data (Figure 1.7).

I quite liked the look of all strings flowing to the center, but I had nine members of the Fellowship to put there. I therefore placed all of the character names in the center, arranged them alphabetically (at first), and had each string flow towards the correct character in the center.

I'm switching from "chords" to "strings" to describe the inner lines from here on out.

However, Figure 1.8 showed me that the inner strings looked a bit odd, especially at the very top and bottom of the visualization (such as the top-left orange strings). It just didn't feel *natural* to me. Thankfully, these string shapes were created as scalable vector graphics (SVG) paths. These are quite flexible and allow you to create diverse types of shapes; they can be straight, circular, or curved. And as I wanted something else than the default that the chord diagram supplied, I had to finally resign myself and learn how to construct these SVG paths myself.

For a short explanation of SVG, please see "Technology & Tools" at the beginning of the book.

Fig.1.8

Making sure the strings end up at the correct vertical location.

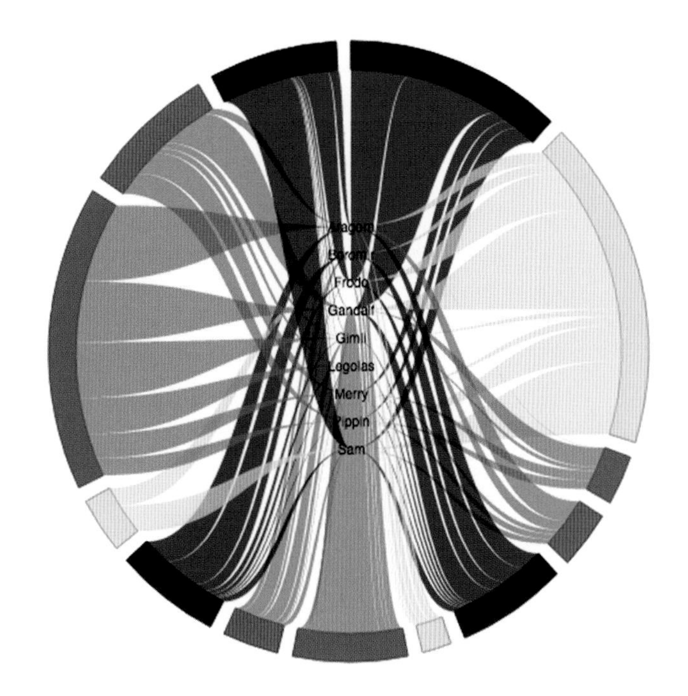

I found a great website[4] that helped me a lot to get the hang of the more complex SVG paths. It showed one thing: a *Cubic Bézier Curve*, and I could move it around while seeing how its path "formula" changed. Using the website, I tried out different shapes and slowly built up an understanding of how to create more elegant S-shaped curves. I went through many, many tweaks of shapes, each one getting a bit closer to a natural looking curve flowing between the outer arcs and inner characters.

A Cubic Bézier Curve is a line that is defined by an end and starting point, and two "handle points" that shape the curved look of the line.

Fig.1.9
(a & b)

Two in-between states while I was trying to make the inner chords look more natural

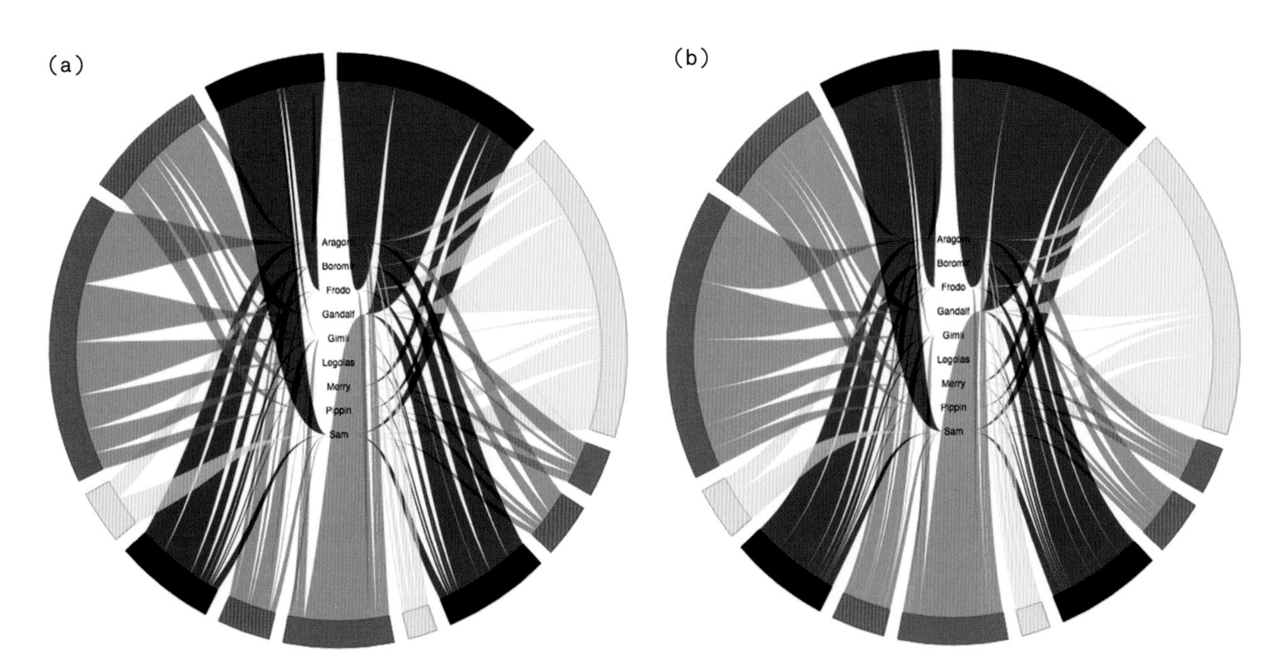

[4] SVG Cubic Bézier Curve Example: http://blogs.sitepointstatic.com/examples/tech/svg-curves/cubic-curve.html

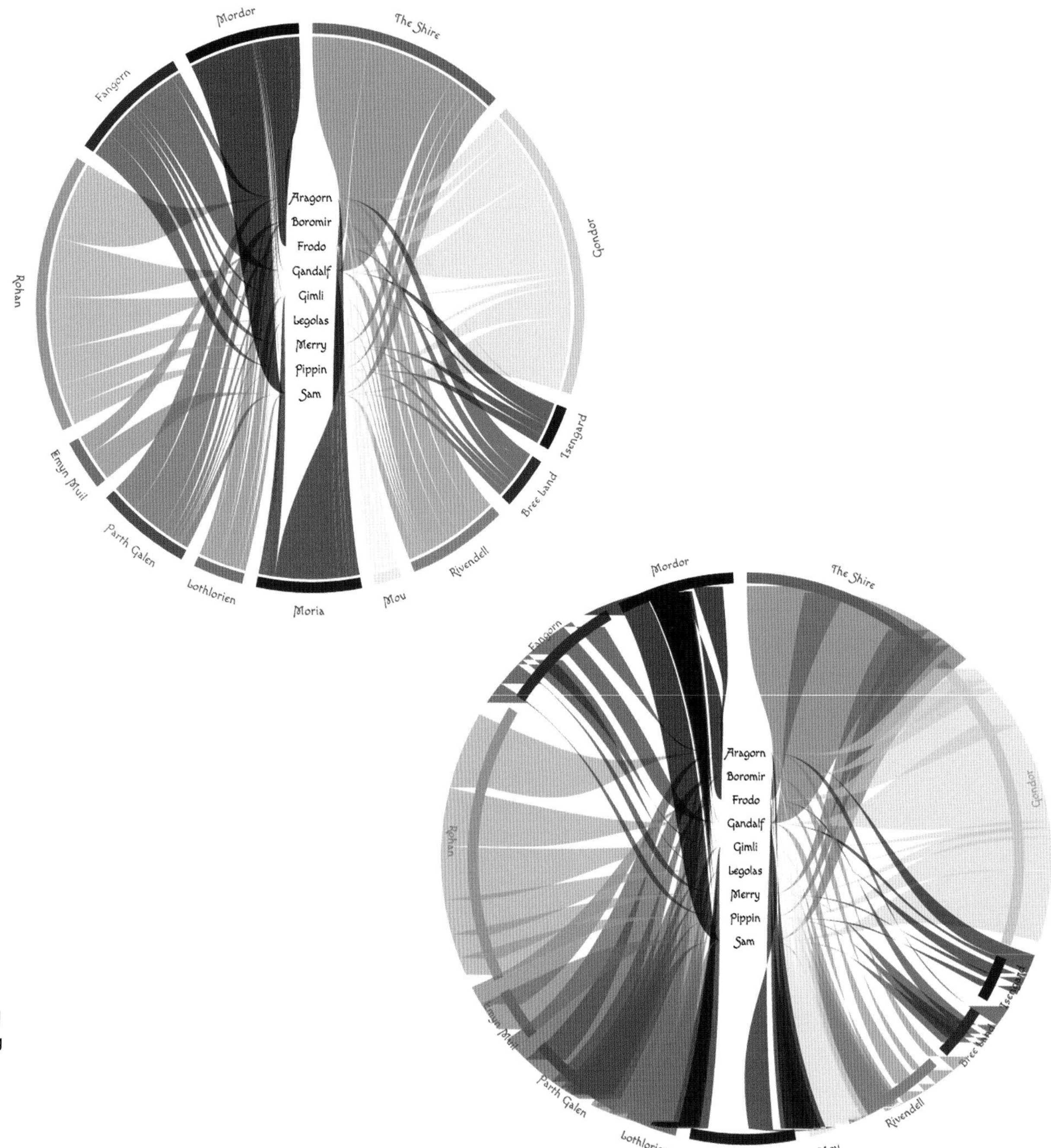

Fig.1.10

Updated color palette and several other design tweaks already make a world of difference.

Fig.1.11

Not quite the expected result while I was trying to move the two circle halves outward.

I was now happy with the resulting shapes and moved on to updating the color scheme. I color-picked several stills of the movie locations to get a color that I felt represented each location, without being the same as another location (a challenge, as there were many green locations) (Figure 1.10).

But even with these changes, the next step that I needed to take was practically screaming at me from the screen: fix the center, which looked way too squished. There was definitely a need to create empty space above and below the inner section. This would not only fix the squished feeling, but also to give the strings in that region the room to really flow in a nice shape.

Moving the two halves outward meant I again had to wrangle those SVG paths, however, this time it was easier, because it mostly amounted to adding or subtracting a horizontal offset to the right or left half, respectively (Figure 1.11).

After I visually stretched the two halves to my liking, I ranked the broad locations clockwise from the top in the order that they first appear in the movie or where the bulk of the scenes are taking place. This thankfully divided up the two halves almost symmetrically. A few shape adjustments and color tweaks later, I ended up with something that could be its own new chart form (Figure 1.12). I even asked around on Twitter for advice on what to name it. I got back some really great suggestions, such as Ginkgo + Leafs, Butterfly + Wings, Labrys + Blades, Bowtie, even the "Eye of Sauron." Eventually I decided to call it Loom + Strings. It seemed the most appropriate, plus it was also suggested by the creator of `D3.js`. (*≧▽≦)

Due to more technical reasons I needed two names, one for the overall shape and one for the inner lines.

The one thing I added at the very end of the project was interactivity. I created a simple mouse hover that gave viewers the option to inspect each location or character in more detail. Shirley offered the suggestion to completely fade out locations where a character had no spoken lines. I also implemented it vice versa; when you hover over a location, the characters without spoken lines there are dimmed. And to help the viewer understand the types of insights that you can get from the visual, I added a short paragraph with some fun insights per character that shows up when you hover over their name.

Finally, I wanted fonts that reminded me of Middle-Earth scripts. Bilbo's handwriting and the Elvish and Dwarfish scripts are so ingrained throughout the movie, and very beautiful, that I had to use it in the visual as well. As for location names, I spent a lot of time searching for the correct translations of the different locations in Elvish (and "Moria" in Dwarfish, of course) as well as the inscription of the One Ring. (*≧▽≦) But I feel that it makes a major impact on the overall feel of the final piece, so it was thankfully worth it!

I'll never know for sure if I got the Elvish translations correct though...

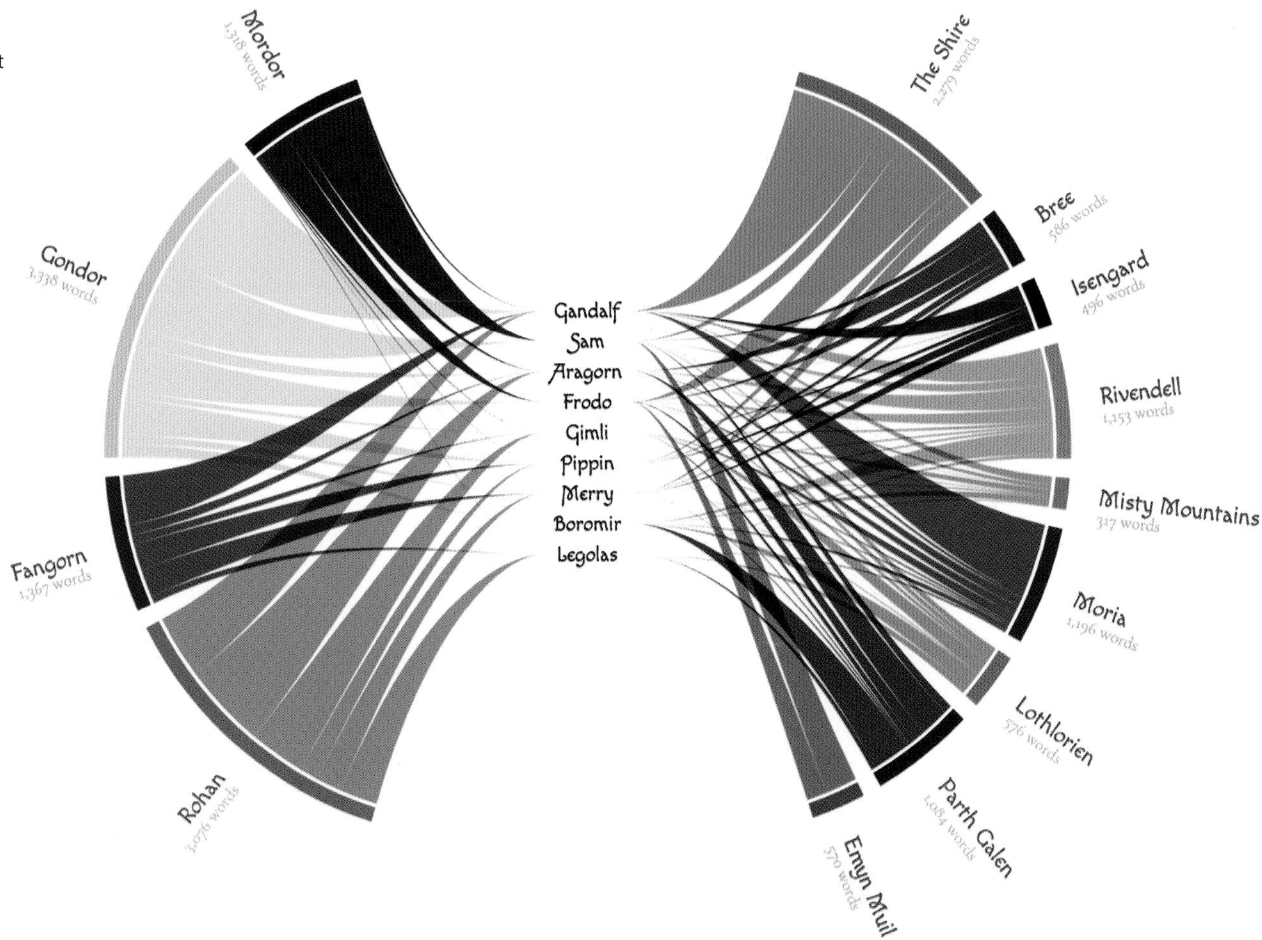

Fig.1.12

The (almost) final result of the visual that had now truly moved away from its chord diagram origins.

↳ Remix What's Out There

When I design my visuals, I always use plain pen and paper (or its "digital pen" equivalent). I don't think about remixing other people's work. I instead try to figure out what would work well for that dataset and the question the visual has to answer. But after looking at my finished designs, I see similarities with visuals that other people have made. It's therefore very rare that I start my JavaScript code from scratch. I typically try to find a code snippet or example that most closely resembles my design and start adjusting that. This can be as simple as the code for a bar chart or donut chart, but in this case I started from the well constructed code of a `D3.js` chord diagram. Having this code as the base made it *much* easier to slowly alter it until it looked like my sketched idea. I literally don't think that I could've created this project without having Mike Bostock's code, which included so many smart ways to set up the arcs and inner chords, to repurpose.

Reflections

It was quite a lot of fun working with a dataset about a topic that I love. I learned several new things about the film from creating the visualization, which is something I always strive for (although that's probably true of most visuals). Also, having successfully adapted a part of `D3.js` code into something different gave me a real confidence boost in my understanding of JavaScript going forward! I did have to hold myself back several times throughout this project from not just laying my laptop aside to watch the whole trilogy again.

The Words of the Lord of the Rings

↳ LotR.VisualCinnamon.com

Who's speaking in Middle Earth

How many words have the members of the Fellowship spoken across Middle Earth during all 3 extended editions of the Lord of the Rings

In more than 11 hours of the LotR trilogy all characters combined speak approximately 32,000 words. The 9 members of the Fellowship alone take up about 17,000 of these words, a bit more than half. In the visualization below you can find out how many words a member has spoken at each general location throughout the trilogy.
The members have been sorted from Gandalf who had the most lines, to Legolas, who spoke even less than Boromir, even though the latter was only in one movie (and some extended scenes).

Hover over characters or locations to get a more detailed overview

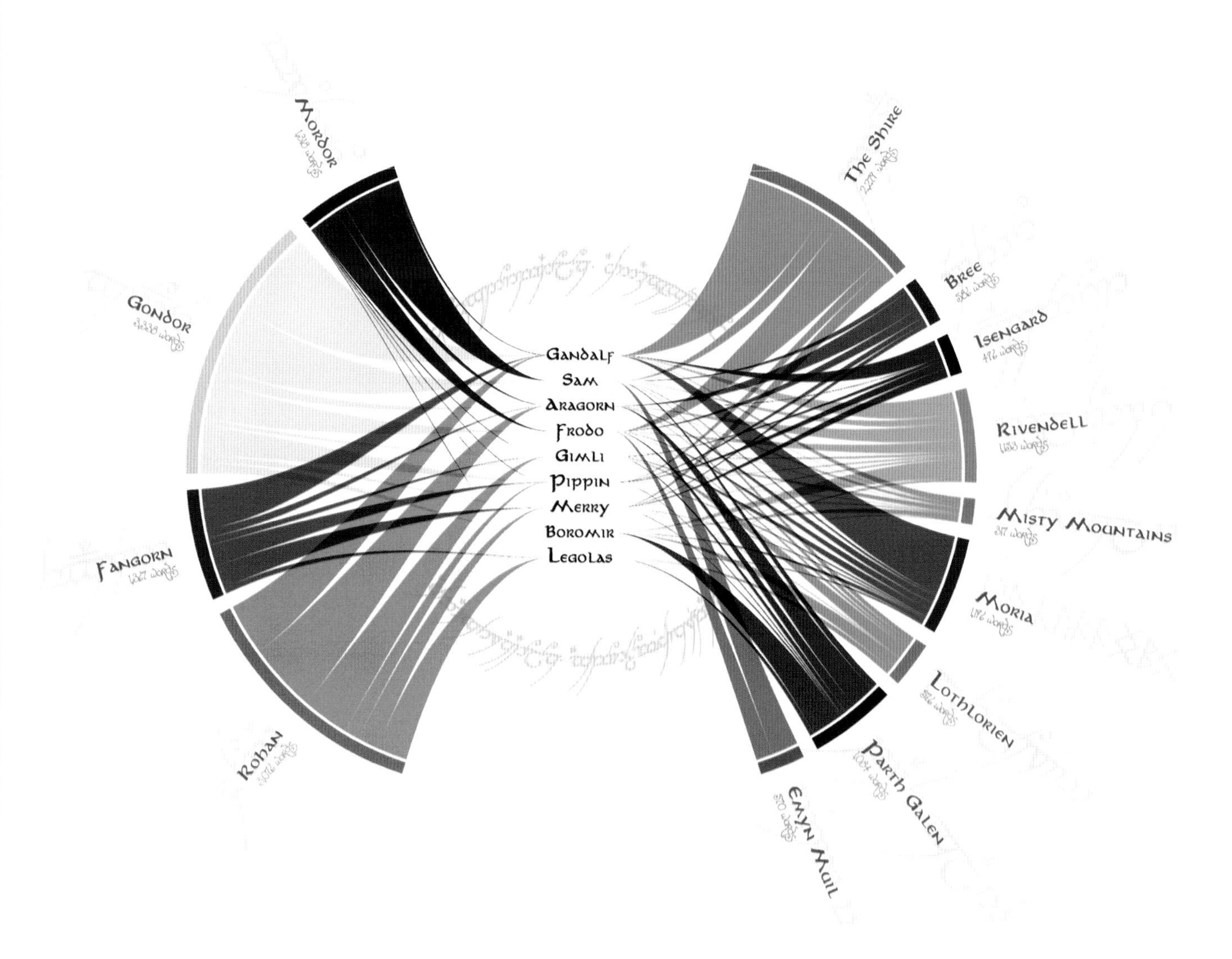

Fig.1.13

The final result of the Lord of the Rings visual with the Elvish or Dwarfish translations behind each location name, and the inscription of the One Ring in the center.

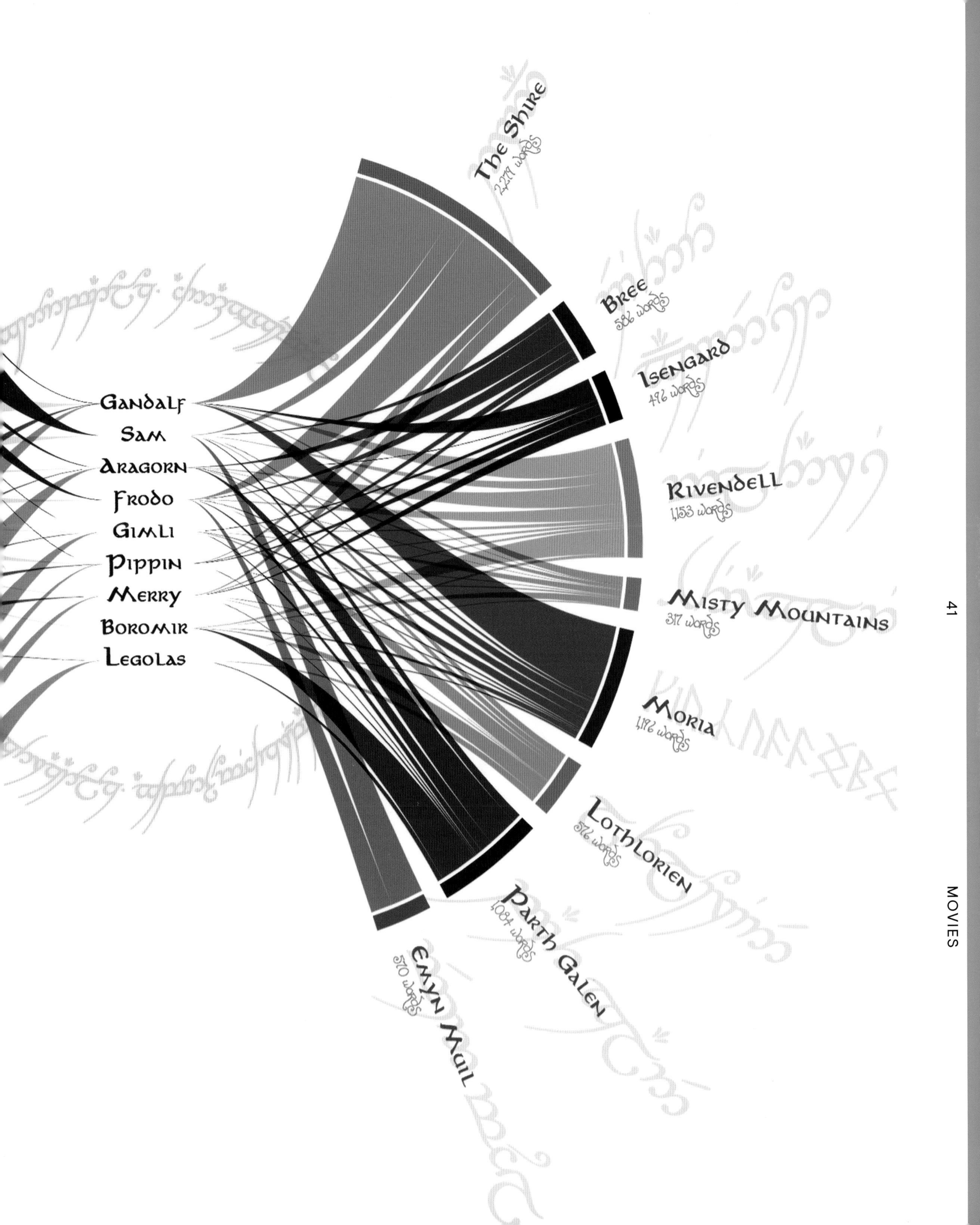
Gandalf
Sam
Aragorn
Frodo
Gimli
Pippin
Merry
Boromir
Legolas
The Shire
2,279 words
Bree
586 words
Isengard
496 words
Rivendell
1,153 words
Misty Mountains
317 words
Moria
1,196 words
Lothlorien
576 words
Parth Galen
1,084 words
Emyn Muil
570 words

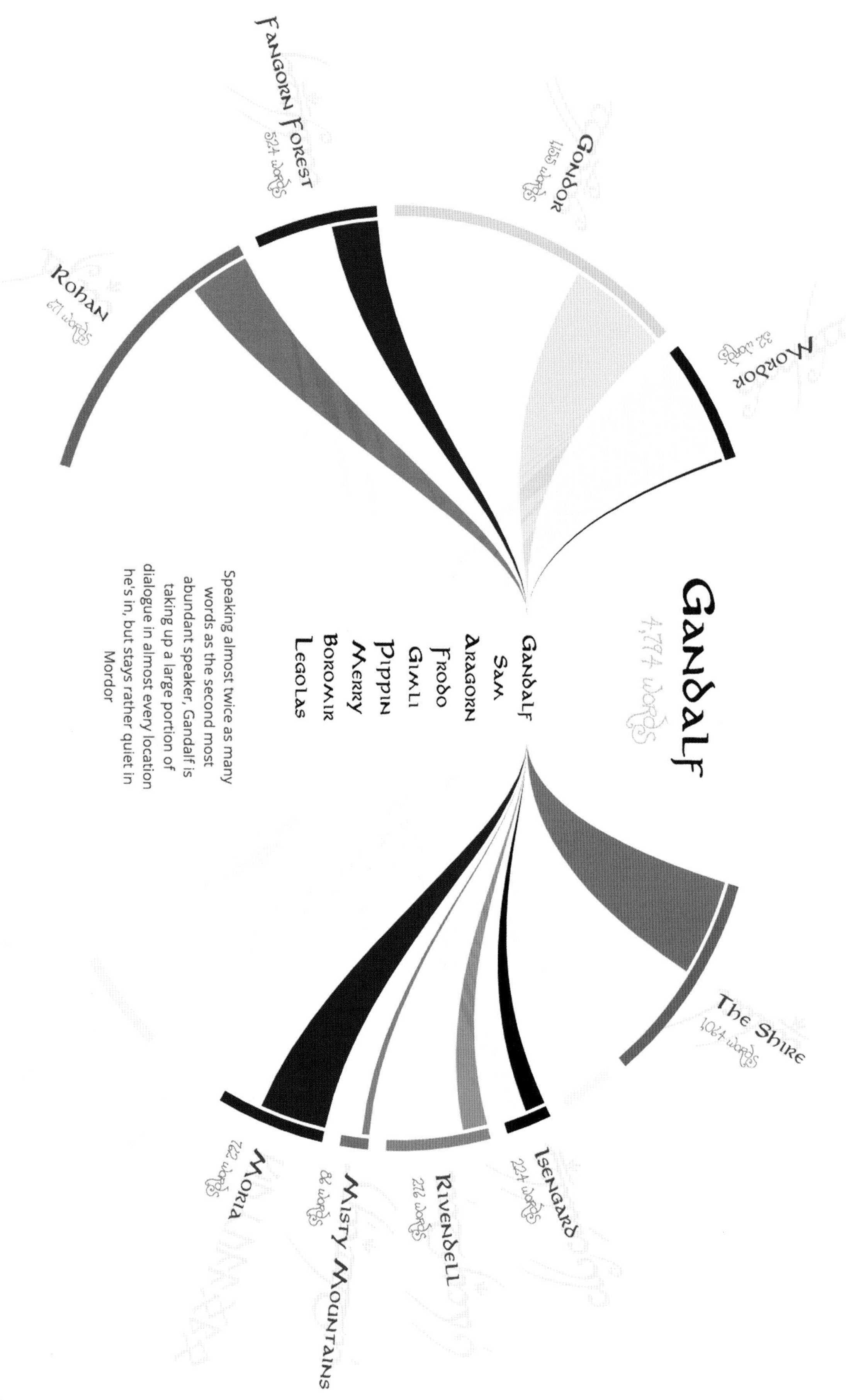

Fig.1.14

Hovering over a character's name shows their strings only, hides the locations where they had no spoken lines, and updates the outer values of words to only theirs.

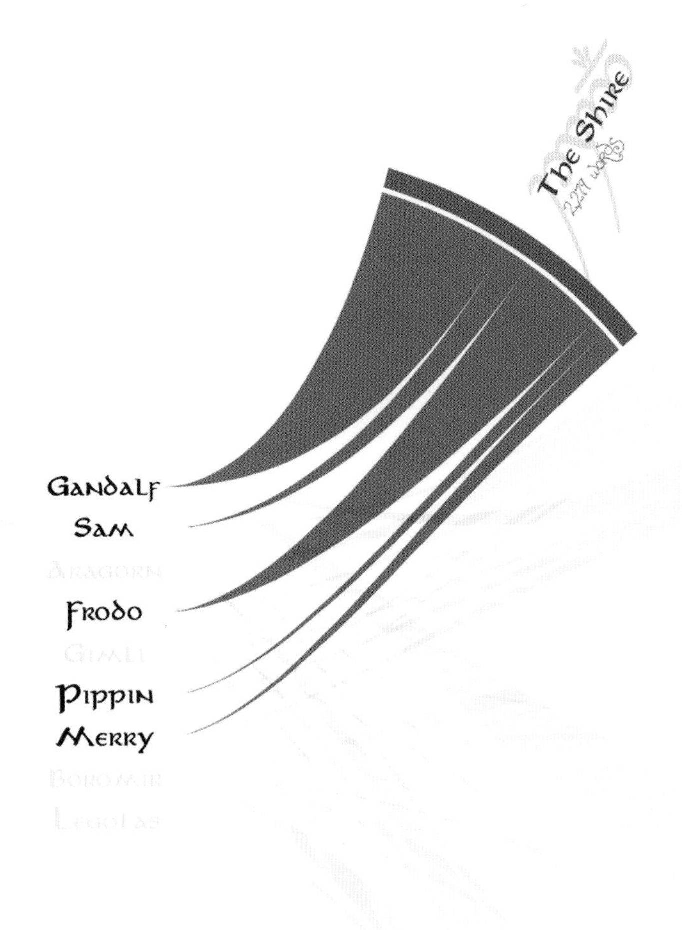

Fig.1.15

Hovering over a location reveals all the Fellowship members that spoke there. With the section in the Shire happening (almost) exclusively before the Fellowship is formed, only the four hobbits and Gandalf have lines there.

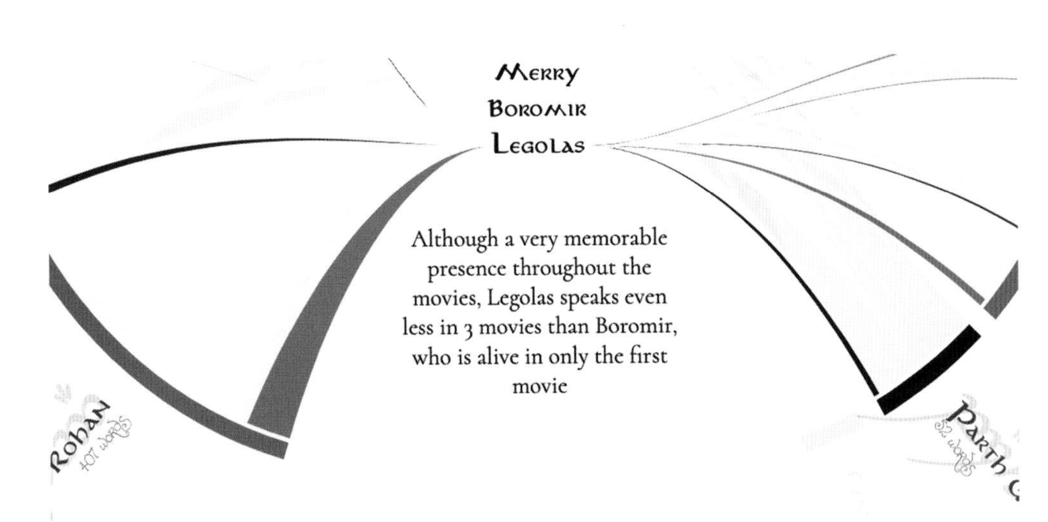

Fig.1.16

Hovering over a character reveals a small section of text with some interesting insight about him, found through this data. (For example, how Legolas talked even less than Boromir, while the latter is really only [truly] alive during the first movie!).

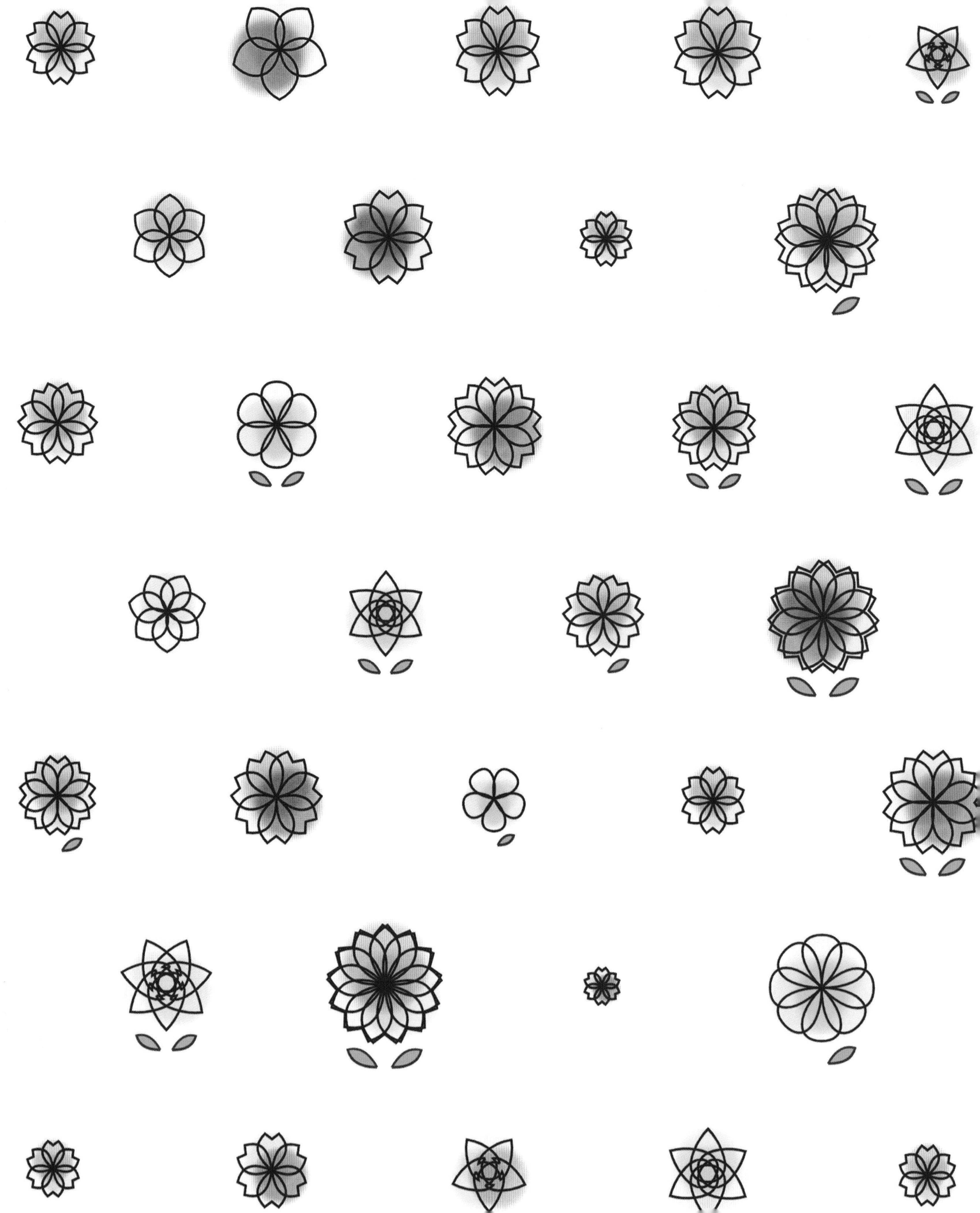

Film Flowers

SHIRLEY

I am absolutely horrible with pop culture references. And while I like to blame the fact that I grew up in various non-English-speaking countries for half of my childhood years (and had my head buried in textbooks for the rest of them), I know it's also because I just didn't watch that many movies growing up.

So I was pretty excited to examine movies for our first topic: how many blockbusters had I seen (or not seen) in my lifetime? And since it was July, I decided to concentrate on summer blockbusters.

Code

The toughest part of coding the flowers was the beginning, where I had to refresh my mind on SVG paths, and the Cubic Bézier Curve command in particular. I needed it in order to draw the flower petal shapes, which I had noted in my original sketches would be mapped to the four parental guidance ratings. At first, I had to draw out the shapes of the petals to work out the commands (Figure 1.4), but after a while, I was able to code the petal shapes directly (Figure 1.5).

For a short explanation of SVG, please see "Technologies & Tools" at the beginning of the book.

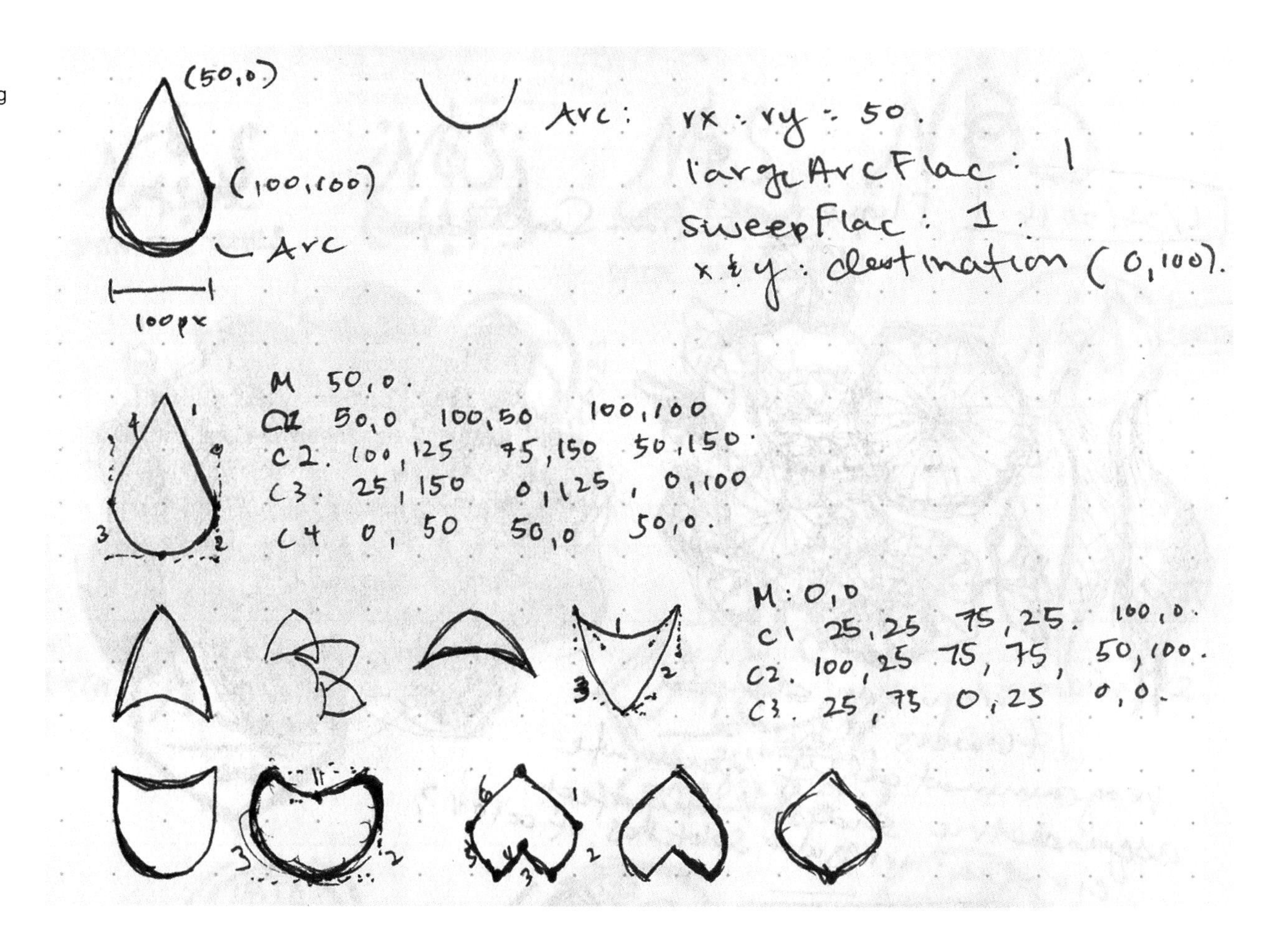

Fig.1.4

My attempts at figuring out Cubic Bézier Curves for the flower petals.

Fig.1.5

SVG paths of each flower petal.

I ended up coding six or seven different petal shapes and making them into flowers—duplicating each petal six times, and rotating them around the center at 60° intervals—to see which petals shapes I liked. I eventually narrowed them down to the four I would need to match each of the four parental guidance ratings to (Figure 1.7). My favorite is definitely the cherry blossom ❀ shape that Nadieh suggested. (ˬ•ᵥ•)✧

Fig.1.7

Each flower petal is duplicated six times and rotated to create a flower.

Once I had the petal shapes I liked, I used `D3.js` to help me map the movies data to the different visual channels of the flower. I used `d3.scaleQuantize()` (takes a continuous numerical input and outputs a discrete number, string, or color) to convert the movie's number of IMDb votes to a discrete number of 5 to 15 petals and `d3.scaleLinear()` (continuous numerical input, mapped to a continuous numerical output) to assign the movie's rating out of 10 to flower petal size:

For a short explanation of `D3.js`, see "Technologies & Tools" at the beginning of the book.

Fig.1.8

The flowers with number of petals and petal size assigned.

I still marvel at how little changed between the mappings I specified in my sketches and what I ended up using in the final. Usually, a lot changes as I prototype.

Drawing with SVG

There are many ways to draw shapes on a web page, but SVG is probably the most straightforward to learn because of how semantically similar it is to HTML.
For example, to draw a circle centered around `[0,0]` with a radius of 10, I would write:

```
<circle cx="0" cy="0" r="10"></circle>
```

And I can manipulate it just like any other HTML element, updating attributes and registering event handlers.

Out of all the SVG elements, I like to use `circle`, `rect`, `text`, `path`, and `g` (group) the most often. And out of those, my absolute favorite is `path`, because with it, I can draw any shape I can imagine.

This is the SVG path I use to draw the cherry blossom petal:

```
<path d="M0,0 C50,40 50,70 20,100 L0,85 L-20,100 C-50,70 -50,40 0,0"></path>
```

In the path string, I use three different commands:

Command	syntax	(how I think about them)
Move to	M x,y	"Pick the pen off the paper, put it back down at x, y."
Line To	L x,y	"Make sure the pen is already on the page and draw a line to x,y" M x,y L x,y
Curve to	C x,y x,y x,y	"Make sure the pen is already on the page and draw a line to the end position. Then take the first control point and the second control point and nudge them until we get the curve we want" M x,y C x,y x,y x,y

It's exactly like drawing curves with the pen tool in Illustrator.

So let's break down the cherry blossom path:

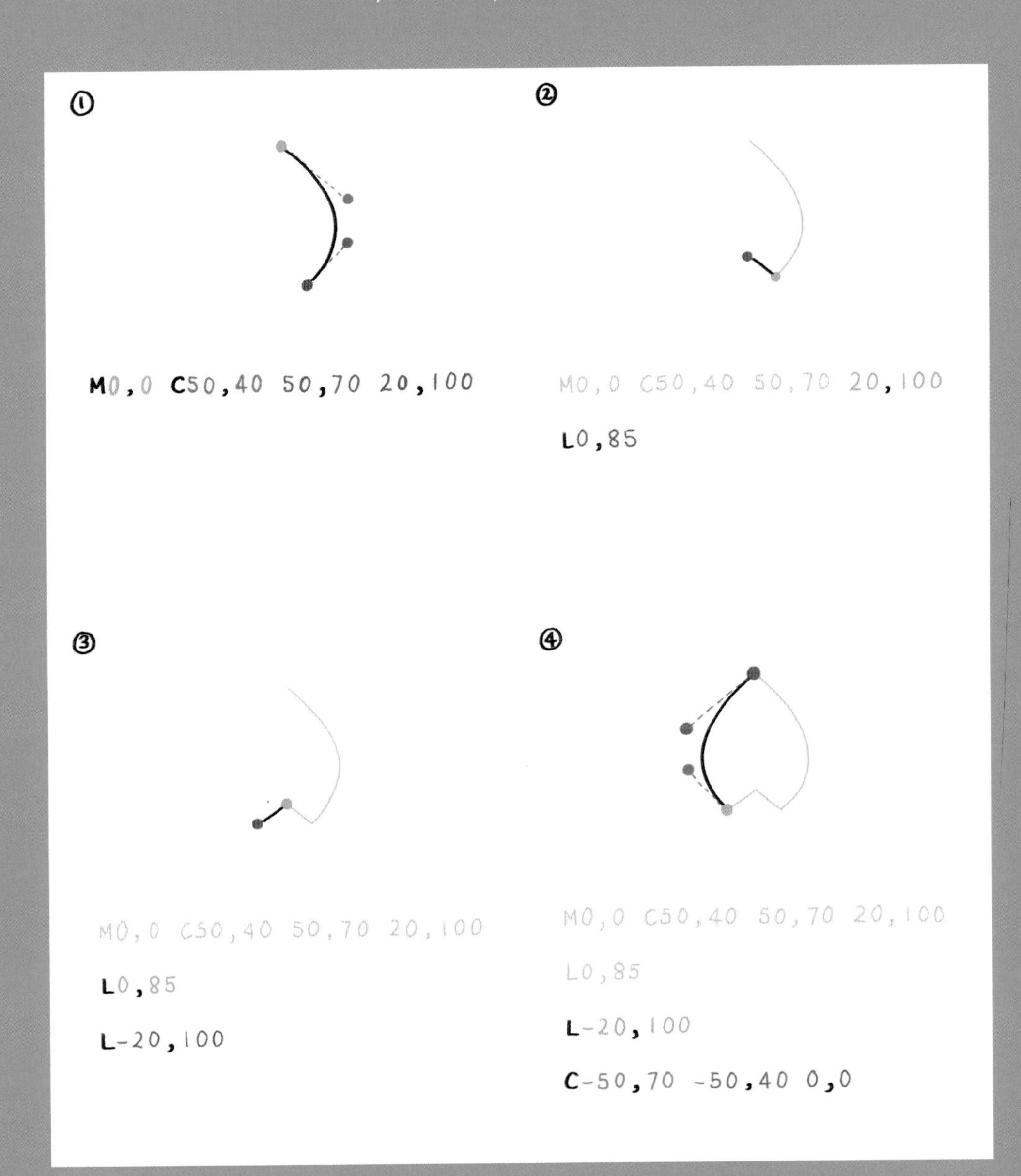

①
M0,0 C50,40 50,70 20,100
②
M0,0 C50,40 50,70 20,100
L0,85
③
M0,0 C50,40 50,70 20,100
L0,85
L-20,100
④
M0,0 C50,40 50,70 20,100
L0,85
L-20,100
C-50,70 -50,40 0,0

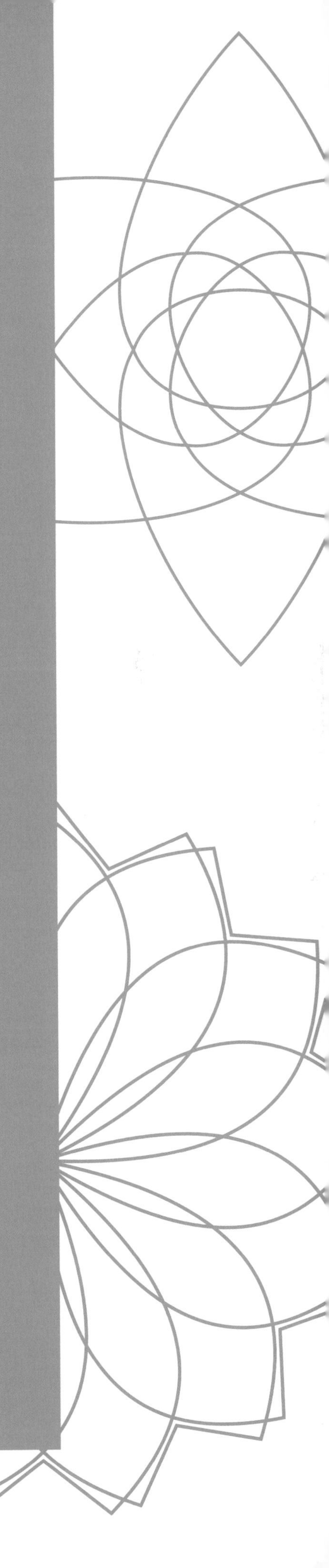

I have to admit: it can be cumbersome to write the path string out by hand, and it's definitely easier (and faster) to draw the shape in Illustrator and export as SVG. But I'm a firm believer in knowing how something works, and I'm so glad I taught myself these path commands. I'm no longer restricted by the shapes that SVG ships with, and I'm not intimidated by the long path strings when I have to inspect or debug.

The most fun part was adding the colors. I used D3.js's `d3.scaleOrdinal()` (discrete input, discrete output) to map a movie's genre to the corresponding color and created an SVG for each color. I liked the effect, but felt that the circle's edges were too harsh, and that's when I remembered Nadieh's set of (brilliant) "SVGs beyond mere shapes"[1] tutorials where she covered different gradients, filter effects, and blend modes. I went through her tutorials on CSS blend modes (`multiply`) and SVG filters (`feGaussianBlur`) to make the colored circles look blended together and the edges feathered.

Finally, I used two SVG `transform` attributes—`translate` and `rotate`—to offset the colors just enough to overlap each other and angle correctly depending on the number of genres for that movie (Figure 1.9). That calculation took me longer than I expected; I went through a lot of back and forth on how I wanted to deal with movies with only one or two genres (for one genre, the colored circle is centered, and for two, they are stacked vertically on top of each other) and spent quite some time nudging the circles together and apart so that they'd fit inside the overall shape of the flowers.

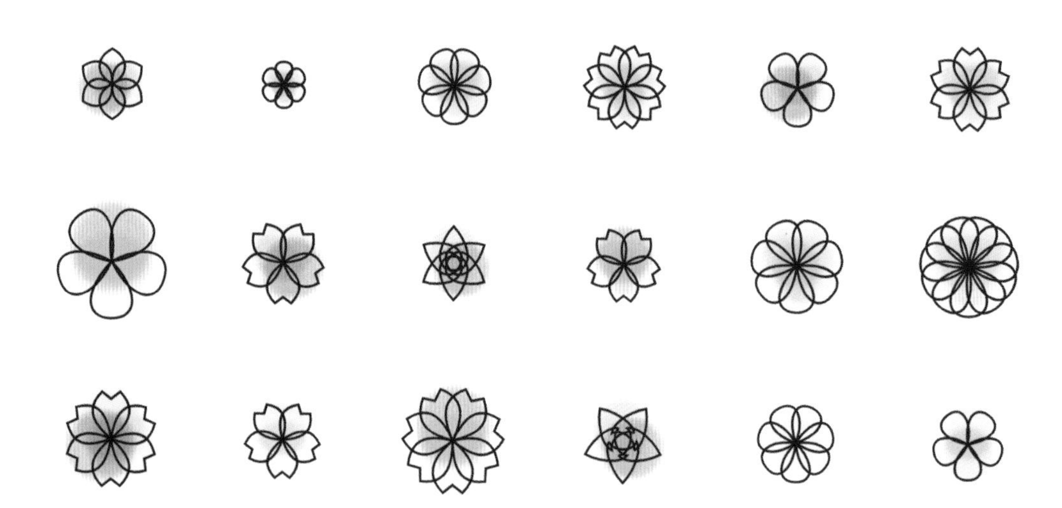

Fig.1.9

The flowers with color assigned and positioned.

Reflections

I was incredibly happy with the end result; I gushed about it to anyone who would listen for a good few days. One of my favorite finds is definitely *Batman & Robin* from 1997, which had an unfortunate 3.7/10 rating on IMDb and is the most adorably tiny speck of a flower. Others, like *Inception*, were marvelously large and complex. But of all the flowers, my favorite is definitely *Harry Potter and the Deathly Hallows: Part 2*, for all the sentimental memories the movie holds for me.

This project is still one of my favorites. I think it's really beautiful in its simplicity, but because there is a clear mapping of the data, I can still find interesting patterns in the data. It gave me a firm grasp on custom SVG paths, which I was previously intimidated by, and started me on flowers, a theme I have revisited many more times since.

[1] Nadieh Bremer, "SVGs beyond mere shapes": https://www.visualcinnamon.com/2016/04/svg-beyond-mere-shapes.html

Scales in D3.js

Scales are one of the most important parts of the D3.js library; I use them in every project. At its base, we need two things for a scale: an input that we pass into *domain* and an output that we specify with *range*. I like to think of scales as mapping from the raw data (*domain*) to the web page (*range*).

Take, for example, a line chart of average temperatures throughout one year. We would want to map the dates to the *x*-axis and the temperatures to the *y*-axis. So we would create two scales:

- A `d3.scaleTime()`, passing in the first and last date as *domain* and mapping those to the left-most and right-most *x*-position, respectively, passing those into *range*.
- A `d3.scaleLinear()`, passing in the minimum and maximum temperatures seen that year as *domain* and mapping those to the bottom and top *y*-positions of the visualization, respectively.

> **In the "Sketch" section, I talked about marks and channels, and about assigning data attributes to visual channels. I think of D3.js scales as the natural next step; it is the code that will turn data attributes into visual channels.**

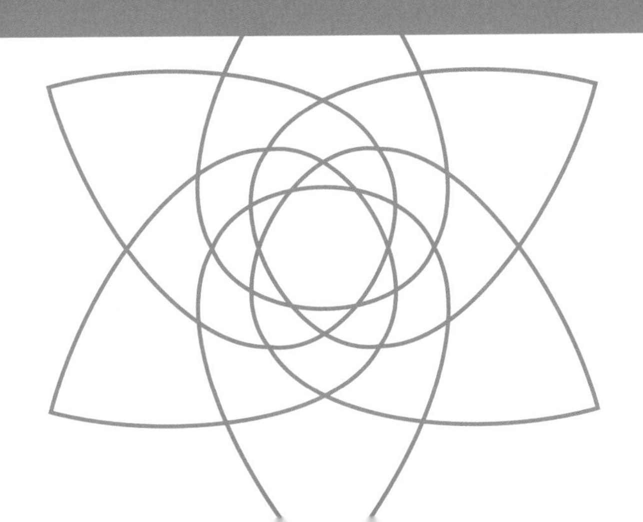

Film Flowers

↳ shirleywu.studio/projects/filmflowers

film flowers

top summer blockbusters
reimagined as flowers

(shirley wu)

G PG PG-13 R

Drama Comedy Adventure Action Other

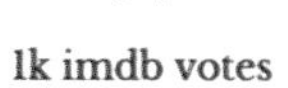

1k imdb votes 414k 827k 1,240k 1,653k

3.7 / 10 5.0 / 10 6.3 / 10 7.7 / 10 9.0 / 10

2016

Now You See Me 2 | The Legend of Tarzan | Independence Day: Resurgence | Central Intelligence | Finding Dory

2015

Terminator Genisys | Straight Outta Compton | Ant-Man | Mission: Impossible - Rogue Nation | Minions

2014

Teenage Mutant Ninja Turtles | 22 Jump Street | Dawn of the Planet of the Apes | Transformers: Age of Extinction | Guardians of the Galaxy

The Dark Knight Rises

Lilo & Stitch

Inception

WALL·E

The Sixth Sense

Slumdog Millionaire

Fig.1.10

A few of the most iconic movies as film flowers.

Harry Potter and the Deathly Hallows: Part 2

Batman & Robin

Toy Story 3

Fig.1.11

My favorite film flowers either because I have very fond memories associated with them, or in the case of Batman & Robin, because it's so tiny and cute!

2012

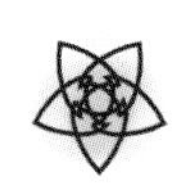
Ice Age: Continental Drift

Ted

Brave

The Amazing Spider-Man

The Dark Knight Rises

2011

Captain America: The First Avenger

Rise of the Planet of the Apes

Cars 2

Transformers: Dark of the Moon

Harry Potter and the Deathly Hallows: Part 2

2010

The Karate Kid

Despicable Me

Inception

The Twilight Saga: Eclipse

Toy Story 3

2009

G.I. Joe: The Rise of Cobra

The Proposal

Ice Age: Dawn of the Dinosaurs

Harry Potter and the Half-Blood Prince

Transformers: Revenge of the Fallen

2008

Slumdog Millionaire

Mamma Mia!

WALL·E

Hancock

The Dark Knight

2007

The Simpsons Movie

Ratatouille

The Bourne Ultimatum

Harry Potter and the Order of the Phoenix

Transformers

2006

Borat: Cultural Learnings of America for Make Benefit Glorious Nation of Kazakhstan

Click

Talladega Nights: The Ballad of Ricky Bobby

Superman Returns

Pirates of the Caribbean: Dead Man's Chest

2005

Mr. & Mrs. Smith

Batman Begins

Charlie and the Chocolate Factory

Wedding Crashers

War of the Worlds

2004

The Village

Dodgeball: A True Underdog Story

I, Robot

The Bourne Supremacy

Spider-Man 2

Fig.1.12

Summer blockbusters from my high school and university years. A personal detail I don't explain in the legend: one leaf means I've seen the movie, and two means I saw it in theatres. Some of my favorite films are from those years, because I started going to the theatres with friends (๑→‿←๑)

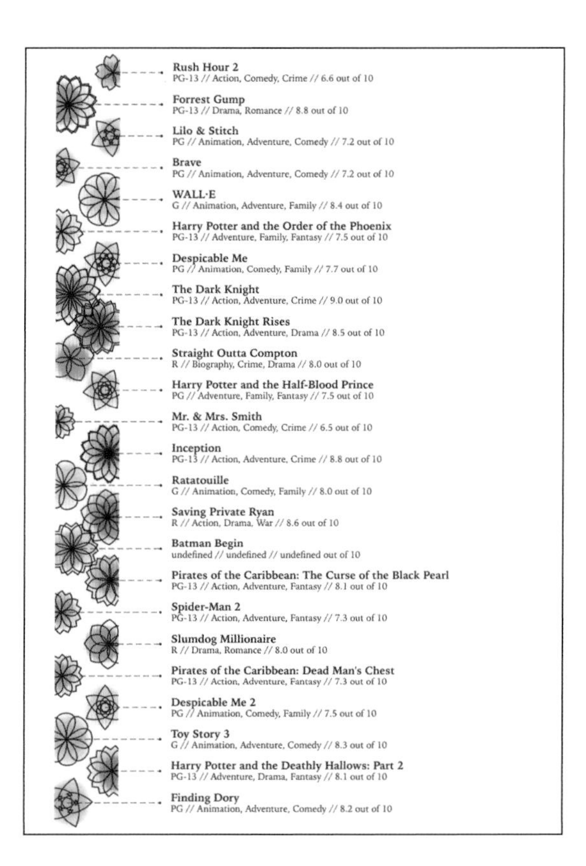

Fig.1.13

I also had fun turning film flowers into washi tape!

Fig.1.14

I wanted to include a legend with the film flowers washi tape I sold. I went through a few iterations until I landed on the final (right).

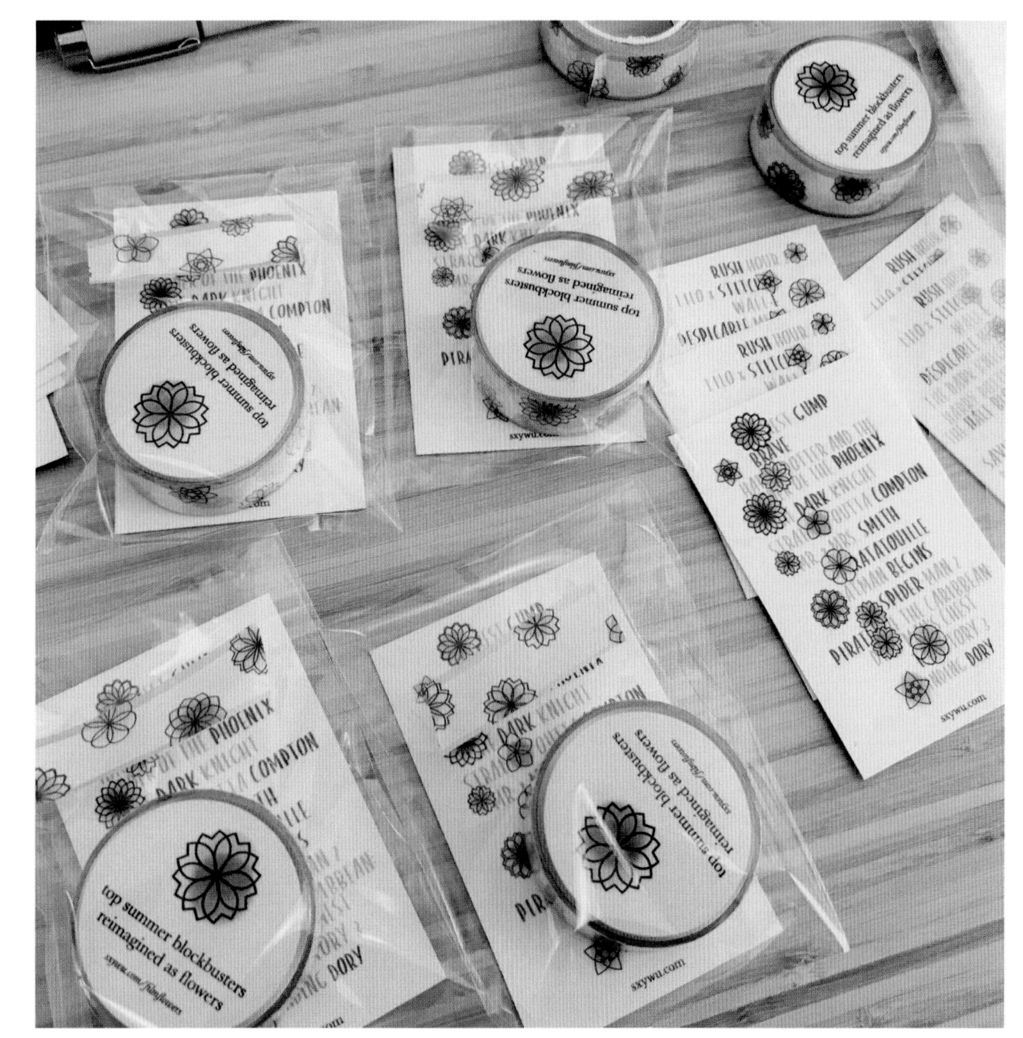

Fig.1.15

The washi tape packed and ready to ship (๑•̀ㅂ•́)و

OLYM

IPICS

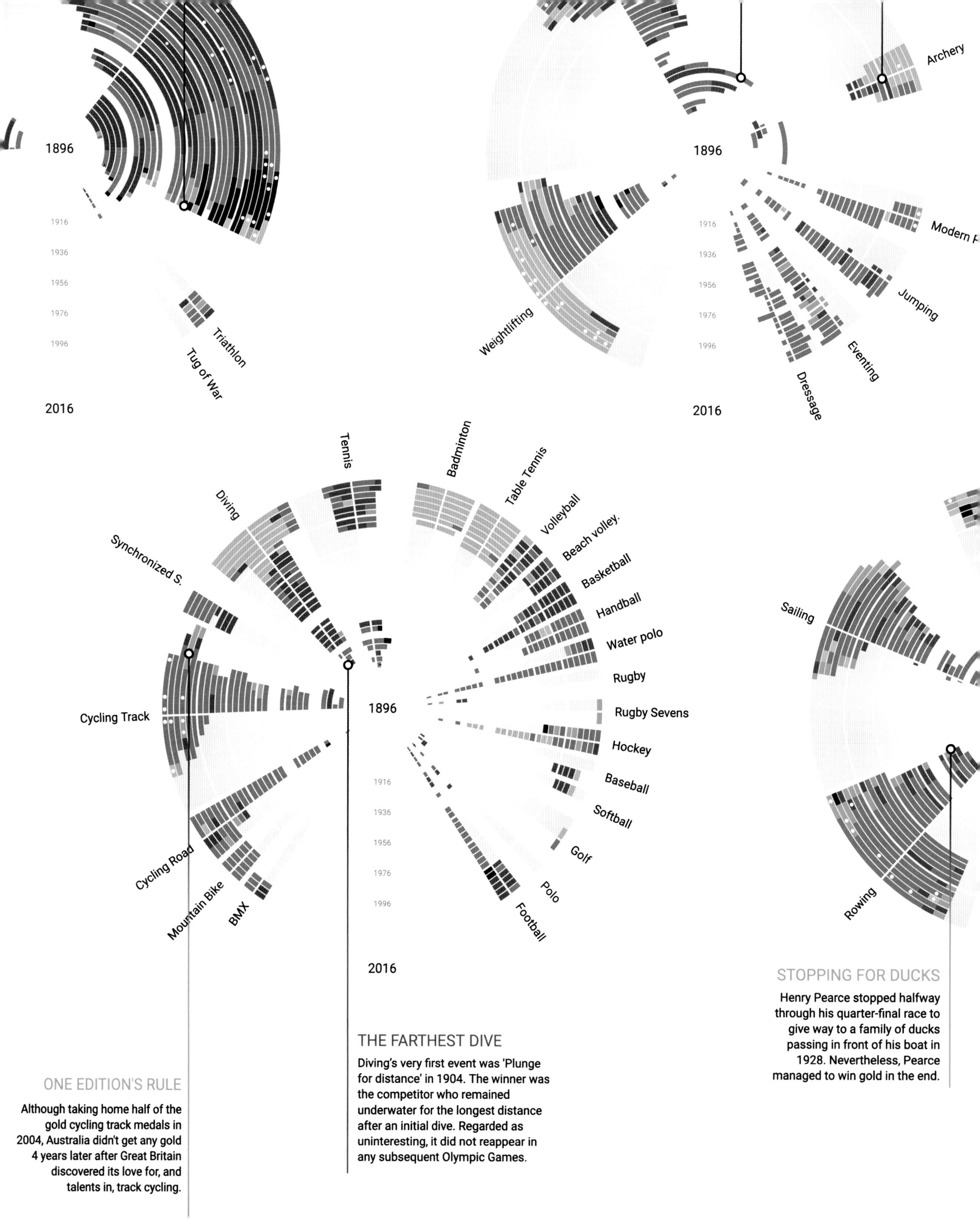

STOPPING FOR DUCKS

Henry Pearce stopped halfway through his quarter-final race to give way to a family of ducks passing in front of his boat in 1928. Nevertheless, Pearce managed to win gold in the end.

THE FARTHEST DIVE

Diving's very first event was 'Plunge for distance' in 1904. The winner was the competitor who remained underwater for the longest distance after an initial dive. Regarded as uninteresting, it did not reappear in any subsequent Olympic Games.

ONE EDITION'S RULE

Although taking home half of the gold cycling track medals in 2004, Australia didn't get any gold 4 years later after Great Britain discovered its love for, and talents in, track cycling.

Olympic Feathers

NADIEH

Before work on this project truly got underway, I already knew what data I wanted to look at: a list of every Olympic gold medalist since the first Olympic Games in 1896. I always love topics that have a rich history; they generally have relatively big and (mostly) diverse datasets. And the Olympics lends itself really well for interesting anecdotes about events that happened in the past, which can be used to point out parts of a visualization.

Data

Thankfully, *The Guardian*, a newspaper in the UK, already created the exact dataset[1] that I needed. It contained a list of every Olympic medalist (gold, silver, and bronze) from 1896 to 2008 (they published the list of 2012 medalists[2] in a separate article). This was a big help because the official website for the Olympics presents the same data in a cumbersome way, making you manually select the edition, sport, and event before you see any results.

The Guardian published these datasets for the 2012 Olympics in London.

After combining the 1896–2008 and 2012 datasets in R, I made two choices to downsize it to something more manageable; I would only look at gold medalists and only count the winning country (and not individual team members) for team events. This would result in a dataset with unique values comprised of the combination of Olympic edition (year), sporting discipline, gender, event, and winner.

For a short explanation of R, please see "Technology & Tools" at the beginning of the book.

However, I noticed the 2012 dataset didn't contain all the medalists. (⊙.⊙) Many team events, such as hockey, were missing and several diving medals weren't in there either. These discrepancies forced me to check each of the 41 sporting disciplines held in 2012 and manually add the information about the missing medalists. This discrepancy made me worry about the 1896–2008 dataset, so I decided to double-check those figures as well. I aggregated the total dataset per Olympic edition to see how many unique gold medals it contained and in how many unique sports and disciplines these were won. Next, I searched through the Wikipedia pages of each Olympic edition to see if the number of events held matched the number of gold medals from my file. This wasn't always the case, but there was always a reason to be found after some more digging. For example, in several years, strangely enough, the names of horses competing in equestrian sports were included in the *Guardian* file (Figure 2.1). This did make for an interesting "quest" to find the horses, but they were easy to locate. (I don't know many other competitors named "Prinzess" or "Lady Mirka"—without a last name—that won gold!) After a few data adjustments, my confidence that all the gold medals were indeed in the dataset was restored.

For some events no gold medal was awarded due to diverse circumstances.

The next step was to prepare an empty dataset for 2016. To do this, I looked at the 2012 events list and a page on the Olympics' website that outlined the differences between the 2012 and 2016 editions. This empty 2016 dataset would make it possible for me to take the data into account during the creation of the visualization. Then, during the 2016 games, I spent 15 minutes each day manually adding every new 2016 gold medalist's name and country to the dataset, and the visualization would be updated automatically.

A major data quirk that I had to reconcile during this project was country names. The map of the world has changed a lot since 1896; countries have disappeared and new ones have been created. Because the five Olympic colors are so well known, and most people think that they represent five continents, I knew from the start that I wanted to color each medal by the continent in which the country lies that won. I therefore had to map each country's National Olympic Committee (NOC) Olympic country code to the official country ISO code, which I could *then* map to a continent using the `countrycode` package in R. For the handful of countries that no longer exist, I manually mapped them to their most obvious current counterpart. I only really needed a correct continent, not the exact country.

Knowing the data visualization colors from the get-go practically never happens for me.

[1] Olympic medal winners dataset from *The Guardian*:
https://www.theguardian.com/sport/datablog/2012/jun/25/olympic-medal-winner-list-data
[2] Olympics 2012 medal breakdown from *The Guardian*:
https://www.theguardian.com/sport/datablog/2012/aug/10/olympics-2012-list-medal-winners

↳ Check Data Accuracy and Completeness

Checking the accuracy of your data is a standard practice in data analysis, and while not the most fun activity, is a lesson we have to constantly relearn. Thankfully, you often don't have to check every value manually. Instead, think about taking sums, counts, and averages and comparing these to common sense values (is the total revenue 10x too big or too small?), or even better, whenever possible, to a different data source (is the total revenue that I get about the same as what was published in the annual report?).

Missing data can actually be harder to find than wrong data. With wrong data you can often see something odd when you look at summary statistics or when you create simple charts. In the Olympics piece I only noticed the missing data because I knew from experience that every recent edition should have exactly one hockey medal for both genders. Once I figured out that the data wasn't complete, I had to think of a different dataset that could show me my data gaps. This doesn't have to be the exact same data of course. Often you don't have a duplicate of the data somewhere else, so you have to think about a proxy instead; it can indirectly show you where something is missing. For this instance that became the number of events that occurred during each edition (data that can be fairly easily found online), which I then compared to the number of gold medals I had in my data.

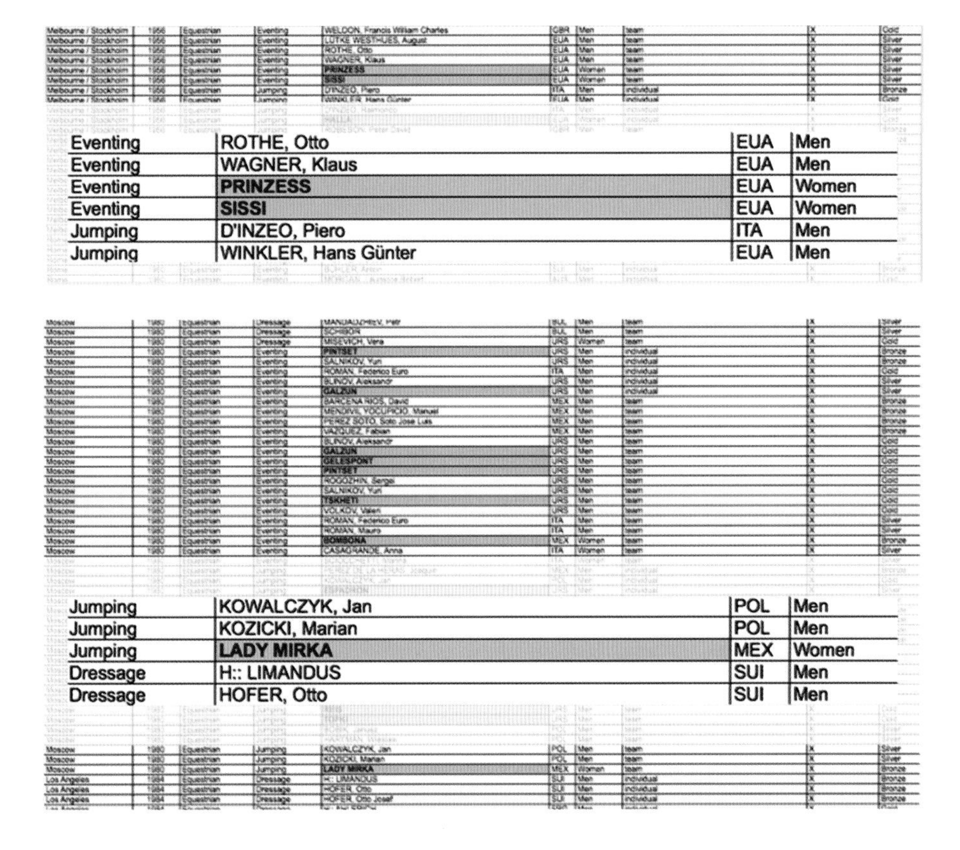

Fig.2.1

For several Olympic editions, the horses were also in the dataset alongside their riders. Not labelled as horses, but as either men or women.

After the Olympics of 2016 ended I added one more level of data to the visual: the current Olympic records and world records. I thought it would be interesting to see which records have been standing for quite some time and which were beaten in 2016.

I couldn't find one nice overview of all the record holders, so I spent 2–3 hours manually searching for and adding these records to my dataset.

Practically all records in athletics and swimming are set by the gold medalists, but this was less often the case in rowing and archery

↳ Add Context Using Remaining Visual Channels

Even if the visualization is mostly complete, I always try to see how I can add more context by visualizing other data variables using remaining "visual channels." Visual channels are the components of a data visualization that we can use to encode data. Position, color, size, and shape are a few examples of channels that can be used. In this case, I added a white circle on top of each medal slice in the cases where they resulted in Olympic records or world records. This can both make the whole more visually appealing and also give the truly interested reader even more ways to dive into and understand the main story that you're trying to convey. But be aware to not go over the top and keep these "extras" to a minimum.

Sketch

The initial inspiration for the visual style of this project came from a very nice design of a peacock's tail made by Ryan Bosse.[3] While admiring the image, I suddenly realized that the peacock's tail feathers could work really well with my data, too. Each sport becomes a "feather," with the symmetry around the middle to distinguish between men's and women's events. Furthermore, the narrow tip at the start that widened towards the outside of the feather could represent each Olympic edition. That way I could place more emphasis on the later editions, using more "pixels" to represent medals won in recent editions versus medals won long ago.

I whipped out my iPad Pro and started sketching some feather shapes (see Figure 2.2). However, I quickly realized that not all sports have exactly one medal per gender. And there were other issues with the sketch, such as the initial feather heights. With different heights, the same years wouldn't align to the same distance from the center, making it more difficult to compare one Olympic edition across all sporting disciplines. But even with all its errors, I felt that the abstract idea was intriguing enough to explore further.

The next morning, I asked a friend and former colleague if he'd like to help me think through the design some more (due to Shirley being on vacation at the time). The process of sketching my idea to him and trying to explain my reasoning helped me realize that there were other logical issues with my design concept (see Figure 2.3). For example, some sporting disciplines have many events in which a medal can be won. Therefore, several feathers would get very wide, such as athletics and swimming. Would that still work as a single feather?

He preferred to just be called "a friend."

Also, the number of events changed a lot throughout the years, with new events being added and others being removed. I thus had to make a choice on what exactly would *define* one medal; would I stretch each edition's number of events to the outside of the feather, or always use the same arc length per medal (i.e., number of degrees)? I went with the latter option, because it felt more consistent across the whole, which created a lot of white space for sports where a large number of events were held during only one edition, such as shooting or archery in 1920. Ultimately, I believed this would give an interesting insight into the history of the modern Olympics.

[3] Image of a peacock tail by Ryan Bosse: https://dribbble.com/rdbosse/projects/230311-Adobe-Creative-Cloud

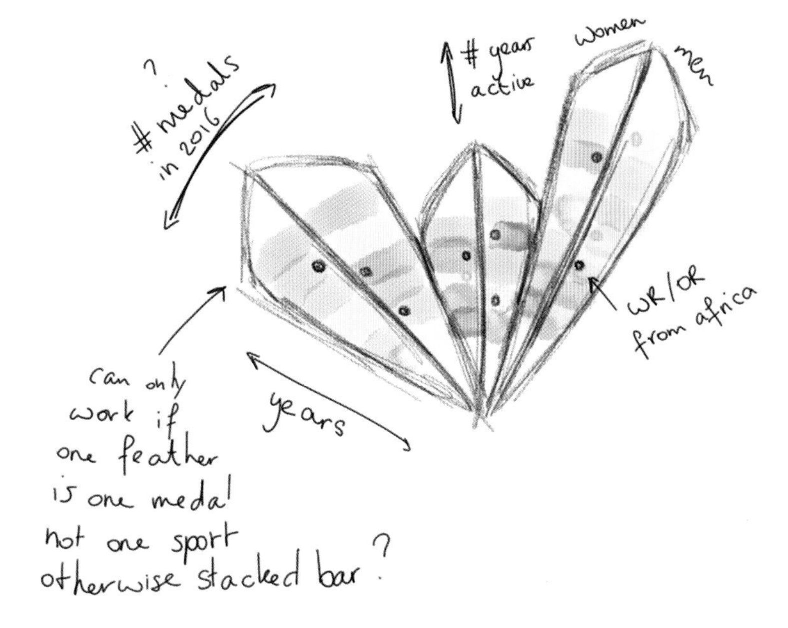

Fig.2.2

The first sketch that I drew where I investigated how the data could be structured as a "feather."

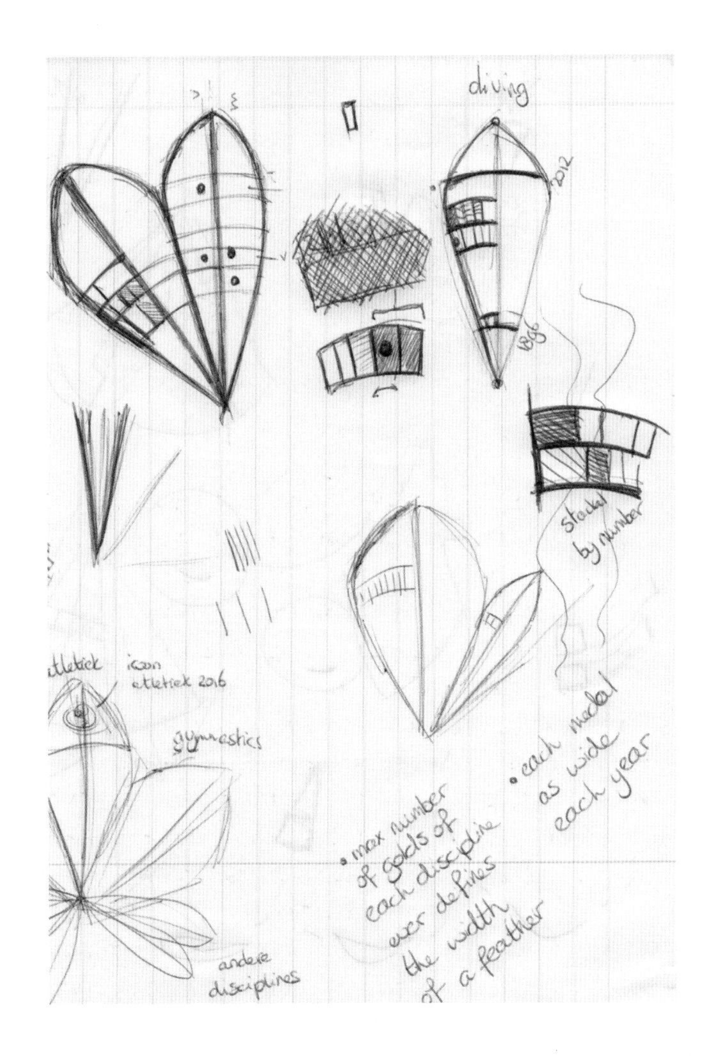

Fig.2.3

Several sketches that arose while trying to explain my idea to a friend and coming across more logical errors in my design.

Now that I decided each medal would be the same arc length and expected that would result in more white space for some sports, I realized it would be a good idea to calculate how "big" a medal would actually become if I placed all 56 sports into one giant circle (which was my initial idea). A few aggregations in R and some math later and I knew for sure that one circle wasn't going to work. Even a medal in the outermost ring of 2016 would barely be a few pixels wide. Using the Olympic logo itself for inspiration, five circles seemed the next best choice.

I sketched out a more detailed version to see how it might look with the new insights I'd gained about the quirks of this particular dataset (see Figure 2.4). At first, I focused on the medals themselves because I wasn't quite sure if they should be sorted in a specific order. But then I decided to sort by continent that won the most medals during that edition to the least, because it added additional layers of insight and context to the data.

After completing this final sketch, it was high time to create it using the actual data to see if it would work with all ±5,000 gold medals.

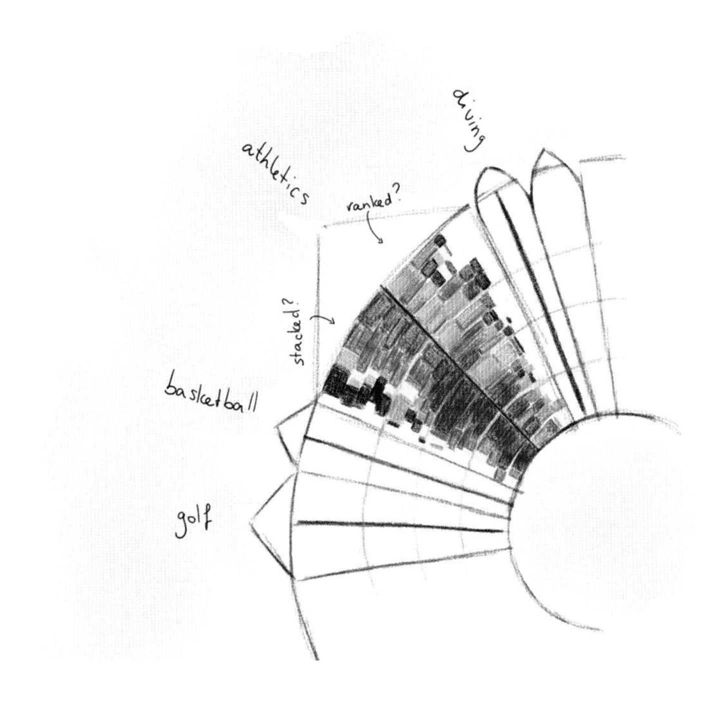

Fig.2.4

Final sketch in which I tested out ordering the continents either always in the same order (stacked) or from the continent that won the most during that edition to the least (ranked).

↳ Sketch to Discover and Remove Thinking Errors

This project required a lot of different sketches for the same basic idea with each version building on lessons learned from the previous drawing. Even when you *think* you have the perfect idea for a visual form in your head, the mind is exceptionally good at glossing over logical faults in the design. It can either be aspects of the data itself that you're not thinking of or that the design is visually not going to work with the particular dataset.

By sketching out the design in your head onto a piece of paper, you'll likely catch any potential issues with your idea early on, saving time later down the line. At this stage, you can iterate on your initial drawing and come up with a new sketch that solves the issues. Don't worry; you don't have to be an artist to sketch out data visualization designs. It's mostly rectangles, circles, and simple curves anyway.

> **If you can't make your design work logically on paper, it's definitely not going to work on the computer with the actual data.**

Code

First, I needed to get those medals on the screen using my carefully prepared (in R) nested medal JSON file. As always, that never happens correctly the first time you think you have enough code typed out to see something decent on the screen. For example, forgetting that, although `rotate` in CSS works with degrees, `Math.cos()`and `Math.sin()` in JavaScript work with radians (Figure 2.5); or when you somehow set the rotation of each sport to the same number (degrees this time) making them overlap each other in Figure 2.6.

Decreasing the medal's opacity helped me figure out what I was doing wrong, until I finally had a rather odd looking (almost abstract) art thing on my screen in Figure 2.7.

Fig.2.5

Five circles where each medal is much too wide, due to using the values for degrees in a sine/cosine function, which stretched each medal out over the entire circle's circumference.

Fig.2.6

Correct medal lengths and placement, but now each sport's slice/feather is rotated to exactly the same degree, making them overplot.

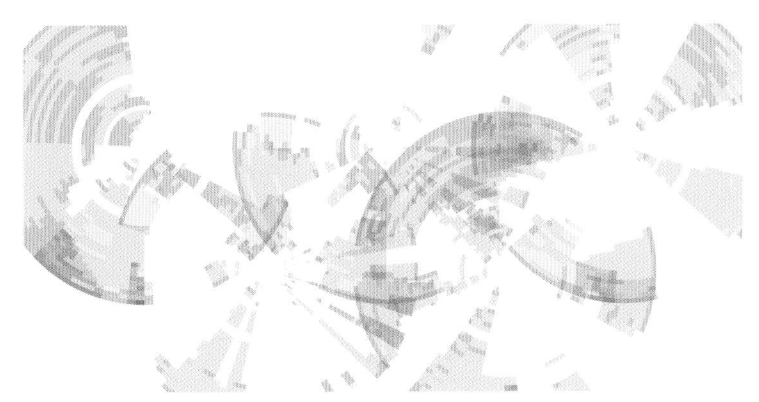

Fig.2.7

Making the medals transparent helped me realize at what point each sport seemed to have been rotated to their expected location.

↳ Precalculate "Visual" Variables

Even though the final visual was made with `D3.js` through JavaScript, it doesn't mean that I had to calculate all the "visual aspects" at the same time (in JavaScript). When dealing with a fixed dataset, which was the case for this Olympics piece, I prefer to precalculate the more complex aspects of the visual because I personally find it easier to perform these calculations in R. But the same applies to any combination of tools. The tool/language that you end up creating your visualization with isn't necessarily the best or easiest tool to figure out aspects such as placements, rotations, sizes, shapes, etc. that are based on the data.

I like to call these types of additions to my data "visual variables." Visual variables have nothing to do with the original dataset, but only apply to how the data will be laid out on the screen. For example, for this project I precalculated the following "visual variables":

- How far each of the five circles as a whole should be rotated to have the opening pointing center down
- How many degrees each of the inner sport feathers had to be rotated, based on the feathers that came before it
- And how far each medal had to be offset from the center of a feather

The only placement variable that I kept calculating "on the fly" in JavaScript was the edition/year scale that decided how far from the center a medal should be drawn. This made it simpler to increase or decrease the circles based on screen width.

These visual variables can either be attached to your original dataset, such as when you've predefined an x and y location per data point, or loaded as a completely separate file and referenced at the right moments. A personal extra benefit in doing this is that it makes my JavaScript a lot more concise. (⌐■_■)

What I love about this type of extremely structured setup, is that once it works for one (nested) object, it works for all of them. Either all ±5,000 medals are still wrongly placed on the screen, or they are all correct after your next code change. You can certainly design with dummy data in a tool such as Sketch or Adobe Illustrator, but it's just impossible to capture the nuance and complexity of seeing your idea applied to *real* data. In Figure 2.8, I finally had all of the medals in place and sporting disciplines labelled.

With the medals in place, I started working on the feather shape. I had not expected that some disciplines would be so exceptionally wide (athletics took up more than half a circle). So I pulled up my now go-to "SVG path playground" website[4] to wrap my head around the math that I needed to mimic the shape of a feather's tip. Luckily, my work with the *Lord of the Rings* visualization from the previous project meant that custom SVG paths were still fresh in my mind and it surprised me how fast I got the feather tip shape to appear (see Figure 2.9).

To make the wider feathers look better, I created multiple tips if a feather was too wide (a simple idea, but not so easy to actually create). A fair amount of iterations later, and having increased my knowledge of custom SVG paths even further, the circles looked a lot better (Figure 2.10).

Fig.2.8

All the medals are finally in their correct location.

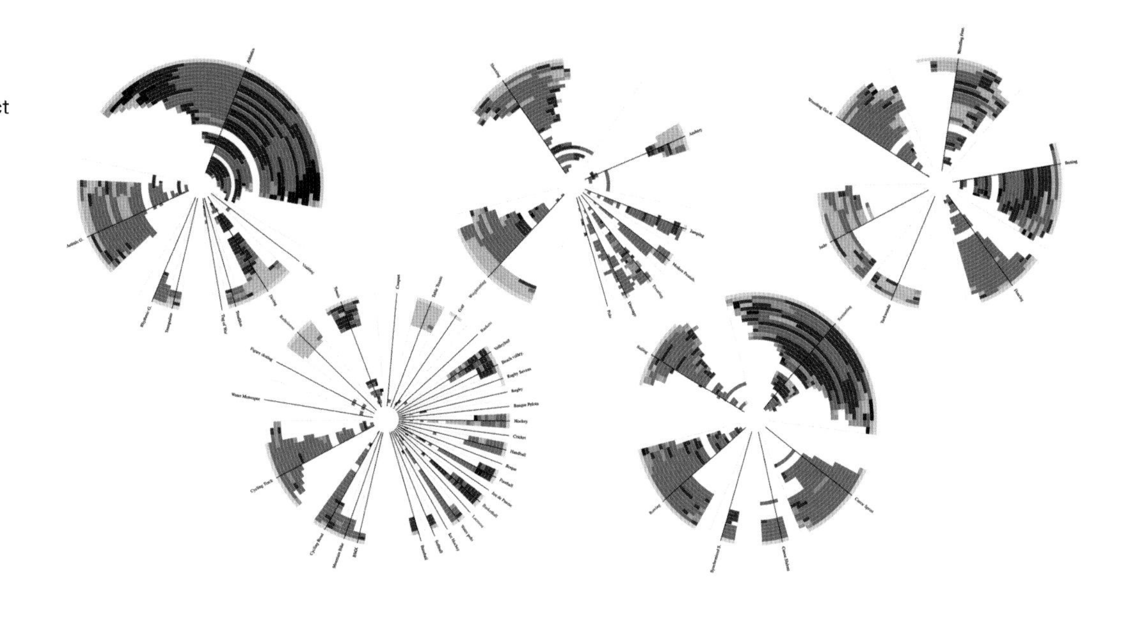

Fig.2.9

Clipping the circles so each sport's feather would end in exactly one "tip."

Fig.2.10

If a sporting discipline was too wide, it got several feather tips.

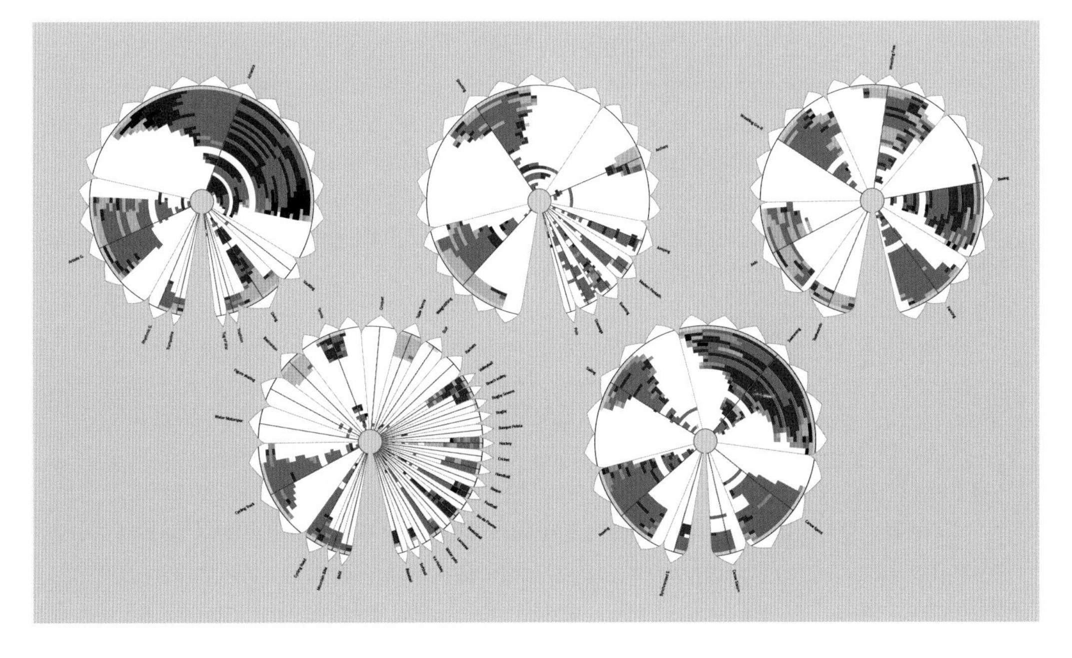

[4] SVG Cubic Bézier Curve Example: http://blogs.sitepointstatic.com/examples/tech/svg-curves/cubic-curve.html

Now that the "broad outline" of the feathers was in place, I focused on changing the details. Figure 2.11 looks a lot like Figure 2.10, but the darker lines were removed completely or turned white. I padded the feathers to make the shapes more pronounced and the feather tips curve just a little bit more smoothly on each end. Finally, I was proud of how the visual looked!

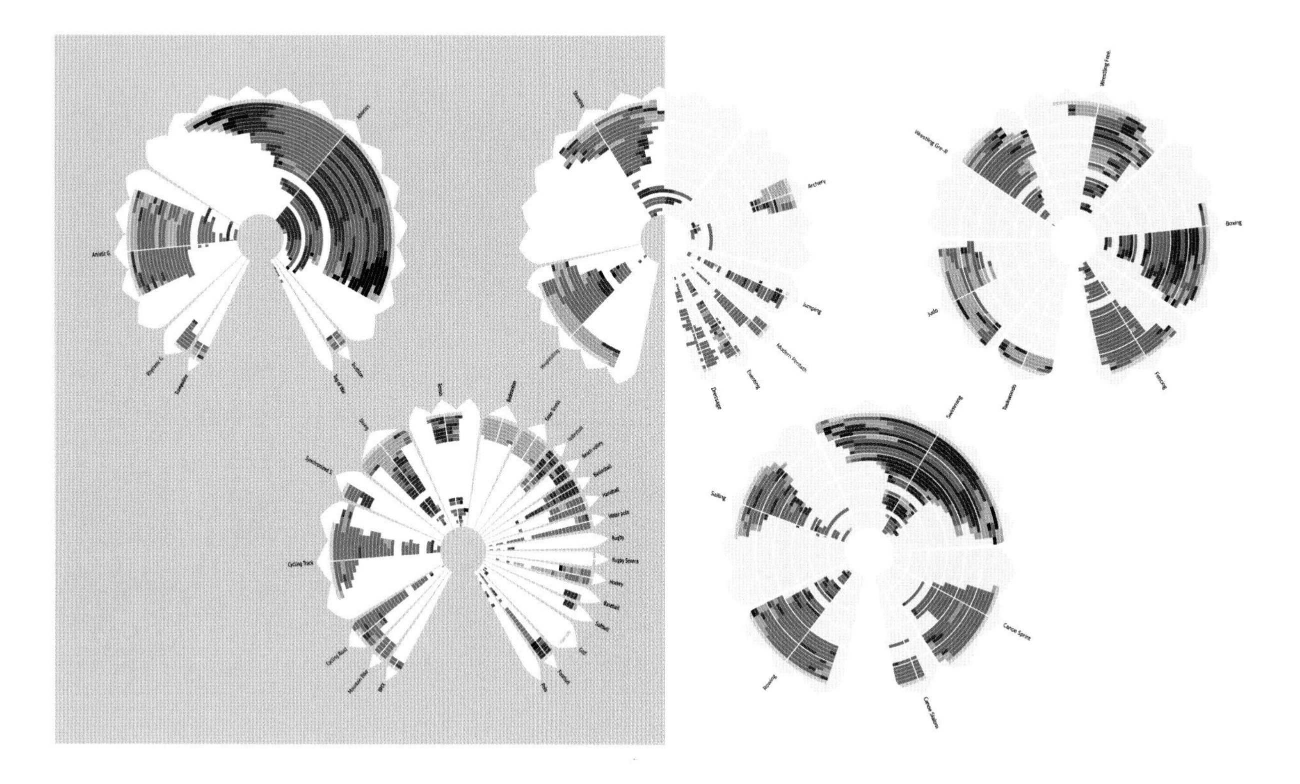

Fig.2.11

Adding several small and subtle changes to the shape of each sport's "feather" to make the whole look a little better.

I sent the image to my friend for feedback. He gently made it clear that the feather tips didn't work with the design and the topic. After a day of denial (all that time spent on the code!), I followed his advice and commented out the code for the feather tips.

That dreary grey background definitely wasn't what I wanted for the final result—it helped to better assess the feather shapes during the development—I wanted to go back to white. However, what should I do with the white inside the circles? Thankfully, my friend came back with a wonderful example that he'd made that used a subtle gradient to fill each circle. The example helped clarify which section of a feather related to the men's and women's events by coloring the backgrounds either bluish or reddish. And it's always a bonus when visual elements are given a data or explanatory function, instead of "just" doing it for visual reasons. After coding the radial gradients into the visual, the feathers looked practically finished (Figure 2.12).

The final big aspect of the overall project was getting the introduction, annotations and legends in there. One textual aspect that I wanted to add were annotations about weird, amazing, or silly things that happened during the history of the games. These types of annotations typically help to "teach" your reader how to actually understand the visual and the data it's showing.

Personally, I prefer the more complex charts that don't show very highly aggregated numbers, but the diversity of the details and even more context around it

For the introduction, my friend showed me a nice example on how to place the title and introductory paragraph side-by-side which was pretty straightforward to reproduce.

As is my usual style, "Olympic Feathers" is definitely not an "understood within five seconds" chart.

Sure, you need to *get* the visual encoding first, but when you do there are many different stories to find and insights revealed. That's just the way I like my visualizations: a little bit of effort to gain a lot of insight.

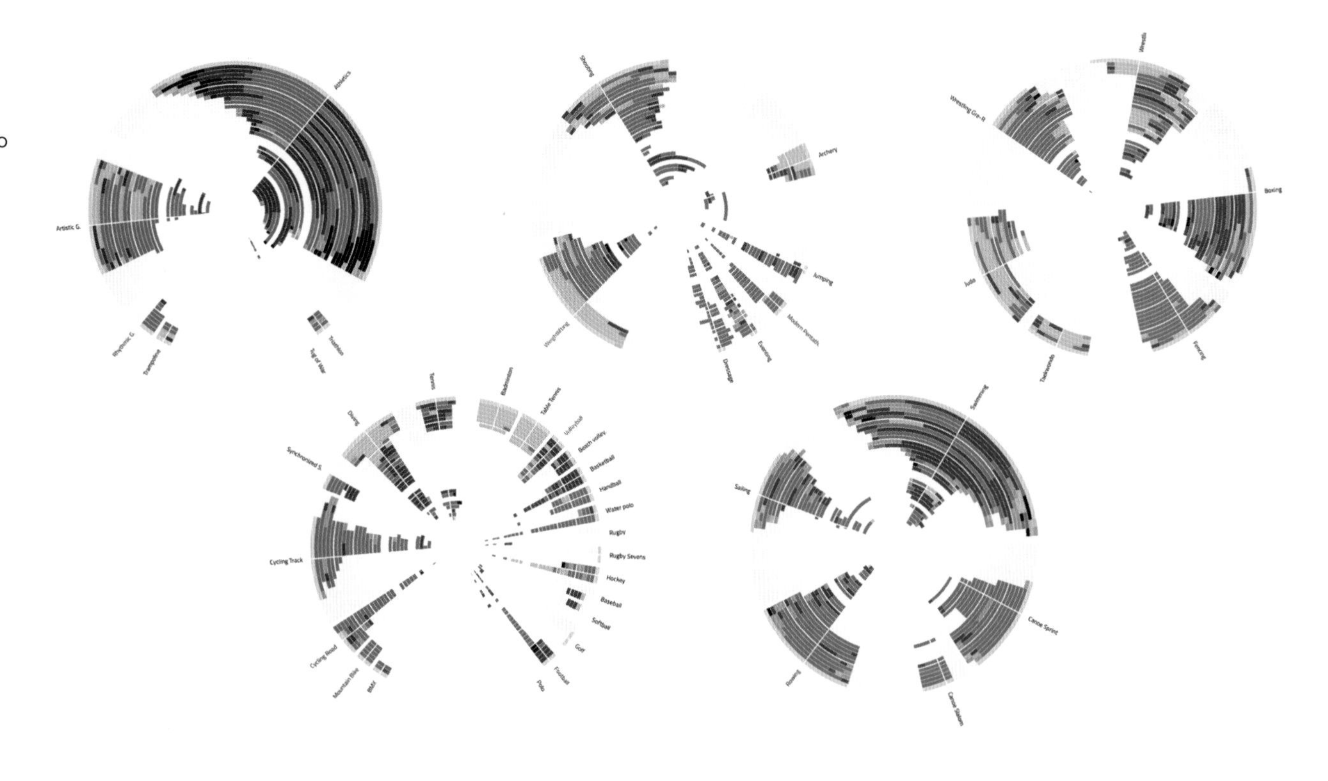

Fig.2.12

No more feather tips and adding subtle radial gradients of two different colors to make the "white space" of the circles stand out and distinguish between the genders.

Reflections

I completed this visual in two weeks using *every second* of free time that I had in my evenings and weekends. Since I wanted to share the final result before the end of the Olympics, data preparation and sketching happened in the week before the Olympics. The programming of the visual occurred during the first week of the Olympics, with daily medal updates for the remainder of Rio 2016 after the visual came out on August 13th. Doing these daily medal updates showed me all kinds of new insights into (the history of) the Olympics and I shared my findings on Twitter with small "annotated" screenshots, such as the ones in Figure 2.13.

August was crazy, hectic, tiring, and fun. I was already a big fan of the Olympics before starting, but after the insights I gained from building and investigating this visual, that only increased.

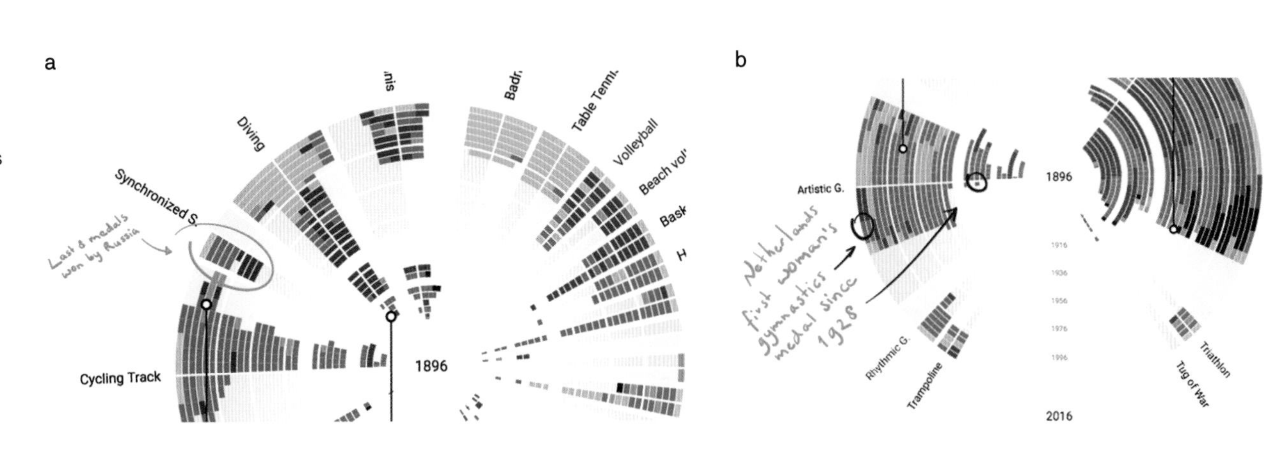

Fig.2.13 (a & b)

Two examples of the annotated screenshots I tweeted to share new insights that I'd found about the Olympic medalists.

Olympic Feathers

↳ OlympicFeathers.VisualCinnamon.com

ALL OLYMPIC GOLD MEDAL WINNERS

summer editions since 1896

More than 5000 Olympic events have had a winner, rewarded with a gold medal from 1904 onwards, in the Summer Olympics since the first games of 1896. Investigate the visuals below to see how each of these medals has been won in the 56 different sporting disciplines that have competed at the games, of which 41 are still held at Rio 2016.

Most of the Olympic sports started out being a men only event. Thankfully this started to change during the 2nd half of the last century. Even the number of medals that can be won for one discipline is slowly becoming the same for both genders. Today at Rio there are 3 disciplines left in which only one gender can compete; the Greco-Roman wrestling, already at the games since the very first edition, is done solely by men. Rhythmic gymnastics & synchronized swimming on the other hand, both at the Olympics since 1984, are only performed by women.

Although Rio could have been celebrating the 31st Olympic Games, 3 editions have been canceled, due to WW I in 1916 and WW II in 1940 & 1944. And yes, Tug of war has truly been part of 5 Olympic Games, from 1900 to 1920. Hover over the medals to see the winning athlete or team or hover over the time-line in the bottom of each circle to find your own interesting stories.

Instead of a medal being represented by a specific width, in these visuals 1 medal always has the same arc length. This makes sure that the more recent the edition of the games, the more emphasis it gets due to the increasing size of the ring

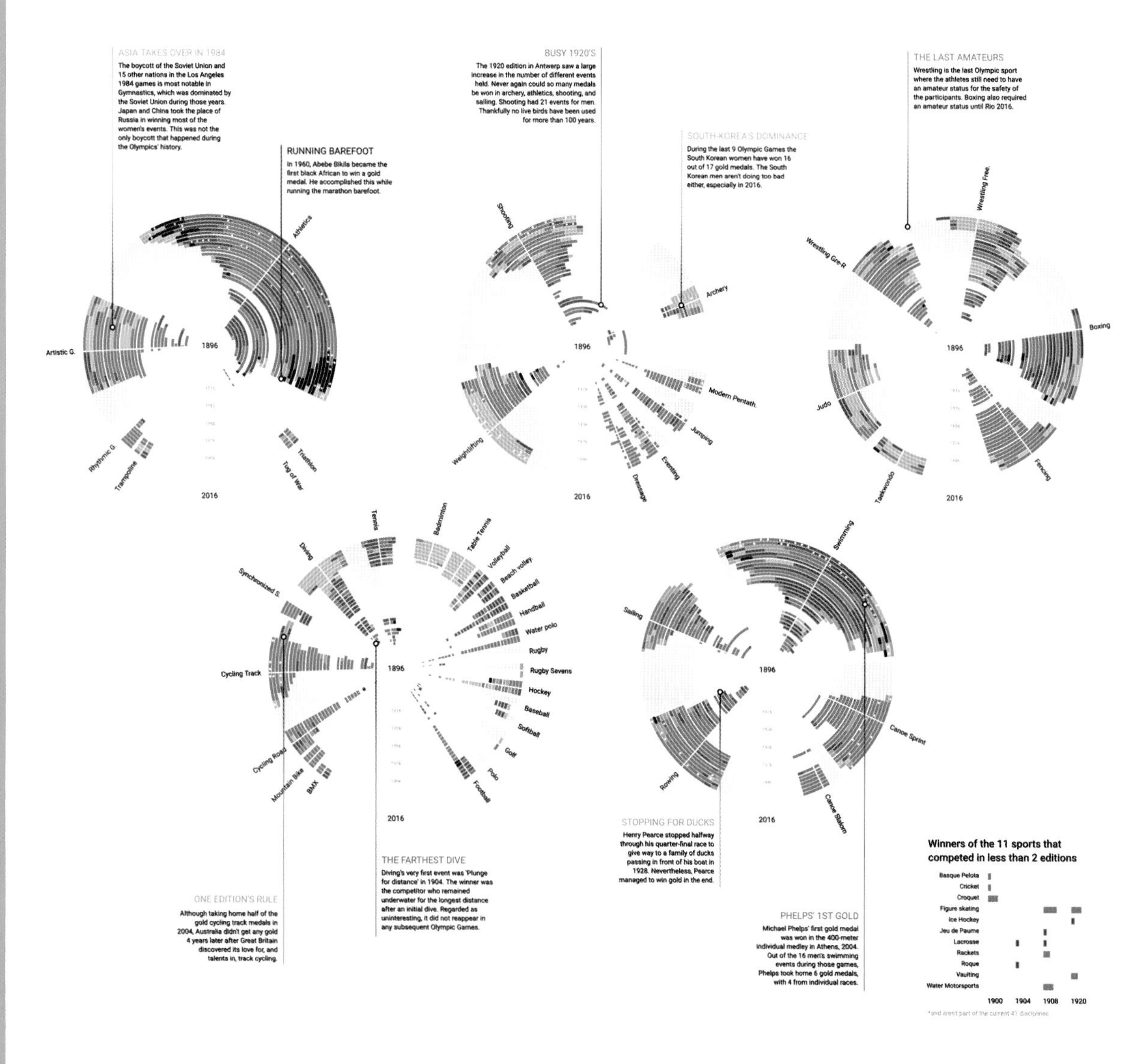

ASIA TAKES OVER IN 1984
The boycott of the Soviet Union and 15 other nations in the Los Angeles 1984 games is most notable in Gymnastics, which was dominated by the Soviet Union during those years. Japan and China took the place of Russia in winning most of the women's events. This was not the only boycott that happened during the Olympics' history.
RUNNING BAREFOOT
In 1960, Abebe Bikila became the first black African to win a gold medal. He accomplished this while running the marathon barefoot.
BUSY 1920'S
The 1920 edition in Antwerp saw a large increase in the number of different events held. Never again could so many medals be won in archery, athletics, shooting, and sailing. Shooting had 21 events for men. Thankfully no live birds have been used for more than 100 years.
SOUTH-KOREA'S DOMINANCE
During the last 9 Olympic Games the South Korean women have won 16 out of 17 gold medals. The South Korean men aren't doing too bad either, especially in 2016.
THE LAST AMATEURS
Wrestling is the last Olympic sport where the athletes still need to have an amateur status for the safety of the participants. Boxing also required an amateur status until Rio 2016.
Athletics
Artistic G.
Rhythmic G.
Trampoline
Triathlon
Tug of war
1896
2016
Shooting
Archery
Modern Pentath
Jumping
Eventing
Dressage
Weightlifting
Wrestling Gre-R
Wrestling Free.
Boxing
Judo
Taekwondo
Fencing
Diving
Tennis
Badminton
Table Tennis
Volleyball
Beach volley
Basketball
Handball
Water polo
Rugby
Rugby Sevens
Hockey
Baseball
Softball
Golf
Polo
Football
Synchronized S.
Cycling Track
Cycling Road
Mountain Bike
BMX
Swimming
Sailing
Rowing
Canoe Sprint
Canoe Slalom
STOPPING FOR DUCKS
Henry Pearce stopped halfway through his quarter-final race to give way to a family of ducks passing in front of his boat in 1928. Nevertheless, Pearce managed to win gold in the end.
THE FARTHEST DIVE
Diving's very first event was 'Plunge for distance' in 1904. The winner was the competitor who remained underwater for the longest distance after an initial dive. Regarded as uninteresting, it did not reappear in any subsequent Olympic Games.
ONE EDITION'S RULE
Although taking home half of the gold cycling track medals in 2004, Australia didn't get any gold 4 years later after Great Britain discovered its love for, and talents in, track cycling.
PHELPS' 1ST GOLD
Michael Phelps' first gold medal was won in the 400-meter individual medley in Athens, 2004. Out of the 16 men's swimming events during those games, Phelps took home 6 gold medals, with 4 from individual races.
Winners of the 11 sports that competed in less than 2 editions
Basque Pelota
Cricket
Croquet
Figure skating
Ice Hockey
Jeu de Paume
Lacrosse
Rackets
Roque
Vaulting
Water Motorsports
1900
1904
1908
1920
*and aren't part of the current 41 disciplines

HOW TO READ A FEATHER

Each circle represents a grouping of several different (but approximately) similar themed sports, such as water or ball sports. Within a circle we find slices. Let's call each slice a feather to make it easier to distinguish as a whole. Each feather represents one discipline.

A feather is split up into 31 sections, radiating outward. Starting from the first Olympic Games in 1896 at the center to the current Olympic Games in Rio in 2016 at the other end. Each discipline is twice as wide as the maximum number of medals that could ever be won during one edition for a gender (men and women get the same width).

The next split is by gender. For the example feather to the right, the small bars going upward on the light red background are gold medals won by women. The bars going towards the bottom, with the light blue background are gold medals won by men.

All the medals have the same arc length and you can see in the bottom (men) section of the example feather to the right how wide 1 medal is for each edition of the Olympics. For medals won by a men & woman team or two gold medals in the same event each person gets 0.5 medal assigned. The medal bars are colored according to the continent in which the country of the winning athlete or team lies. Furthermore, for each edition and gender, the bars are stacked from the continent that won the most medals to the least.

Finally, some sport disciplines have Olympic records, such as athletics and swimming. As an extra level of detail, the events in which the gold medalist reached a currently standing Olympic record (after Rio 2016) are marked with a white dot. You can see the record when you hover over the medal.

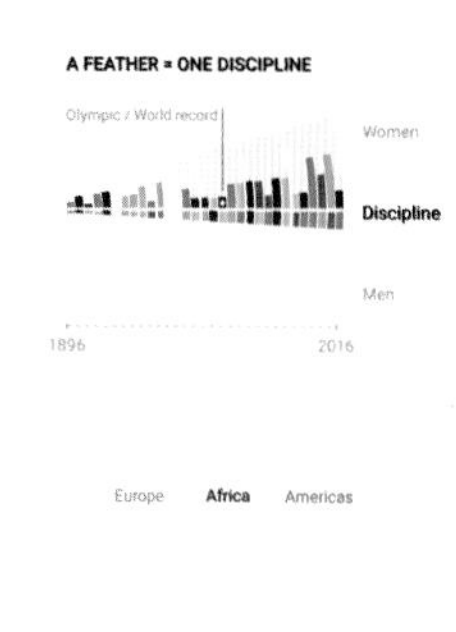

Fig.2.14 a

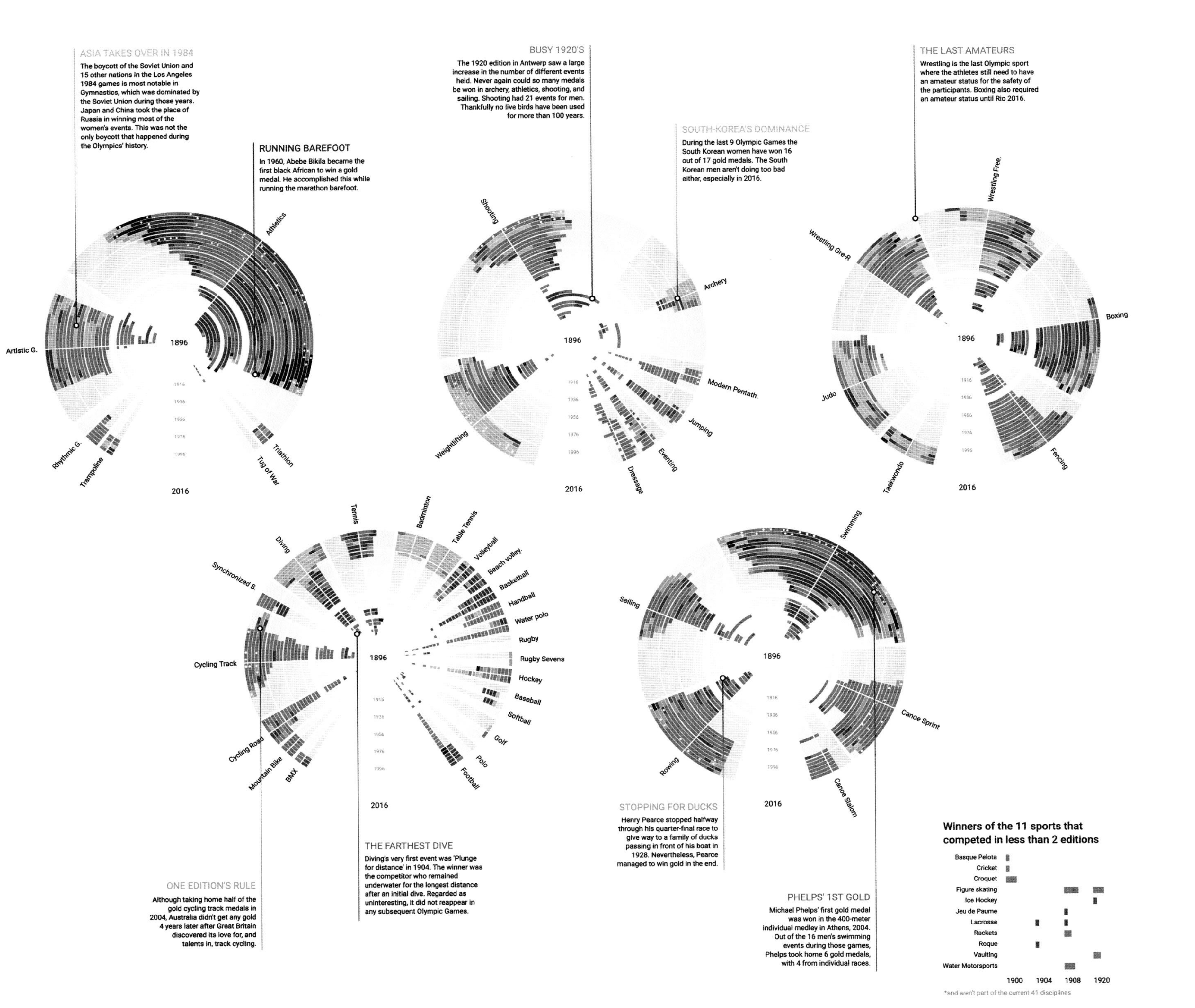
ASIA TAKES OVER IN 1984
The boycott of the Soviet Union and 15 other nations in the Los Angeles 1984 games is most notable in Gymnastics, which was dominated by the Soviet Union during those years. Japan and China took the place of Russia in winning most of the women's events. This was not the only boycott that happened during the Olympics' history.
RUNNING BAREFOOT
In 1960, Abebe Bikila became the first black African to win a gold medal. He accomplished this while running the marathon barefoot.
Athletics
Artistic G.
1896
Rhythmic G.
Trampoline
Triathlon
Tug of War
2016
BUSY 1920'S
The 1920 edition in Antwerp saw a large increase in the number of different events held. Never again could so many medals be won in archery, athletics, shooting, and sailing. Shooting had 21 events for men. Thankfully no live birds have been used for more than 100 years.
SOUTH-KOREA'S DOMINANCE
During the last 9 Olympic Games the South Korean women have won 16 out of 17 gold medals. The South Korean men aren't doing too bad either, especially in 2016.
Shooting
Archery
Modern Pentath.
Jumping
Eventing
Dressage
Weightlifting
THE LAST AMATEURS
Wrestling is the last Olympic sport where the athletes still need to have an amateur status for the safety of the participants. Boxing also required an amateur status until Rio 2016.
Wrestling Free.
Wrestling Gre-R
Boxing
Judo
Taekwondo
Fencing
Tennis
Badminton
Table Tennis
Volleyball
Beach volley
Basketball
Handball
Water polo
Rugby
Rugby Sevens
Hockey
Baseball
Softball
Golf
Polo
Football
Diving
Synchronized S.
Cycling Track
Cycling Road
Mountain Bike
BMX
THE FARTHEST DIVE
Diving's very first event was 'Plunge for distance' in 1904. The winner was the competitor who remained underwater for the longest distance after an initial dive. Regarded as uninteresting, it did not reappear in any subsequent Olympic Games.
ONE EDITION'S RULE
Although taking home half of the gold cycling track medals in 2004, Australia didn't get any gold 4 years later after Great Britain discovered its love for, and talents in, track cycling.
Swimming
Sailing
Rowing
Canoe Sprint
Canoe Slalom
STOPPING FOR DUCKS
Henry Pearce stopped halfway through his quarter-final race to give way to a family of ducks passing in front of his boat in 1928. Nevertheless, Pearce managed to win gold in the end.
PHELPS' 1ST GOLD
Michael Phelps' first gold medal was won in the 400-meter individual medley in Athens, 2004. Out of the 16 men's swimming events during those games, Phelps took home 6 gold medals, with 4 from individual races.
Winners of the 11 sports that competed in less than 2 editions
Basque Pelota
Cricket
Croquet
Figure skating
Ice Hockey
Jeu de Paume
Lacrosse
Rackets
Roque
Vaulting
Water Motorsports
1900
1904
1908
1920
*and aren't part of the current 41 disciplines

Fig.2.14 b

A zoom in on the five circles.

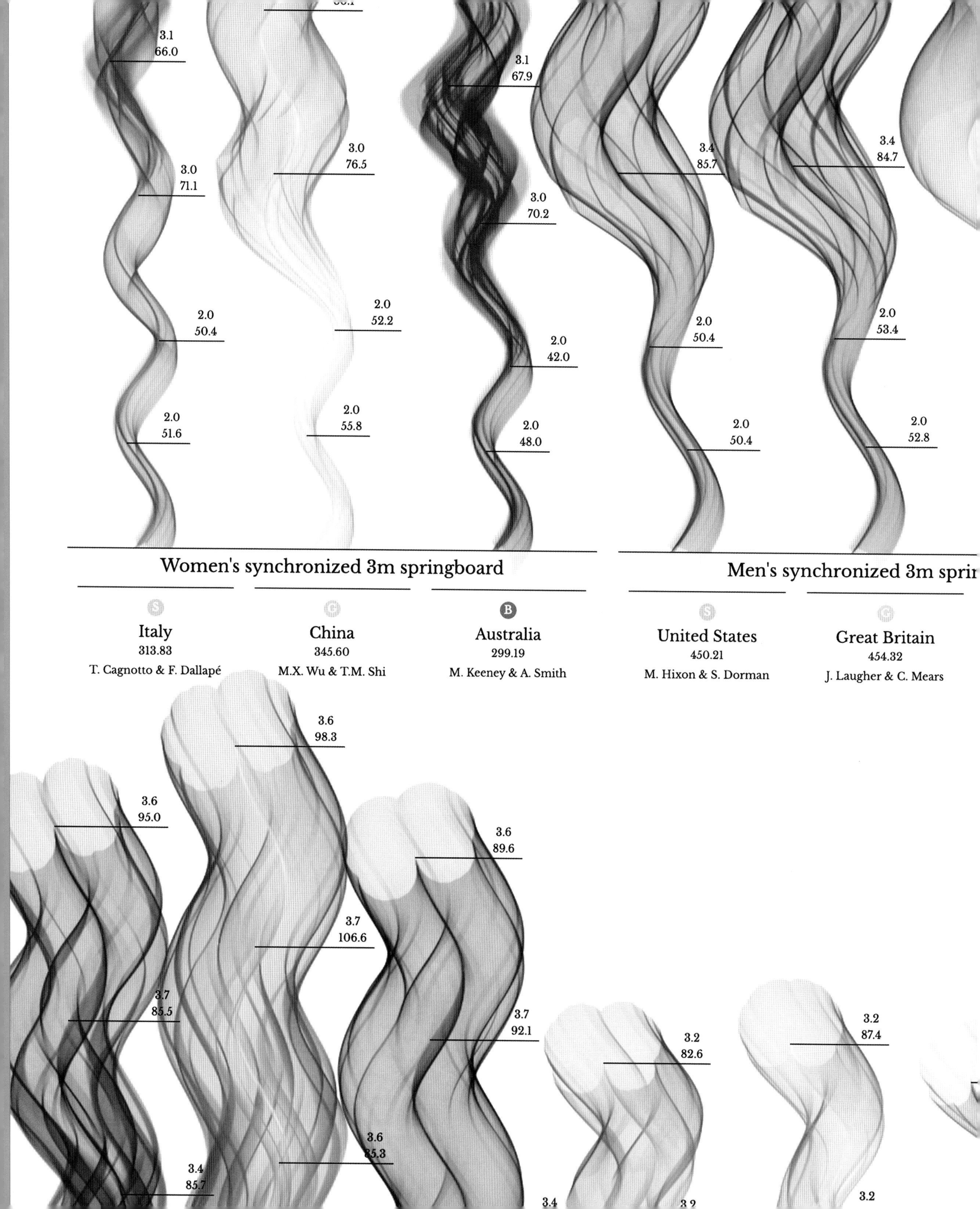

3.1
66.0
3.0
76.5
3.0
71.1
2.0
52.2
2.0
50.4
2.0
55.8
2.0
51.6
3.1
67.9
3.0
70.2
2.0
42.0
2.0
48.0
3.4
85.7
2.0
50.4
2.0
50.4
3.4
84.7
2.0
53.4
2.0
52.8
Women's synchronized 3m springboard
S
Italy
313.83
T. Cagnotto & F. Dallapé
G
China
345.60
M.X. Wu & T.M. Shi
B
Australia
299.19
M. Keeney & A. Smith
Men's synchronized 3m sprir
S
United States
450.21
M. Hixon & S. Dorman
G
Great Britain
454.32
J. Laugher & C. Mears
3.6
98.3
3.6
95.0
3.6
89.6
3.7
106.6
3.7
85.5
3.7
92.1
3.2
82.6
3.2
87.4
3.6
85.3
3.4
85.7
3.4
3.2

Dive Fractals

SHIRLEY

Ever since I was little, I've loved watching the summer Olympics, and out of all the events, I especially loved gymnastics and diving. And even though that's most likely because I come from a Chinese household and they were the only Olympic sports we watched, I also loved them for their grace and beauty.

I binged both sports during the 2016 Olympics and ended up enjoying synchronized diving the most. That event wasn't around when I was a kid, and I marveled at the trust and respect that each partnership showed and at what it must take to be so in sync with each other. The divers seemed to be driven by a responsibility to do well for their partner and by a fear of letting each other down. When the teams medaled, their tearful embraces were beautifully emotional.

I wanted to recreate that beauty somehow.

Data

When I first started brainstorming on the data, I thought I could do something with their scores. The women's diving events have five rounds and men's have six. Each round consisted of a difficulty score, six execution scores, and five synchro scores. I wanted to get the scores of the top three teams for each event.

After many Google searches, I could only find the overall scores and none of the breakdowns online, so I decided to gather them myself. It was absolutely painful; I re-watched all four synchronized diving events and manually entered all of the scores into a JSON file (Figure 2.1). Not only did I have to keep pausing and rewinding to make sure I didn't miss any of the teams' scores for a round, but I had to be really careful about putting the array brackets in the right places (the numbers all started to blend together after a while).

For a short explanation of JSON, or JavaScript Object Notation, please see "Technologies & Tools" at the beginning of the book.

Fig.2.1

The JSON data for a single team for a single event.

```
{
  "event": "Women's synchronized 3m springboard",
  "event_key": "event1",
  "date": "Aug 7, 2016",
  "country": "Italy",
  "athletes": ["T. Cagnotto", "F. Dallapé"],
  "total": 313.83,
  "gender": "W",
  "medal": "silver",
  "breakdown": [
    [2, 51.6, [8.5, 8.5, 8, 8.5, 8.5, 8.5], [9, 8.5, 9, 5.5, 8.5], "Back Dive"],
    [2, 50.4, [8, 8, 8.5, 8.5, 8.5, 9], [8.5, 8.5, 8.5, 8, 8.5], "Reverse Dive"],
    [3, 71.1, [7.5, 7.5, 8, 8, 8, 8], [8, 7.5, 8, 8.5, 8], "Forward 2½ Somersault 1 Twist"],
    [3.1, 66.03, [8, 7.5, 7.5, 6.5, 7, 7], [7, 7, 7, 7, 7.5], "Forward 3½ Somersaults"],
    [3, 74.7, [8, 8, 8, 7.5, 8, 8], [8.5, 8.5, 8.5, 8.5, 8.5], "Inward 2½ Somersaults"]
  ]
}
```

I originally decided on a JSON file because I thought I needed a nested structure for the score breakdowns. But in retrospect, I should have used a flat structure with a column for each score and Excel or Google Sheets to manage my data gathering process. I could then programmatically get the data into the structure I needed (nested or not), and I wouldn't have had to struggle so long to make sure all the array brackets and decimal points were in the right places.

One of the most important lessons I learned from Nadieh during my work on *Data Sketches* is to do my manual data gathering and cleaning in Excel or Google Sheets; they make the whole process go faster.

It was only after I was almost done with the data gathering process that I finally found a page on the Rio Olympics website with the score breakdowns I was looking for.

I'll be honest, I threw a fit , I threw a fit over that one (ノಠ益ಠ)ノ彡┻━┻

Sketch

Because I was traveling for the first three weeks of the month, I knew I didn't have the time to come up with different sketches. Thankfully, sometime during the middle of my trip, a friend of mine sent me a link to Yuri Vishnevsky's experimentation with algorithmic art and his attempts at generating a set of unique business cards (Figure 2.2). Some of them in particular looked like the top-down view after the divers entered the water, and I knew I wanted to have a similar effect for my visualization.

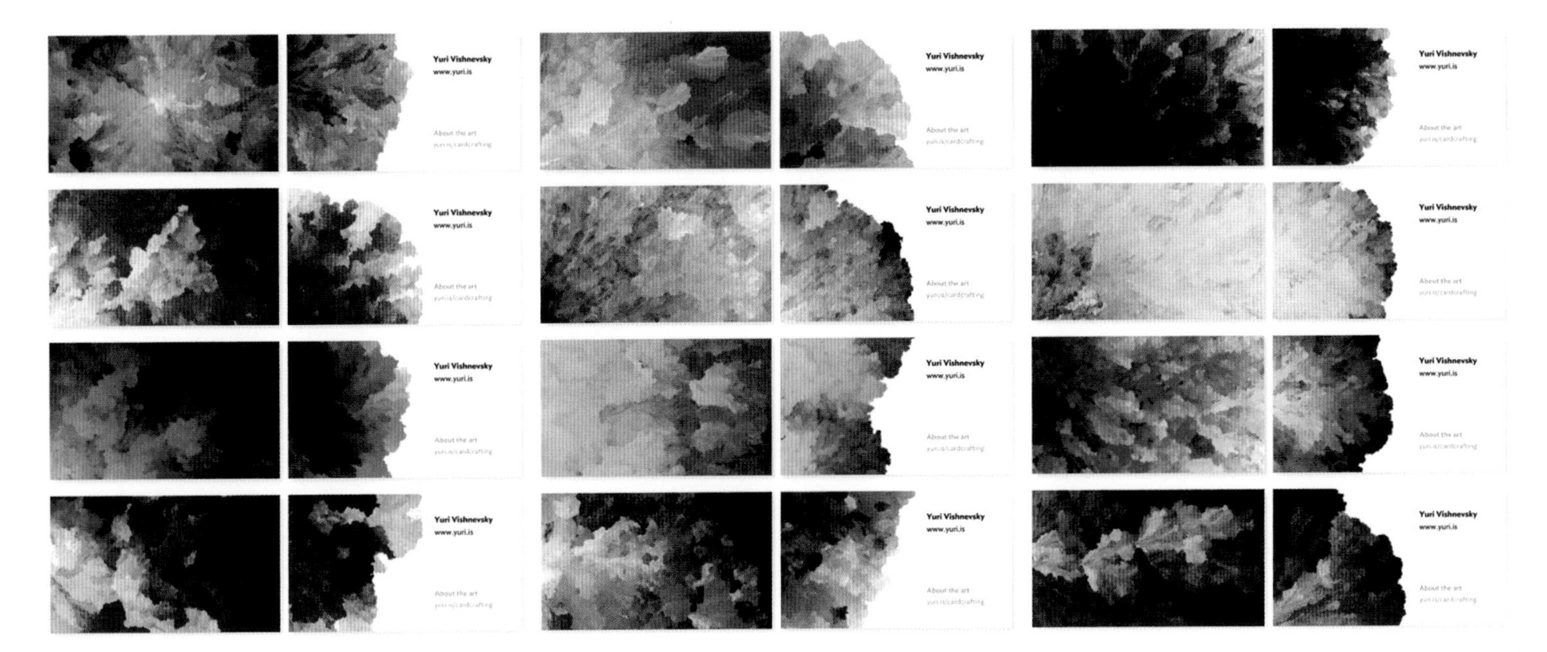

Fig.2.2

Yuri Vishnevsky's algorithmically generated business cards.

In the write-up for his project, Yuri didn't provide a link to his code, but he did mention the term "fractal search algorithm." A quick search for "fractal search algorithm javascript" led me to Dan Gries' tutorial for "Sweeping Fractal Lines" (Figure 2.3).

Gorgeous. And though it didn't look like the water splash in Yuri's project, it reminded me of silk flowing through water or of ink in water. I thought: how beautiful would it be if I could make the divers' data look like this?

Fig.2.3

Dan Gries' "Sweeping Fractal Lines."

Code

The first few days were spent just trying to understand Dan's code. At only ±250 lines, it was beautiful in its brevity. As with any piece of code I wanted to understand, I printed out all the lines and went through them line by line:

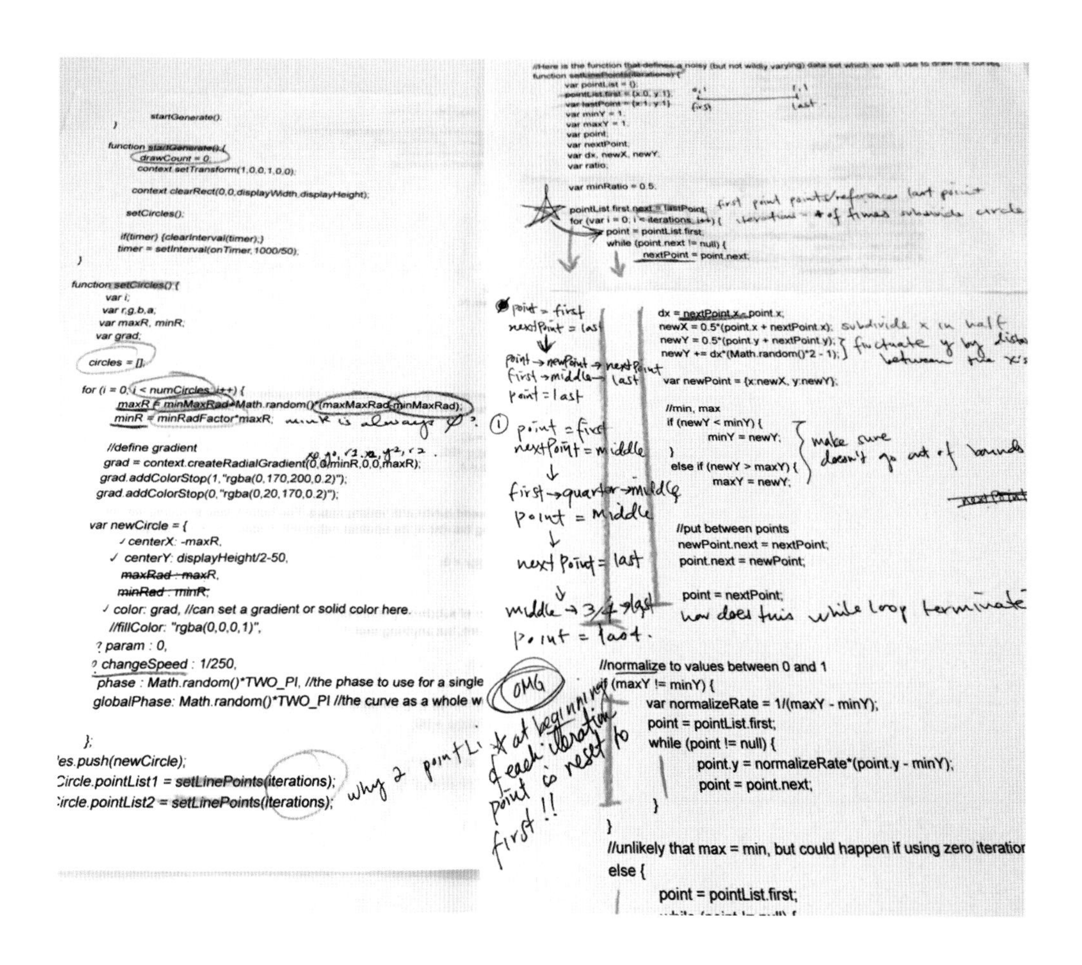

Fig.2.4

A printout of Dan's code with my notes. (I can't say that my notes are the most coherent.)

The key was to understand enough of what was going on to be able to replace the randomly generated data with my diving data. After the first day, I was able to boil it down to two key functions:

- `setLinePoints` implements the "fractal subdivision process," which takes a line, recursively divides it in the middle, and varies the *y*-position of that midpoint. This programmatically creates a "squiggly" line.
- `onTimer` is called on every page repaint (triggered by `requestAnimationFrame`, a browser method for animation) and does most of the heavy lifting: it is passed two of the "squiggly" lines, one for starting, and one for ending. On every repaint, it slowly "morphs" (interpolates) from the starting to ending line, drawing to canvas as it goes. And as soon as it hits the ending line, it assigns the ending line as the new starting lines, generates a new ending line with `setLinePoints`, and starts over. This happens until it hits the edge of the screen.

This is very similar to how I generated the watercolor effect in my "Nature" chapter.

From there, I knew how I wanted to map my data:

- Each round would be a distinct starting line generated with `setLinePoints`
- The radius of each circle is denoted by the difficulty score of that round's dive
- The height between each starting and ending line is the score the divers received for that round
- The colors are the two primary colors on that team's country flag

For guidance on interpolators in animation, check out the lesson "Custom Animations" on page 269 of my "Nature" chapter. For translating lines to circles, refer to the lesson "Trigonometry" on page 109 of my "Travel" chapter.

I thought this mapping would help me compare both the difficulty of each dive (radius) and overall score (height) of each team.

My first attempt at adapting Dan's code took a while and didn't turn out quite as I expected. I understood `setLinePoints` quite well and thus was able to use the divers' execution scores to generate the "squiggly" circles, but didn't understand some of the key parts of `onTimer`. And because `onTimer` took care of actually drawing the circles, I ended up with disjointed lines (Figure 2.5, left).

Not quite what I was expecting. I wasn't grasping how Dan was interpolating the shapes of the curves, creating the weird discontinuity between sections. I knew then that I had to re-implement Dan's code in a way that I could understand. So I kept most of Dan's code from `setLinePoints`, but completely rewrote and separated `onTimer` into three separate functions: the first function to transition from each starting line to the next using a custom `D3.js` *interpolator*, the second to convert the "squiggly" line to a circle, and a third to draw the circle to canvas. This worked marvelously (Figure 2.5, right).

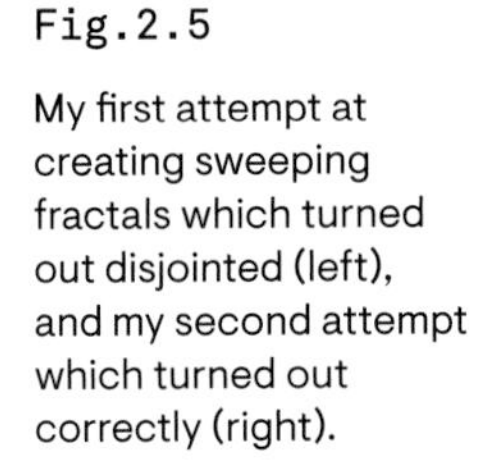

Fig.2.5

My first attempt at creating sweeping fractals which turned out disjointed (left), and my second attempt which turned out correctly (right).

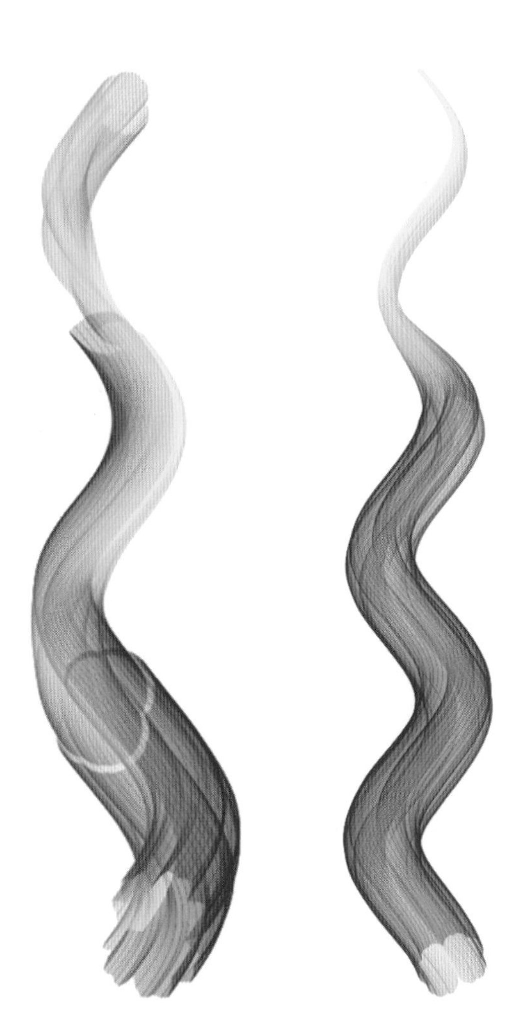

SVG vs. Canvas

SVG and the Canvas API are the two most common ways of drawing shapes on a web page, and they both have their pros and cons—mostly around usability and performance.

With SVG, each shape is represented as an individual DOM element in the browser. I can set attributes to position and size the shape, set styles, and register event listeners (hover, click, etc.) on it.

I think of DOM, or Document Object Model, as the way the browser keeps track of all the elements (`divs` and `spans` being the most common) that need to be rendered or styled in a webpage.

> **SVG is the best for visualizing a dataset that doesn't need to update and registering some basic user interactions like click and hover.**

Once I have to update the visualization to reflect updates in the data, using SVG becomes more complicated. I have to keep track of which new SVG elements need to be created and inserted into the DOM, which need to be removed, and which elements' attributes have to be updated; `D3.js`'s "enter-update-exit" pattern helps manage this, but even that can get complex and error-prone. In those situations with a smaller dataset but complicated updates, I like to use SVG with `React.js` or `Vue.js` managing the DOM. But once I have a bigger dataset and need to draw and animate more than a thousand SVG elements in the DOM, I like to use the Canvas API instead.

For an introduction to SVG, see the "Drawing with SVG" lesson on page 50 of my "Movies" chapter.

For a short explanation of React.js and Vue.js, see "Technologies & Tools" at the beginning of the book.

Canvas, unlike SVG, is just one DOM element (which makes it more performant and recommended for large datasets). We draw shapes within canvas with a series of JavaScript commands, and once they're drawn, we no longer have access to the shapes. This is because canvas has no notion of individual shapes (as the shapes aren't individual DOM elements) and instead keeps track of each pixel and its RGB values to draw the image.

And even though canvas requires more code to get started (especially when drawing more complex shapes), once we have the code to draw the initial dataset, updating is much easier; we clear the canvas and redraw with the new dataset. On the other hand, because canvas doesn't track individual shapes, handling animations and user interactions is trickier (but still doable).

I think of them as SVG is like Illustrator and canvas is like Photoshop.

For guidance on how to implement animations in canvas, see lesson "Custom Animations" on page 269 of my "Nature" chapter. For how to handle user interactions, see lesson "Canvas Interactions" on page 181 of my "Books" chapter.

> **In general, I prefer canvas when I'm creating complex, interactive visualizations that animate because it's so much more performant.**

	SVG	Canvas
Pro	Easy to get started Easy to register user interactions Easy to animate	Very performant Easy to update
Cons	Potentially complex DOM updates Not performant for large number of elements	More work to get started More work to handle interactions Have to write custom animations

And just for fun, I also made it animate like Dan's did, but with `d3.timer()` (an abstraction on top of `requestAnimationFrame`) instead:

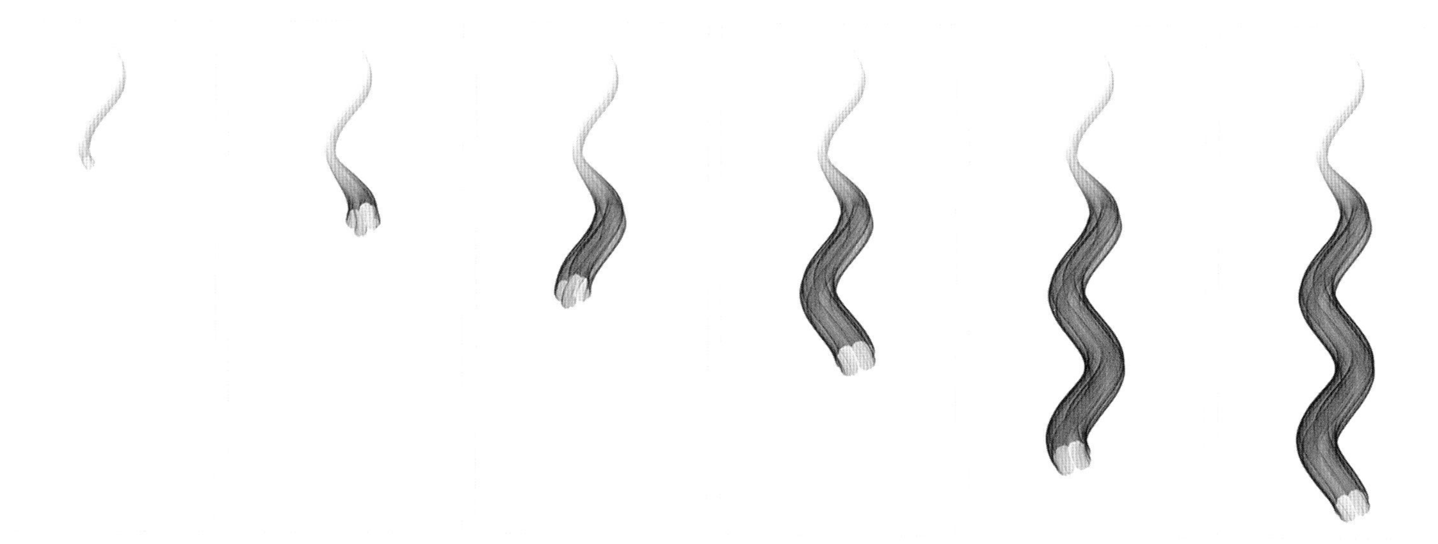

Fig.2.6

My version of the sweeping fractal lines, animated.

I then refactored my code to be able to show and animate multiple flows instead of the single one I was prototyping with, with each flow representing one team's data (Figure 2.7). My happiest moment was getting the animation working for all of the flows, and having the animation's duration depend on the height of the flow instead of a set duration (I'm all about these small wins. (*•̀ᴗ•́*)و)

Fig.2.7

Sweeping fractal lines for each of the teams and events.

Up until this point, I had been using Block Builder to prototype the visuals. But now that it looked how I wanted, it was time to add some structure and polish. As a front-end engineer, I tend to think in terms of components, so I moved my code into `React.js`. I put each diving event (with three teams, ordered by their position on the podium: Silver, Gold, and Bronze) into a component so that I could annotate them together with event name, team information, and score for each round (Figure 2.8).

Block Builder is a now deprecated in-browser code editor for creating and sharing `D3.js` examples.

To finish, I added a title, description, and a hover interaction where I displayed more detailed score breakdown for that round.

Fig.2.8

The sweeping fractal lines for the women's teams, annotated.

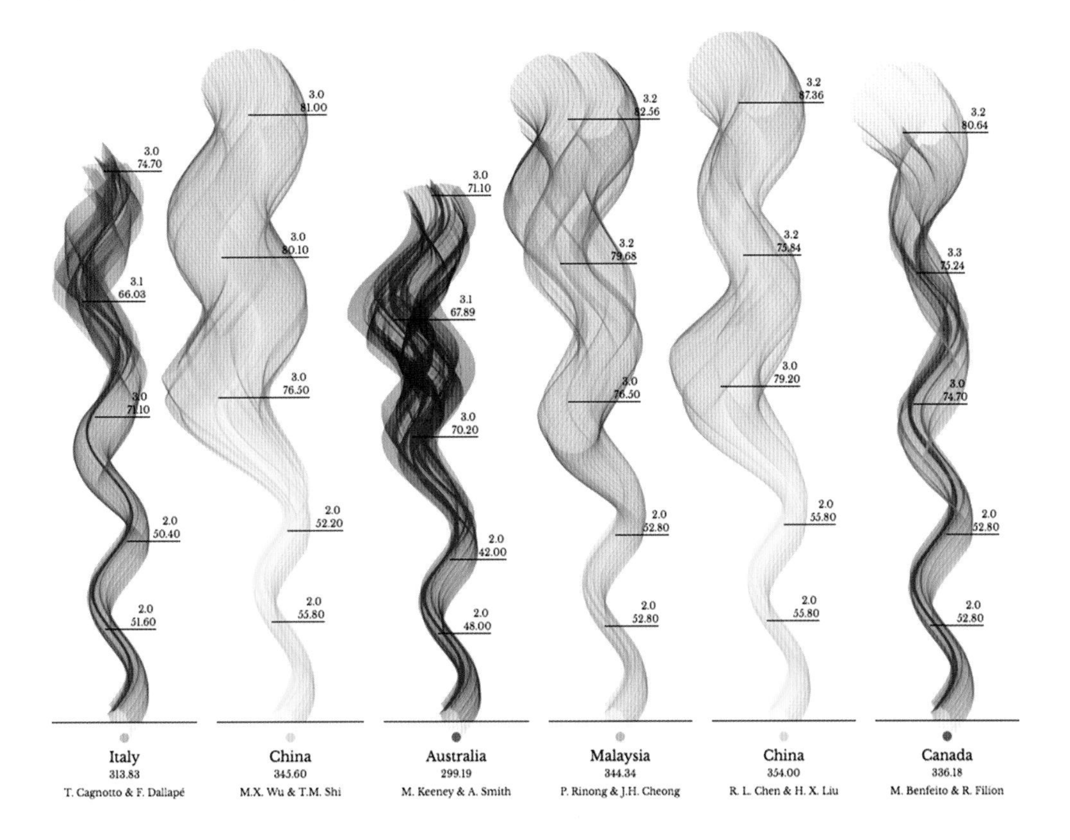

Reflections

When I first finished this project, I felt it left a lot to be desired. I wanted to add more annotations to highlight incredible and emotional moments in each event, because those were what made the events beautiful to watch for me. I had links to videos of those moments, but ran out of time to incorporate them. I was also hoping the visual would look like silk flowing in the water, but I got something that to me looked more like smoke billowing in the wind. As a data visualization, it was hard to compare anything other than height (and thus, overall score) between the different teams; it lacked detailed insights. And as a web project, it was hardly performant: there was a delay whenever I hovered over an annotation.

Now I like it a lot more. It taught me a very important lesson: we'll have good projects and bad projects, but the act of finishing is an important one (that piece of wisdom is from Nadieh). I pushed through the technical challenges and finished, and the aesthetic that bothered me so much back then—the billowing smokestack—has grown on me. I now think it looks beautiful! It was also, unknowingly, one of my first forays into data art, and I'm glad I challenged myself to make something more unorthodox, because it has since given me the courage to keep experimenting.

Dive Fractals

↳ shirleywu.studio/projects/olympics

dive fractals:

synchronized diving in the olympics

Ever since I was a kid, I've loved watching the summer Olympics, and out of all the events, I especially loved gymnastics and diving. And though it was mostly because I come from a Chinese household and it was all we watched, I also loved the two for their grace and beauty. In this visualization, I was inspired by Dan Gries's Sweeping Fractal Lines, and wanted to recreate the beauty and fluidity of the divers in the flowing lines.

Each line represents a **synchronized diving team**, and its **height** represents its total score. There are a total of six rounds of diving for the mens' event, and five for the women, which explains the disparity in height between the two events. Diving is scored based on a **difficulty score** (ranging from 2.0 to 3.9, in these cases), and six **execution scores** (out of 10). I've used the difficulty and execution scores to determine the radius of the lines at each round, and a wider section indicates a difficult dive executed well. Conversely, a narrower section may have a low difficulty rating, or even a high difficulty with low execution.

Each round is denoted by its difficulty as well as the score for that round. Hover to see all the execution scores for that round, and click to see all dives from that round.

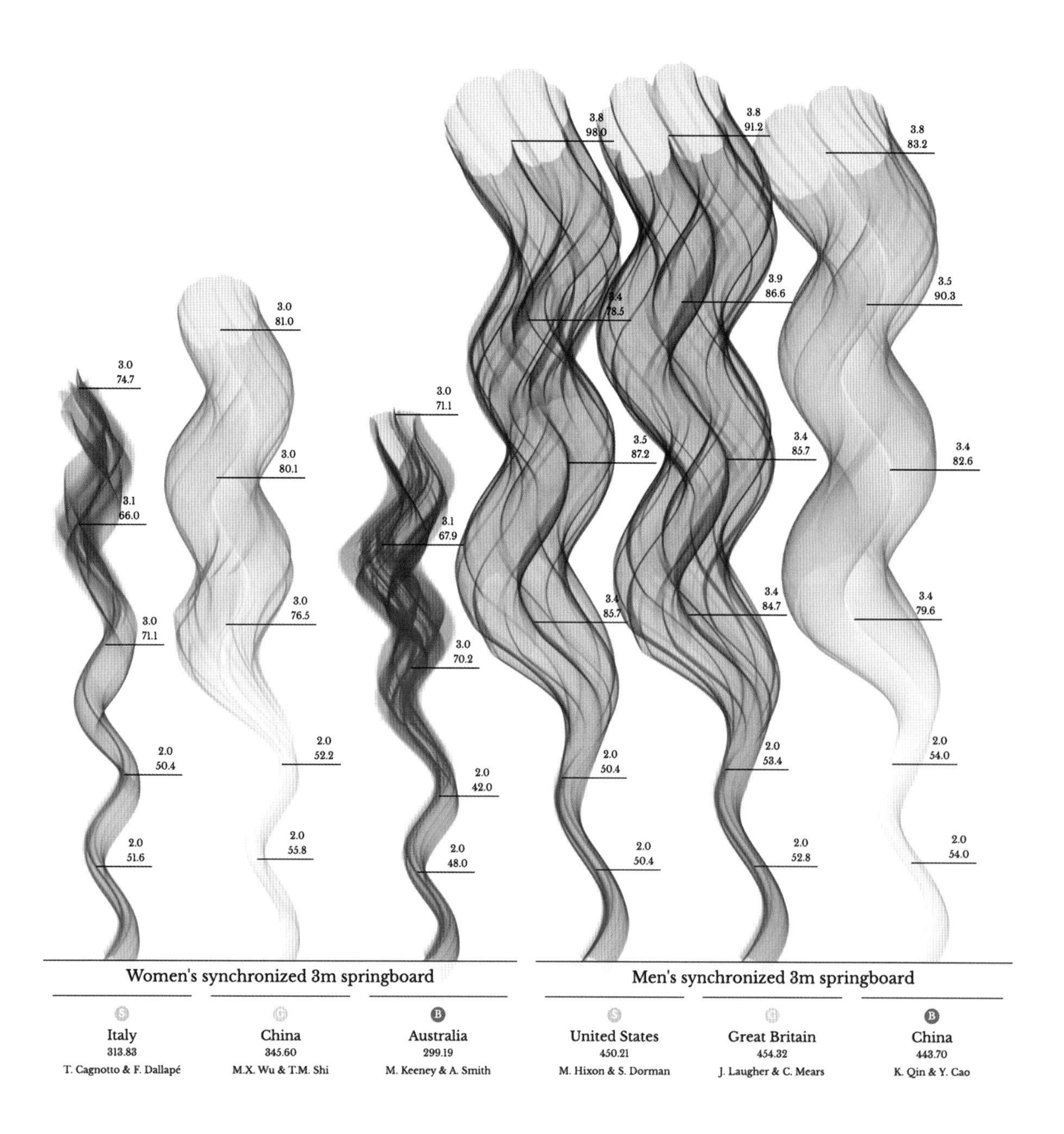

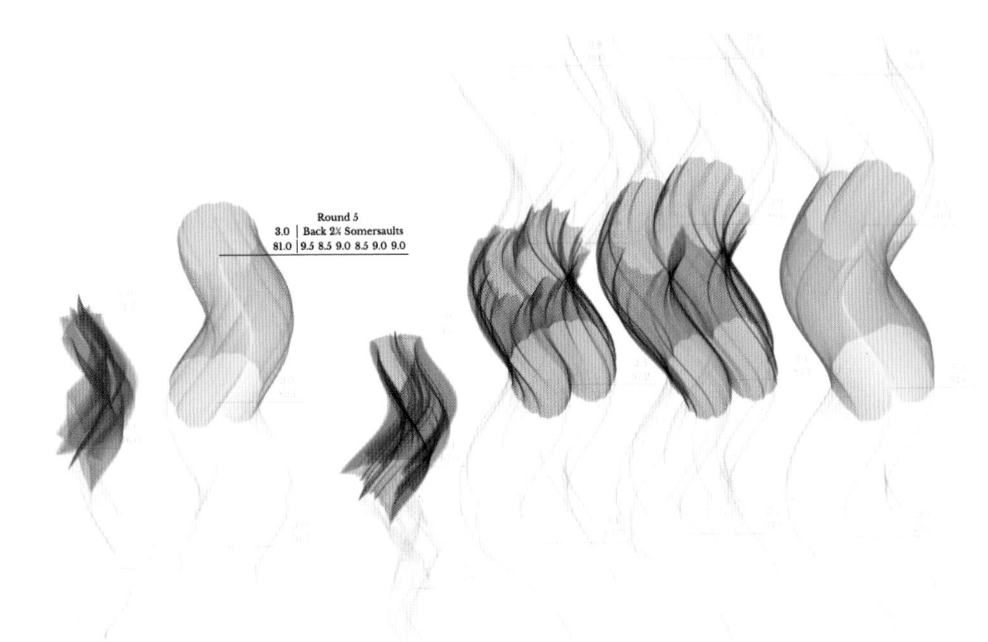

Fig.2.10

Hovering displays more detailed score breakdown, and clicking filters all flows by that round.

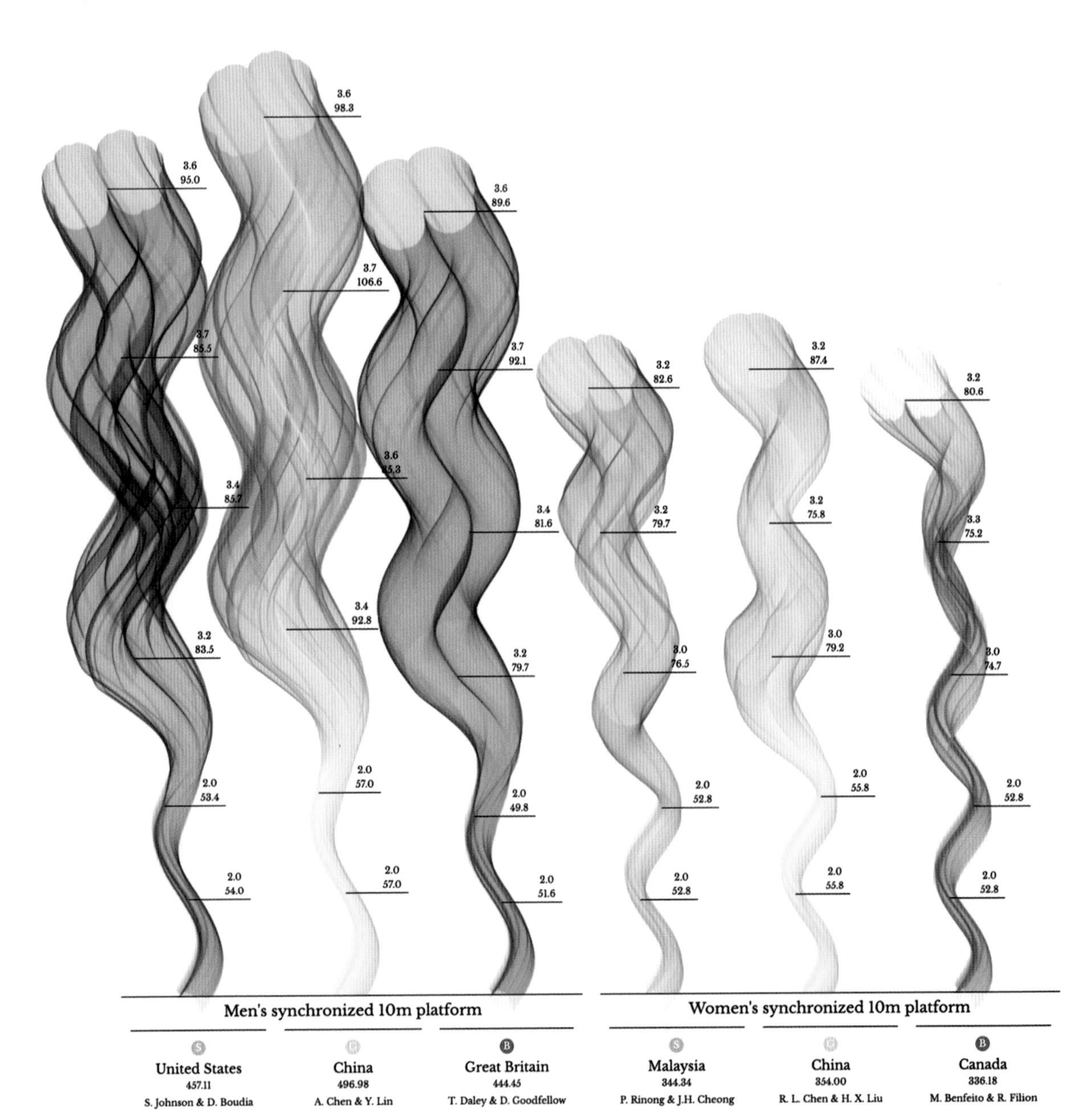

by Shirley Wu for August data sketch|es.

Fig.2.9

The final visualization of the synchronized dives. The women's events are shorter because they only have five rounds instead of the mens' six. Each sweeping fractal line is colored by the team's country flag. I love seeing how close the scores ended up being for the men's 3m springboard and the women's 10m platform.

TRA

VEL

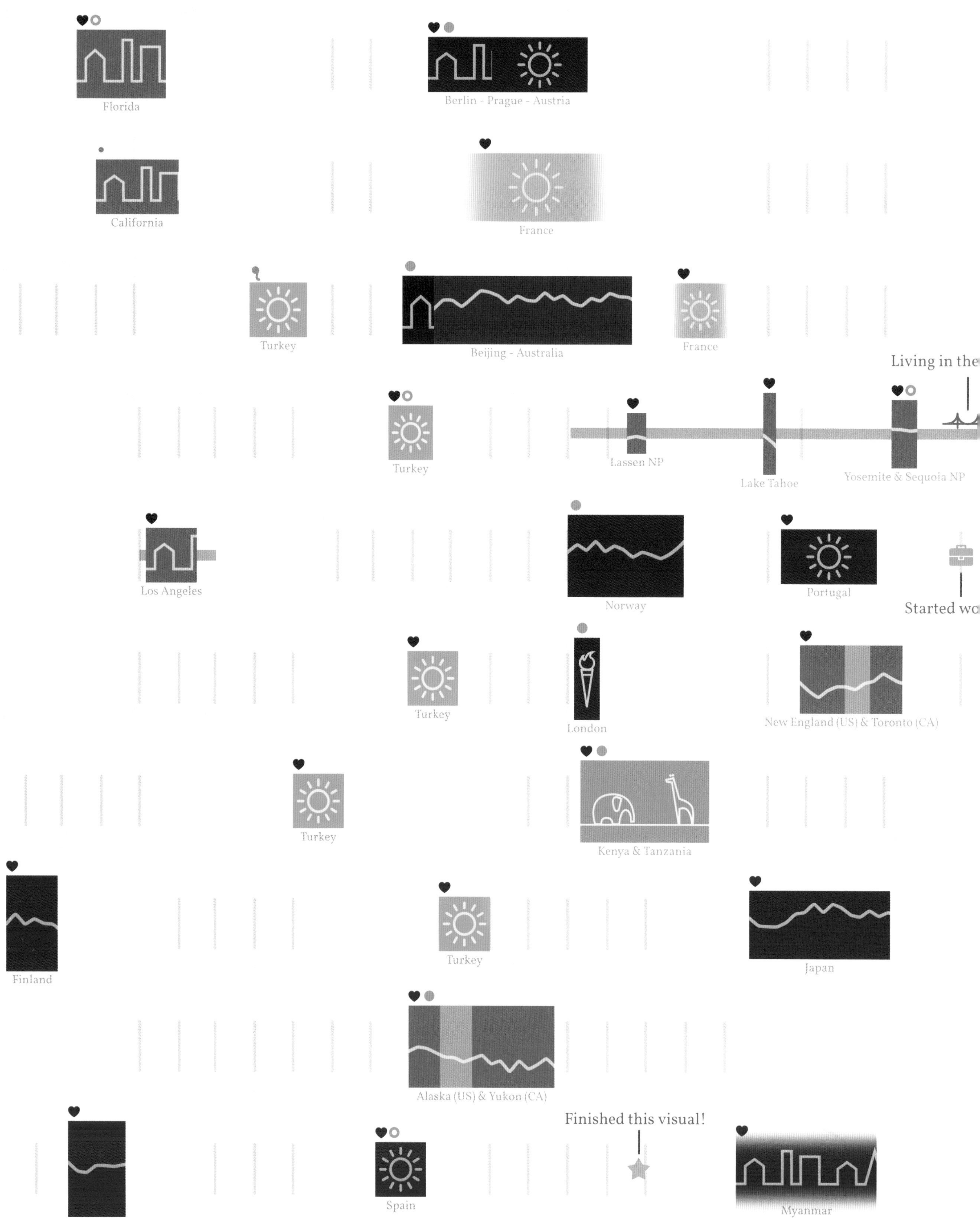

Florida
Berlin - Prague - Austria
California
France
Turkey
Beijing - Australia
France
Living in the
Turkey
Lassen NP
Lake Tahoe
Yosemite & Sequoia NP
Los Angeles
Norway
Portugal
Started wo
Turkey
London
New England (US) & Toronto (CA)
Turkey
Kenya & Tanzania
Finland
Turkey
Japan
Alaska (US) & Yukon (CA)
Finished this visual!
Spain
Myanmar

My Life in Vacations

NADIEH

For me, this topic became more about the data gathering aspects than the visual. I truly *love* being able to travel and see the wonders of our world and I'm extremely grateful that I have the opportunity to do so. I spend *hours upon hours* planning beforehand—searching for the most beautiful, unique, and amazing places to sleep—and I book them weeks in advance. (Vacations are really the only thing that I splurge on, and we all need something to splurge on right?) Even before I was old enough to plan my vacations, for the first 10 years of my life my parents took me along each summer in our cute caravan to enjoy the warm summers of France. The idea of mapping out my vacations (when, where, and with whom) was immediately fixed in my mind.

Data

Since it's (sadly) natural to forget things from your own childhood, I first needed to consult a reputable source to create the data for the first ±10 years of my life: my mother. She has the long-term memory of an elephant. It was quite fun to discuss distant memories from my childhood with her, and together we managed to get a good starting vacation list. I then went to my dad's place to browse through old vacation photos. It took me approximately 4–5 hours to sort all the photos my dad had by year and month (the printing month and year were thankfully written on the backs of most photos), and another four hours to cross-reference the vacation photos with the list my mom and I put together, and put them in a folder corresponding to year. And as an added bonus, this project has now resulted in a nicely sorted box of my childhood photos (Figure 3.1).

Yup, my pre-teens were still in the mostly analog world.

From age 12, I started to keep journals from these trips, writing in whatever notebook I could find. By combining the journals and the ticket stubs of as many museums and tours I could find, I was able to pinpoint exactly where I had gone and when (Figure 3.2).

We got a digital camera in 2004 and I could use each photo's metadata to identify exactly when the photo was taken. I eventually ended up with a file containing all kinds of metadata about each of my vacations.

I included vacations where I didn't stay at home and that weren't business trips abroad.

a

b

Fig.3.1 (a & b)

The different sorting stages of getting through my childhood photos.

Fig.3.2

Part of my collection of travel diaries and journals which helped me find exact travel dates and destinations.

↳ Manually Add New Variables to Your Data

As this was a project related to me personally, all of the variables had to be manually created. It took some effort to gather all of the information I needed, but it did provide me with a dataset that I was totally enthusiastic about representing visually!

Sketch

My idea this time was a very simple one. Each year since I've been born would be one row and the periods in which I went on vacation would be a colored rectangle and annotated. However, I wanted to play around with different kinds of annotations to make it easier to see trends; where I went, with whom, how we got there, what was the main purpose, things like that. The sketch in Figure 3.3 was my first attempt at creating these annotations.

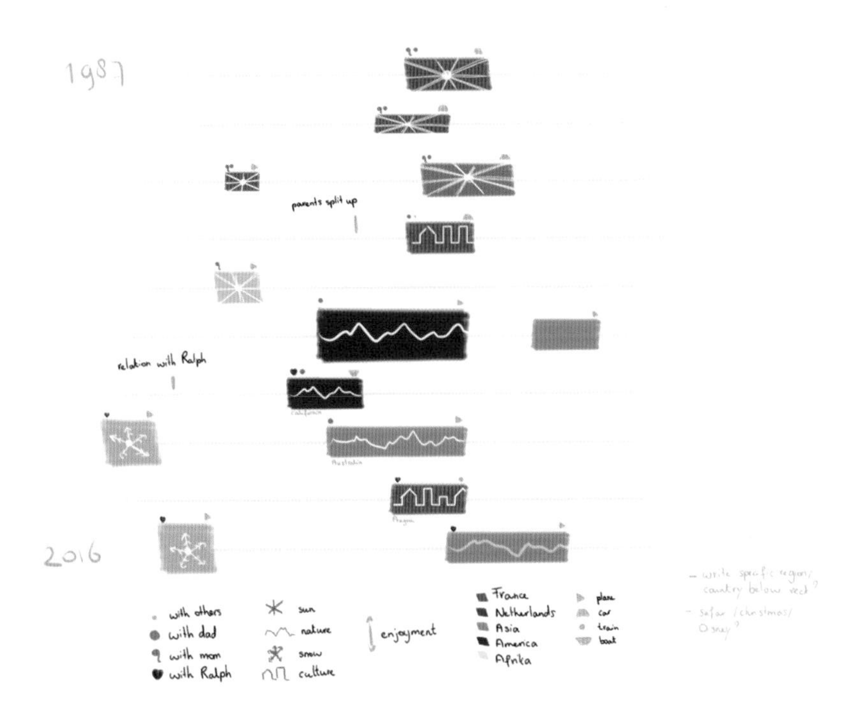

Fig.3.3

A row per year with colored blocks during the periods that I was on vacation.

But I quickly realized that what I was sketching was a poor reflection of reality. There are 52 weeks in a year, and typically I'm on vacation during a maximum of five of those weeks—a lot! I know, but objectively it's less than 10% of a year. Therefore, if I were to create this visual, less than 1/10th of a row would be colored, translating to a visualization with a lot of white space and a few tiny colored blocks in between.

I started thinking of ways to give the vacation weeks more space and I was reminded of the way time zones run across the Earth: from North to South pole, and *not at all* in a straight line. Maybe I could also squeeze the months in which I wasn't on vacation in that year (Figure 3.4)? Yes, that would mean that the months wouldn't align perfectly anymore, making it pretty darn difficult to compare exact months across the years. But I was more interested in seeing trends between vacations than getting a perfect overview of the periods in which I was away.

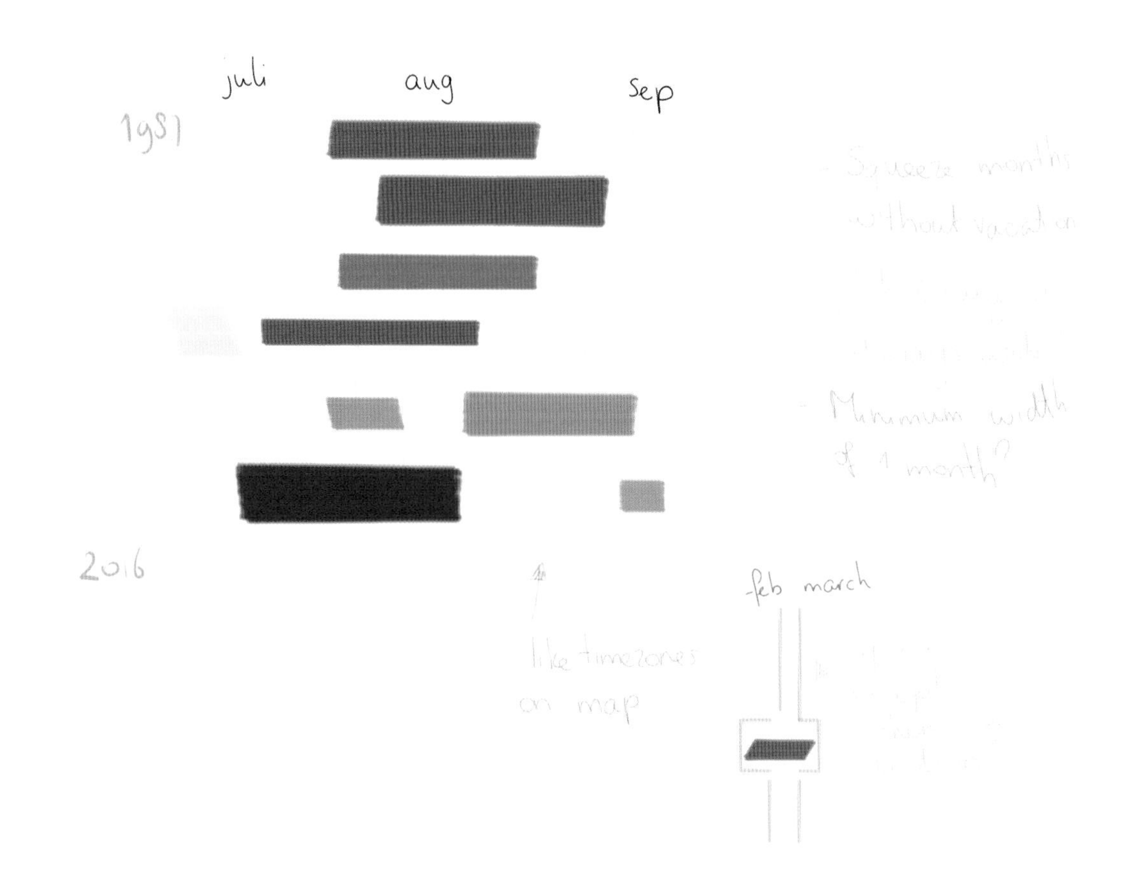

Fig.3.4

Sketching the idea of squishing a month in which I hadn't been on vacation.

↳ Sketch to Discover and Remove Thinking Errors

Starting off your visualizations with a simple paper-and-pen sketch can help you spot whether your idea will actually work with your dataset and allow you to iterate quickly. By sketching out my simple design of colored blocks per row I could quickly assess that my initial idea wouldn't represent the data in a good way visually. Furthermore, sketching also helped me to come up with the new approach of squishing some months.

Code

Even though the idea was simple in theory, the data collection and execution required more time in practice, with a lot of custom code. The first step was getting those vacation rectangles in the right location. Due to the introduction of the "squeezed months" concept, and centering the year around August (my typical vacation month), it took me at least 30 iterations before I had the math and loops in my code working correctly (Figure 3.5).

Connecting all the months by a line was less trivial than I expected. It took many iterations before I was happy with the curving of the lines in between the years. (The one in Figure 3.6, although not useful, was definitely my favorite.)

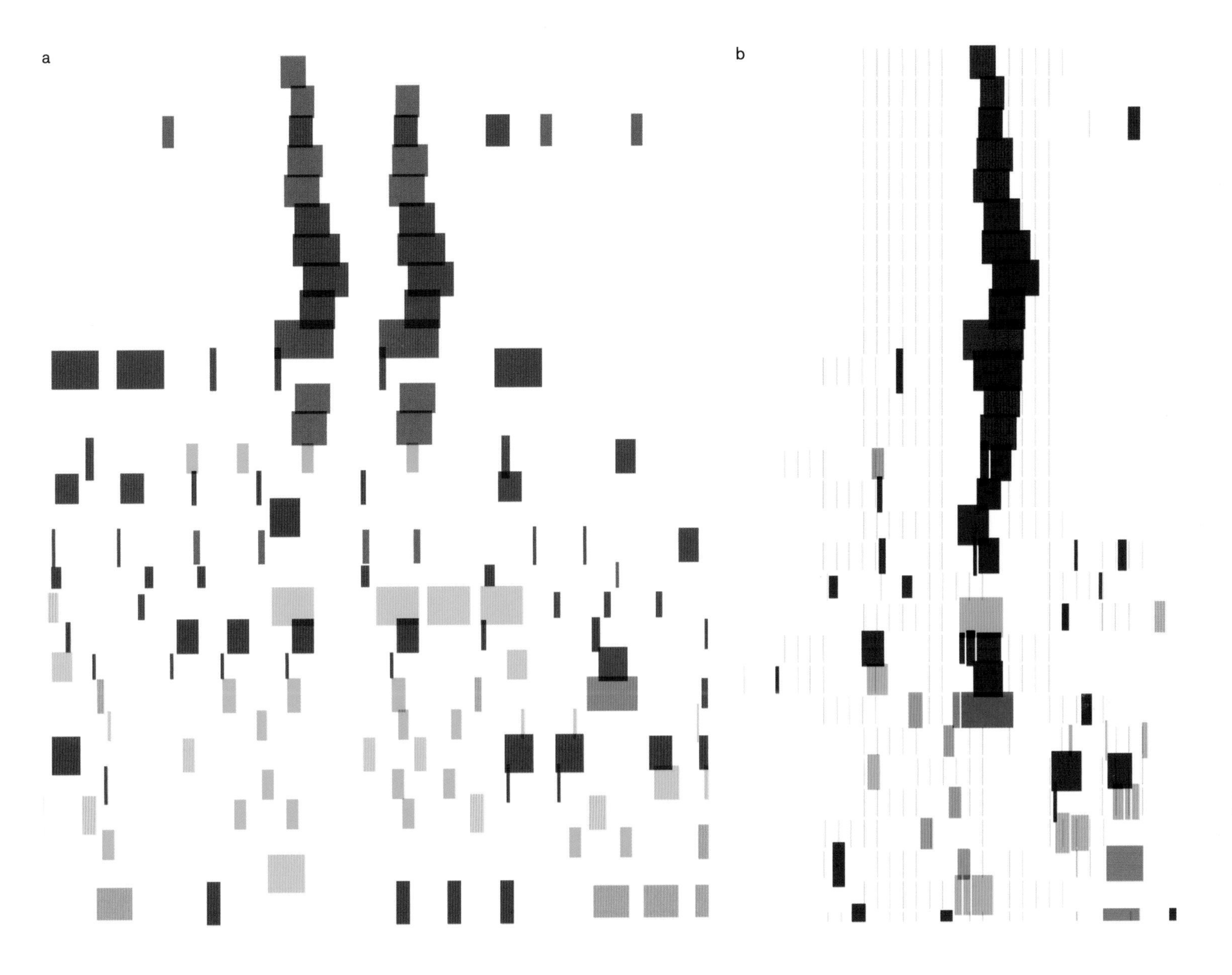

Fig.3.5
(a & b)

Two iterations of getting the vacation rectangles in the correct horizontal location.

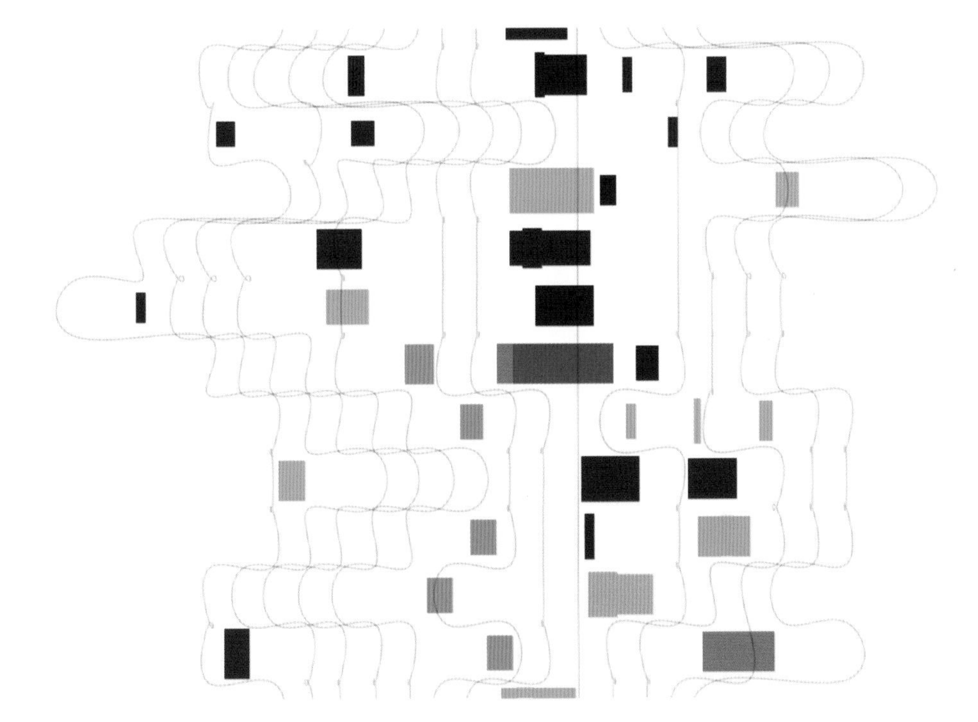

Fig.3.6

Some wonky results for the month dividing lines.

Next up were the shapes inside the vacations representing the purpose of my travels, such as "culture" or "nature." I was trying to go for easy-to-decipher patterns that would be understandable without referring to the legend. After trying out a few different techniques I found it easiest to create a small line chart or icon on top of each rectangle. For example, the pattern on top of "nature" vacations is nothing more than a randomly drawn line. It's different for each rectangle and on each refresh. The house-like pattern for my "cultural" vacations always uses the same setup of lines until it reaches the end of the rectangle. The "sun" and "snow" icons were made from basic shapes, a bit of trigonometry, and a for-loop. However, the "Mickey Mouse," "Olympic torch," and "safari animal" icons were too complex to make with code. I made them in Adobe Illustrator, saved the tiny SVGs, and loaded them onto their respective rectangles.

I made the pattern of lines for the "culture" vacations long enough to fit the longest rectangle.

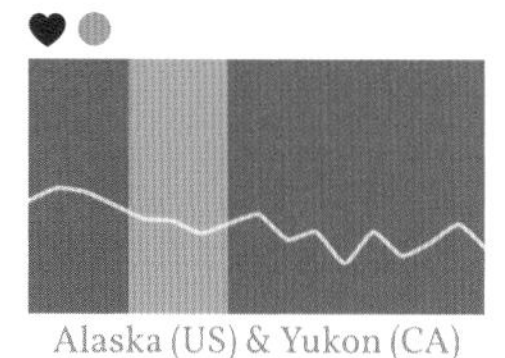

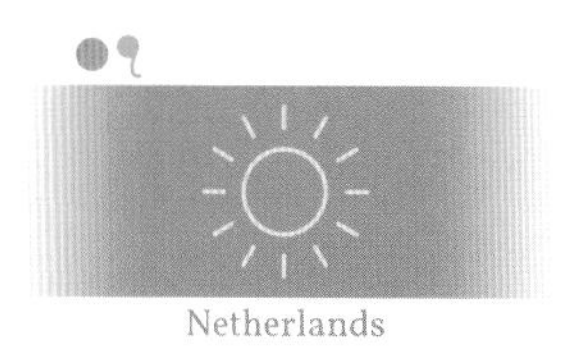

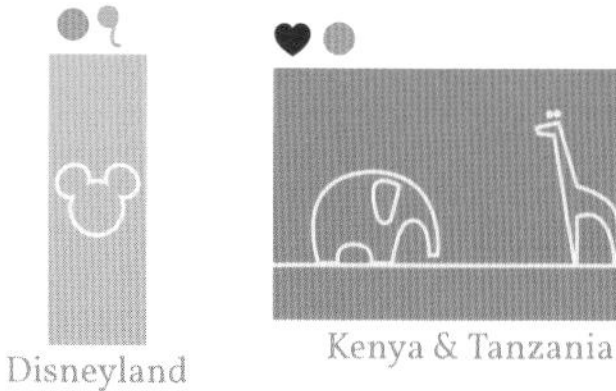

Fig.3.7

The different types of vacation patterns and icons inside the rectangles and people icons on top of the rectangles.

Moving on to the annotation of "who I was with" on top of each rectangle, the final icons are actually pretty similar to my first sketch: my boyfriend is represented by a heart, my dad is a blue circle, and my mom a pink small circle with a downward swoosh. Since this entire visual is already very personal, I didn't mind the exact icons relating to something that only I would understand.

Initially, I was using a gradient for the vacations I didn't have exact dates for in order to fade the ends out a bit. However, I wasn't sure what to use for vacations where I couldn't remember how much I enjoyed it, being too young at the time. The problem is that you can only apply *one* gradient to an SVG element. That's until I read one of Shirley's write-ups again and she talks about the motion blur filter, and I was like: of course, I wrote *a whole tutorial on that*—how could I forget! So I replaced the gradient with the blur filter. The nice thing is that you can set it up to blur in either horizontal, vertical, or both directions. I did find that the vacations in my early youth, where both the duration and enjoyment are blurred, are a bit like vague spots (Figure 3.8), but then Shirley said that this was a nice metaphor to how I remember them, just blurry pieces (or nothing at all), which is absolutely true indeed.

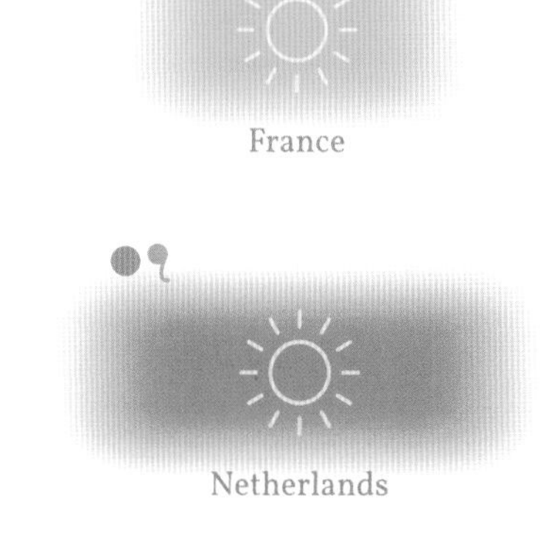

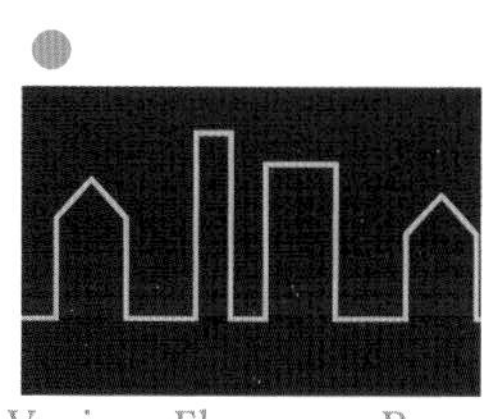

Fig.3.8

The different types of blurry filters applied to the vacation rectangles to denote either an unknown start and end date, not remembering how much I enjoyed the vacation, or both.

The final visual touch, apart from a legend, were the textual and icon-based annotations along the sides to explain the visual and a few major life events (Figure 3.9). Due to all of these custom touches—such as the icons, the patterns, the legend, the annotations to help people understand my life through vacations more—I finally came to the point where this project had more lines of code than my previous project about the Olympics. Not such a simple project after all! (╥﹏╥)

I kept the interactions to a minimum, limiting it to help the viewer understand the most difficult aspect of the viz: differentiating between months, due to all the squeezing going on. There's only a small hover that highlights the same month throughout the years in light grey.

In the end the final visual became quite long, a bit like those long-form infographic monsters that were all the rage a few years ago. But it also reminds me a bit of a children's book; simple with bright colors, easy icons, and a bit of text every now and then.

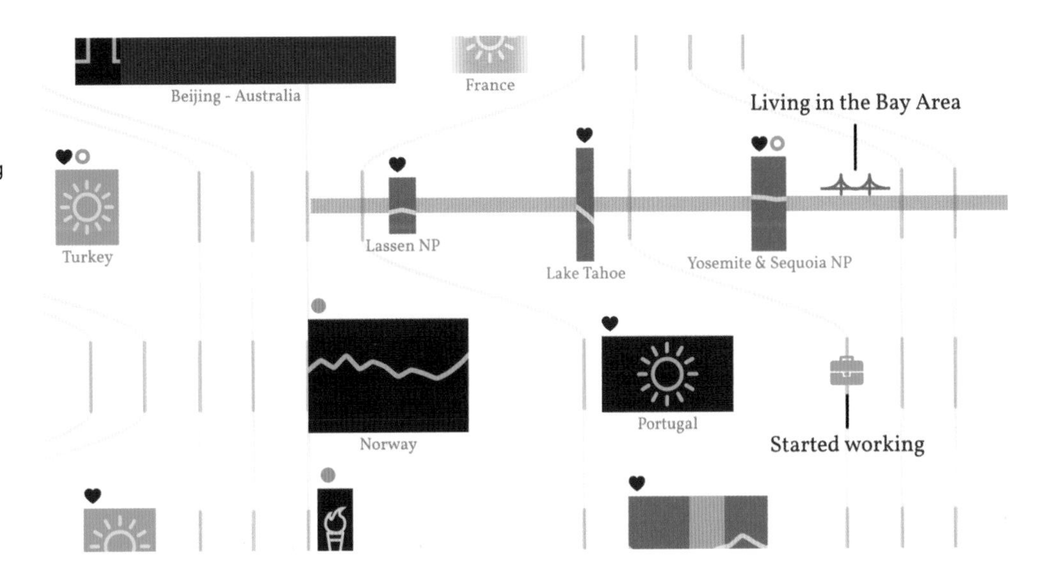

Fig.3.9

Two of the small icon-based annotations that I added to the visual to denote the big events in my life: doing an exchange at the University of California Berkeley and living in the Bay Area, and starting to work after graduating.

Reflections

With the amount of code and hours that this project still required, I wonder if I should've gone for a more complicated sketch/idea that would've looked more visually appealing (but would still take the same amount of time). By comparison, this one just looks simple. But as Shirley stated in her previous project, we wouldn't be super proud of every visual we created during *Data Sketches*. I really wanted to try out this idea where you got a more general sense of time rather than seeing time as fixed, and give the vacations themselves more room to exist. I'll leave it up to you to decide if that was a good choice in the end.

What I do like is that I've never made a visual with such personal data before. So maybe it's not that interesting visually for other people to look at, but for me personally it has a special place in my heart. (˘ ³˘)ノ°゚°❍

My Life in Vacations

↳ Vacations.VisualCinnamon.com

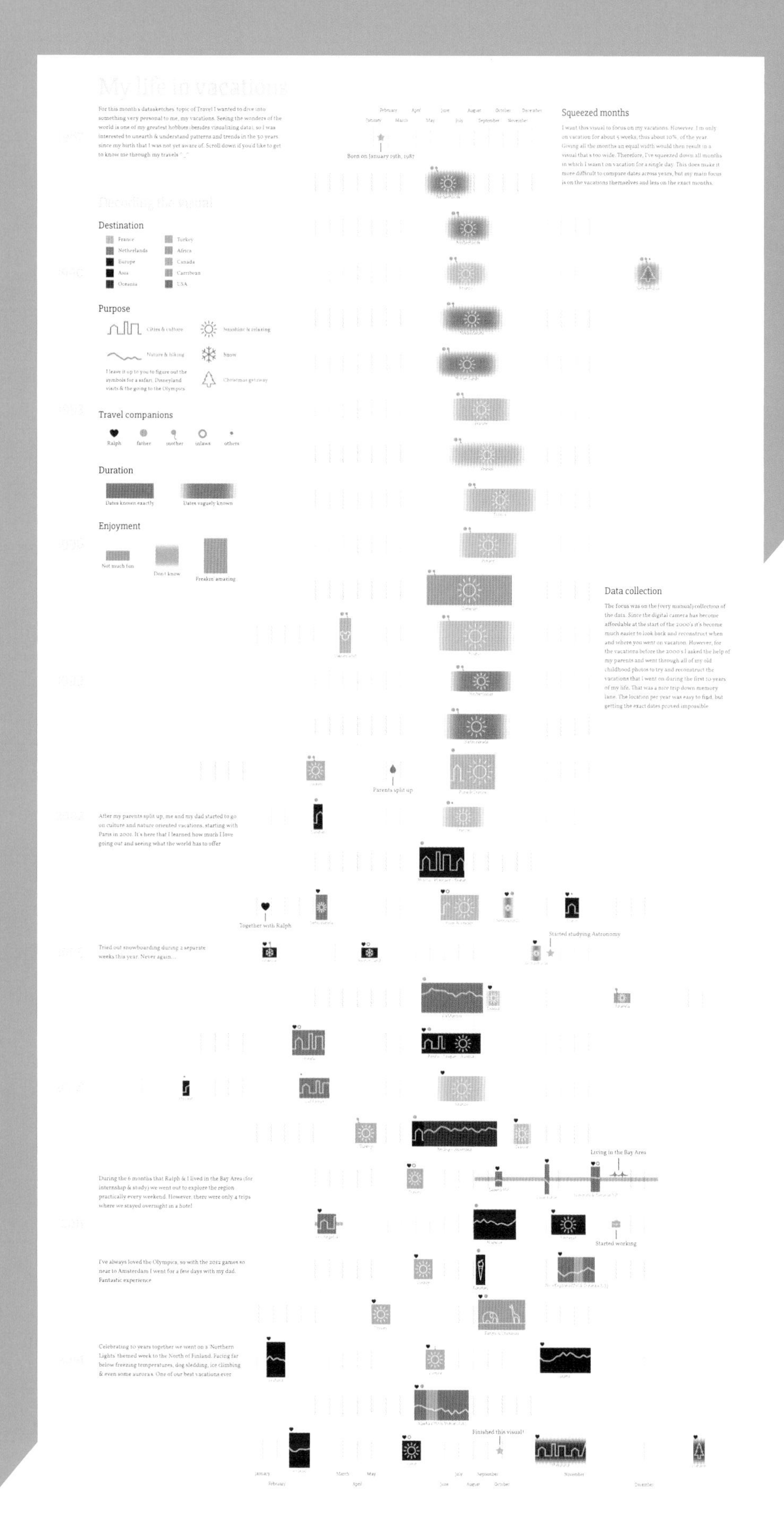

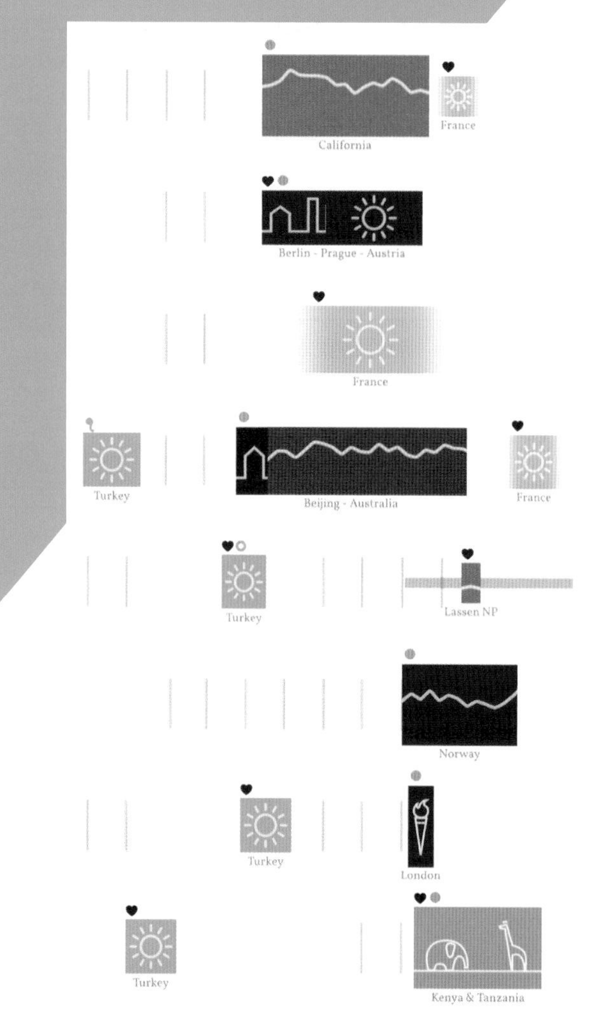

Fig.3.11

Zooming in on a section generally between June and September, from 2006 and 2013.

Fig.3.10

On a hover, the same month is highlighted across all years with a grey area

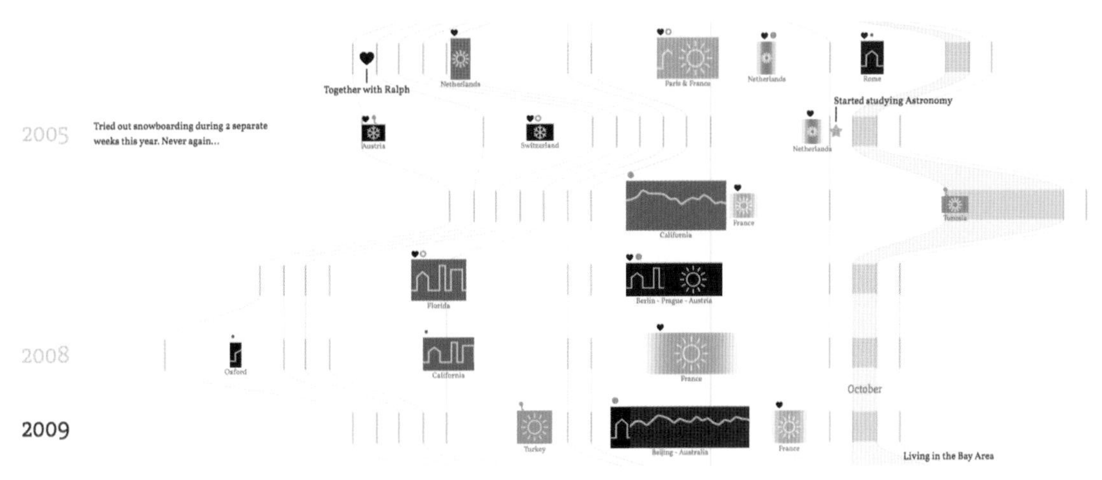

Fig.3.12

On a hover, the same month is highlighted across all years with a grey area (July in this case).

Fig.3.13

I had two vacations planned for the months after finishing this visual and therefore didn't know how much I would enjoy them

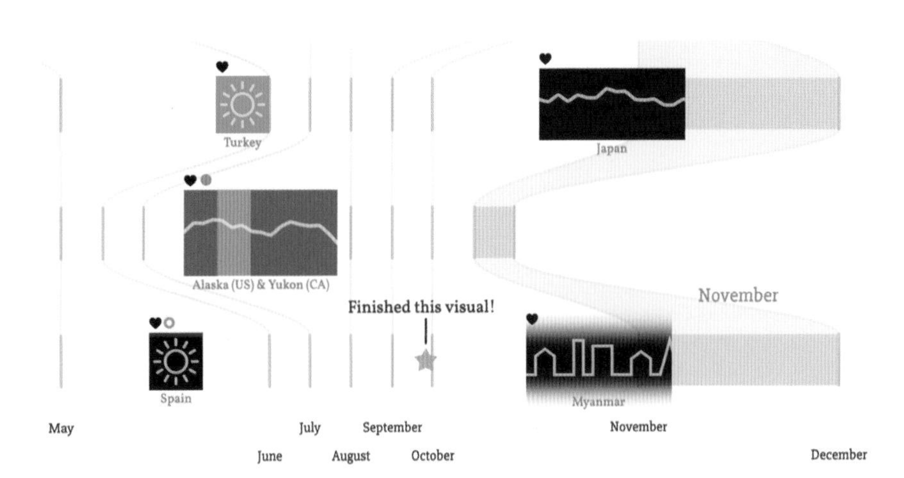

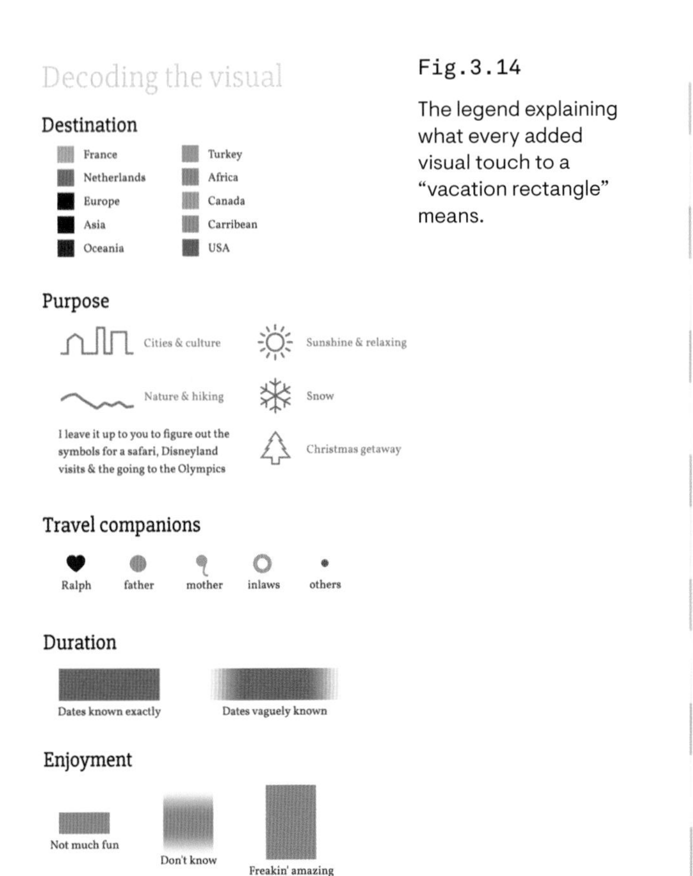

Fig.3.14

The legend explaining what every added visual touch to a "vacation rectangle" means.

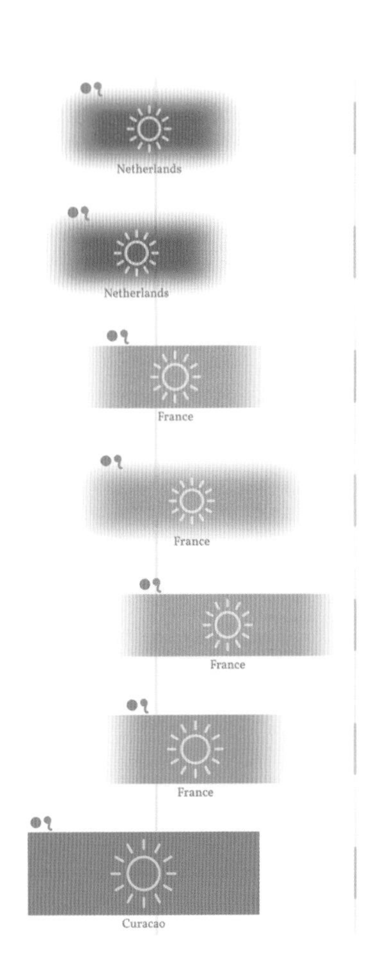

Fig.3.15

A sun vacation during July/August was the only type of vacation that happened in the first 10 years of my life.

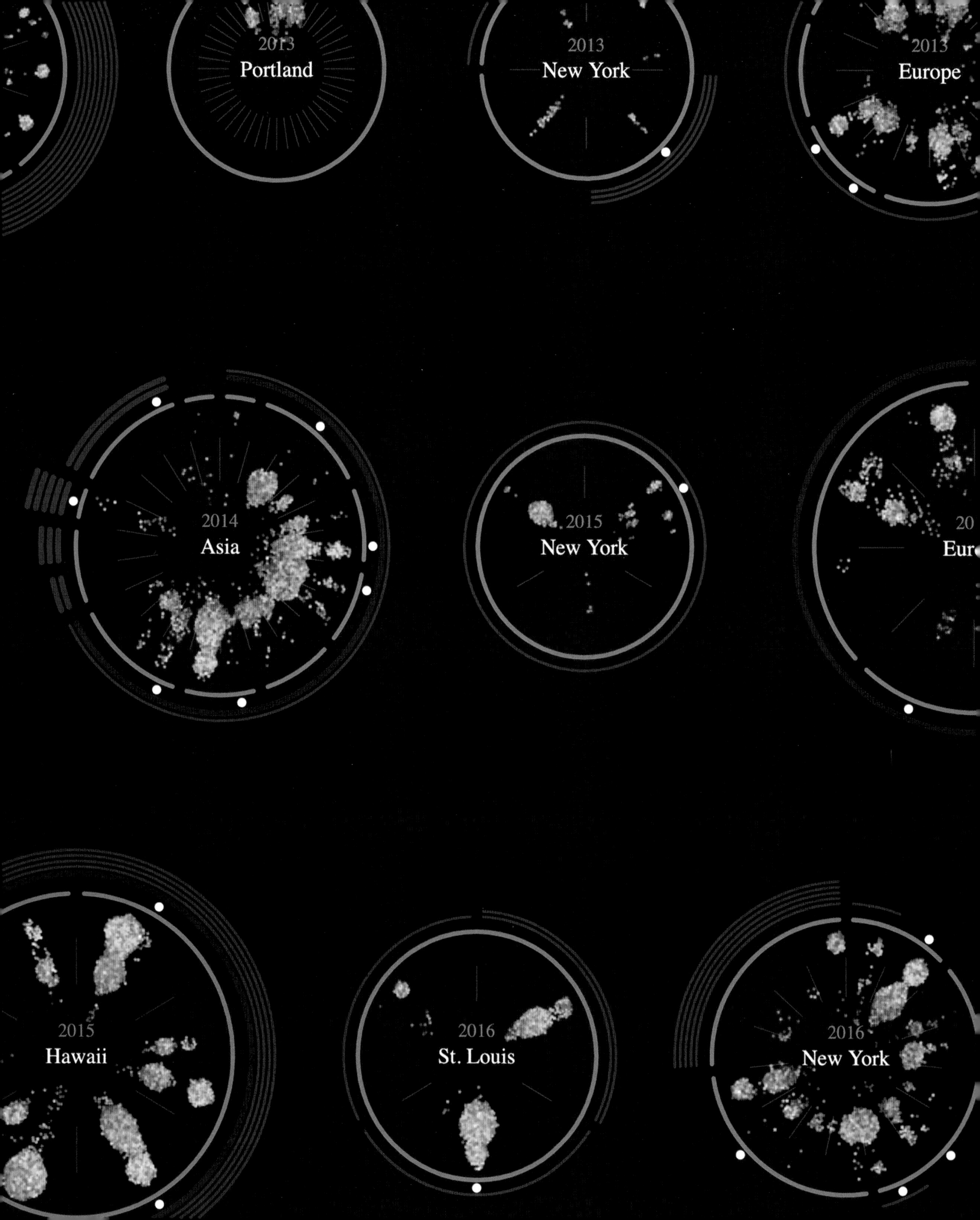

2013
Portland
2013
New York
2013
Europe
2014
Asia
2015
New York
2015
Hawaii
2016
St. Louis
2016
New York

Four Years of Vacations in 20,000 Colors

SHIRLEY

Traveling is one of my absolute favorite activities, and since graduating college, I've tried to go abroad at least once a year. (I moved around a lot as a kid, so I get restless when I'm in the same place for a long period of time; traveling helps alleviate that.) Since I take so many pictures when I'm traveling, I thought it'd be a really fun idea to gather all the photos I took on my trips and visualize them.

Data

I started by going through all of my photos from the last four years (thank you iPhone and Dropbox), and grouped them by year and airport code. I learned two things from this exercise: I like going to the same places, and I take a ridiculous amount of pictures. On a two-and-a-half week trip to Japan and China in 2016, I took more than 1,000 pictures.

Admittedly, the amount has increased as I've gotten newer phones with more storage, which might also be interesting to analyze.

After I sorted through my images, I played with `gm`, a Node.js wrapper for `GraphicsMagick` (a collection of tools for reading, writing, and manipulating images). My first idea was to reduce the images down to their top three colors, then blur and resize them. I experimented with every fifth photo from the ±1,000 images I took on my latest trip (because 1,000 is just ridiculous), and displayed them in the browser in chronological order to look for any interesting color patterns (Figure 3.1). Unfortunately, I didn't make out anything noteworthy with the colors (it also wasn't very performant and took a few seconds to load). For my second attempt, I decided to go through every fifth image from every trip instead of just the most recent. I resized the images (Figure 3.2), read the images' metadata, and saved the dates they were taken as well as the latitude/longitude of where they were taken (when available).

This came out to about 800 images, which was just enough to notice some lag in loading all the images, but not enough to do anything particularly exciting with them.

As using all the images (±4,000 photos) would be way too memory intensive, I decided to analyze the colors instead. I searched around for how to get an image's primary color(s), and found `get-image-colors`, a Node.js package that returns an array of the most frequently appearing colors in a given image. Unfortunately at this point, I did something minorly stupid; I passed in the *fully-sized images* (2,500x3,200 pixels and approximately 3MB each) to get the primary image colors.

Fig.3.1

Every fifth image from my latest trip and their top three colors, blurred and resized.

Fig.3.2

Every fifth image, resized.

Around 30 minutes and 120 images later, I realized that I could have just passed in my resized images (13KB) to get the colors instead, which would've been so much faster. D'OH!

Once I figured that out and fixed the code to take in the resized images, the package worked wonderfully; it returned more than 4,000 colors for all 800 images in a matter of seconds:

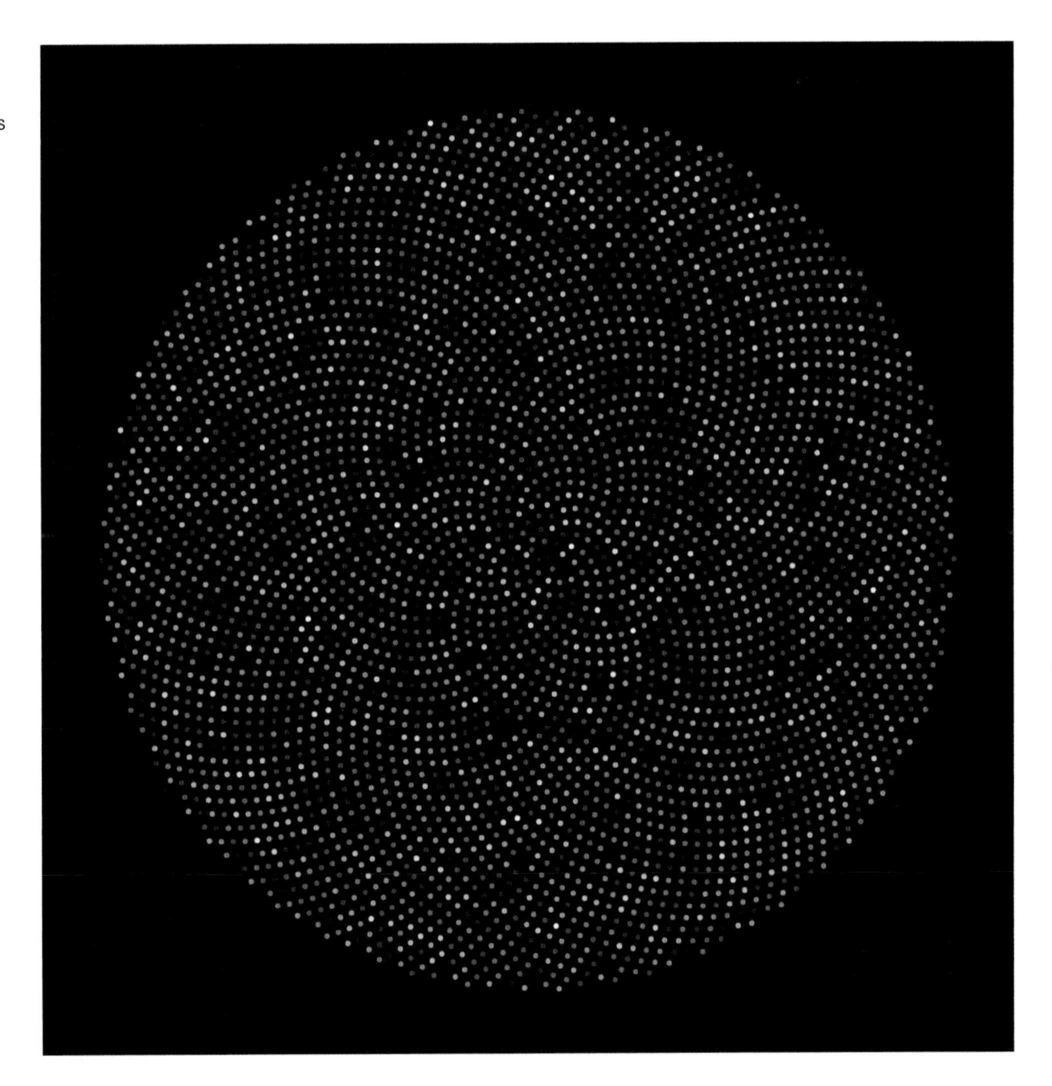

Fig.3.3

The top five colors for each of the 800 images

Sketch

When I started brainstorming ways to visualize the data, I immediately thought of food (my travels are always dictated by food). I remembered the ice cream I ate in Japan the month before, and thought how cool (so punny!) it would be to make an ice cream image out of all my photos.

I started looking for inspiration on The Noun Project, a searchable website dedicated to icons, and found a super cute set of ice cream and beverage icons to base my design on.

I decided to have each trip's visualization be in the shape of a food from that location or region (Figure 3.4). I'd place each color along the icon path, with the *x*-axis being time of day the photo was taken and the *y*-axis being the *n*th day of the trip. The food would be sized according to the number of photos that trip had, and I would apply the "gooey effect" in SVG to blend the colors together. Finally, I would clip the colors with the icon so that they would have clean edges.

The gooey effect is achieved in SVG by applying a blur filter and upping the contrast, so that shapes look like they stick and blend together like gooeys.

I was really excited about this idea, but given the time crunch I decided to explore another idea instead.

The second idea came when I was driving out of a parking lot after dinner. I saw a string of lights along a building which reminded me of autumn and the holiday season. I thought it would be so pretty to visualize the colors as a string of lights or stars in the night sky (Figure 3.5).

The plan was to place the colors in a scatterplot, with the year the photo was taken on the *y*-axis and the time of day or the location on the *x*-axis. Then, I wanted to use the `d3.forceSimulation()` layout to get rid of overlap and use motion blur in both directions to give a glow effect. Finally, I wanted to run a string through all the lights/colors from the same trip.

I decided to try out this idea of lights in the sky, but when I plugged my data into it, it didn't turn out as I expected (Figure 3.6).

It's interesting looking back on it now how my sketch/design process has changed; when I did this project three years ago, I concentrated a lot on how to technically implement the design, whereas now, I concentrate a lot on the end user experience, on creating good visual metaphors, and bringing end users delight.

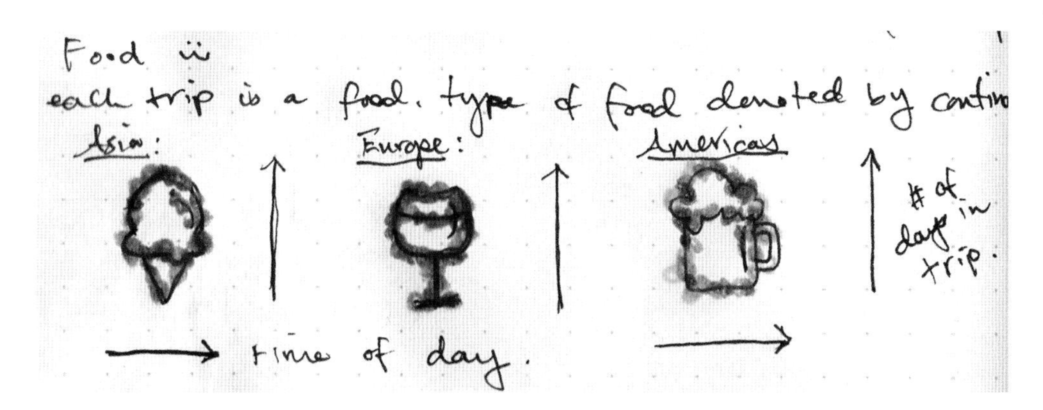

Fig.3.4

Idea with food and beverage icons

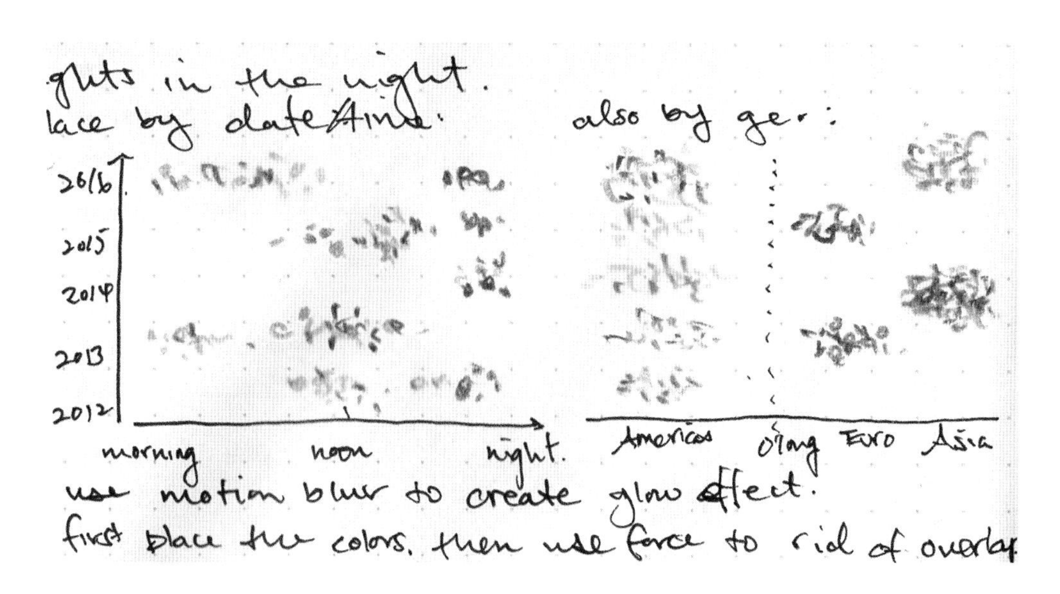

Fig.3.5

Idea with lights in the night sky

Fig.3.6

I call this piece "Man Peeing into Puddle." (*≧艸≦)

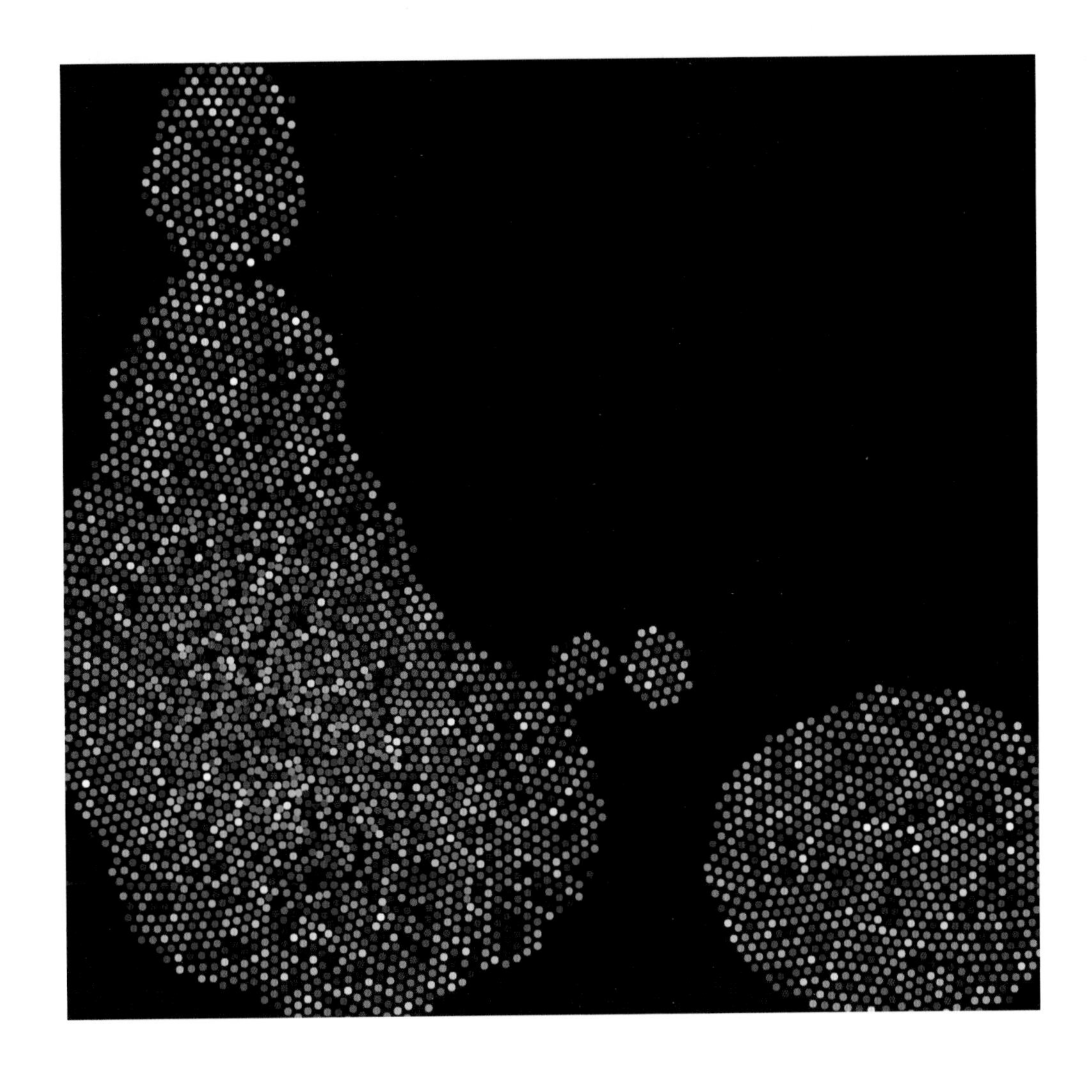

This is essentially what happens when I sketch before actually exploring the data. For "Movies," I had the time to explore my data before really sketching and deciding on my idea, and with "Olympics," I just happened to luck out with the data and the idea I went with. So I got complacent for this project (and plus, I had already done so much work to get the data to where it is) and convinced myself that *this idea will work*! It obviously did not. Important lesson learned: always *explore the data first*.

↳ Exploring Data: List Attributes

Because I don't have a data background, I've had to learn on the job, and over the years I've developed some guidelines for myself.

The first step of this guideline (after I've gathered and cleaned the data), is to always identify the "lowest" unit of data and list out its relevant attributes. When I list the attributes, I also like to note if it's **quantitative, categorical, ordinal, temporal,** or **spatial**; this really helps me when I'm designing and thinking through what visual channels might be appropriate. For this project, I gathered my vacation photos and extracted the top colors per photo. Therefore, the "lowest" unit was the individual colors, and the attributes were the RGB and HSL for the color (quantitative), the time the photo was taken (temporal), and the trip that it was taken on (categorical).

Once I have the attributes listed, I highlight the ones I think are interesting and use those to formulate questions and hypotheses. Then, I plug the data into a charting library to test my hypotheses, explore, and get a better understanding.

For more detail, refer to the lessons "Explore Data: Ask Questions" on page 304 of my "Culture" chapter, and "Explore Data: Use Charting Libraries" on page 336 of my "Community" chapter

Code

Around the time I was going to start coding, the short list for Kantar's "Information Is Beautiful" Awards[1] had been released and I came across Nicholas Rougeux's absolutely gorgeous "Off the Staff" project[2] (Figure 3.7).
It inspired me to try out a radial format, something I never worked with before. So I grouped the colors by trip and positioned them in a grid (Figure 3.8, left), then used the time and day the photo was taken as the angle of the colors around the center (Figure 3.8, right).

I used trigonometry to convert each color's designated angle (the time that photo was taken) into *x*- and *y*-positions. After I found the correct positions, I used the new `D3.js` *force* module to actually place the colors. In particular, I used `d3.forceX()` and `d3.forceY()` to nudge the colors toward the *x*- and *y*-positions I had previously calculated, and then `d3.forceCollide()` to specify a radius for each node within where other nodes cannot enter (thus getting rid of overlap but making sure the colors are as close to each other as possible).

The force layout is a graph layout algorithm that calculates the positions of nodes in a graph. It does this by making sure that nodes repel each other so as not to overlap, while also placing nodes that are connected closer together (but not overlapping)

Now that I had the colors sorted by time, I could see that on some trips I took photos consistently and almost daily, while on others, photos (and thus the colors) were bunched into a few days with none in between. But I wanted to see if I could get more out of the data. In particular, I wondered if I could sort color to map to a visual channel. After some research, I found the blog post,"The incredibly challenging task of sorting colours" by Alan Zucconi.[3] It went into incredible detail about the different ways of sorting, but for me the section on HSV (hue, saturation, value) sorting sounded the most promising, and I decided to map the hue to the radius (Figure 3.9).

Fig.3.7

Nicholas's gorgeous "Off the Staff" project

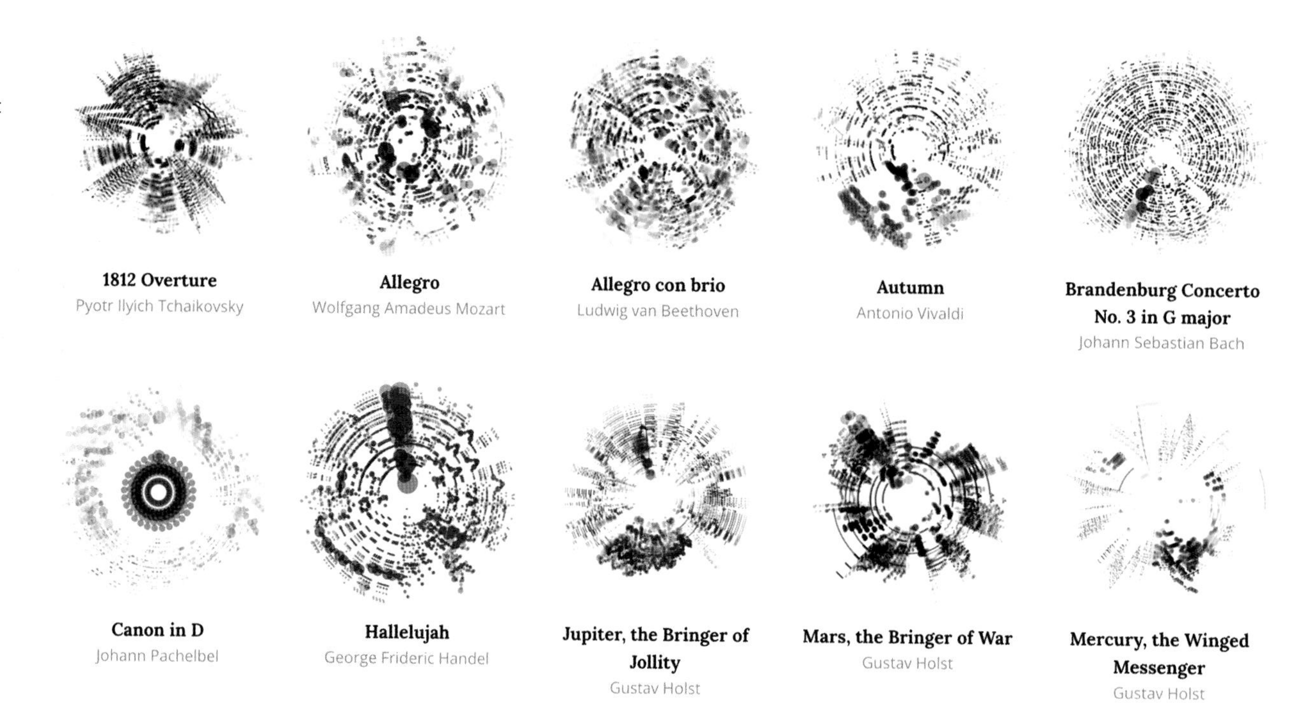

[1] Kantar "Information Is Beautiful" Awards: https://www.informationisbeautifulawards.com
[2] Nicholas Rougeux, "Off the Staff": https://www.c82.net/offthestaff/
[3] Alan Zucconi, "The incredibly challenging task of sorting colours": http://www.alanzucconi.com/2015/09/30/colour-sorting/

Fig.3.8

Going radial.

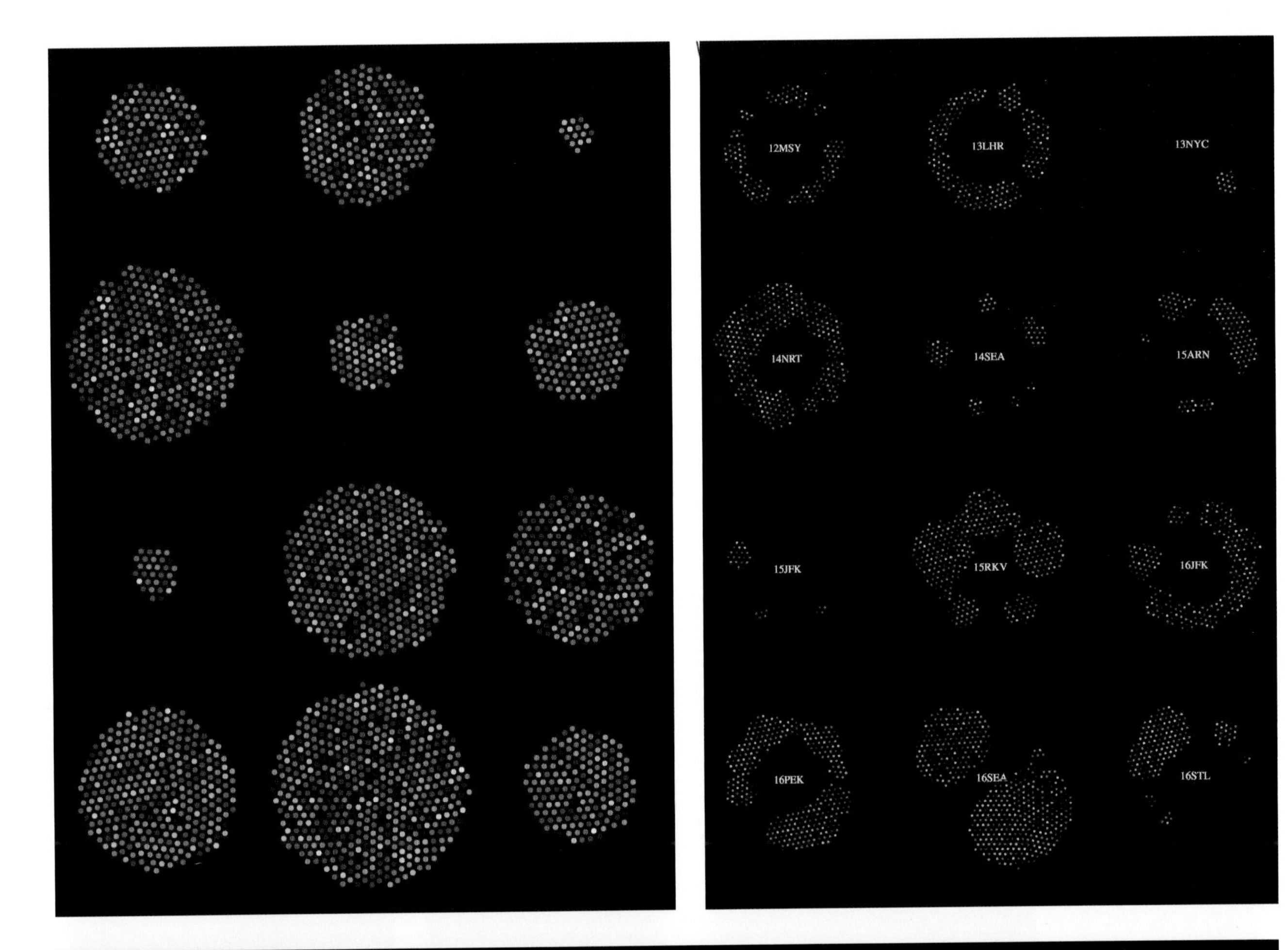

Fig.3.9

Sorting by hue.

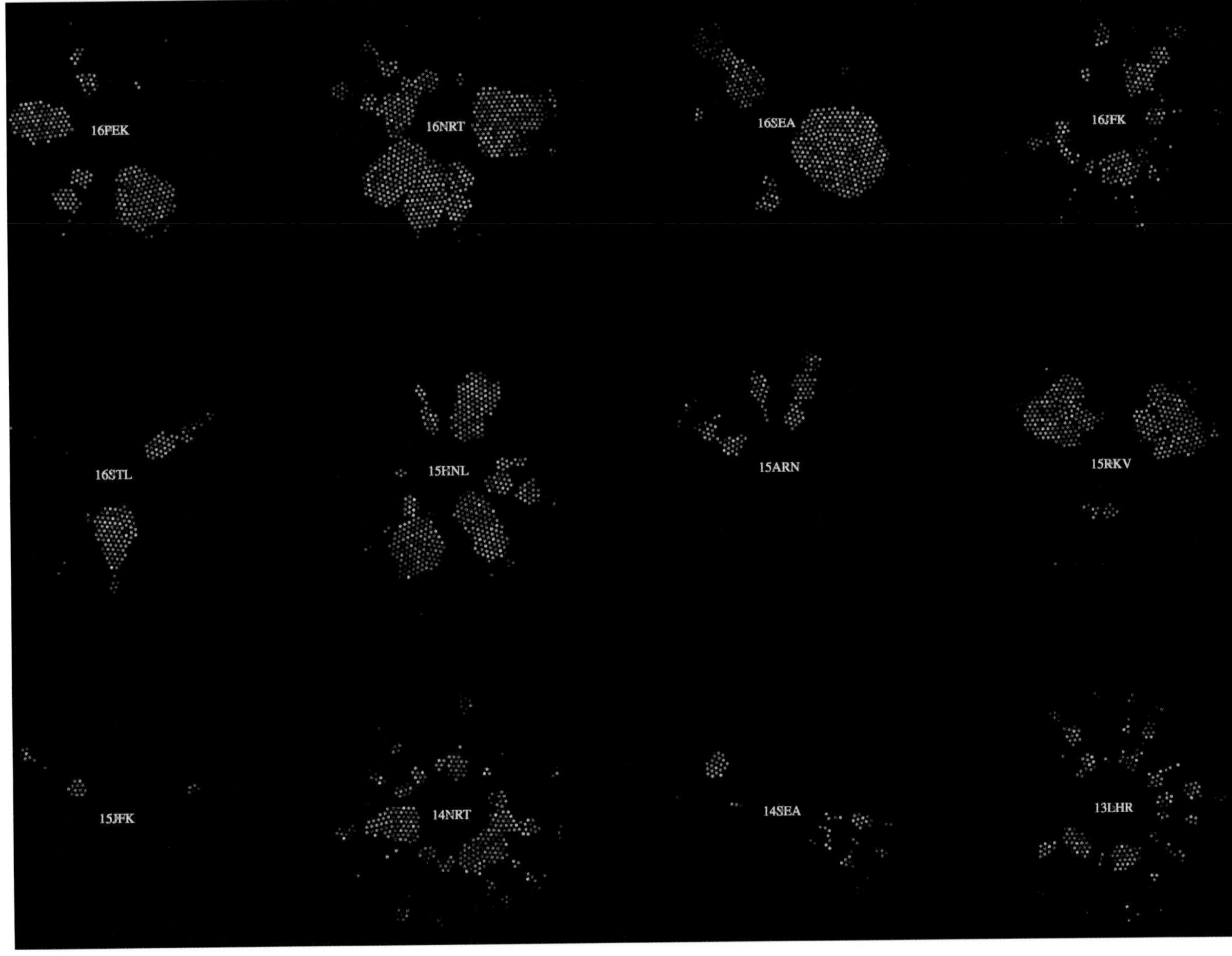

From this, I started to see patterns of which days and which trips had which colors. Unsurprisingly, I had loads of green and blue from my pictures of scenery; I love taking photos of nature, of oceans and mountains and lakes and forests. But I didn't know what the reds and oranges were until I looked into the underlying photos ... and found that they were mostly food photos!

My 2016 trip to New York and 2014 trip to Japan have so many reds and oranges (¯ — ¯;.

At one point, I also tried to switch the encoding of mapping time to *angle* and hue to *radius* (Figure 3.10, left) to the other way around (Figure 3.10, right). I found it really beautiful, but it looked too much like Nicholas's "Off the Staff" project, and it was also harder to read. I wanted to focus on what kind of photos I took across the days, but mapping time to *radius* meant that the first day of any trip was much smaller in circumference than the last day, and that was very inaccurate. It was a good experiment to switch and see what it looked like (and it was an easy, two line switch) but I ultimately decided to go back to my original encoding.

`Fig.3.10`

Switching the encoding of time and hue

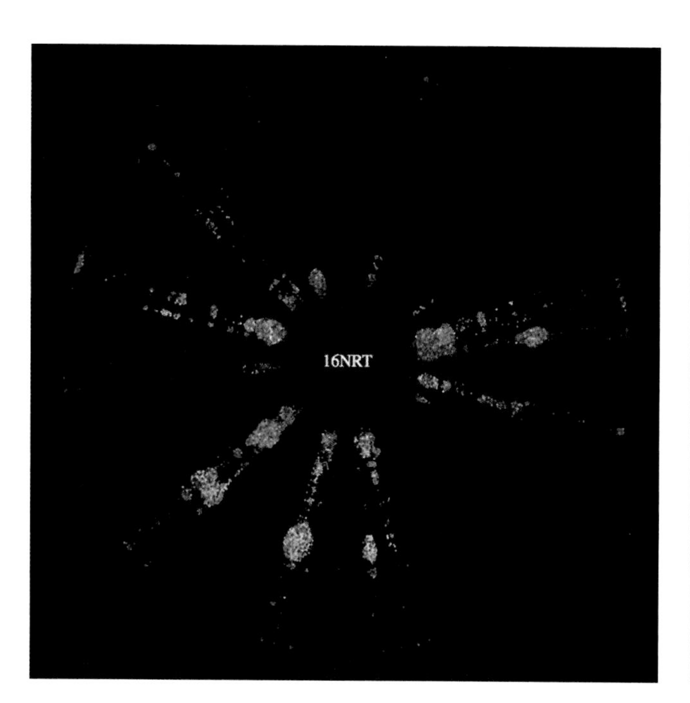

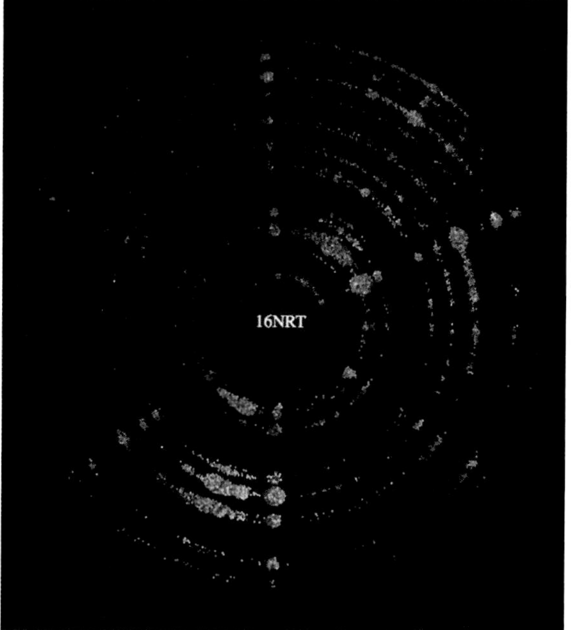

All of the code until this point only took a day. It took three more days to gather additional metadata on my trips (where I stayed, what I enjoyed, who I traveled with and visited), figure out the annotations for that metadata (Figure 3.11), and work through the interactions on how to show that information.

> **When people say that 80% of the work for a dataviz is in the annotations, they're really not kidding!**

Trigonometry

When I was in high school, I distinctly remember thinking to myself: when would I ever use trig in real life? I had my answer seven years later, when I created my first data visualization.

I use trig functions all the time now, especially when calculating radial layouts. I'll map the *angle* and *radius* to attributes in my data (in this project, I mapped *angle* to the time the photo was taken and *radius* to the hue of the color) and get the corresponding *x*- and *y*-positions:

```
x = radius * Math.cos(angle) // make sure angle is in radians
y = radius * Math.sin(angle)
```

And I can use the resulting *x*/*y* to position my marks.

If I want to go the other way—perhaps I have the *x*/*y* of a mark, and I'm trying to get the *angle* of its rotation—I will use `Math.atan2`:

```
angle = Math.atan2(y, x) // make sure to pass in y first!
```

A word of caution with `Math.atan2`: because it only returns values between $-\pi$ to π, if you want a number between 0 and 2π, it will require additional calculations after.

Finally, the distance formula is just super all-purpose useful:

```
dist = Math.sqrt(Math.pow(x2 - x1, 2) + Math.pow(y2 - y2, 2))
```

These are just the formulas I use most often. There are, of course, many other use cases worth brushing up on trigonometry for. ✧(○ •ᵥ•)

For an explanation of marks, see the "Marks & Channels" lesson on page 47 of my "Movies" chapter.

A nice trick to remember that y comes first when using `atan2`: tangent is opposite (y) over adjacent (x).

Fig.3.11

The final annotations.

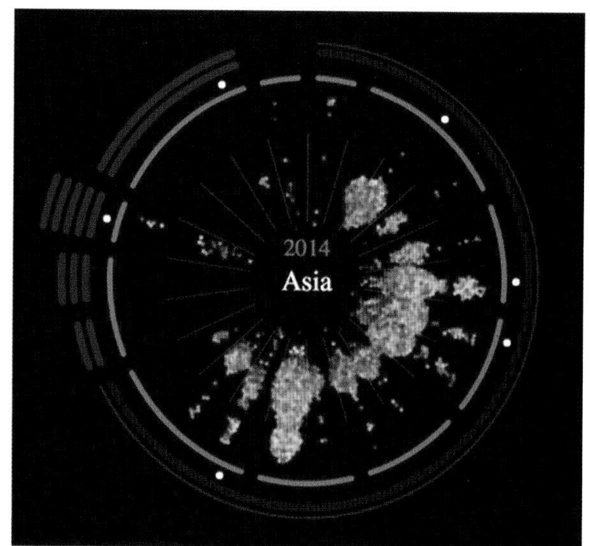

In particular, the interactive elements were hard to get down; originally, hovering over a line or a dot would show the information for that particular element. But since the lines and dots were all so small and thin, the flicker was unbearable whenever I accidentally unhovered or tried to move to the next line. After many iterations, I landed on something I really liked: placing the hover over a whole day (instead of the lines and dots within that day) would display all the information—where I was, who I was with, if I enjoyed my time—from that day.

Reflections

I'm really happy with how this project turned out, especially with how much I learned about color and trigonometry and the opportunity to try out a layout I haven't before. I am also happy that I pushed through with the annotations and legend portion and got them to a point that I could be satisfied with, since that's what I usually despise working on (discipline!).

This project is one of my favorite *personal* projects, because of how much I learned about myself (and how I like to travel) from it. It's also been a great way to commemorate and recall some of my favorite memories with my favorite people. (•ᵥ• ू)✧

Four Years of Vacations in 20,000 Colors

↳ shirleywu.studio/projects/travel

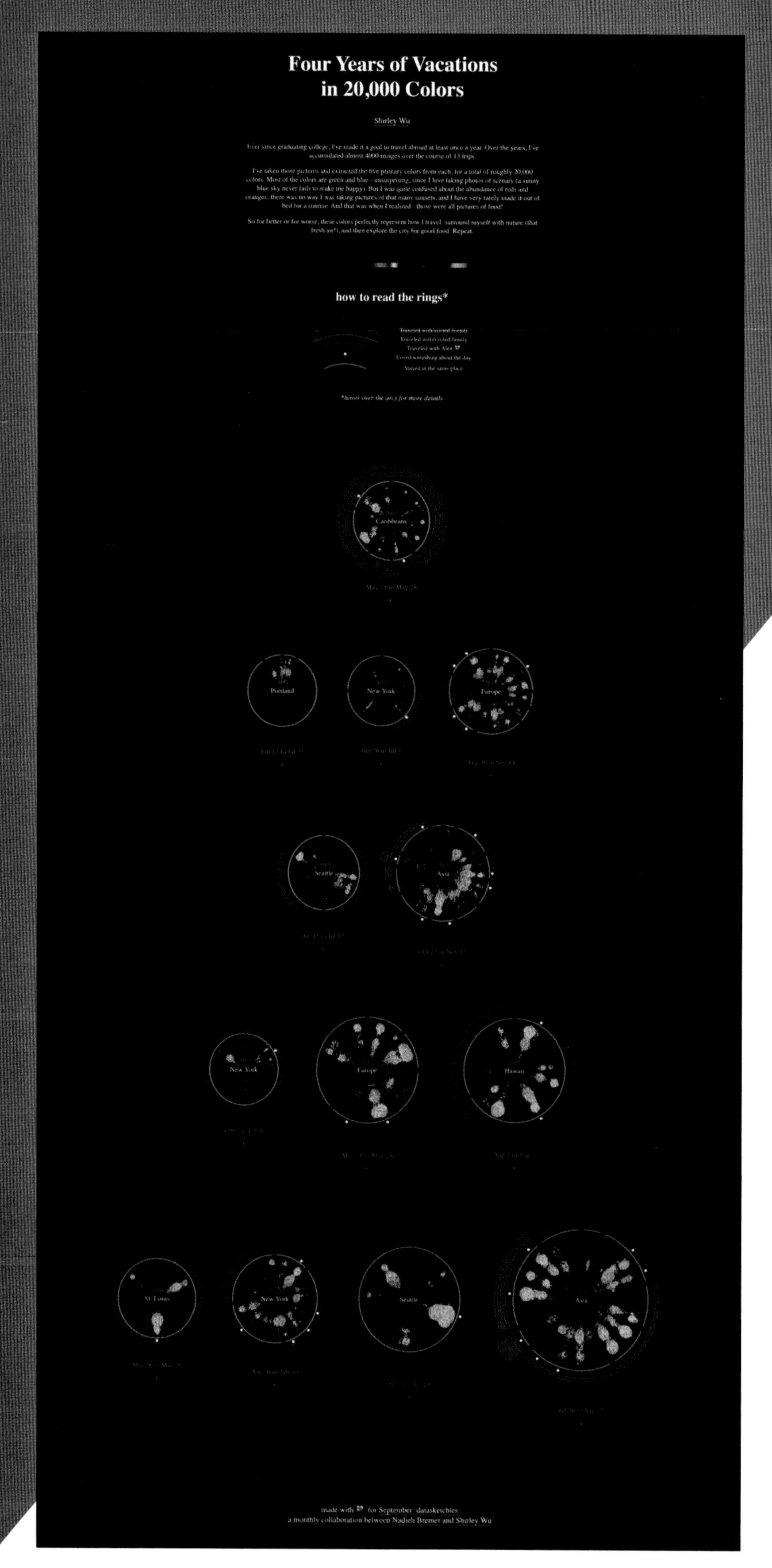

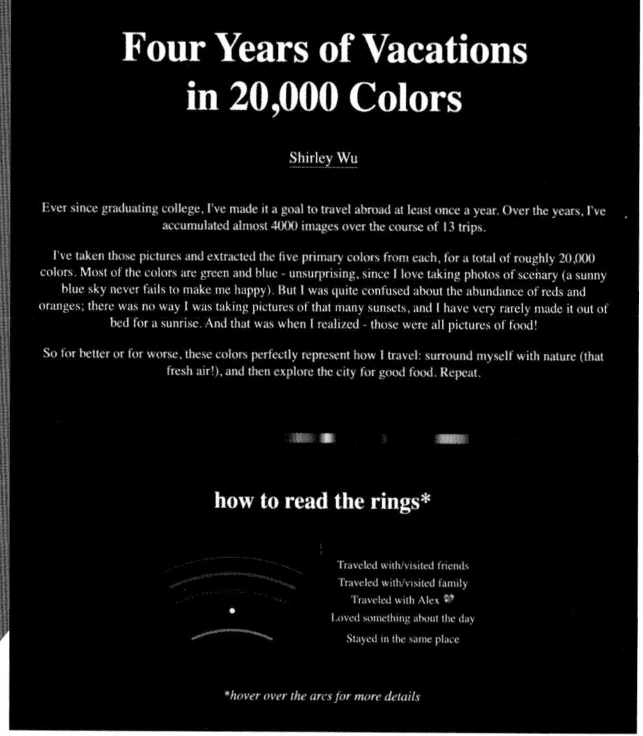

Four Years of Vacations
in 20,000 Colors
Shirley Wu
Ever since graduating college, I've made it a goal to travel abroad at least once a year. Over the years, I've accumulated almost 4000 images over the course of 13 trips.
I've taken those pictures and extracted the five primary colors from each, for a total of roughly 20,000 colors. Most of the colors are green and blue - unsurprising, since I love taking photos of scenary (a sunny blue sky never fails to make me happy). But I was quite confused about the abundance of reds and oranges; there was no way I was taking pictures of that many sunsets, and I have very rarely made it out of bed for a sunrise. And that was when I realized - those were all pictures of food!
So for better or for worse, these colors perfectly represent how I travel: surround myself with nature (that fresh air!), and then explore the city for good food. Repeat.
how to read the rings*
Traveled with/visited friends
Traveled with/visited family
Traveled with Alex
Loved something about the day
Stayed in the same place
*hover over the arcs for more details

Fig.3.12

The final visualization with legends, annotations, fourteen trips, and 20,000 colors. My favorite is definitely the 2014 trip to Japan and China, I made so many great memories. ᘳ(ↄ´ω`)ↄ✧

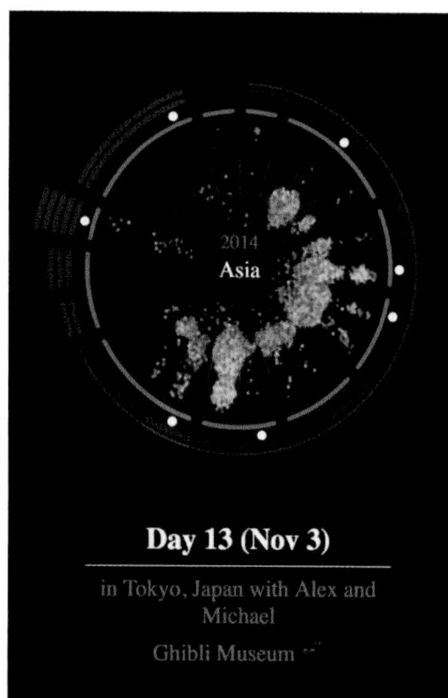

Fig.3.13

Hovering over a particular day shows detailed information from that day. I used a `d3.arc()` generator to create the pie slice shape used to highlight that day; it's still one of my favorite interactions.

Jun 12 to Jul 25

Jun 28 to Jul 1

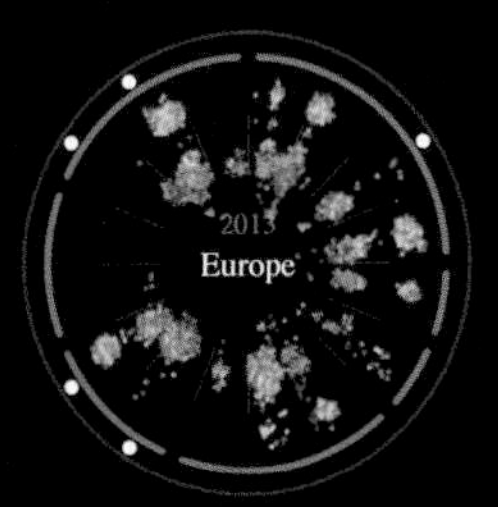

Aug 30 to Sep 14

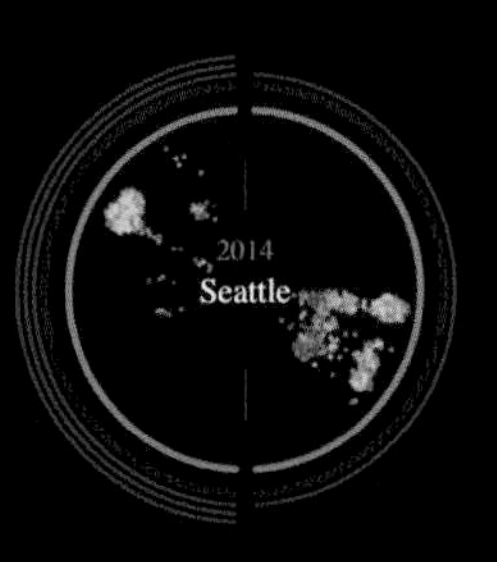

Jul 11 to Jul 12

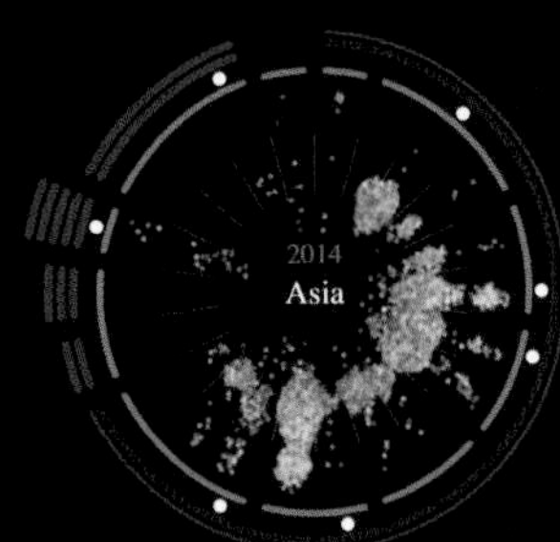

Oct 22 to Nov 12

Feb 6 to Feb 8

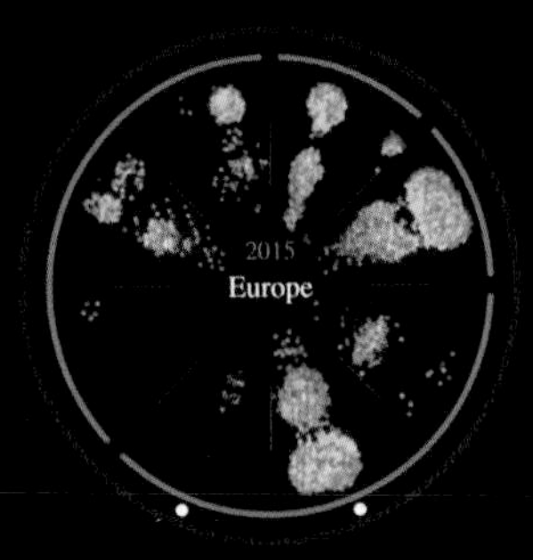

May 21 to May 28

Aug 2 to Aug 7

Mar 26 to Mar 28

Apr 16 to Apr 30

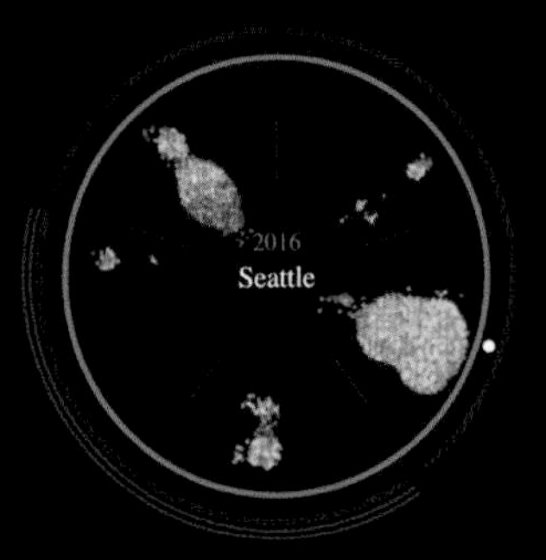

Jul 21 to Jul 25

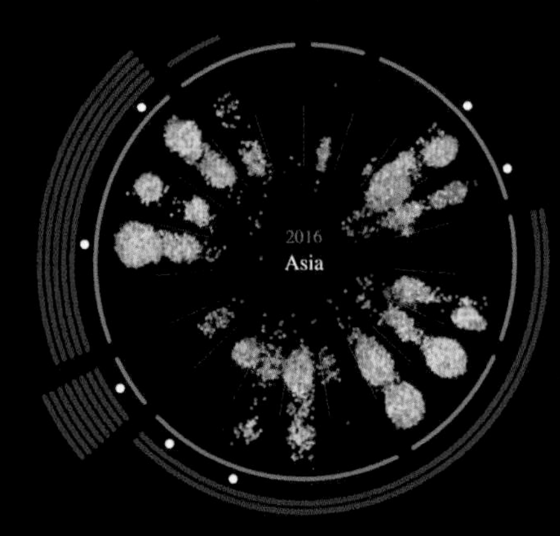

Jul 30 to Aug 17

PRESI
& RO

DENTS

YALS

Diana
II
anuel II
Nicholas II
William II
Victoria
Pauline of Wurttemberg
Marie Antoinette
Louis XIV

Royal Constellations

NADIEH

Talk about timing. Our original project idea was going to be around the US elections, but with the way that turned out, we decided to look for something different. As a European, focusing on American presidents didn't interest me much, but I felt that the European royal families would be a good equivalent to how Americans might see the presidential families.

The first thing that came to mind when I thought about the royals was their bloodlines. Due to all of the intermarriages in the past, I was sure most royals in Europe were related. The question then was: how *close* of a relative? Were they all cousins twice removed, or perhaps even closer than that?

Data

As with other projects, I thought that such a popular topic would yield many nice datasets with information about the royal bloodlines, but the reality was much different. Sure, many people had drawn the family trees of various royal families, but was there a dataset that offered a connected look at all of Europe's royals?

Luckily, I found a file online[1] that seems to be *the* information source of royal genealogy. It contains genealogical information of 3,000 individuals, all somehow connected to royalty in Europe! (/◕ヮ◕)/*:･ﾟ✧

a

```
6021    0 @I659@ INDI
6022    1 NAME Juliana of_Netherlands //
6023    1 TITL Queen
6024    1 SEX F
6025    1 BIRT
6026    2 DATE 30 APR 1909
6027    2 PLAC The Hague,Netherlands
6028    1 FAMS @F243@
6029    1 FAMC @F242@
6030    0 @I660@ INDI
6031    1 NAME Bernhard of_Lippe- Biesterfeld//
6032    1 TITL Prince
6033    1 SEX M
6034    1 BIRT
6035    2 DATE 29 JUN 1911
6036    2 PLAC Jena
6037    1 FAMS @F243@
6038    0 @I661@ INDI
6039    1 NAME Beatrix of_Netherlands //
6040    1 TITL Queen
6041    1 SEX F
6042    1 BIRT
6043    2 DATE 31 JAN 1938
6044    2 PLAC Soetdijk,Palace,Netherlands
6045    1 FAMS @F443@
6046    1 FAMC @F243@
```

b

```
25638   0 @F242@ FAM
25639   1 HUSB @I658@
25640   1 WIFE @I657@
25641   1 CHIL @I659@
25642   1 MARR
25643   2 DATE  7 FEB 1901
25644   2 PLAC The Hague,Netherlands
25645   0 @F243@ FAM
25646   1 HUSB @I660@
25647   1 WIFE @I659@
25648   1 CHIL @I661@
25649   1 CHIL @I1204@
25650   1 CHIL @I1205@
25651   1 CHIL @I1206@
25652   1 MARR
25653   2 DATE  7 JAN 1937
25654   2 PLAC The Hague,Netherlands
25655   0 @F244@ FAM
25656   1 HUSB @I613@
25657   1 WIFE @I662@
25658   1 CHIL @I663@
25659   0 @F245@ FAM
25660   1 HUSB @I663@
25661   1 WIFE @I664@
25662   1 CHIL @I665@
25663   0 @F246@ FAM
```


Fig.4.1 (a & b)

The left side shows how individuals are described in the original data. The right side shows a data snippet of how these individuals are related through family connections.

However, it was created in 1992 and therefore missed the recent additions to the royal families. And although the file contained lots of people related to royalty, the list wasn't complete. I therefore spent some time checking connections to famous people and added (grand)parents or (grand)children if needed. In the end, I manually added ±150 people to the file.

Family trees are a perfect match for a visualization type known as a *network* or *graph* that reveals interconnections between a set of entities. Therefore, the odd format of the original data had to be transformed into a "nodes" and "links" file. A *nodes* dataset contains one row per entity—a person in this case—with a person's characteristics in different columns, such as name, title, and date of birth. There's also a unique ID to be able to reference this person in the *links* file.

[1] www.daml.org/2001/01/gedcom/royal92.ged

A *links* file shows how these people are connected, with each row representing a connection between two people (from the *nodes* file). You can also add more attributes to these rows, such as the type of connection (e.g. "parent-child," "wife-husband").

At the beginning, all I did was get the nodes and links files in a usable shape. But while wrangling my way through the "force layout," which is what you use in D3.js to create a network visual, I thought about new variables to add or values to pre-calculate to get more insights from the structure of this gigantic family tree. Another addition came in the form of trying to fill in missing information. The original file has several characteristics for individuals: a unique ID, the name, their royal title, birth date, and death date (and location). Birth dates especially seemed like a wonderful way to create some more structure in the network, but these weren't known for everybody in the dataset. Thankfully I didn't need an exact date; I just needed a date that was in the right half of the century.

I wrote a small script in R that estimated birth dates either by looking if the death date of a person was known and then subtracting 60 years, or by looking at the birth or death date of the spouse, children, or parents, in that order, and again subtracting or adding some years to account for generational differences. This gave me an estimate for 95% of the people with unknown birth dates and I eventually filtered the remaining people out when I only kept the people born after 1000 AD.

My final addition to the data was to calculate which of the 10 royal hereditary families each person was most closely related to and within how many steps. I used this in the final network to make each individual want to be, horizontally, near their closest royal family connection.

This will probably make more sense once you've read the final parts of the next section and see the final visual.

Sketch & Code

In my initial sketches, I couldn't go beyond drawing some abstract circles connected by lines; I didn't actually know what kind of network the data would give me. I first needed to know the insights that were hidden in the "shape" of the data and thus had to visualize the network first before I could think about the more broader design.

> Instead of sketching with pen and paper, I sketched with code

The first challenge was just getting something visible on the screen. Being new to the updated force layout in D3.js, and not fully understanding all of its settings, I found basic examples on the site "Block Builder" which helped me visualize circles and lines.

Fig.4.2

09:40 The first successful result on my screen, a collection of circles and lines that were quickly expanding and moving out of my screen.

Fig.4.3

09:53 Slightly updating the network settings left me with a "hairball."

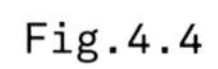

Fig.4.4

10:08 Making the canvas 36× bigger in size finally created enough space for the network to reveal some structure.

However, they just kept on expanding outward right outside of my screen (Figure 4.2), which wasn't very helpful...

Adjusting some settings of the network so the circles wouldn't keep excessively pushing themselves away gave me an even more useless "hairball" as a result (Figure 4.3). What the network really needed was some space. I therefore converted back to my initial settings and zoomed the page out very far on my browser so it would fit a square of 6,000 pixels wide, instead of the 1,000 pixels I had before. Everything became very tiny, but *some* structure started to appear (Figure 4.4). Some time and many iterations later I managed to squish it into a square of "only" 3,500 pixels.Even with some structure visible, I quickly came to the conclusion that I really needed more context to find insights. As mentioned before, I felt that birth year would be the most interesting variable that brings sense to the connections in this network; chronology is quite inherent to a family tree. I therefore colored the nodes by year of birth from dark blue for people born the longest time ago, all the way to yellow for the youngest person.

A network is called a "hairball" when it looks like a circular clump showing no structure. It's pretty much the worst possible outcome.

At this point I was still relying only on those individuals with known birth dates. People with missing dates remained grey.

Fig.4.5

11:42 Coloring everybody by birth year (when known).

The underlying algorithm that is used in D3.js to create these networks is based on the idea of gravity and charges. The circles try to push away from other circles, but the lines between them work a little like (elastic) strings making sure connected circles don't move away from each other (too far). In short, it has its roots in actual physics. But thankfully, in a computer environment we can "mess" with the laws f nature and play with how "gravity" works.

Fig.4.6

12:50 Pulling the network apart by year of birth, but the 40% of individuals with unknown birth dates represented as grey circles are pulled towards the center.

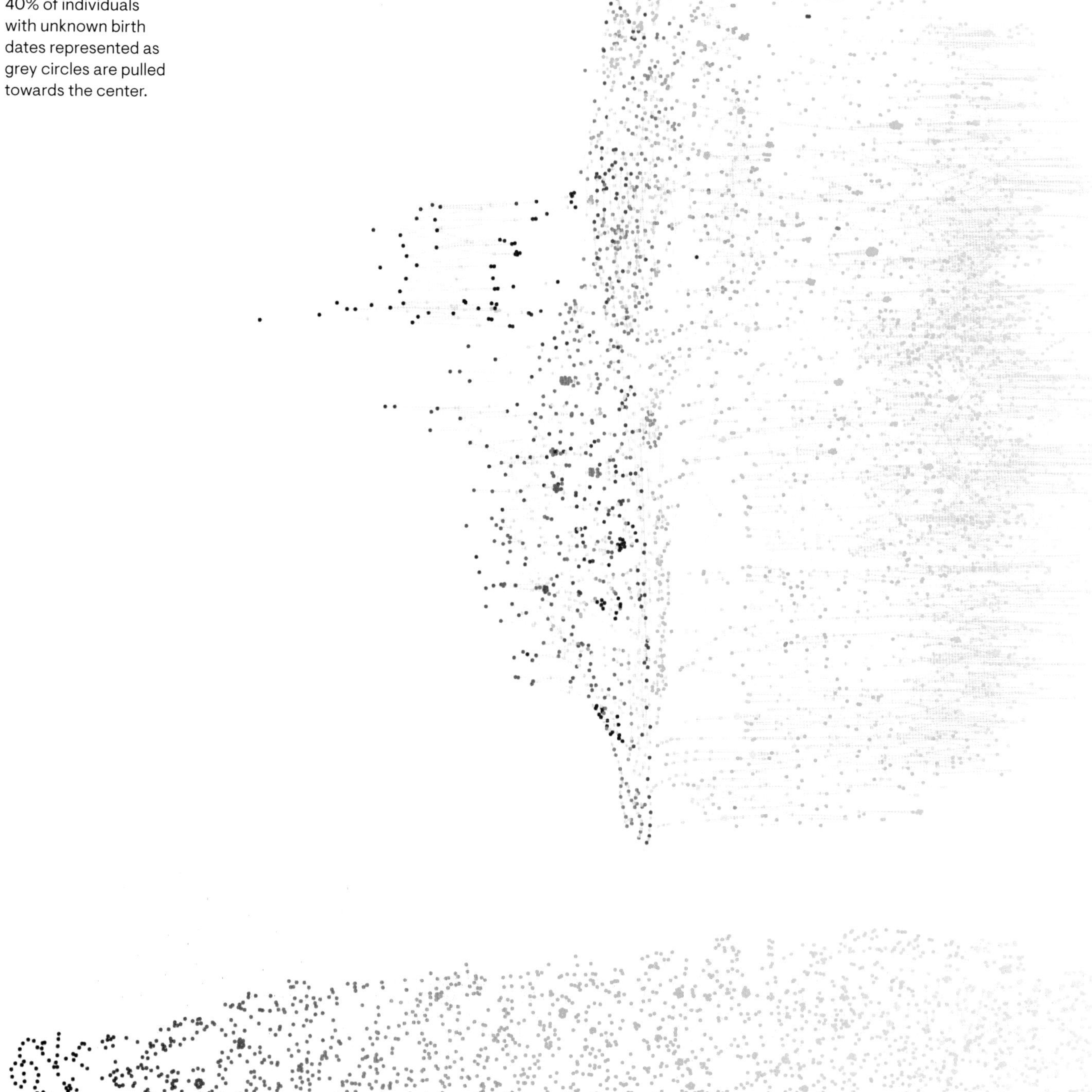

Fig.4.7

15:51 After estimating a birth year for the people with unknown years of birth the network was finally truly "pulled apart" chronologically..

The change that had the biggest impact on the structure came when I started using the year that an individual was born to define which *horizontal* position that person's circle would be placed in the visual.

Although I've since fallen in love with this technique of applying a data variable to the gravitational pull, it has one major caveat: you can't have unknown data for the variable that you use. Just like you can't place a circle in a scatterplot if you don't know the value for both the x- and y-axis, you need to know where each circle needs to be pulled. (And my dataset didn't have birth data for about 40% of the people.) That didn't stop me from first trying the most straightforward "fix" to at least get a sense of the result: everybody with unknown birth dates would be pulled towards the center. It created the rather odd looking shape from Figure 4.6, which I felt showed enough potential to investigate further.

That's when I decided that I *had* to estimate a birth year for all the unknowns. A few hours later, I was looking at a network that was organized in a way that's normal for a family tree: chronologically, from left to right in this case (Figure 4.7).

I explained how I estimated the birth dates in the "Data" section.

And although I first thought that I had gotten to the point where I could get some insights, when I truly started looking for info, the network was still rather uninsightful; it didn't tell me anything about recent royal history, or what has led to there now being only 10 royal families in Europe these days. Having spent more than six hours on this in one sitting (what are weekends for otherwise...) I stopped to try again the next day.

To get my mind off the network structure itself I started thinking more about the general design. As an astronomer, many things remind me of the night sky and network visuals often look like constellations to me. I therefore switched to a deep dark blue background and made the circles look like little stars by adding a touch of yellow glow.

Now that I was more pleased with the aesthetic look, I remembered the initial question that I wanted the visual to convey: how intermarried and interconnected were all of the royal families? To answer this, I knew that I needed to make distinctions between the different "branches" of the current royal families and see how intertwined they remained and/or became.

First, I made the current royal hereditary leaders big and noticeable. Then I spread them out evenly along the vertical axis, which would hopefully pull apart the network, placing closely related royal leaders near each other and the most loosely connected monarchs on the outside (Figure 4.8).

Fig.4.8

Fixing the 10 current royal hereditary leaders, the big circles, along a vertical line started to pull apart the separate family trees.

Continuing on the idea of connections and focusing on the current monarchs, I calculated the "distance" of each person to a current monarch, in other words, how many hops did it take to go from that person, through family connections, to each royal. I used that number to set the circles' opacity. Individuals closest to a royal would be very visible, diminishing to almost completely transparent for those more than six steps away from all of the royal leaders. And that's when I finally started to see some potential!

The entire network is connected; each and every person is related to everybody else.

To separate the network even more into the current family trees I used the "shortest distance to a royal" in another way as well: By generating a slight pull of the individuals to their closest current royal descendant. And also (non-linearly) squishing the date scale the farther back in time it goes. This makes the visual more focused on the (most interesting) last two centuries.

A vertical pull in the previous landscape style screenshots, and a horizontal pull in the newer portrait-style version.

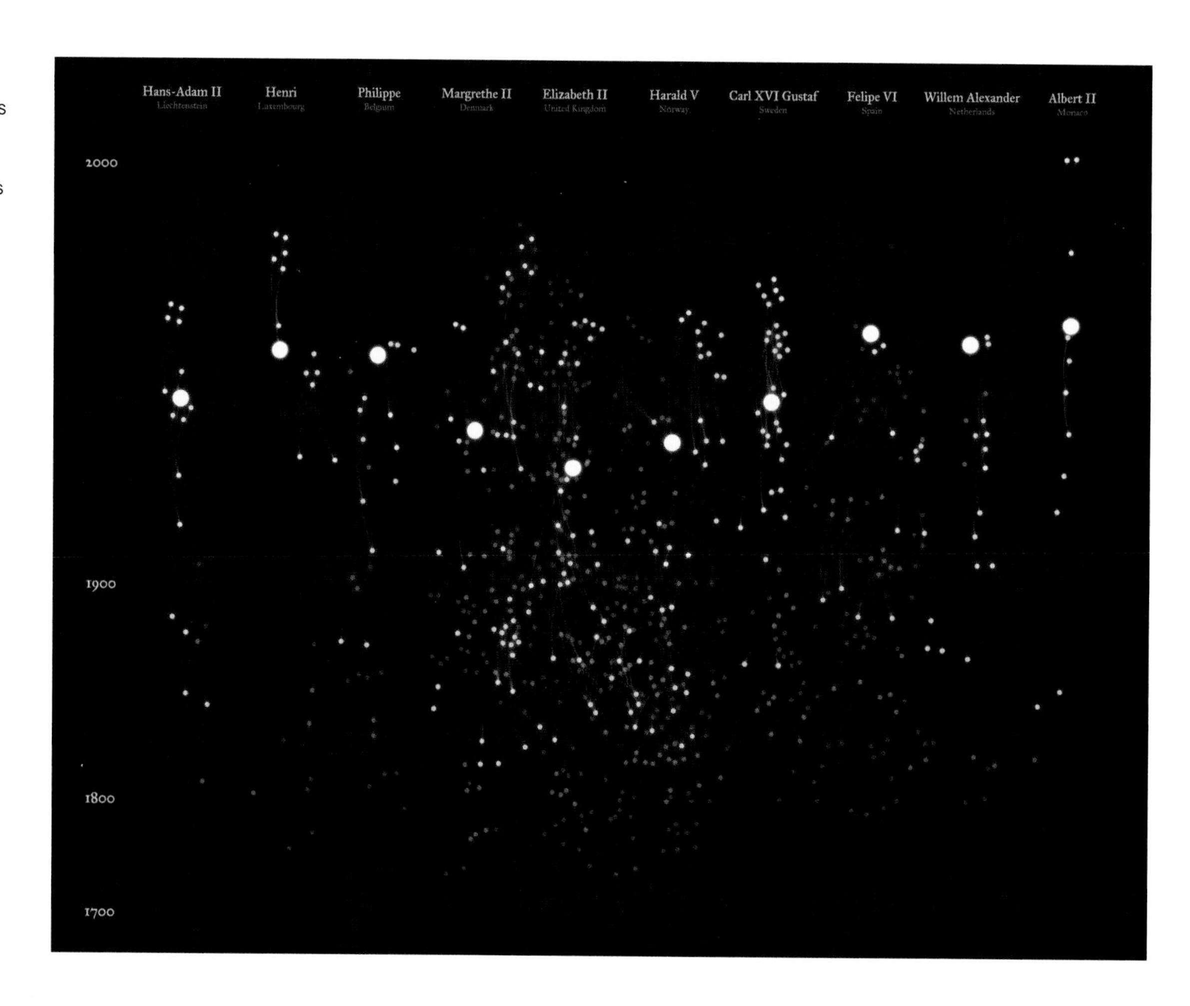

Fig.4.9

Pulling the family trees more apart by making each individual be slightly pulled towards their closest current relative that's at the head of a royal family.

And that's when I ended up with a network structure that I was happy with. It now revealed insights into how connected the current monarchs are. For example, the Monaco line (all the way on the right side of Figure 4.9) separated from the rest of Europe more than 200 years ago—the most recent connection to the rest of the network happens at around 1800—but Denmark, England, Norway, and Sweden, which are closer in relation, are all in the center.

This genealogy seems mostly focused on the English bloodline the farther back it goes.

Next up, I wanted to investigate how to use interactivity to make the closest connection between two people more apparent. After some brainstorming I came up with two main things to achieve: first, to show how far "six degrees of separation" spread from a hovered-over person into the network. The second thing was to show the shortest path between any two people.

In this case I mean that as six generations backwards and forward in time.

↳ Design with Code

Unlike more general graphic design, the effectiveness of a data visualization design really stands or falls with the data that goes into it. You can have an amazing looking pixel perfect design based on dummy data, but when the actual data behaves in an unexpected way, you'll have to restart your design process. That's why I prefer to only sketch out the rough abstract shape that I want to fit my data into, and then move towards making the idea into a reality by using the actual data. If that idea doesn't work with my data I can easily sketch out some other rough idea and try it again.

Afterwards I continue to design with code, where I test out other aspects of a typical design, such as details, colors, fonts, and effects, by programming it. That way I can see the effect of, say, a different color palette appearing on *all* of the data after just one refresh of my page.

Although you can assess pretty well how a chart would look if your dataset only has a dozen or so datapoints, the more data that you need to visualize, the faster you need to go from design to working with the actual, and total, dataset!

I typically start out with very simple and typically slow code, which makes it quick for me to determine if I'm happy with the visual result. Only when I saw that the "six degrees of separation" hover showed interesting aspects across different people, did I implement a faster method.

Due to the overall "space" theme, I decided to scale the colors that appear on a hover to the temperature scale of actual stars.

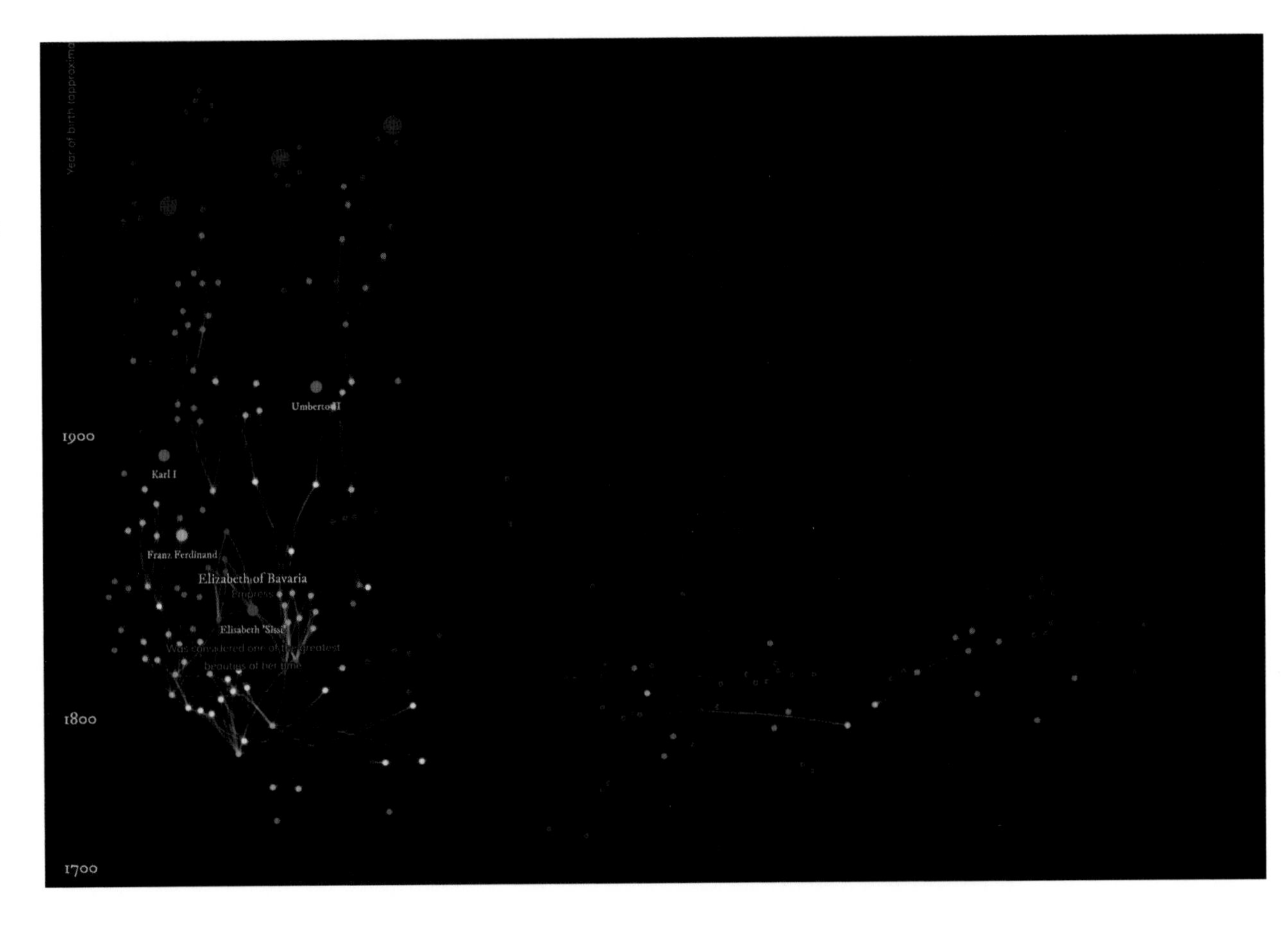

Fig.4.10

Hovering over Sissi shows which other people she's connected to within six generations; including three current monarchs.

At first I was afraid that with almost 3,000 circles in the network that a "shortest path calculation" would take too much time. I nevertheless first wanted to make it *actually* work and then see if it was something to keep and optimize. After a short search I found a script on GitHub[2] I adjusted to my needs. Surprisingly enough it performed amazingly well, returning a shortest path before I could blink! Adding some transitions to fade out the other circles and highlight the shortest path, and I was done much faster than expected.

I'm quite happy with the small dotted rings that rotate around the two selected people. It's the small victories.

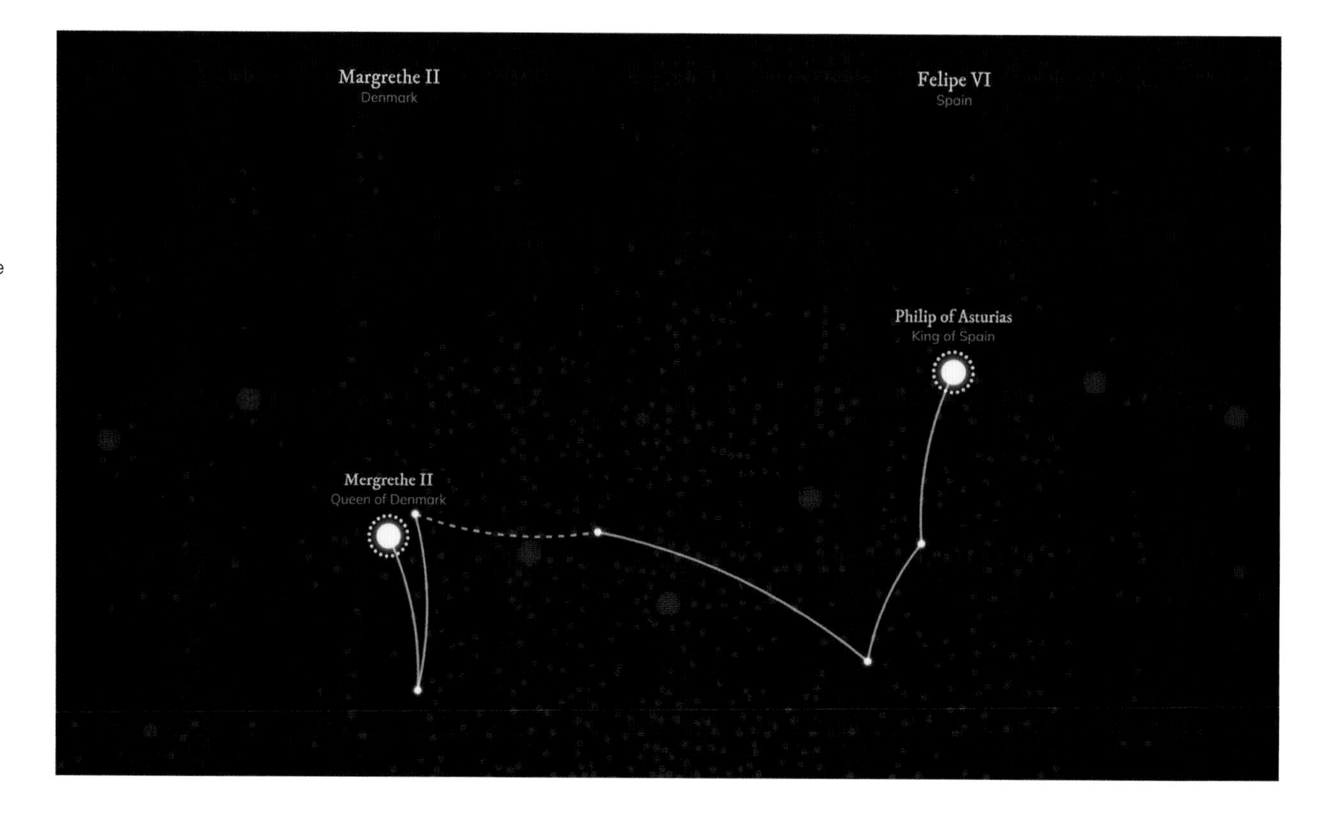

Fig.4.11

The visual representation of the shortest path between two clicked people. A solid line is a connection by blood, a dotted line represents a marriage connection.

Finally, to add a little bit more of a hierarchy and visual appeal to the visual, I highlighted several "interesting" royals throughout history, from Henry VIII to Marie Antoinette to the last monarchs of several European countries that are no longer a monarchy. A quick introduction and legend at the top, and I was done with my "Royal Constellations."

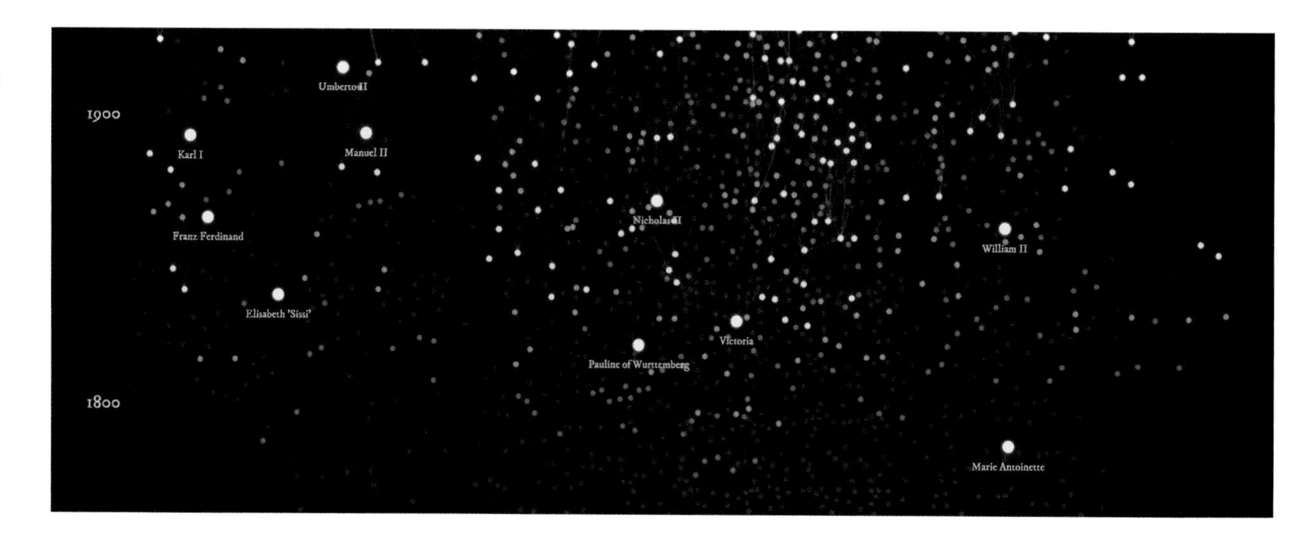

Fig.4.12

Several famous royals from the past are highlighted as well.

Reflections

I completed this project in just two weeks due to an upcoming vacation. With more time, I would've added other things, like a search box where you could find any royal in the visual by name. But I'm happy with what I managed in two weeks. While the interactive experience could have been improved upon, it accomplishes exactly what I wanted: to show how interconnected the European royal families really are. Plus, I learned a *heck* of a lot more about creating network charts with `D3.js`!

A year later, I finally felt comfortable enough with canvas that I rebuilt and replaced the entire visual in canvas instead of SVG, making it *a lot* faster.

[2] ShortestPathCalculator script by @julianbrowne:
https://github.com/julianbrowne/dijkstra/blob/master/javascripts/ShortestPathCalculator.js

Royal Constellations

↳ RoyalConstellations.VisualCinnamon.com

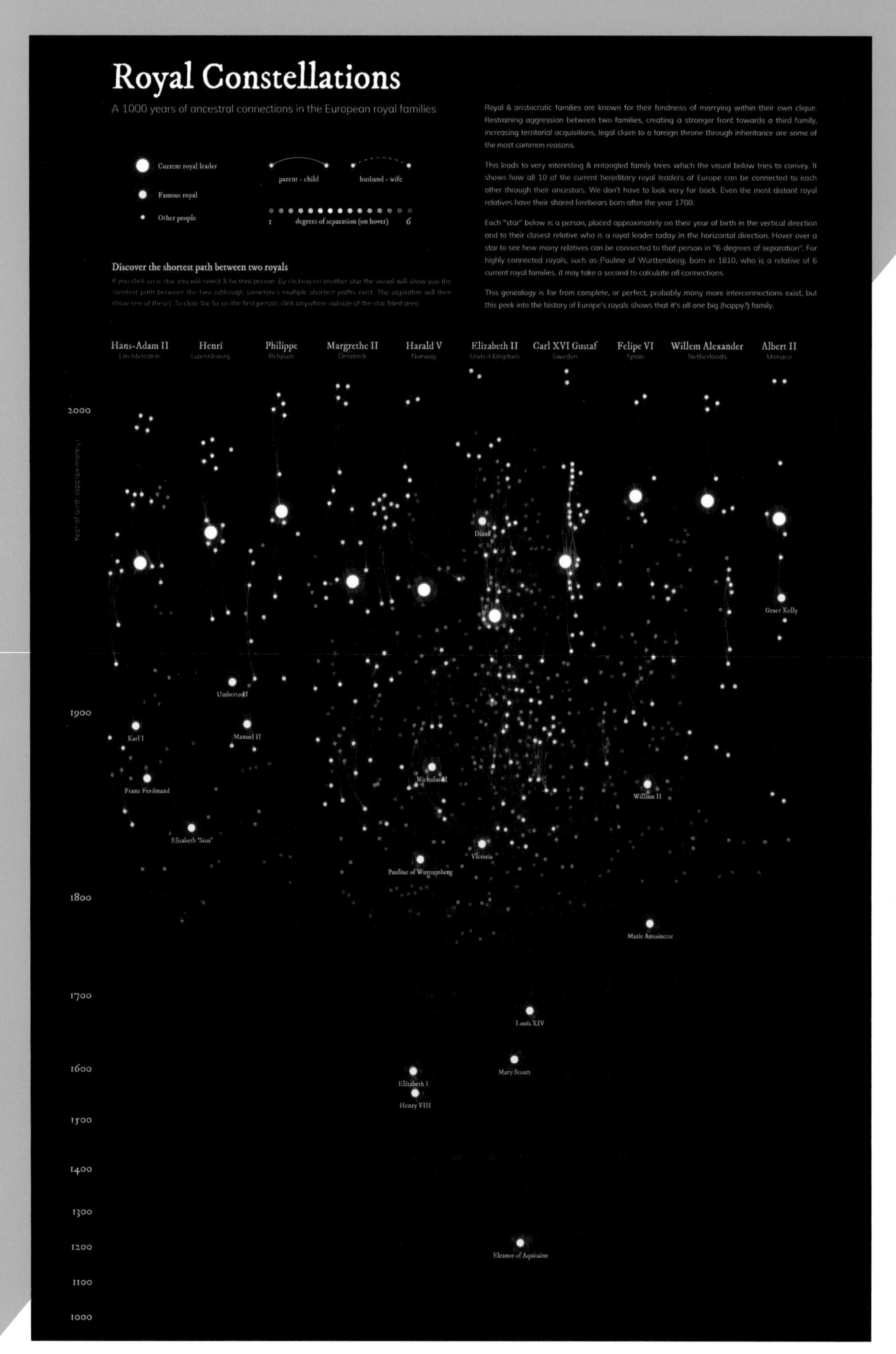

Fig.4.13

The final result of "Royal Constellations."

Royal Constellations

A 1000 years of ancestral connections in the European royal families

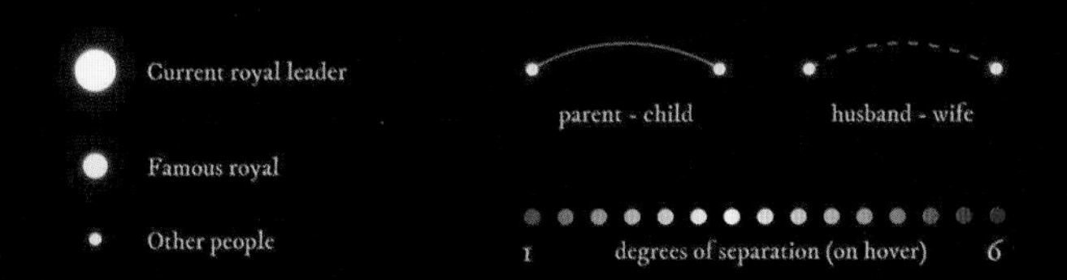

Discover the shortest path between two royals

If you click on a star you will select & fix that person. By clicking on another star the visual will show you the shortest path between the two (although sometimes multiple shortest paths exist. The algorithm will then show one of these). To clear the fix on the first person, click anywhere outside of the star filled area.

Royal & aristocratic families are known for their fondness of marrying within their own clique. Restraining aggression between two families, creating a stronger front towards a third family, increasing territorial acquisitions, legal claim to a foreign throne through inheritance are some of the most common reasons.

This leads to very interesting & entangled family trees which the visual below tries to convey. It shows how all 10 of the current hereditary royal leaders of Europe can be connected to each other through their ancestors. We don't have to look very far back. Even the most distant royal relatives have their shared forebears born after the year 1700.

Each "star" below is a person, placed approximately on their year of birth in the vertical direction and to their closest relative who is a royal leader today in the horizontal direction. Hover over a star to see how many relatives can be connected to that person in "6-degrees of separation". For highly connected royals, such as Pauline of Württemberg, born in 1810, who is a relative of 6 current royal families, it may take a second to calculate all connections.

This genealogy is far from complete, or perfect, probably many more interconnections exist, but this peek into the history of Europe's royals shows that it's all one big (happy?) family.

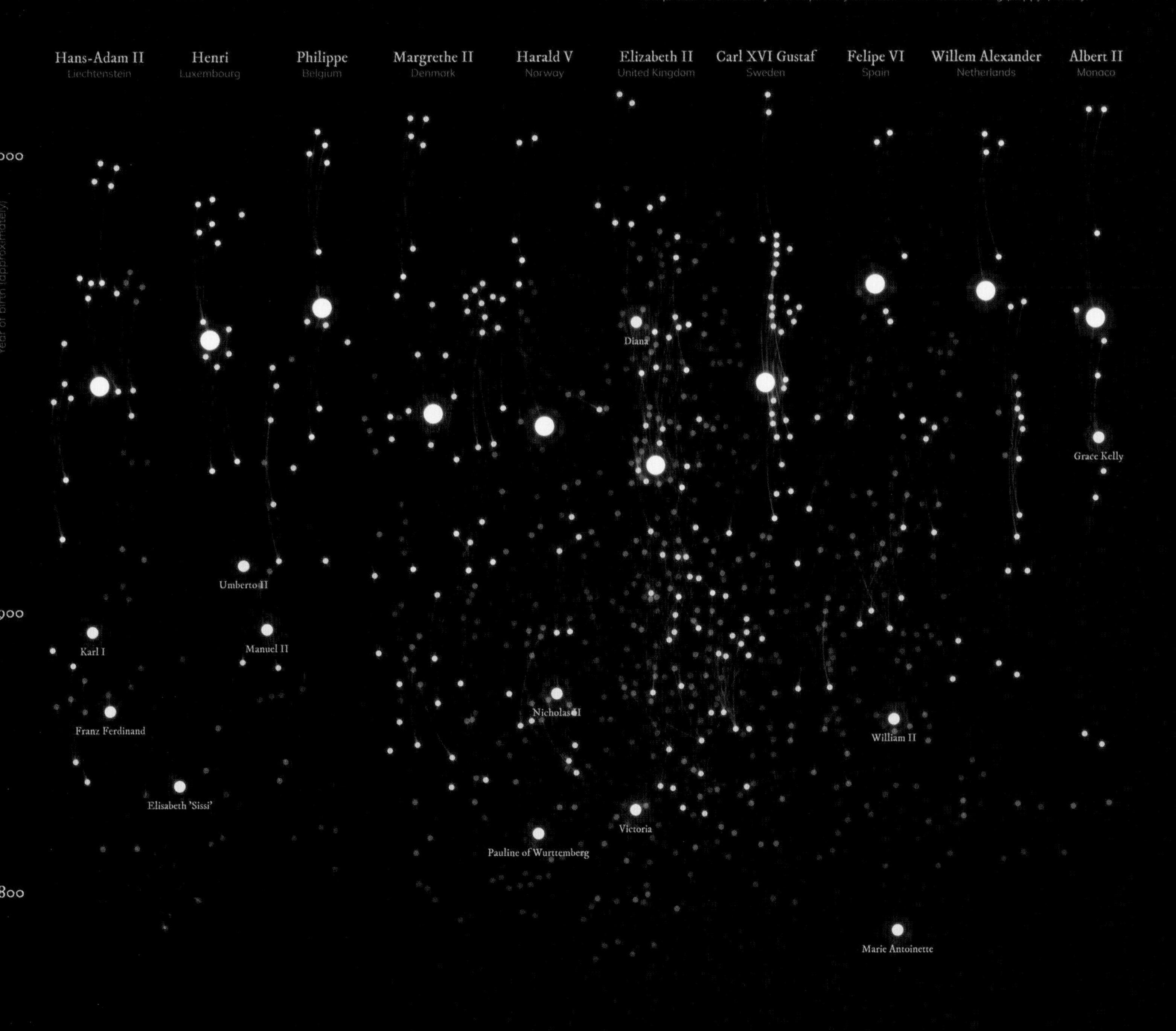

700

Louis XIV

600

Elizabeth I

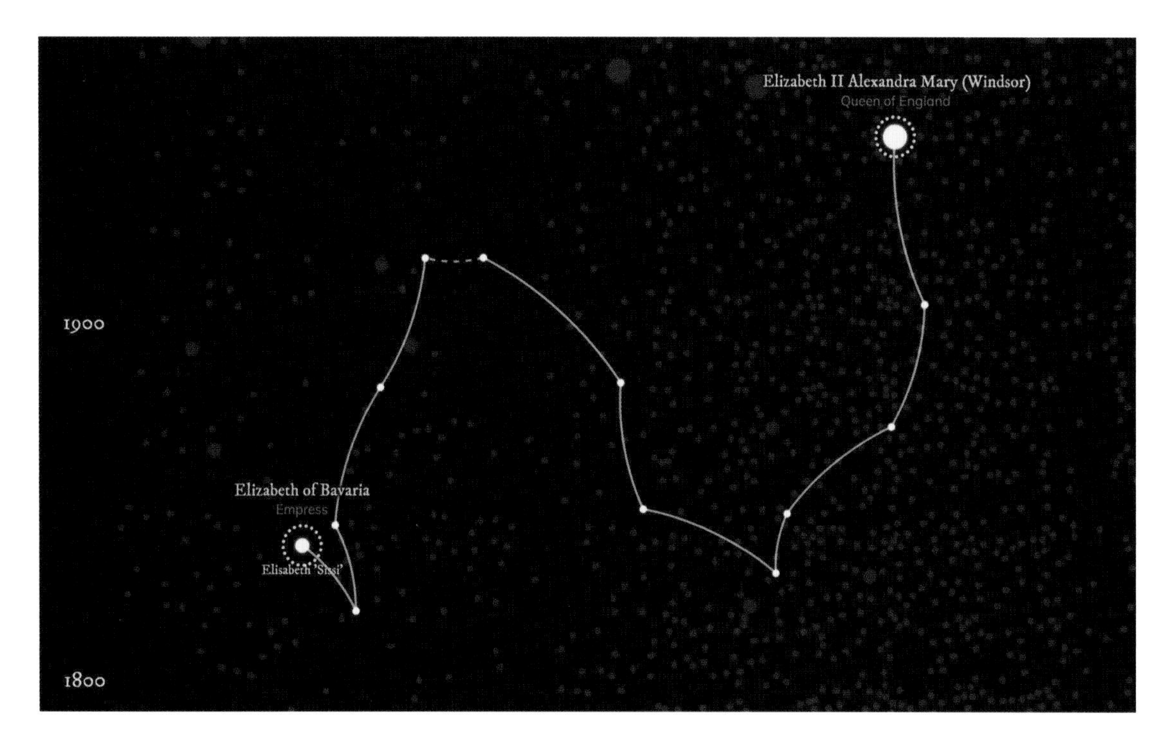

Fig.4.14

The shortest path between Queen Elizabeth II of the UK and Elizabeth of Bavaria (also known as Sissi), Empress of Austria and Queen of Hungary in the 1800s, is a bit of a distant one. However, all connections except one are blood relations: the other is a husband-wife connection between a previous king of Belgium (Leopold III) and princess of Sweden (Astrid).

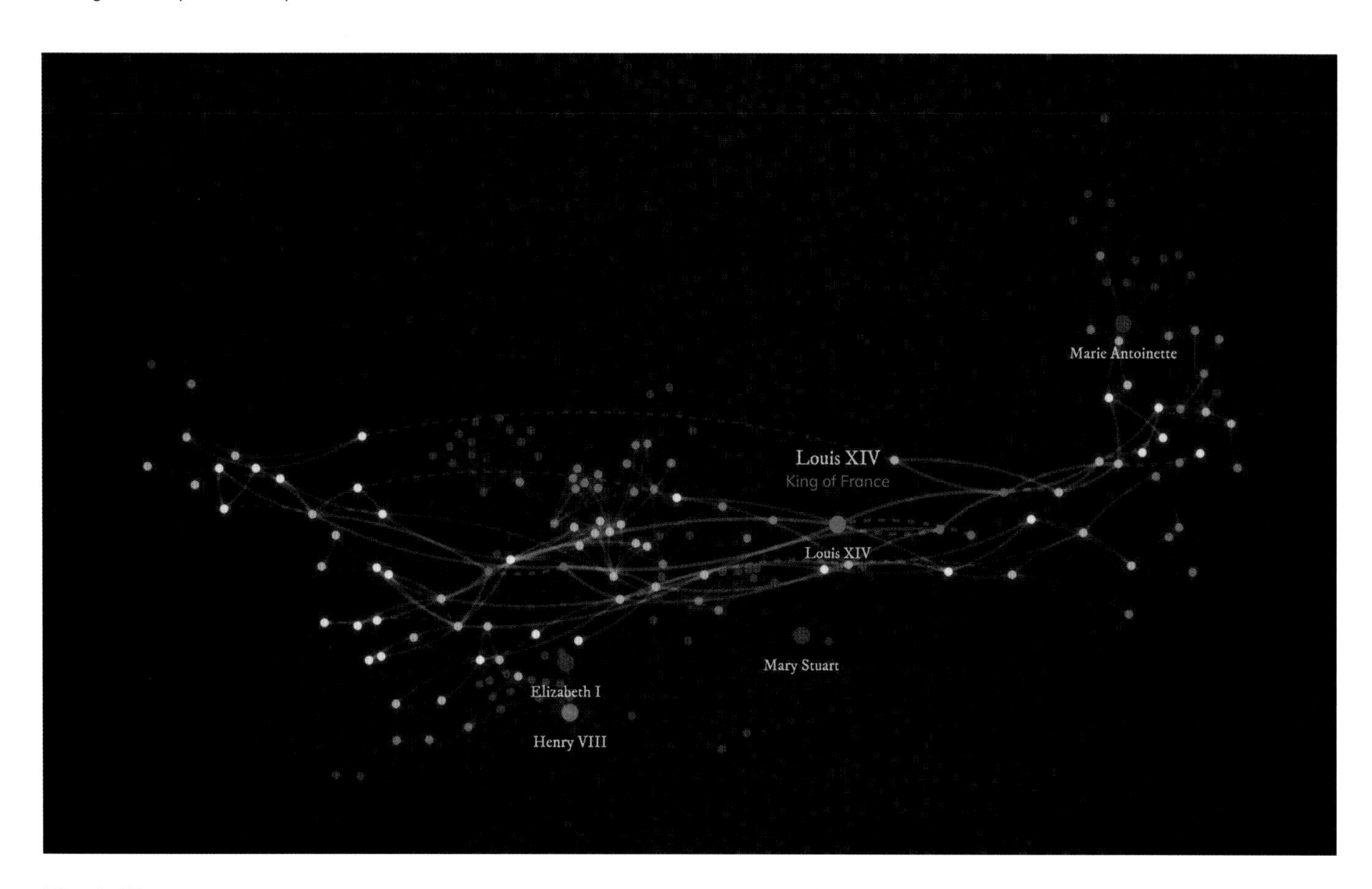

Fig.4.15

Hovering over Louis XIV of France, also known as the Sun King. With 72 years, he has been the longest reigning monarch in European history.

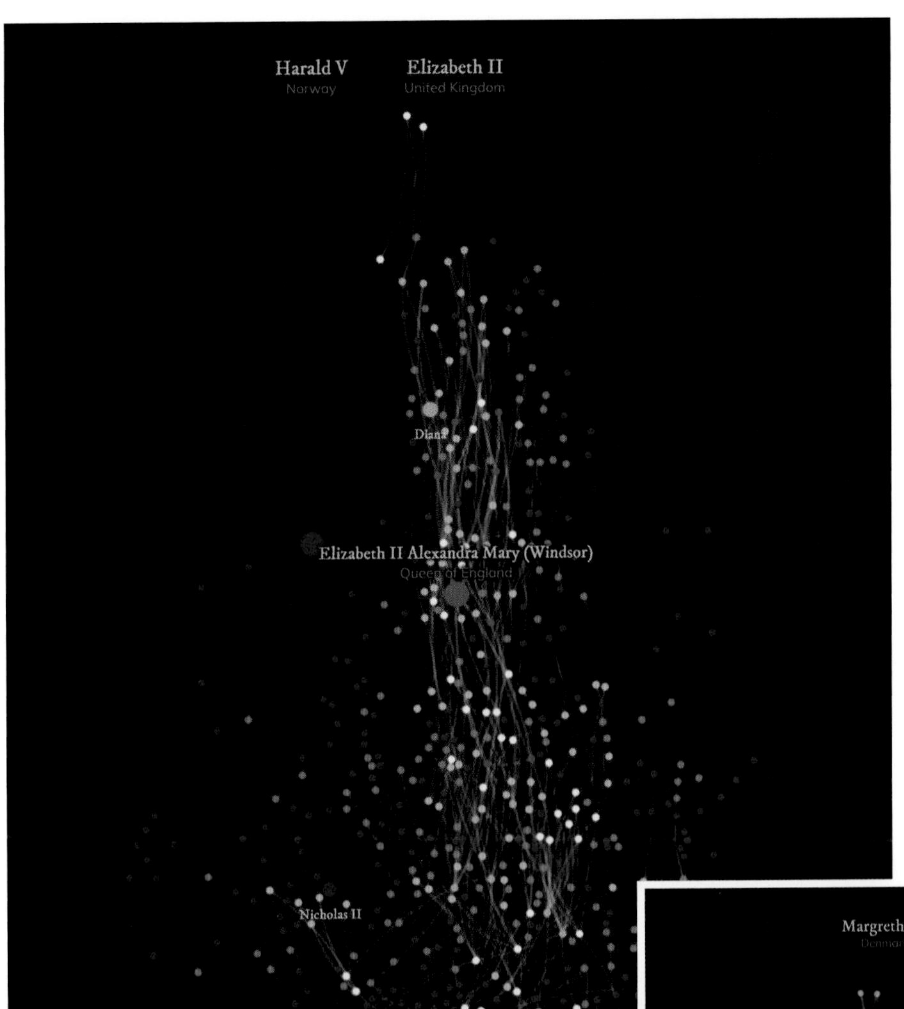

Fig.4.16

Hovering over Queen Elizabeth II shows one connection within six steps to another monarch: King Harald V of Norway, seen as the red large circle to the left of Elizabeth's blue large circle. He is Queen Elizabeth's closest relative who is also a monarch, and they are second cousins. (As a side note, Queen Margrethe II of Denmark, King Juan Carlos of Spain and King Carl XVI Gustaf of Sweden aren't much farther away: they are her third cousins.)

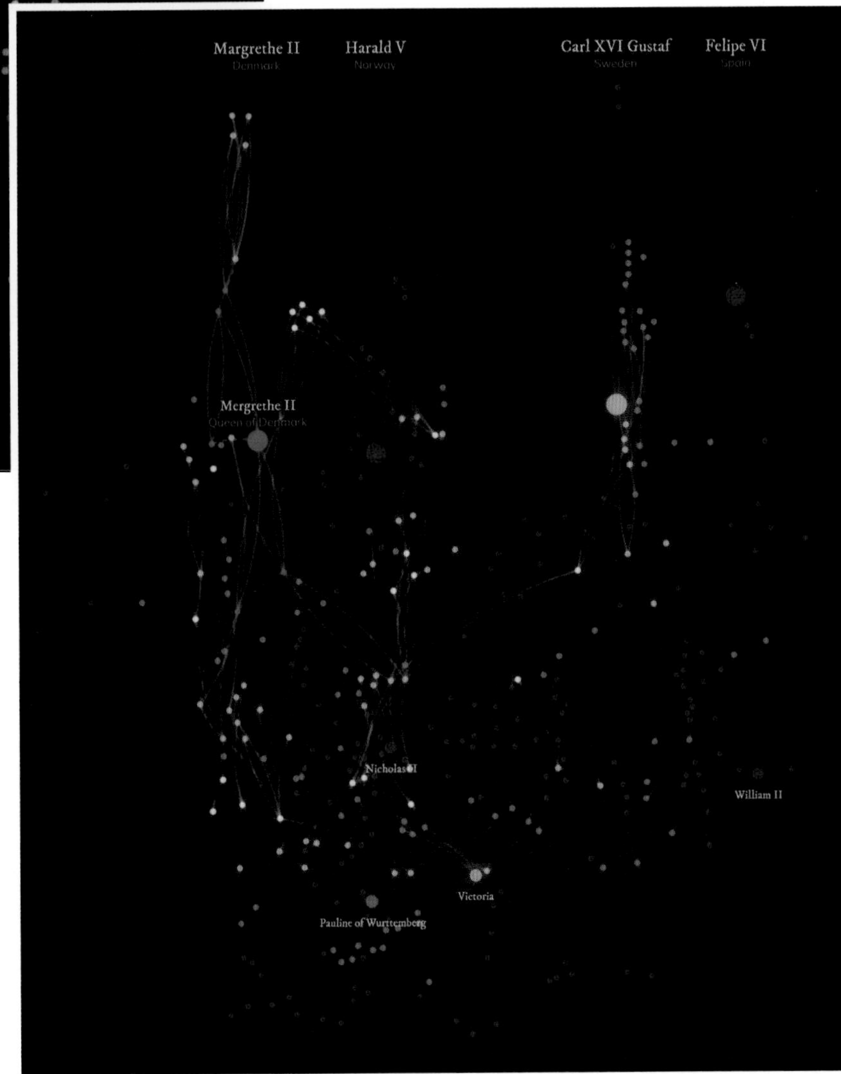

Fig.4.17

Hovering over Queen Margrethe II of Denmark, the big blue circle in the top/middle-left.

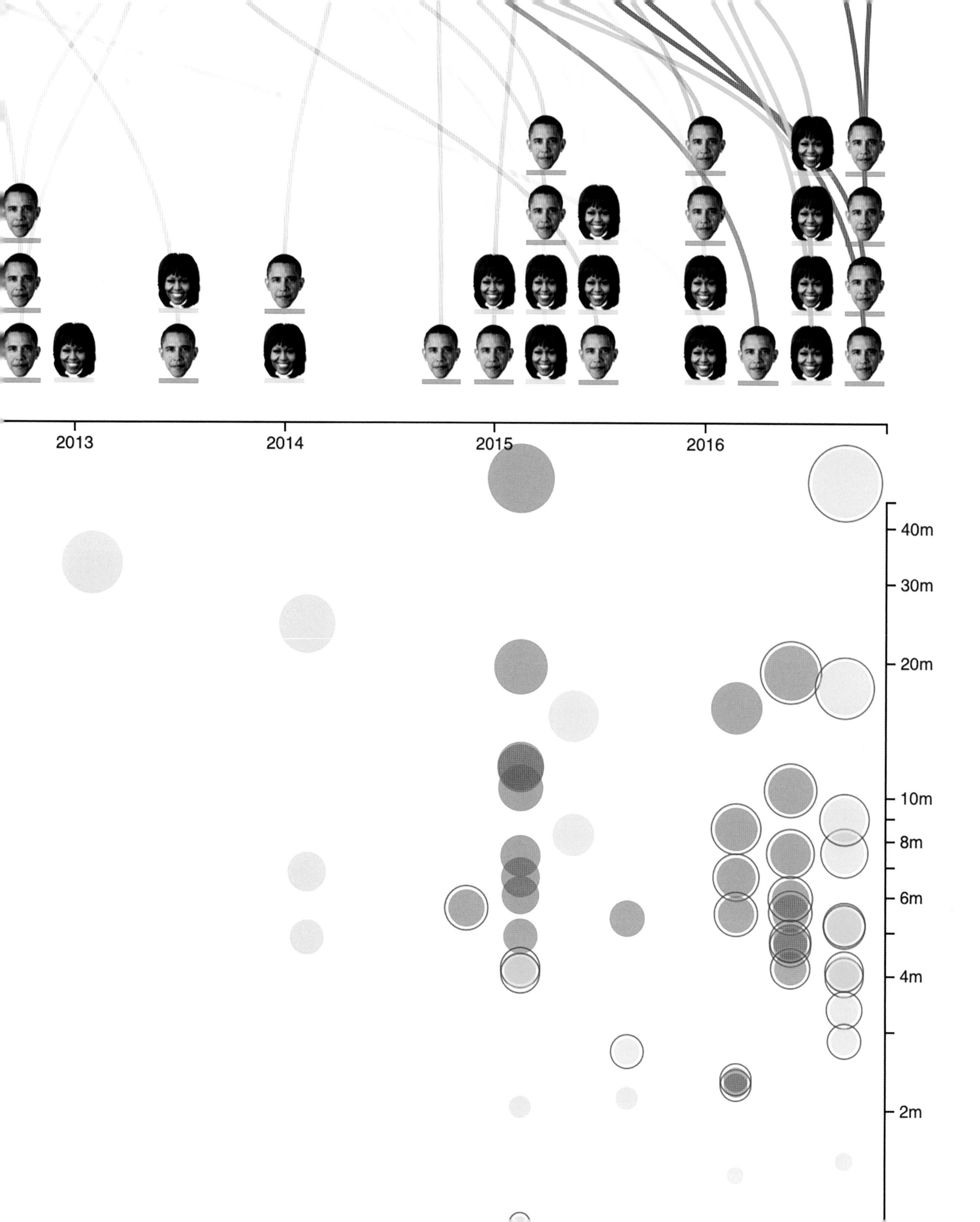
2013
2014
2015
2016
40m
30m
20m
10m
8m
6m
4m
2m

Putting Emojis on the President's Face

SHIRLEY

Originally, Nadieh and I slotted October for the US presidential elections, in order to have something ready by Election Day. But as October approached, I realized more and more that I wanted absolutely nothing to do with the 2016 election (other than to vote). We mused on what to do instead, and I realized: a future president getting voted in also means a current president leaving the White House.

While we might not all agree on what the Obama administration did or didn't do politically, I think most of us can agree that the Obamas are one cool couple. I thought of them as the slightly dorky parents that I wouldn't mind associating with if they were my friends' parents. Like that time when Mr. President recorded a video of himself talking to a mirror so that he could plug healthcare.gov, or when Madame First Lady *Carpool Karaoke'd* with James Corden and she was so hilariously relatable. This made me realize that I would, in some way, miss them being our POTUS and FLOTUS.

Data

I wanted to do something silly and light-hearted to commemorate the Obama's eight years in office. And the first thing I could think of was all of their appearances on late-night talk shows I watched on YouTube (that "Slow Jam the News with President Obama" was so good). I started digging around to see if there was a list of talk show appearances for both Barack and Michelle—and it was IMDb to the rescue again! Both of them had their own IMDb pages, so I cross-referenced Barack Obama's 214 credits and Michelle Obama's 123 with a Wikipedia article of late-night American network TV programs. I was able to narrow down the credits to 24 late-night appearances for Barack and 22 for Michelle. I then compiled a list with the date of their appearance, name of the late-night talk show, and the show's YouTube channel if available.

I actually didn't realize until I published it and someone pointed it out: I included The Ellen DeGeneres Show but that's a daytime TV show! The errors that happen in our data gathering. (¯ — ¯ ;

Using that information, I was able to use the YouTube Search API to search for keyword "obama" and filter the results down by `channelId` and a publish date within five days of the interview date. Unfortunately, because I wasn't sure how many videos were published for each interview (if at all), I set the `maxResults` to 15. This meant that I would get back videos within those days that had nothing to do with the Obamas, and I had to manually go through all of the videos to weed out the irrelevant ones (there were 244 videos, and 186 were ultimately filtered out, leaving only 58).

The list was unfortunately incomplete, as there were interviews with past shows and hosts like the *Late Show with David Letterman* and the *Tonight Show with Jay Leno* that weren't on YouTube. I chose to keep them out instead of trying to find the videos on other websites or on unofficial channels, in the hopes that this would make for cleaner and more consistent data.

Around that time, I attended the annual `D3.js` unconference. I told some friends there about the dataset I had gathered and how I wasn't sure what to do with it. They were enthusiastic: "You should get the captions for each video and do something with the words!"; "Wouldn't it be cool if you could run facial detection on the video and correlate their emotions with what they're saying?" And even though I was quite worried about how long it would take me, the idea took root in my mind.

The sort-of annual d3.unconf started out in San Francisco and is an event very dear to my heart.

After some research, I realized there wasn't a financially affordable way for me to pass whole videos into any facial recognition software, but thankfully I was still surrounded by some very resourceful friends at the conference. They suggested an alternative: take screenshots of the video and upload the images to Google Cloud Vision API for analysis. Here's what I did (and the Node.js packages I used):

Thank you Ian Johnson and Erik Cunningham!

1. I took the list of all videos, and downloaded them (and their captions, if available) with `youtube-dl`.
2. I used `vtt-to-json` to convert the captions into JSON and got the timestamp every time someone talked, and used that timestamp to screenshot the video with `fluent-ffmpeg`.
3. For each screenshot, I uploaded it to Google Cloud Vision API, which gave back data of any faces it found in the image, the *x*/*y* bounds of the face, and any emotions detected on the face.
4. I saved the videos' captions into one JSON file.
5. I joined the captions with the annotated image data (including emotions) from the Google Cloud Vision API.

This entire process took a few days, and once I finally had all the data cleaned I started thinking about the design. I learned my lesson from "Travel," and made sure to explore the basic shape of the data, as well as all the different types of data I had on hand before starting to sketch.

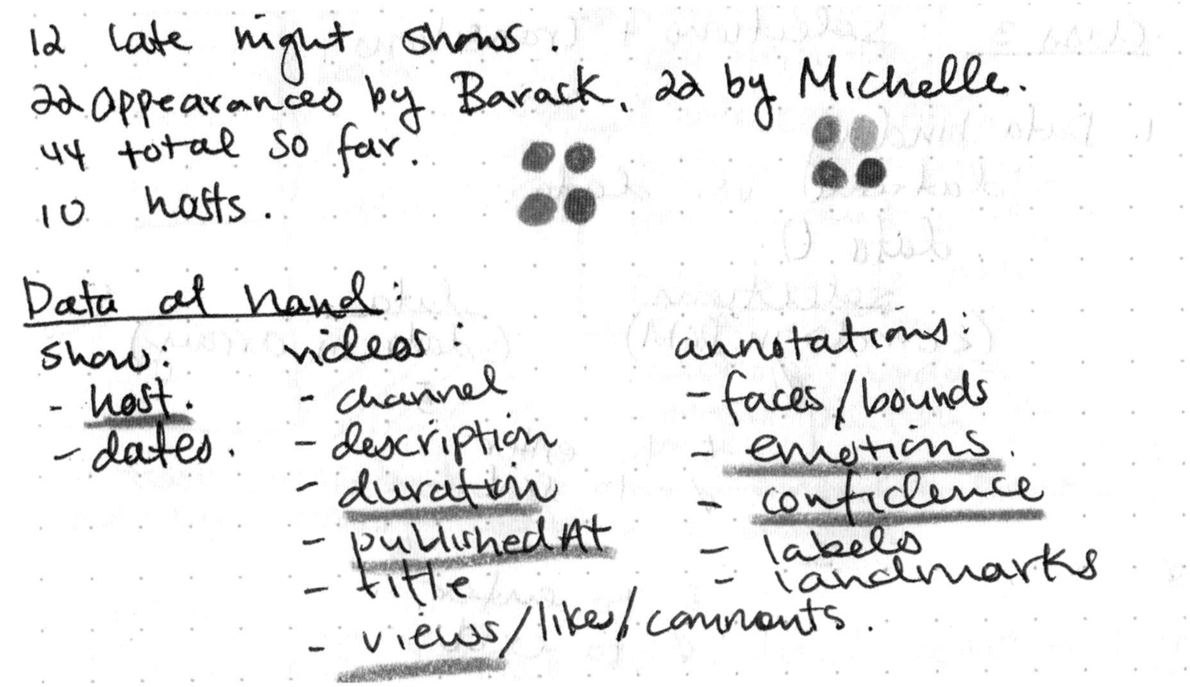

Fig.4.1

List of the data I had.

↳ Data Can Be Found in Many Different Ways

When I start on a data visualization project, I always start with a curiosity, a topic I'm interested in—whether that's the Obamas' late-night talk show appearances, recurring themes in *Hamilton* the musical, *Dance Dance Revolution*, or summer blockbusters. I find that having that curiosity keeps me motivated and gives me direction when looking for datasets.

Once I have a curiosity, I start with a Google search. Sometimes I'm lucky, and I find a dataset that's already downloadable in JSON or CSV format (*very* rare). Oftentimes, I'll find a website where the data is in plain text on the page; in those instances, I can get the data with their API or by scraping the site. For scraping, I like using Node.js modules (mostly `http` to programmatically request the webpage), and for APIs, I've found that there's usually a corresponding Node.js package that does all the hard parts (like OAuth authentication for signing into a service) for me.

If I can't piece the data together from resources online, then I have to manually gather and enter the dataset. It's tedious work, but I've learned (the hard way) that Excel makes the process so much easier. One time, I found the data I wanted in image form, and was able to get the underlying raw data by just ... asking the person who ran the website. It was amazing.

For a short explanation of JSON and CSV, see "Technologies & Tools" at the beginning of the book.

Whenever I think, "Somebody must have done this already" while coding, somebody usually has!

You can read about this story in my "Music" chapter. (*≧▽≦)

Sketch

Once I had the metadata listed, I started thinking about what I wanted to convey. I knew it was ambitious, but I wanted to make full use of such a rich dataset and also get across why even though there were 44 appearances I only had screenshots for 29 videos. I'd also been wanting to try my hand at "scrollytelling" to test my technical skills and decided to create a multi-section project. I mapped out my first few sections to introduce the dataset by show and host (Figure 4.2, ①) and then by date (Figure 4.2, ②). I would display all the videos I had (Figure 4.2, ③), highlight the videos with captions (Figure 4.2, ④), and for the videos with captions, plot the emotions detected throughout the video on a timeline (Figure 4.2, ⑤).

"Scrollytelling" is a popular technique where we tie progression of the story and its corresponding visualizations to user scroll.

And even though I was still blurry on the latter half of the sections, I decided to take a few days to build out those initial sections first. This helped me get a better feel of the video data, which I hadn't explored in as much detail as the hosts and appearances data.

I'm glad I did that because I originally envisioned something where all the videos would be lined up in rows, and a timeline would show a bar graph of both word frequency based on the captions and all the times that POTUS and FLOTUS laughed. But after working on the initial sections, I realized that that was *way* too much data to try and show on one screen. I came up with another sequence of sections to introduce the videos instead and decided to only show the selected video's timeline of captions and laughs (Figure 4.3).

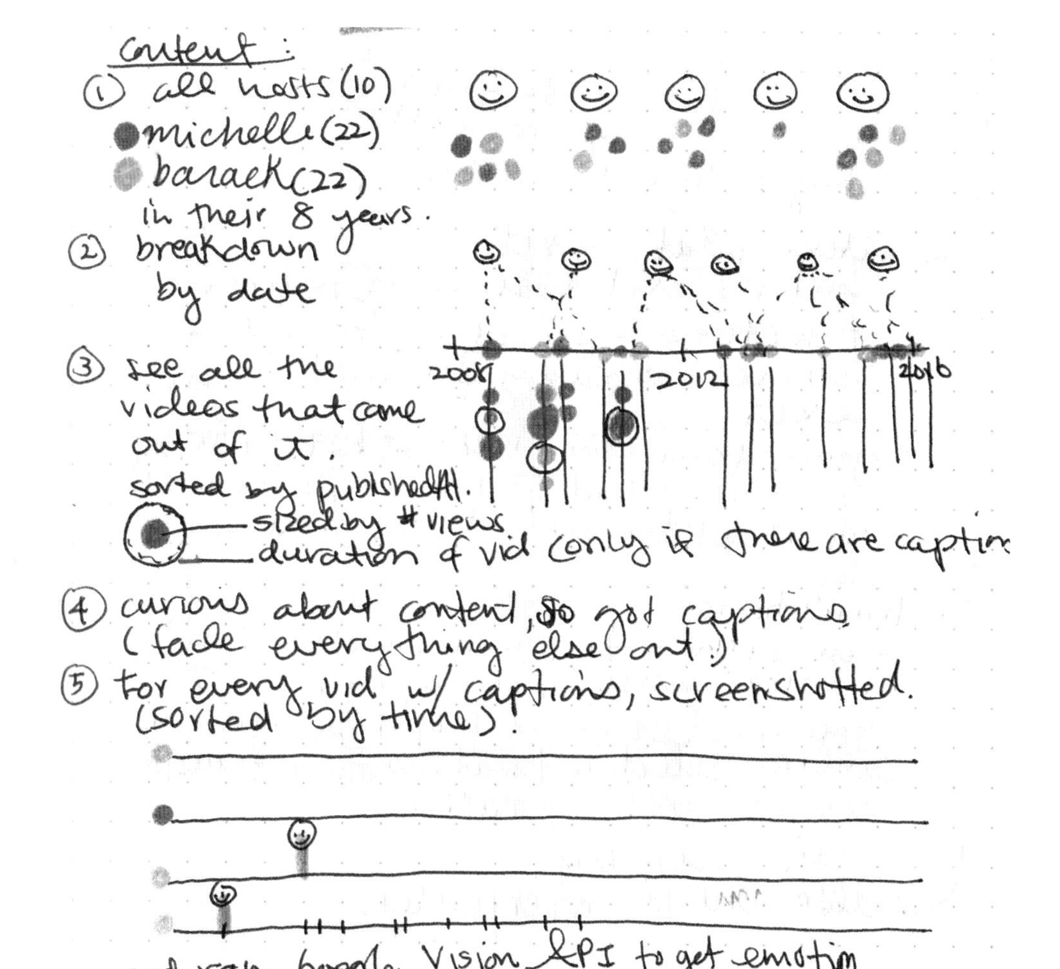

Fig.4.2

Sketch of the first few sections.

Fig.4.3

Revised sketch to introduce the videos and the emotions I detected in them.

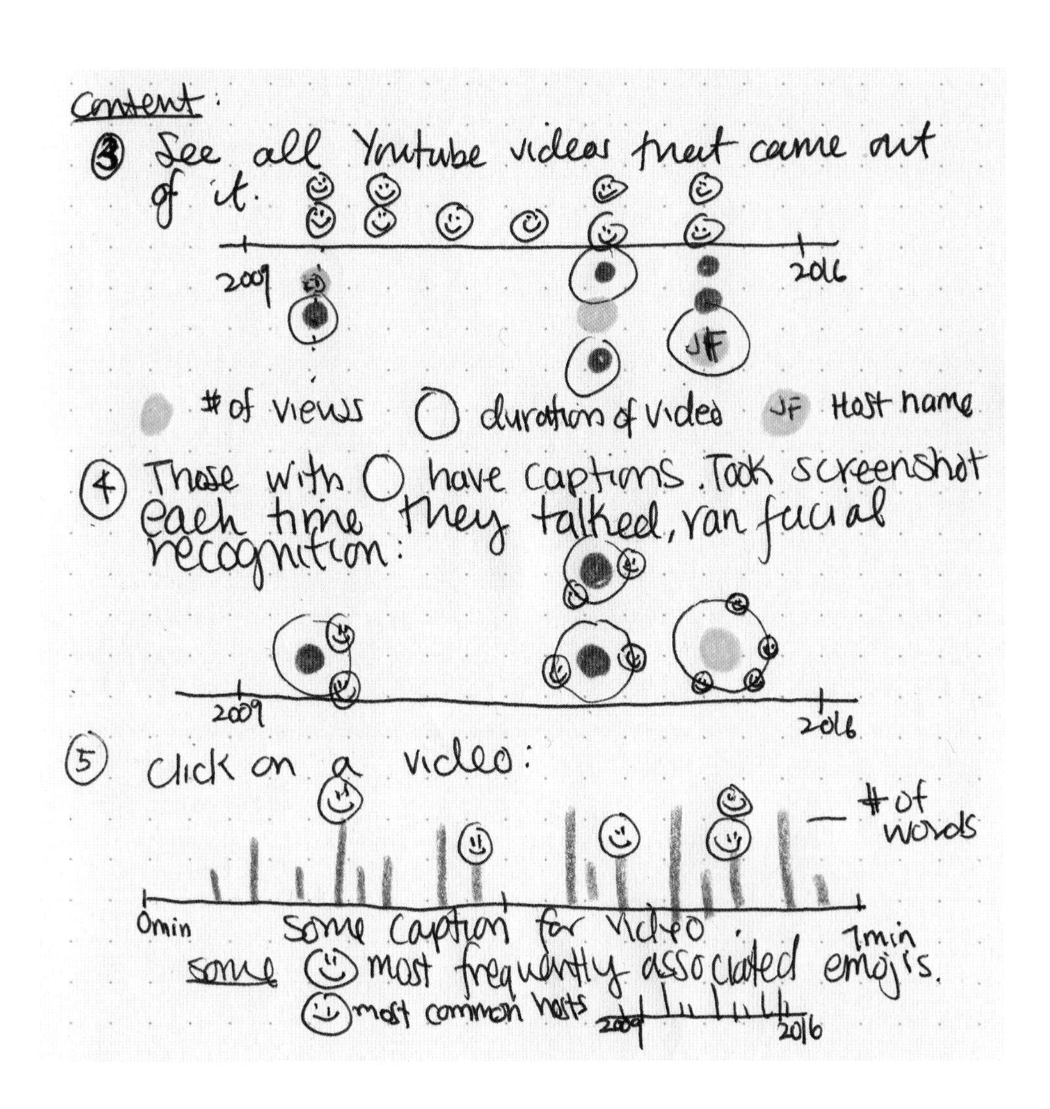

Code

As soon as I figured out that the Google Cloud Vision API gave me the *x*/*y* bounds of faces and heads, I knew I had to put emojis on their faces (unfortunately I couldn't include these screenshots for copyright reasons, but please make sure to visit the finished project online[1] to see them instead).

And boy, do I not regret it at all; it brought me so much joy. The rest of the visualization was basically just a build up to the end tool for exploring all the photos with emojis on their faces.

Silliness aside, this project's implementation actually brought me quite a few first-time technical challenges:

- Animation based on scroll
- Cropping SVG images
- Horizontal fisheye to expand hovered images
- Fitting all the visualizations in mobile (⊙ _ ⊙)
- Making scrubbing the fisheye smooth on mobile (╬ಠ益ಠ)

[1] Shirley Wu, "Putting Emojis on the President's Face": http:// shirleywu.studio/projects/obamas

And because I was just trying to get it all done, I didn't take many screenshots of my progress (which *I do* regret). But *I did* spend quite a bit of time finding headshots of the Obamas and cutting just their faces out to use as icons throughout the visualization (Figure 4.4). ヽ(0 ∀ 0 *★)° *

The first thing I wanted to do was to implement animation tied to scrolling like Tony Chu's and Stephanie Yee's "A Visual Introduction to Machine Learning"[2] (an incredible piece of scrollytelling where all the animations are tied to scroll), because it always seemed like a fun technical challenge. I was able to work out the logic with the help of Tony's "Small Scroll-linked Animation Demo"[3]: The key is to leave a subset of the section static, where scrolling results in *no* movement in the visualization (so that the reader can actually read the text), and then tie scrolling through the rest of the section with animating the visualization from that first section to the next (Figure 4.5).

For example, in the top two sections, each headshot of the Obamas represented a late-night TV show appearance. In the first section, I grouped the appearances by the show host and laid the headshots out in a grid under the show host's headshots (Figure 4.6, top). For the second section, I arranged their appearances along a timeline of their eight years in office and linked them to the corresponding show host (Figure 4.7, bottom).

There are a lot of explainers out there now on how to do scrollytelling, but I really found Tony's helpful when I was getting started.

It's really important to make sure not to move too many things at once. You don't want to have text moving at the same time the visualization is animating. Too much movement makes people unsure where to concentrate on and might even make them nauseous.

Fig.4.4

Headshots of the Obamas that I spent a long time cutting out with Photoshop.

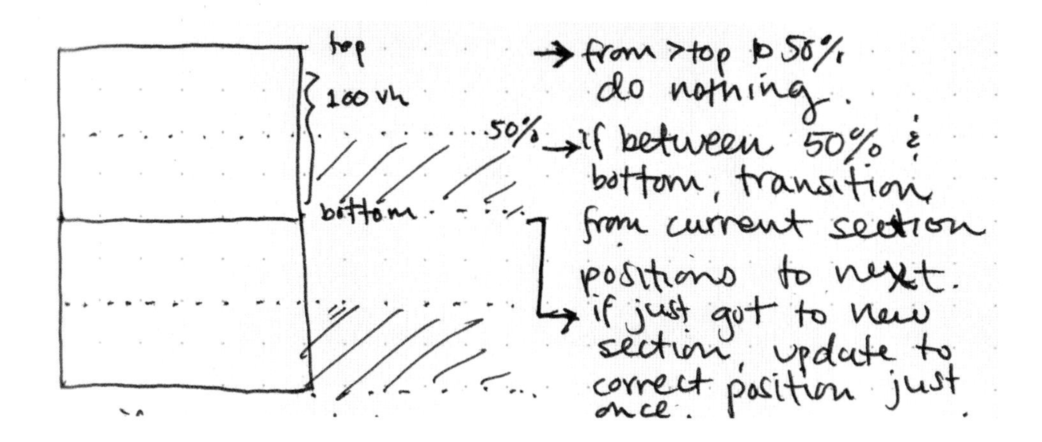

Fig.4.5

Logic for scroll-based animation.

[2] Tony Chu and Stephanie Yee, "A Visual Introduction to Machine Learning": http://www.r2d3.us/visual-intro-to-machine-learning-part-1/
[3] Tony Chu, "Small Scroll-linked Animation Demo": http://bl.ocks.org/tonyhschu/raw/af64df46f7b5b760fc1db1260dd6ec6a/

Fig.4.6

First two sections with annotations of where scroll does or does not trigger animations.

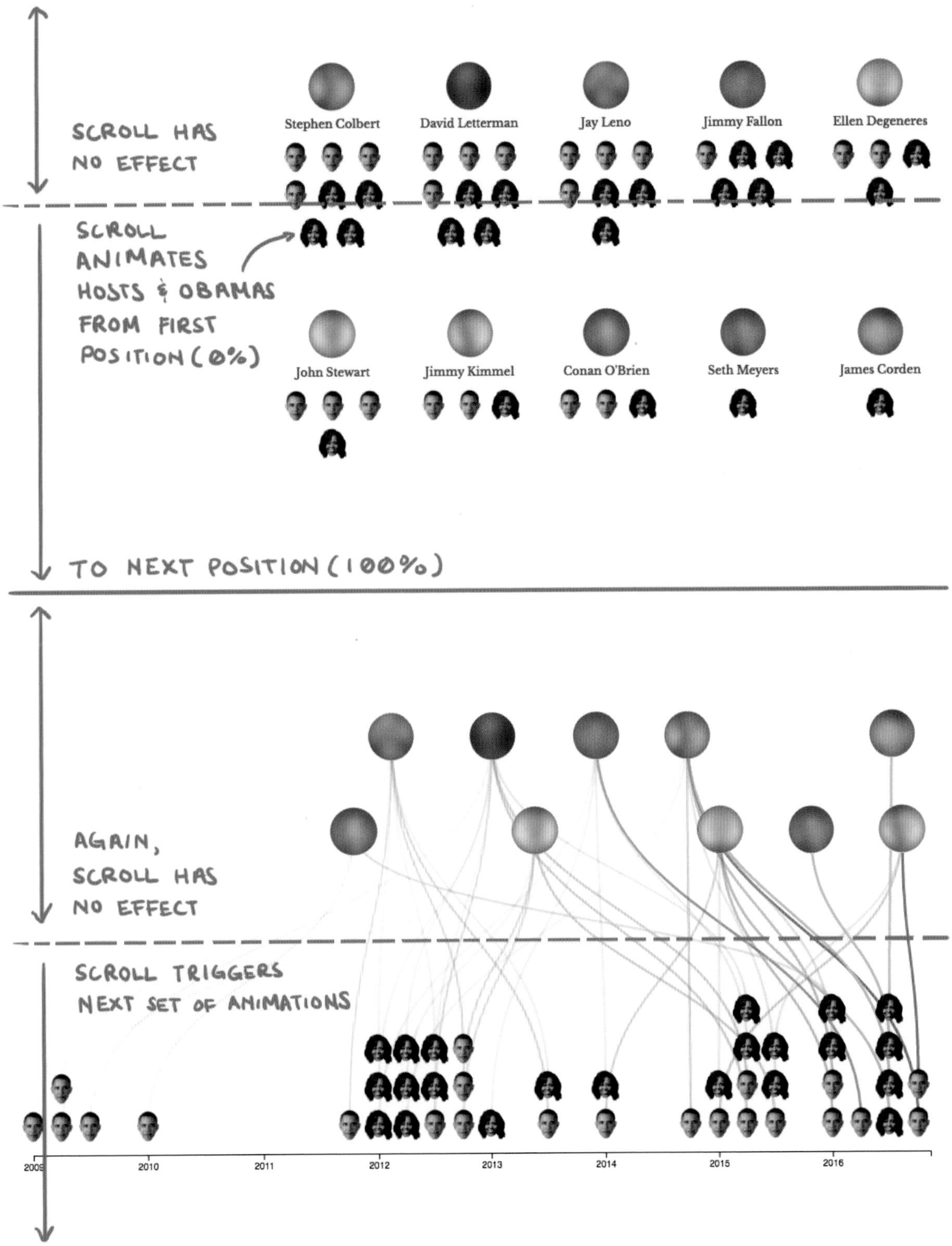

I passed the Obamas' and show hosts' positions in the first section as starting points into `d3.interpolate()`, and the positions in the second section as the ending points. I then passed the amount scrolled between the sections into the interpolator to get the correct in-between *x*/*y* positions for the Obamas and show hosts and animate them down the screen. I repeated this for every pair of sections.

Fig.4.7

For section two, hovering a host shows corresponding guest appearances. It was a very simple implementation, but it made me really happy because it helped a lot in making the tangle of links easier to navigate.

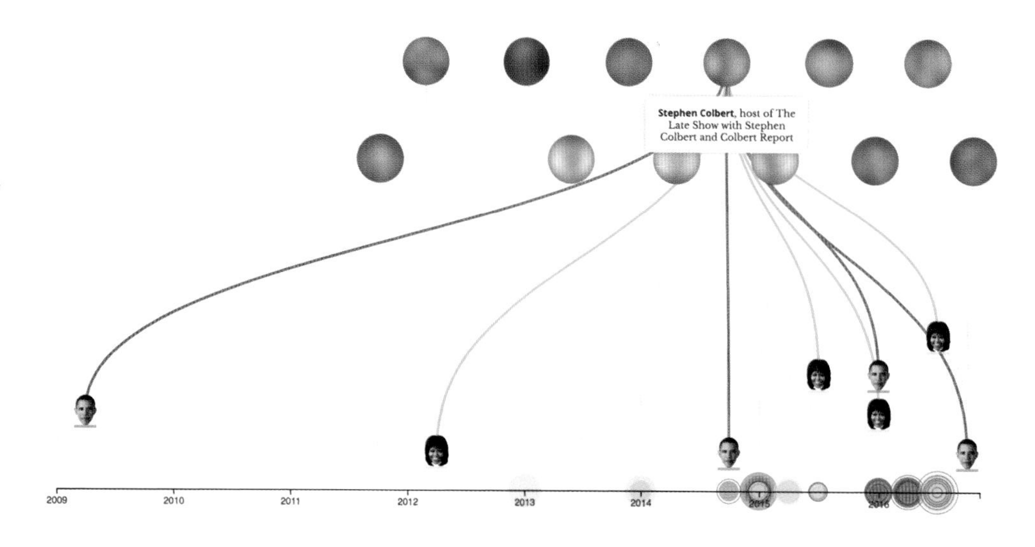

In the third section, I introduced the videos I downloaded from YouTube, with the *x*-axis using the same timeline as the previous section (I love this little detail) and the *y*-axis being the number of views. In the fourth section, I filtered down to just the videos with captions; the filled circle radius is the number of views, and the outer ring radius is the duration of the video. It's fun to see that some of the videos got a lot of views despite being shorter in duration, or vice versa. The small dots on the ring represented the times when Google Cloud Vision API said there were expressions of joy, so we can see whether the laughter was present throughout the video, or if it was concentrated in specific blocks (Figure 4.8). My favorite part about this was actually in the description, where I calculated the number of times the POTUS and FLOTUS laughed and found out that the FLOTUS had significantly more laughter!

I like the intent behind the dots and outer ring, but it's definitely some of my more experimental work—I'm not sure if I'll do the same thing now, since it is quite overwhelming and confusing.

Fig.4.8

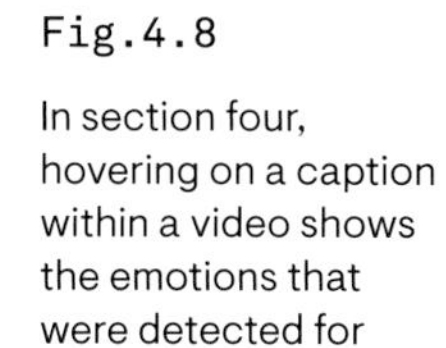

In section four, hovering on a caption within a video shows the emotions that were detected for that caption.

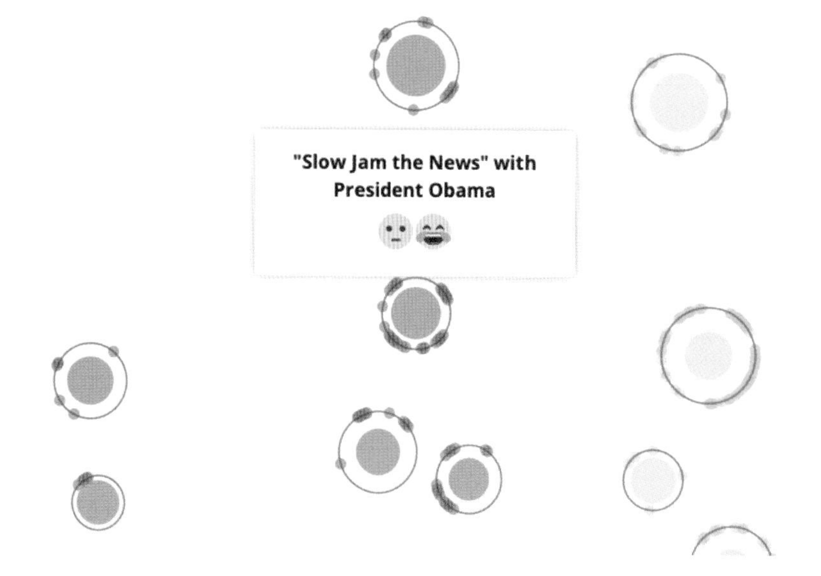

For the final section (that I spent the whole project building up to), my goal was to create an interface that would encourage exploration of the emoji-filled photos. It centers around a timeline of the captions and emotions detected in the selected video, and hovering the timeline shows the full screenshot for that scene with emojis placed on any detected faces.

How Scrollytelling Works

I use scrollytelling when I want to explain different parts of the data or break down interesting visual insights step-by-step to a more general audience. It's great because I can control exactly how the reader interacts with the visualization, and all the reader has to do is scroll.

Scrollytelling is when a user scrolls past a series of "containers" on the page—usually `divs`—and each container triggers a change in the visualization. At the bare minimum, there are four things that scrollytelling has to do:

1. Add event listener for scroll. I also usually use `lodash.js`'s `debounce` to make sure that the callback function isn't called more than once every 64–200ms, depending on how computationally heavy the visualization update is.

2. Within the scroll callback function, calculate which container we're in given the scroll position and each container's top and bottom *y*-positions.

3. Trigger corresponding changes to the visualization. If there are animations and those animations are tied to scroll (the animation progress is directly mapped to the scroll progress, rather than automatically started upon entering the container), calculate the difference between the scroll position and the top of that container.

4. Divide the difference by the height of the container, and pass that decimal value to the interpolator.

I implemented my first scrollytelling project from scratch because I wanted to understand how it worked, but I've since relied on JavaScript libraries to help me manage the scrolling. My current favorite is `Scrollama.js` from The Pudding, because of its super straightforward API and focus on performance; I highly recommend it. (And let's face it, scrollytelling isn't something you ever want to implement twice!)

A callback function is a common type of function used in web development that is run ("called back") when some other action—in this case, scroll—has occurred.

For how animations and interpolators work, see the lesson "Custom Animations" on page 269 of my "Nature" chapter.

`Scrollama.js` aims to be more performant by offloading the computations in step 2 (which can get expensive if done on every scroll) to the Intersection Observer API, which can detect when an element enters the viewport—I'm such a fan!

As soon as I implemented the timeline, I knew I needed to implement the *fisheye* effect—a way for users to hover and expand specific slices to better preview them. I had seen the *New York Times* implementation of the horizontal fisheye for navigating through fashion week images[4], and I knew Nadieh had adjusted that code for canvas before. It took a bit of digging to find an SVG equivalent for cropping centered images (use `preserveAspectRatio` and set the `viewBox` width the same as the image width), and a few passes through Nadieh's and the original fisheye code to get the subset I needed (mapping an *x*-coordinate to its new position based on the distortion) to get it working in my own code.

I found my SVG image cropping solution on Sara Soueidan's blog[5]; she has the best explainers. (ﾉ≧∀≦)ﾉ

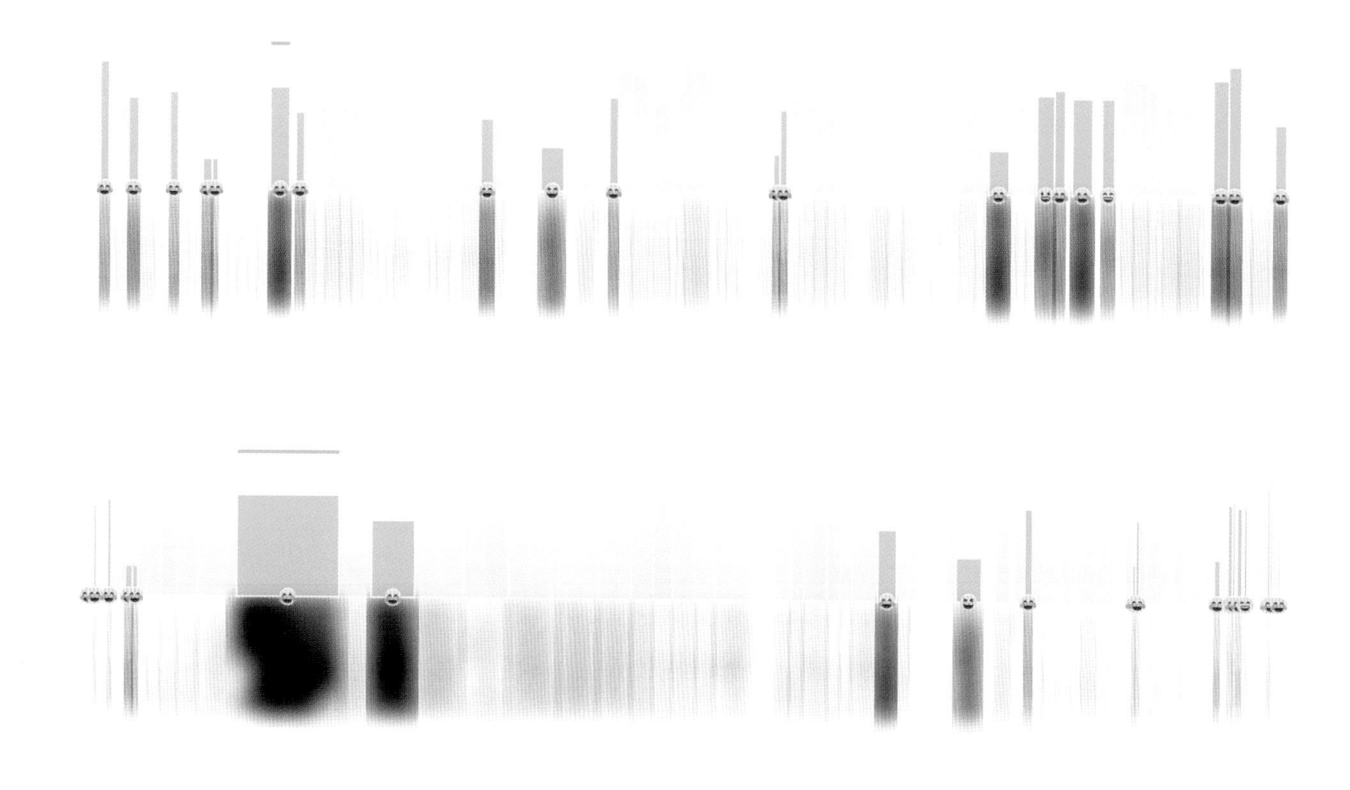

Fig.4.9

The timeline as-is (top) and the timeline with the fisheye effect on hover (bottom).

One of the final things I had to do was make sure that all the visualizations worked in mobile. It was the first time I really tried to be mobile friendly, and after some research, I ended up using `ismobilejs`, a JavaScript package that detects what device the user is on (and I've used this package in all my projects ever since). If the package detected the user was coming from mobile, I passed in a narrower width for all my visualizations, resulting in very squished visualizations (but they did fit the screen!). My happiest mobile moment, though, was getting that fisheye effect working in mobile. When Iinitially used the browser's built-in `touchmove` event listener, scrubbing was extremely buggy and finicky. After spending an entire day thinking of a different interaction I could make with the timeline on mobile, I realized that `D3.js` already had a great touch implementation with the *drag* module; the end result was so smooth it was magical butter.

For examples on designing visualizations specifically for mobile, see Nadieh's "Dataviz For Mobile And Desktop Isn't Always About Scaling" lesson on page 290 of her "Culture" chapter.

[4] Front Row to Fashion Week, February 2014: http://www.nytimes.com/newsgraphics/2014/02/14/fashion-week-editors-picks/index.html
[5] https://www.sarasoueidan.com/blog/svg-coordinate-systems/

Reflections

This project took much longer than all my previous *Data Sketches* projects, but I'm so happy with myself for overcoming all the technical challenges I faced and figured out. I'm especially happy that I implemented scrollytelling from scratch because I now know exactly how it all works and use that knowledge regularly in my work. One of the best things, though, is that it gave me a sense of technical fearlessness, where I really felt that if there was something I wanted to implement, that I'd be able to figure out how to do it, given enough time and the right Google searches.

Sadly, the response to this project was less enthusiastic than I was hoping for, considering the amount of effort I put into animating the visuals and the overall amount of time I ended up spending. I think this was because my text merely recited the facts in the data—which wasn't very memorable for anyone. (This was in contrast to my "Book" project about *Hamilton* the musical, where I really focused on what I learned in my analysis and wrote an article that really made the *Hamilton* fans *feel* for the musical; that was much more successful.)

But that in turn also taught me a great lesson; how I feel about my own projects shouldn't be dictated by external validation, but rather how much fun I had building it. And I'm really proud of everything I was able to accomplish; I think it really set me up well technically for all my subsequent projects. I like my silly, fun, Obamas project very much. 。゜✶.＼(*^∀^*)ノ.✶゜。

Putting Emojis on the President's Face

↳ shirleywu.studio/projects/obamas

Putting 😐😀😄😁😆😂s on the President's Face

BY SHIRLEY WU

In 2009, President Obama became the first sitting president to appear on a late-night talk show. First Lady Michelle Obama followed suit in 2012, appearing on *The Tonight Show with Jay Leno*. They've been on many more talk shows since, promoting HealthCare.gov and urging America to get moving. From this plethora of comedic gold arose a great opportunity: to put emojis on the POTUS and FLOTUS's face.

You may be wondering, "why do something so ridiculously silly"? Simple: it's been a long and exhausting election season, and *we'll all need a pick-me-up come Election Day*.

The Tonight Show Starring Jimmy Fallon
The Late Late Show with James Corden

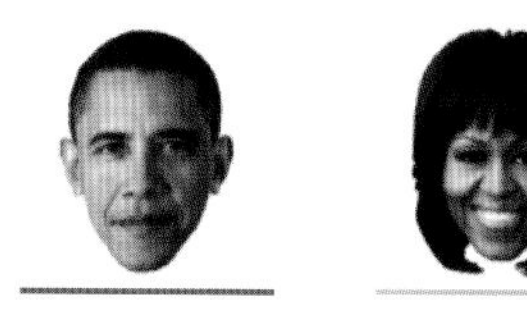

Start
↓

Since his first time on *The Tonight Show with Jay Leno*, the President has made **24** late-night appearances on 11 shows with 10 different hosts. Impressively, the First Lady has very similar numbers despite a three-year late start: **22** appearances across 12 shows, also with 10 hosts.

They seem to favor hosts David Letterman and Stephen Colbert over the years, appearing four times each on both shows. Over the past half year however, the POTUS and FLOTUS have both appeared on newer shows hosted by Seth Meyers, James Corden and Samantha Bee.

(*Hover over the images* for more detail on the host or appearance.)

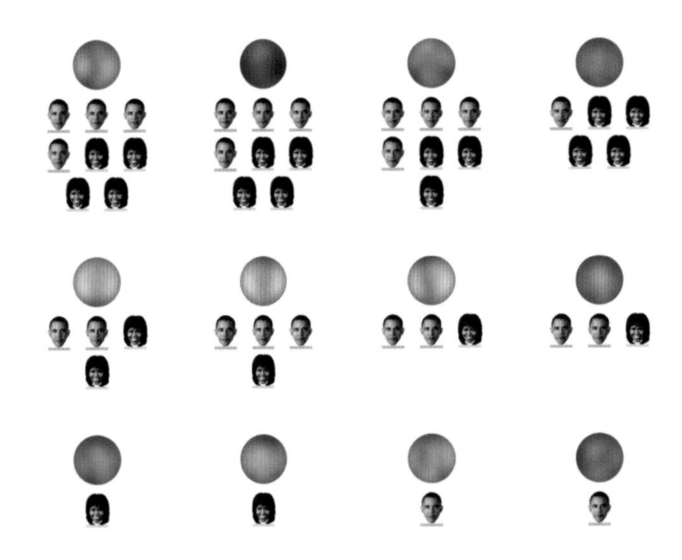

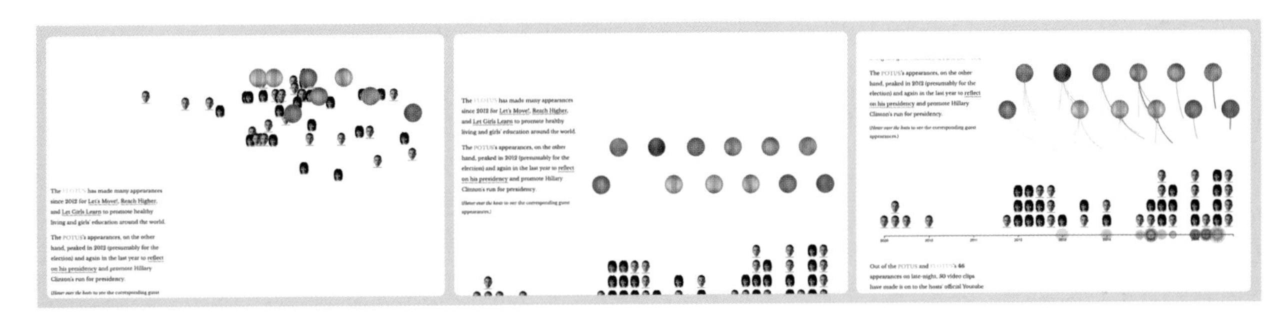

Fig.4.10

Scroll-based animation from first section to the next.

The FLOTUS has made many appearances since 2012 for Let's Move!, Reach Higher, and Let Girls Learn to promote healthy living and girls' education around the world.

The POTUS's appearances, on the other hand, peaked in 2012 (presumably for the election) and again in the last year to reflect on his presidency and promote Hillary Clinton's run for presidency.

(*Hover over the hosts* to see the corresponding guest appearances.)

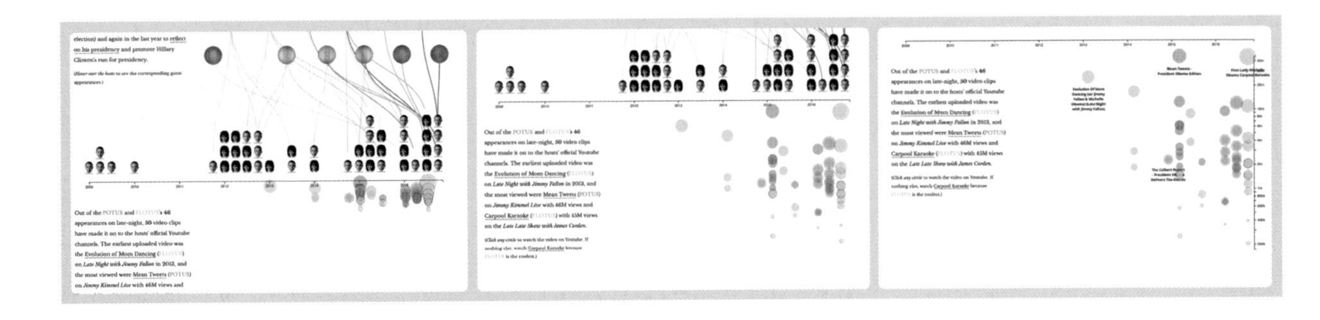

Out of the POTUS and FLOTUS's **46** appearances on late-night, **50** video clips have made it on to the hosts' official Youtube channels. The earliest uploaded video was the Evolution of Mom Dancing (FLOTUS) on *Late Night with Jimmy Fallon* in 2013, and the most viewed were Mean Tweets (POTUS) on *Jimmy Kimmel Live* with 46M views and Carpool Karaoke (FLOTUS) with 45M views on the *Late Late Show with James Corden*.

(*Click any circle* to watch the video on Youtube. If nothing else, watch Carpool Karaoke because FLOTUS is the coolest.)

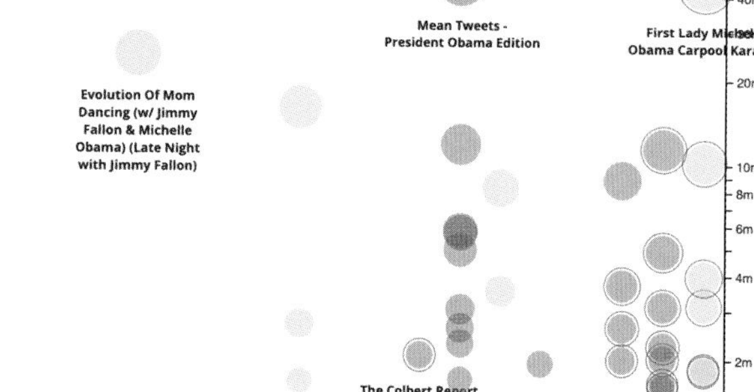

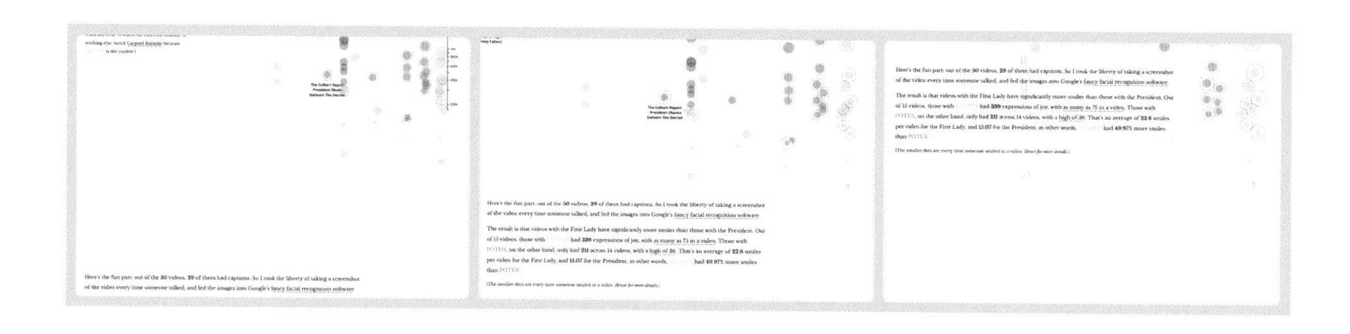

Here's the fun part: out of the **50** videos, **29** of them had captions. So I took the liberty of taking a screenshot of the video every time someone talked, and fed the images into Google's fancy facial recognition software.

The result is that videos with the First Lady have significantly more smiles than those with the President. Out of 15 videos, those with FLOTUS had **339** expressions of joy, with as many as 75 in a video. Those with POTUS, on the other hand, only had **211** across 14 videos, with a high of 36. That's an average of **22.6** smiles per video for the First Lady, and **15.07** for the President; in other words, FLOTUS had **49.97%** more smiles than POTUS.

(The smaller dots are every time someone smiled in a video. *Hover for more details.*)

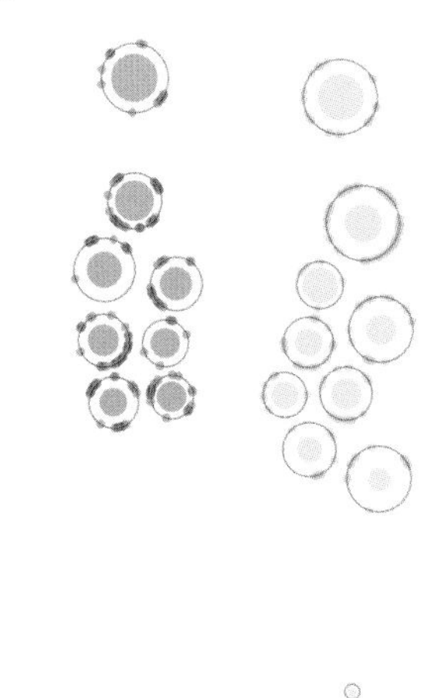

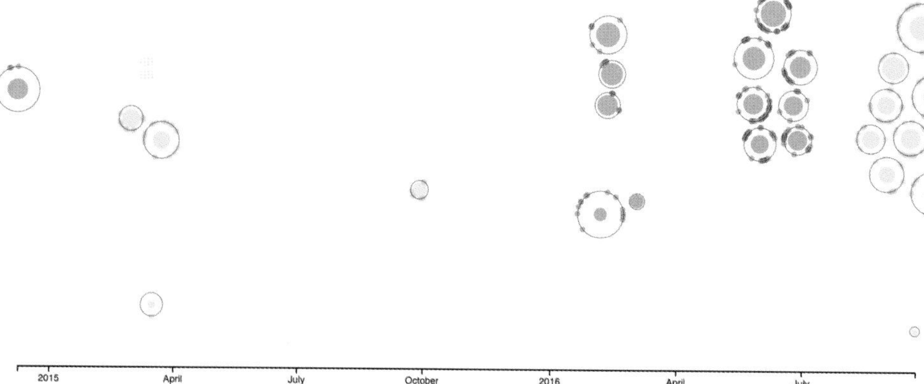

And finally, the pick-me-up: below are circles representing the **29** videos, each with a screenshot for every time someone talked. *Select a video* to see the images, and *hover over the timeline* to read the corresponding captions. *Click on the timeline* to see 😂's on the President's face. Then, *click on the image* to jump to that moment of the video.

First Lady Michelle Obama Carpool Karaoke

on The Late Late Show with James Corden, July 20, 2016

I'M ON BREAK.

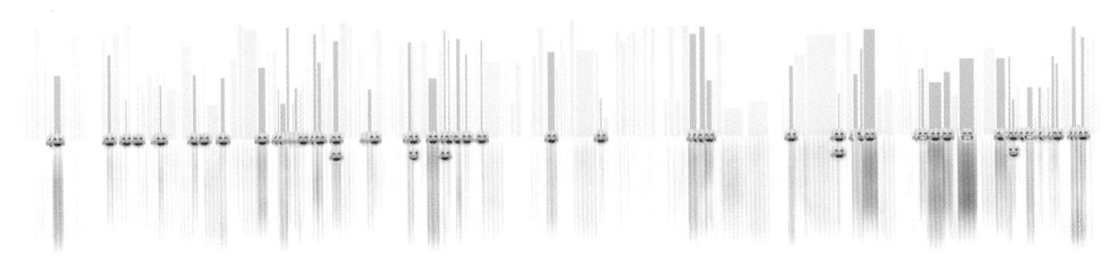

No matter what happens on Tuesday,
I hope this put a 😂 on your face.

If it did, consider sharing the ❤

OKS

A Storm of Swords
The Name of the Wind
Storm Front
The Two Towers
Wizard's First Rule
Nature
Good Omens: The Nice and Accurate Prophecies of Agnes Nutter, Witch
Small Favor
Prisoner of Azkaban
Number
The Gathering Storm
The Wee Free Men
The Light Fantastic
Words of Radiance
Good
Place
The Ocean at the End of the Lane
The Man in the High Castle
Location
Ice
A Breath of Snow and Ashes
Water
Neverwhere
The Sandman, Vol. 3: Dream Country
Battle
Object
Misery
Fighting
The Sandman, Vol. 2: The Doll's House
Guards! Guards!
The Gunslinger
Jewel
Bad
A Clash of Kings
Harry Potter and the Sorcerer's Stone
A Game of Thrones
Harry Potter and the Cursed Child - Parts One and Two
Government
The Final Empire
Blood
ssion
Harry Potter and the Half-Blood Prince
Royal
The Lord of the Rings
Outlander
The Shining
Interview with the Vampire
The Way of Kings
The Queen of the Damned
Light
The Hobbit
The Wise Man's Fear
Magic
Dragonfly in Amber
Quality
Old Thomas
Animal
Fire
The Vampire Lestat
Harry Potter and the Goblet of Fire
Harry Potter and the Chamber of Secrets
A Dance with Dragons
Do Androids Dream of Electric Sheep?
Dead
Mort
Pet Sematary
Coraline
Carrie
The Dragon Reborn
The Way of Shadows
Color
White Night
Name
Harry Potter and the Order of the Phoenix
Dark
The Green Mile
The Color of Magic
Blue Moon
Harry Potter and the Deathly Hallows
Stat
American Gods
Religion
The Eye of the World
Dead as a Doornail
Salem's Lot
Fates Fate
The Return of the King
Happening
A Feast for Crows
New
Space
Attraction
The Shadow Rising
Dead in the Family
All Together Dead
Club Dead
Time
The Fellowship of the Ring
The Hero of Ages
Summer Knight
Drums of Autumn
Night Watch
Group
Anansi Boys

Magic is Everywhere

NADIEH

I spent a large portion of my youth reading comic books, especially "Donald Duck" and "Asterix." But at age 11, I picked up Terry Goodkind's *The Wizard's First Rule* from our local library and was instantly hooked on the fantasy genre. I was so grasped by the feeling of disappearing inside these strange and magical worlds that authors created through their words. And this hasn't changed; fantasy is still the only fiction genre that I enjoy reading. So for this topic, I knew that I wanted to focus on fantasy books.

As diving into the words of a book or series itself could potentially be a copyright issue (say, my favorite one, *The Stormlight Archive* by Brandon Sanderson), I looked for a different angle. I've always felt that the titles of fantasy books are somewhat similar: either something about magic (makes sense), or some "name/object" of some "fantasy place," such as *The Spears of Laconia*. Maybe I could dig into the trends and patterns of these titles?

Data

I had to get my hands on a whole bunch of fantasy book titles and scraping the web looked like the fastest way to do this. On Amazon I found a section that showed the top 100 fantasy authors from that day. I wrote a small web scraper function in R with the `rvest` package that automatically scanned through the five pages on Amazon (20 authors per page) and saved their names. However, I couldn't find an easy way to get their most popular books and the Amazon API seemed too much of a hassle to figure out.

Most big company APIs seem like too much hassle to me for that matter...

Luckily, Goodreads has a very nice API. I wrote another small script with help from the `rgoodreads` package to request information about the 10 most popular books per author, along with information about the number of ratings, average rating, and publication date for each of the 100 authors that I had gotten from the Amazon list.

↳ Data Can Be Found in Many Different Ways

While I didn't come across a single public dataset with a bunch of fantasy book titles, I knew that there are other ways of finding data than simply looking for structured CSV or JSON files. Instead I figured out what *was* available: the Amazon author list to get popular fantasy authors and the Goodreads API to retrieve information about those authors. Combining those resources, I was able to create the dataset of fantasy book titles.

Next came the trickiest part; I had to do text mining on the titles, which in this case consisted of text cleaning, replacing words by more general terms, and clustering of similar titles. For the text cleaning I made a few choices. For one, I only kept the authors that had a *median* number of ratings per book that was above 20. Furthermore, I wasn't looking for any omnibus—a collection of several books—or books written by many people. For this (although not perfect) I looked at how often the exact same book title appeared in my list and took out those that appeared more than twice. Furthermore, I removed all books with terms such as "box," "set," or "edition," making sure to manually check all the deletions. Finally, I scrapped books with no ratings. This left me with 862 book titles from 97 different authors.

I also removed some very specific words to this particular dataset of books, such as "Part."

Now the data was ready for some text cleaning by removing digits, punctuation, and stop words (which are some of the most common words in the language, such as "a," "is," "of," and carry no meaning to interpreting the text). I did a quick word count after the title cleaning to get a sense of what words occur most often in book titles. As these are words, I couldn't resist visualizing the results as a word cloud (see Figure 5.1). The bigger the size of the word, the more often it appears in titles (the location and angle have no meaning). I was very happy to see how often the word "magic" occured!

Fig.5.1

The words occurring most often in the 862 fantasy book titles. The bigger a word's size, the more often it occurs.

I wanted to look for trends in these words. However, for a standard text mining algorithm, the words "wizard" and "witch" are as different as "wizard" and "radio," even though we humans understand the relationship between these words. I first tried to automatically get hypernyms of each noun in the titles, but that sadly didn't give me good enough results, the terms weren't general enough or already overgeneralized. I therefore set about doing it manually and replaced all ±800 unique words across all titles by more general terms, such as "name," "magic," "location," and so on.

A hypernym is a word that lies higher in the hierarchy of concepts. Like "fruit," which is a hypernym of a "banana."

↳ Manually Add New Variables to Your Data

Manually enriching your data, because either doing it perfectly with the computer isn't possible or takes too long, or because the extra data is unstructured, is something that you need to embrace when doing data analysis and creating data visualizations.

In this case, after trying an automated route, I *manually* converted each unique word from all the titles into a more general term. This variable in turn became the main aspect that defined the location of the books, thus it became quite important and worth the time investment!

I loaded this curated list back into R and replaced all the specific title words with their general ones. The final data preparation step was to turn the set of fantasy book titles into a numerical matrix, which could then be used in clustering analyses. I won't go into the details of how this was done, but if you're interested, you can google for "Document Term Matrix."

I tried several clustering techniques on the books, such as K-means, Principal Component Analysis, and tSNE, to see which result would visually give back the most insightful result. My goal with the clustering was to get an x and y location for each book, where similar titles were grouped or placed together in a 2D plane. Inspecting the resulting visuals for each technique, I found that tSNE gave back the best grouping of titles; books were spread out nicely and there were clear clusters of different topics.

I placed the ±40 most occurring words/terms on top of the tSNE plot, in their "average" location when looking at all the positions of the books that contained that term. While not perfect, this gave me a sense of where certain topics were located. A hotspot of books were present in most terms, but a few stragglers in other locations pulled most terms toward the middle.

I'll start using the word "terms" now to denote the most common replaced words of the book's titles, such as "magic," "name," or "nature."

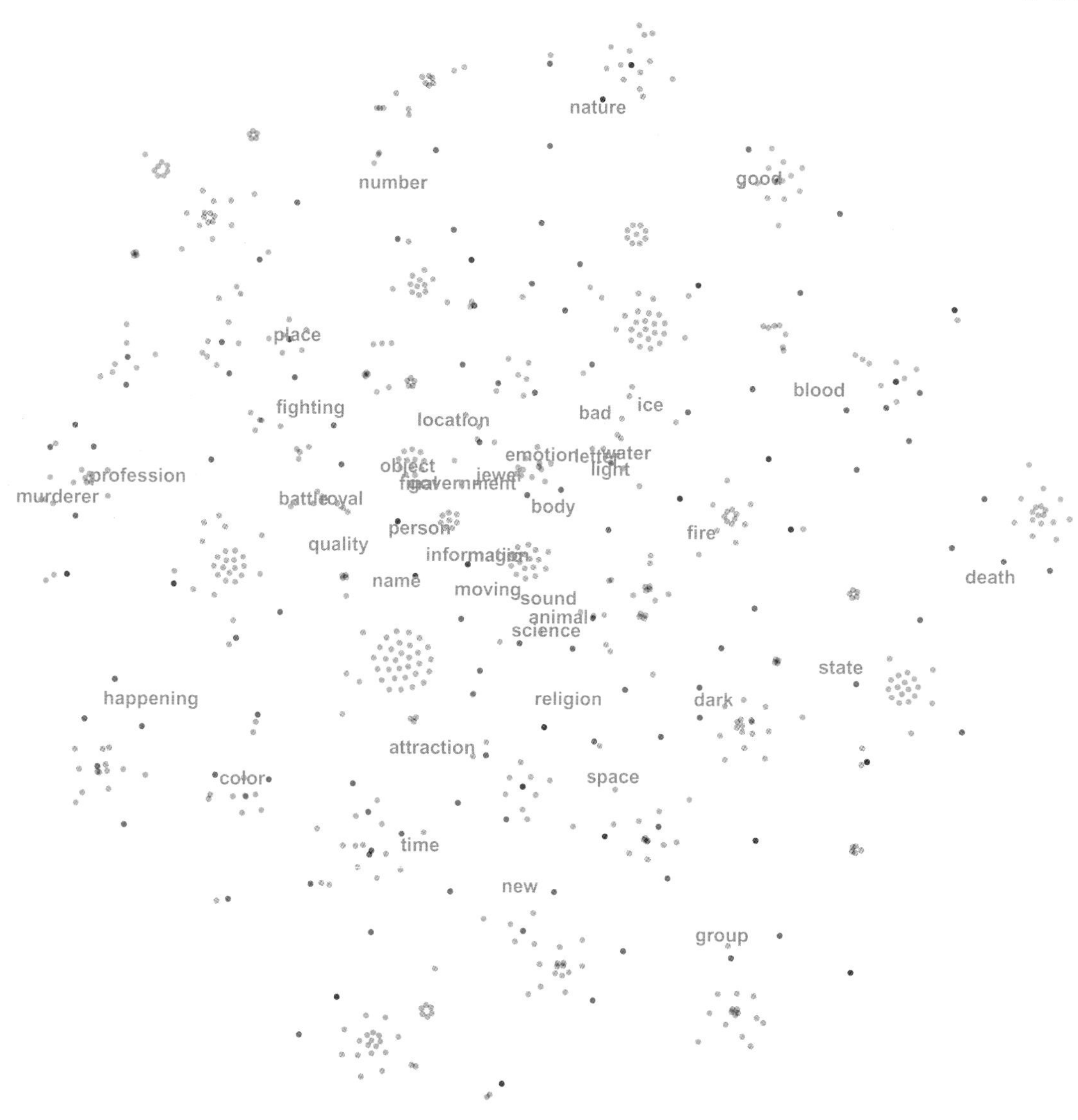

Fig.5.2

The clustering result of running a tSNE on the book title words, with the "central" locations of the ±40 most common words plotted in pink.

My final data step was to prepare an extra variable that would be used to draw a path between all books from the same author. Many books didn't have a publication year from the Goodreads data, so I couldn't do it chronologically. The next best thing was to draw the shortest path between the books, so the length of the lines in the final visual would be minimal. Thankfully, I could use the "Traveling Salesman Problem" approach (imagine a salesman wanting to travel between cities in the shortest distance possible). With the TSP package in R, I calculated the order in which the books should be connected.

Sketch

Throughout the data preparation phase I was thinking about how I wanted to visualize the results. From the get-go I already had the idea to put the books in a 2D plane, placing similar titles together. But how could I get an interesting shape for the books, other than just a boring circle?

Since I was looking at titles, I thought it would be fun to somehow base the resulting "shape" of a book on the title as well. I could split the 360 degrees of a circle in 26 parts, one for each letter in the English alphabet, and then stack small circles on the correct angles, one for each letter in a title. I would then connect all the letters from a word in the title with curved lines, sort of "spelling it out." In Figure 5.3, you can see where I was still deciding if I wanted the lines connecting the letters, to only go around the outside or through the middle of the circle as well (the top right part of the sketch).

Fig.5.3

Figuring out how to visualize the book "marks" themselves in an interesting manner.

After having had so much fun with SVG paths during previous projects, I wanted the lines between the letters to follow circular paths. One of the elements that I had to figure out for these SVG arc paths was the `sweepflag`. Not too difficult, but much easier to figure out if you draw it (see Figure 5.4).

The `sweepflag` determines whether you want the curved line between two points to be drawn *clockwise* or *counterclockwise*.

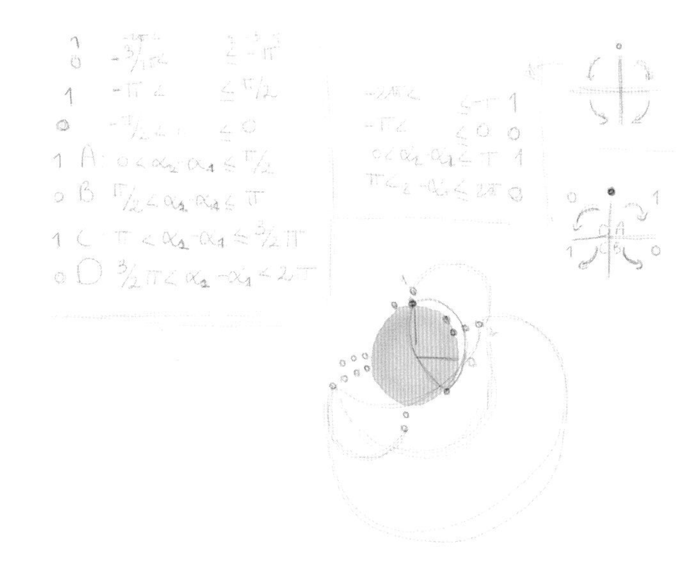

Fig.5.4

Trying to figure out how to get the SVG arc sweepflag signs correct for starting positions in each quadrant of the circle.

Skipping ahead a bit to a point where I had already started on a simple visualization of the books (plotting the book circles, nothing fancy), I looked into the most common terms of the titles again, such as "magic" or "royal." From my explorations during the data phase I knew that placing them in their exact "average" location just wasn't quite right. I wanted to have them more on top of their hotspot and not be pulled towards the center by a few books in other locations. Therefore, I created simple plots in R that showed me where books relating to a certain term were located. See the pink circles that belong to books with the term in their title that's above each mini chart in Figure 5.5, such as a clear grouping of books whose titles relate to "fighting" on the left side of the total circular shape.

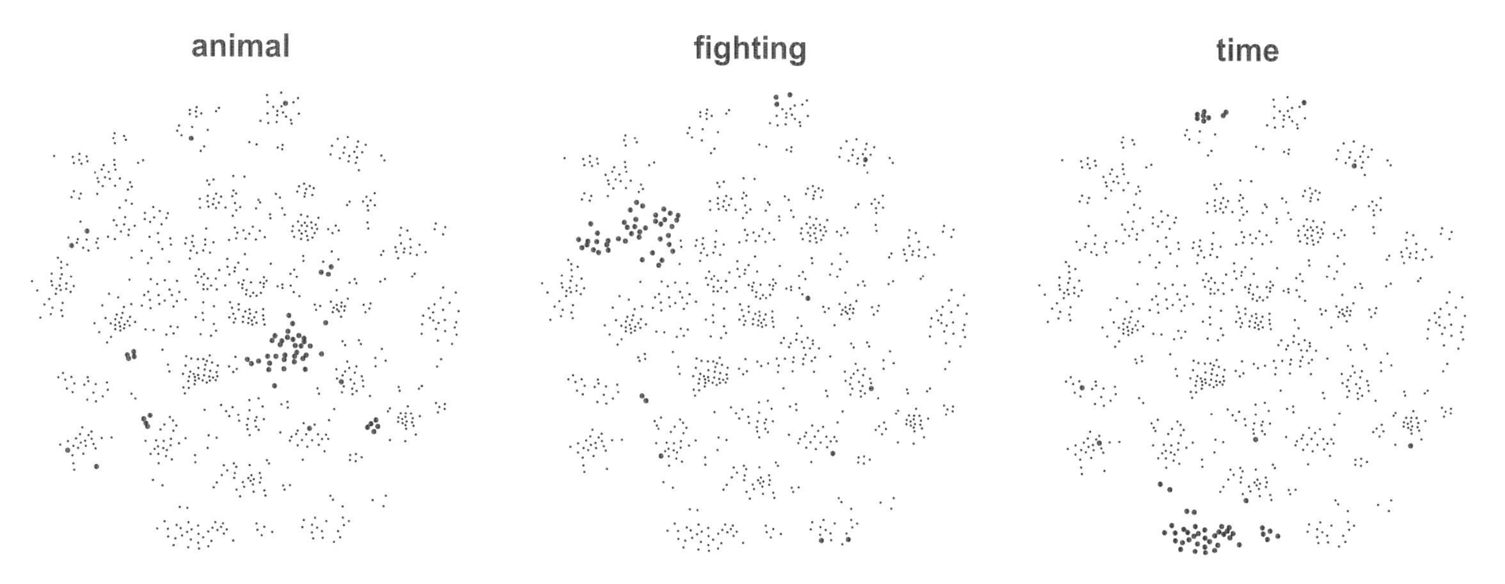

Fig.5.5

Seeing where books that relate to a certain theme/term fall within the whole.

I then took the tSNE plot with all the books into Adobe Illustrator, and together with the charts from R, I manually (yes, again) drew ovals for the top 35 terms that had a clear hotspot. This resulted in the arrangement that you can see in Figure 5.6, which I could use as a background image behind the book circles.

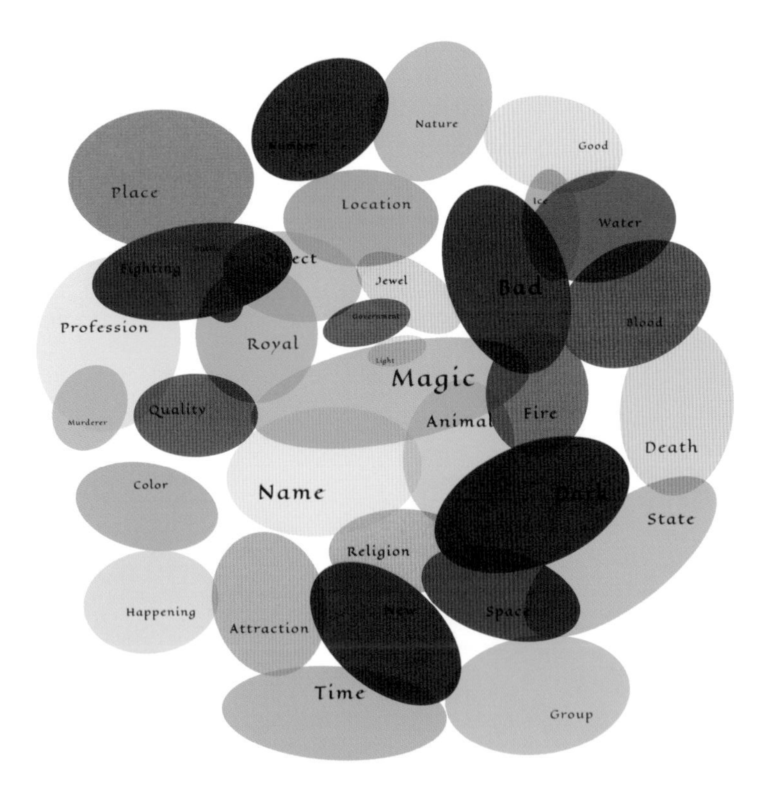

Fig.5.6

The landscape that reveals itself when looking at the hotspot locations of the most common ±30 terms

What this exercise also taught me was that "magic" is found practically everywhere throughout the circular shape (Figure 5.7), hence the title of the final visual.

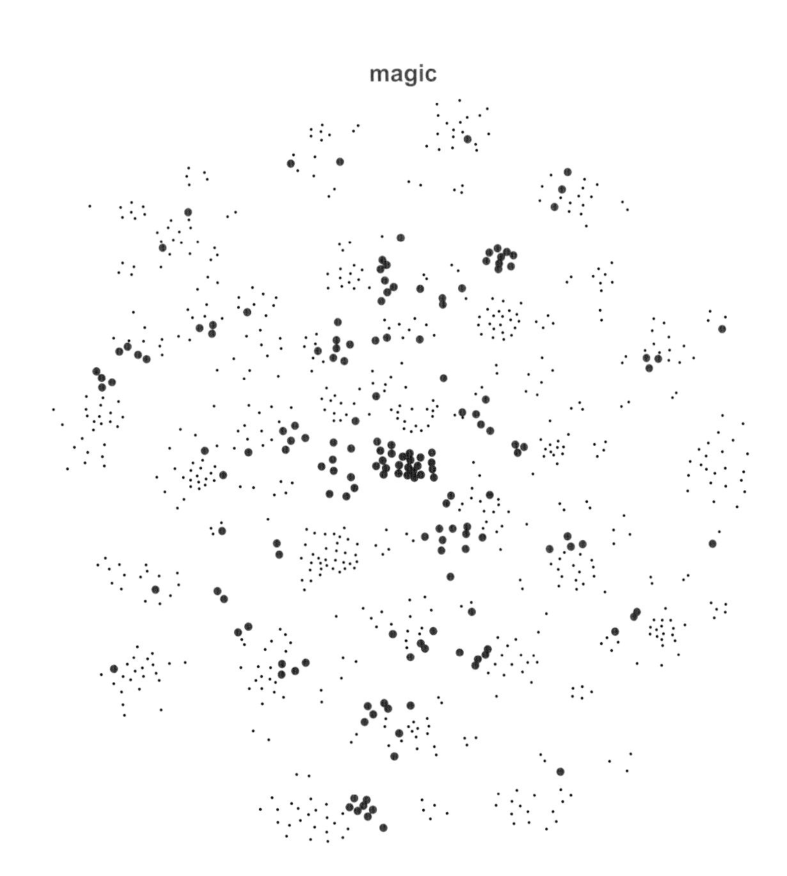

Fig.5.7

Books that have a title that refers to "magic" are found practically everywhere in the tSNE result.

Code

With the x and y locations of the tSNE result finally attached to each book title, I could begin to code the visual with `D3.js`. Thanks to past experience with creating circular paths between two locations (such as the lines in between two circles in my previous project about European royalty) it started out rather painless. A simple addition to the visuals was to size the circles in their area according to how many ratings the book had gotten. Furthermore, I used the thickness of the path that connected books from the same author to denote the author's rank in the Amazon top 100; the thicker the path, the higher the rank.

People on Goodreads definitely like J.K. Rowling; those circles became huge!

Fig.5.8

Placing the books in their tSNE locations in the browser while sizing the circles and line thickness by number of ratings and author rank, respectively.

Fig.5.9

Coloring the circles and connecting lines of some of my favorite authors and book series.

To make the visual a bit more personal, I chose five authors (my three favorite authors, plus two other authors that I enjoyed a particular series from) and marked these with colors, both in terms of their circles and paths (Figure 5.9).

Those authors are Brandon Sanderson, Patrick Rothfuss, and J.K. Rowling, plus Terry Goodkind and Brent Weeks.

Next, I focused on the book "shape," adding the small circles, one for each letter in a title, around the main book circle and then connecting the letters of one word with a path. I had some fun with these arcs, trying to make them swoosh a bit. However, the paths were getting much too big, obscuring titles and other books, so I tuned it down a bit. (Figure 5.10 is still my favorite visual result.)

As you can see from the previous images, many titles overlapped even though the tSNE algorithm did a great job of separating the books in terms of main themes. Since it was important to be able to read the title of a book, I had to adjust the positions of the circles to reduce overlap.

Fig.5.10

Creating each book's visual mark with the title "spelled" out with the smaller circles around the main one, connecting each word in a title with swooshing arcs.

So, yes, for the third time in this project alone, I took the manual approach. I wrote a small function that made it possible to drag each book around, saving the new location in the data. I've since come to love the fact that you can actually save data variables into your local browser so that, even after a refresh, the books would reappear on their moved locations! (Search for `window.localStorage.setItem()`)

Using the handy `d3.drag()` it's not very difficult to move SVG groups across the page.

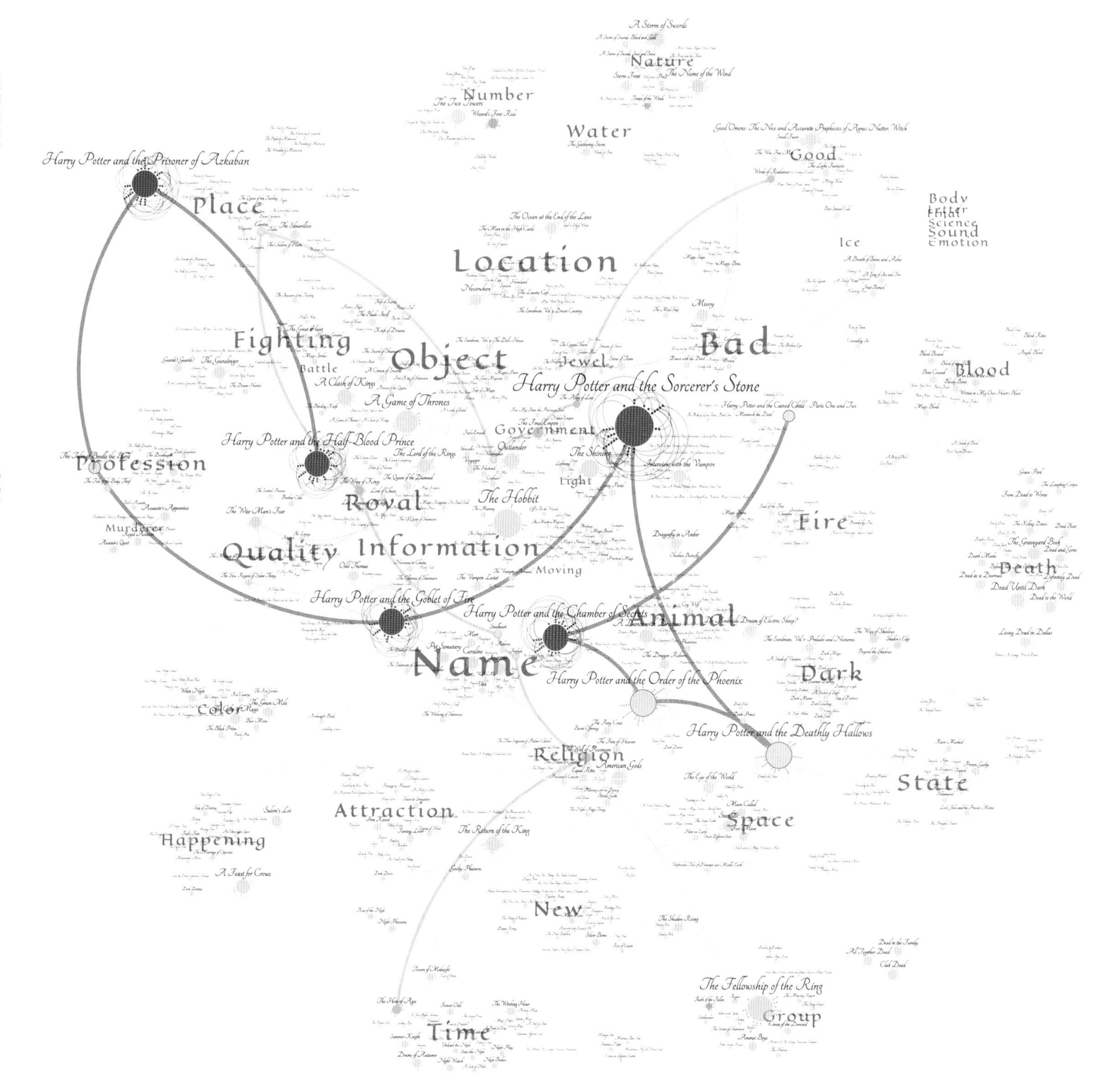

Fig.5.11

Having moved the books to reduce the worst textual overlaps.

↳ Precalculate “Visual” Variables

For this project I precalculated “visual variables” to add to the dataset, the x and y pixel locations, first gotten from the tSNE clustering and then fine-tuned by dragging them around to prevent (title) overlap. This made it possible to immediately place each book’s circle onto its final location. However, I also precalculated in what order the books by the same author should be connected to have the shortest line. It would’ve made no sense to have each viewer’s browser have to calculate almost 100 shortest paths, if the outcome is always the same.

Surprisingly, it only took me about an hour to slightly shift the ±850 books into practically non-overlapping locations. Taking these updated x and y positions into R, I created a new CSV file that became the dataset that is used for the final visual result.

With the books done, I focused on the background locations of the most common terms, such as “magic,” “blood,” and “time,” using the hotspot-oval SVG from Figure 5.6. The advantage of using an SVG image is that I could still make changes to the ovals with JavaScript and CSS. I felt that blurring all the ovals extensively to merge the colors around the outsides would be the way to go (Figure 5.12). Thankfully, I found that it looked even better than I had imagined. (/◕ワ◕)/*:・ﾟ✧

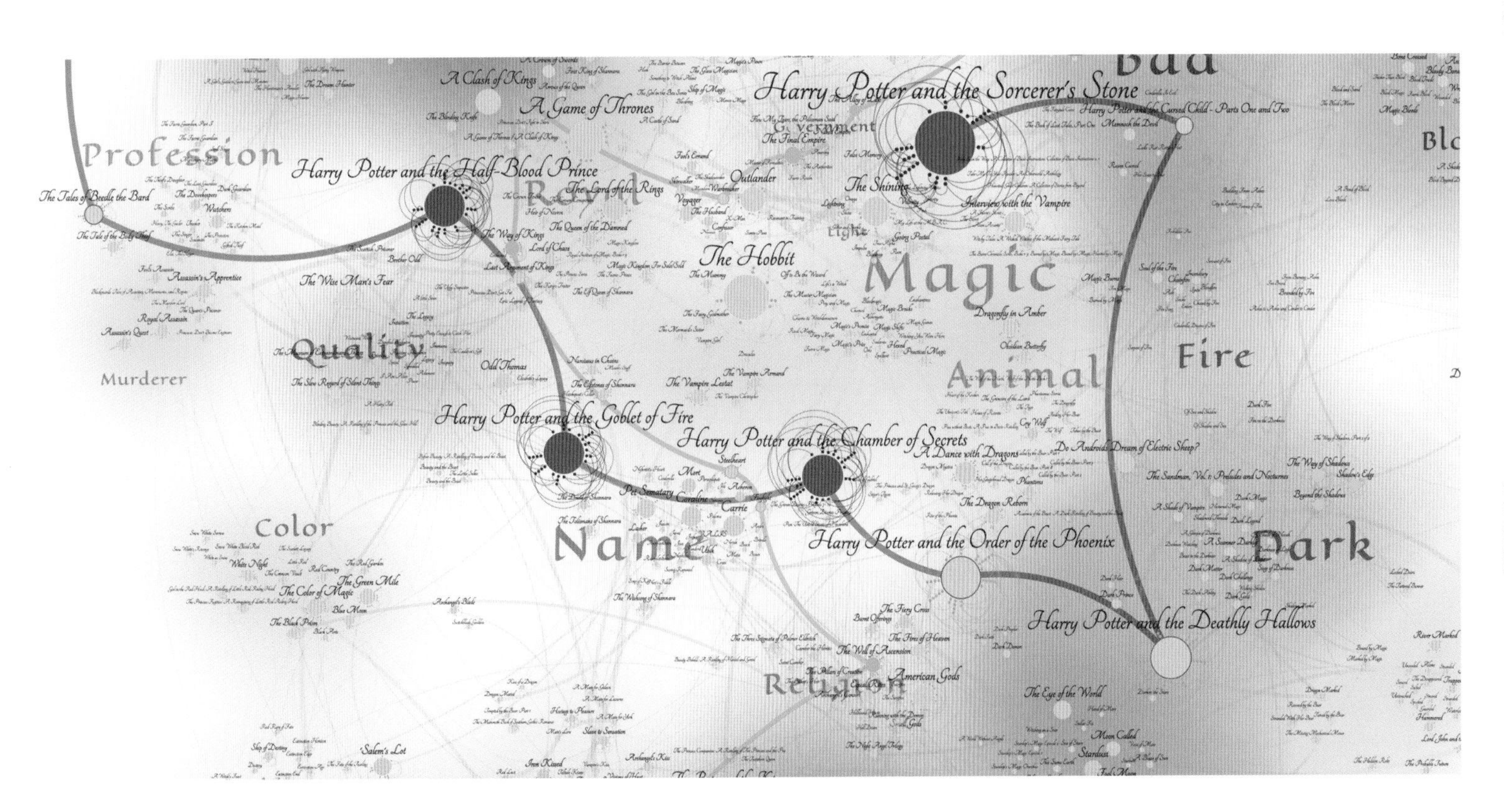

Fig.5.12

Using SVG blur filters to make the background ovals smoothly blend into each other.

As a bonus, having this (blurred) background gave me the option to make the book circles, swooshes, and lines a nice crisp white, instead of the boring grey they were before.

Fig.5.13

To explain what the smaller dots and swooshes around each book's circle meant, I created an animation that shows a book's title being "spelled out" (Figure 5.14)

Finally, I only added a minor interactive element; when you hover over a book, it highlights all the books by that author. With the online page done, I also turned the visual into a static print version, updating the layout to one I felt more fitting for a print.

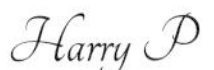

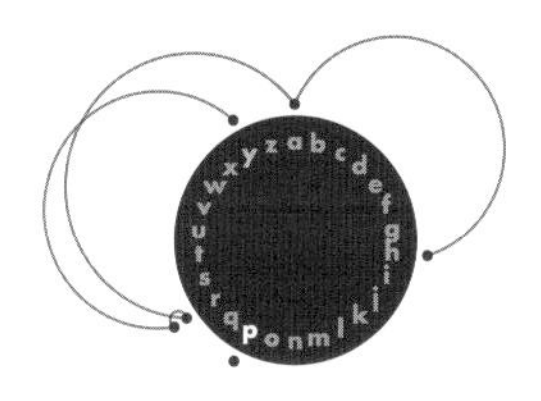

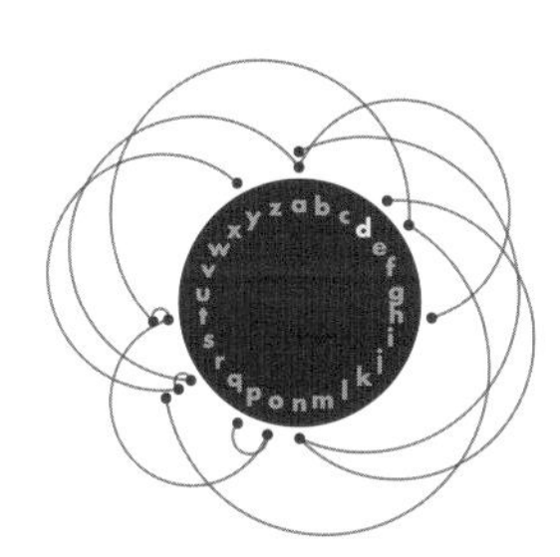

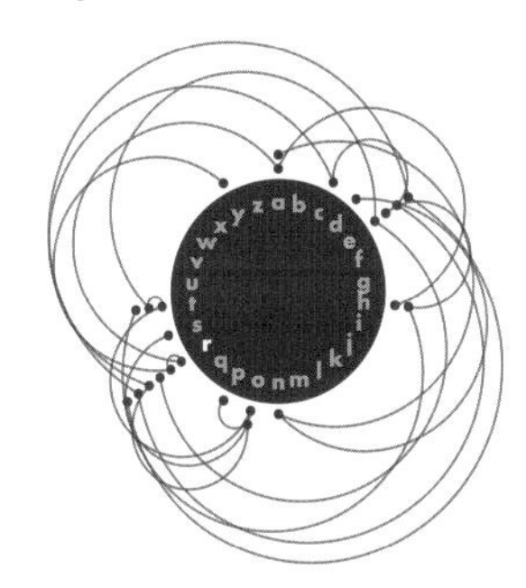

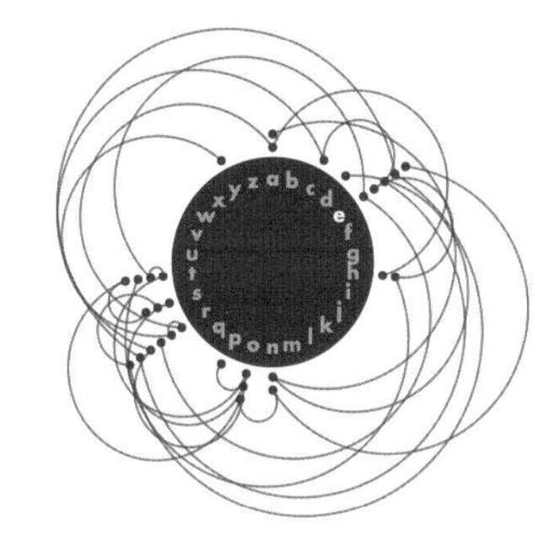

Fig.5.14

A few different moments from the animated legend explaining how to interpret the smaller dots and swooshes around each book's main circle.

Reflections

I'm very happy with the end result; I love rainbow colors, and the blurry aspect reminds me of fairy dust. Perhaps it's more data art than dataviz though. (*^▽^*)ゞ The code part was thankfully not very difficult this time. The most intricate things to program were the swooshes. The whole project was really more about going back and forth between the data in R, other elements in Adobe Illustrator and the main visual in JavaScript, and `D3.js` taking more time than expected.

During the entire process of creating the visualization I noticed that there are far more terms that relate to bad things, such as "blood," "death," and "fire," than things relating to good aspects, such as "light." Maybe references to evil and bad things sell better/create more interesting stories?

Also, I had expected that many authors would probably be fixed within a certain region of the map, all their books following the same title themes. However, that turned out to be false. Most authors are actually spread across the map. Only a few really stick within one location; Charlaine Harris is quite fixed on including "death" in her titles, for example.

Finally, although the interactive hover makes it easier to explore this visual, in the end I prefer the static poster version. It's large enough so the small details of the swooshes and tiny dots around each book's main circle can clearly be seen, and even the smallest book titles are legible. It's both nice to look at and hopefully invites you to dig in and find insights into the world of naming a fantasy book.

Magic is Everywhere

↳ MagicIsEverywhere.VisualCinnamon.com

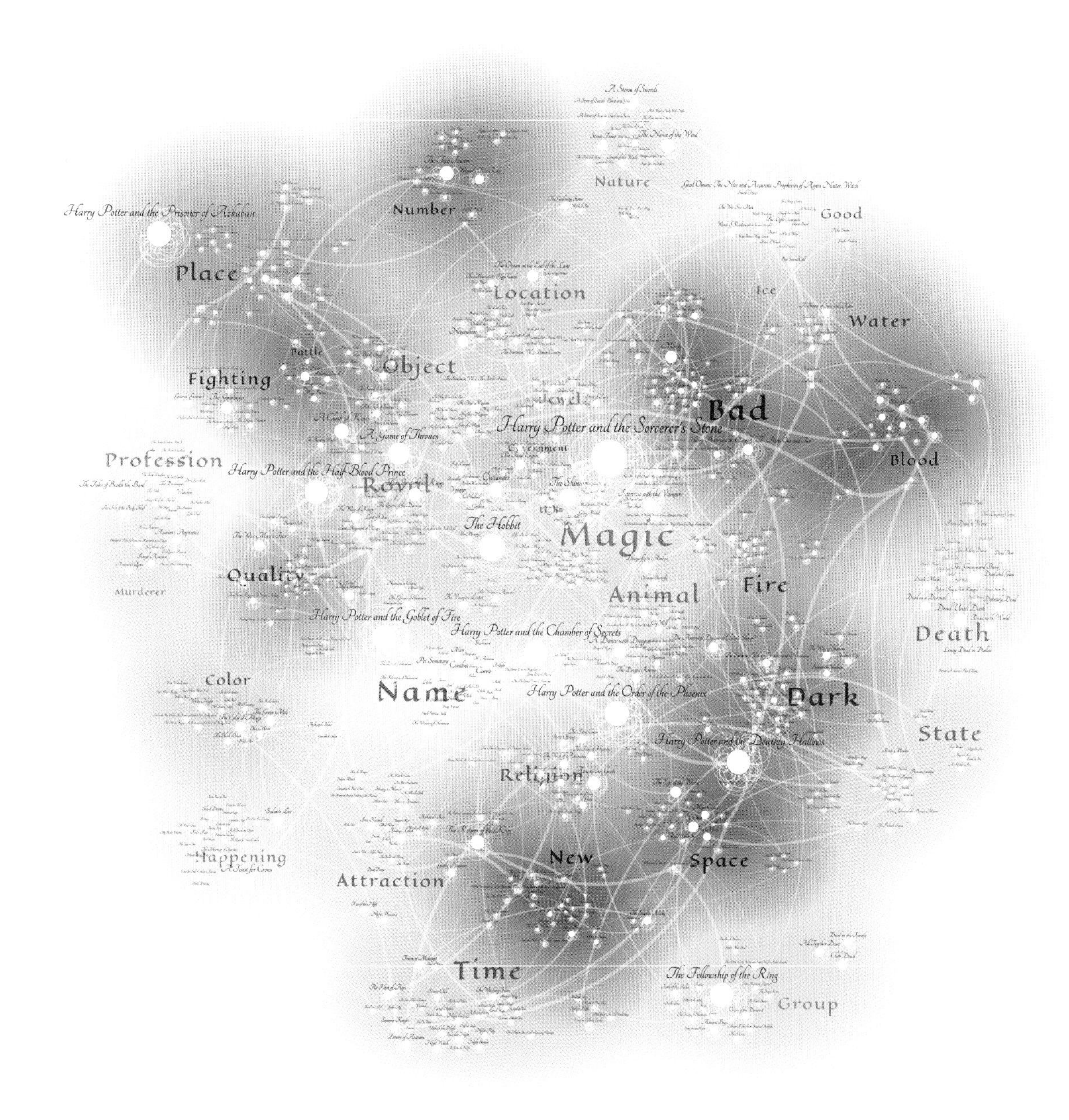

Magic is everywhere

Investigating patterns in Fantasy book titles

The titles from the top 10 books of the top 100 best-selling fantasy authors on Amazon were collected. Using text-mining the titles were analyzed for general subjects or terms, such as **fire**, **royal**, **time**, & more. Finally, these titles were clustered in a 2-dimensional plane, which placed books with similar themed titles together.

Created by Nadieh Bremer | Visual Cinnamon

Harry Potter and the Sorcerer's Stone

Ha *Harry* *Harry Potter*

Books by the same author
Linked by a line

Number of ratings on Goodreads
The area of each large book circle

Rank in Amazon's fantasy top 100
The thickness of the lines connecting books

Fig.5.15

The final poster/print based version of "Magic is Everywhere."

A Storm of Swords
A Storm of Swords: Blood and Gold
Storm Front
The Name of the Wind
The Two Towers
Wizard's First Rule
Nature
Good Omens: The Nice and Accurate Prophecies of Agnes Nutter, Witch
Azkaban
Number
The Gathering Storm
The Wee Free Men
Good
The Light Fantastic
Words of Radiance
The Silmarillion
The Ocean at the End of the Lane
The Man in the High Castle
Location
Ice
A Breath of Snow and Ashes
Water
Neverwhere
Battle
Object
Misery
The Sandman, Vol. 2: The Doll's House
ghting
The Gunslinger
Jewel
Bad
A Clash of Kings
Harry Potter and the Sorcerer's Stone
A Game of Thrones
Harry Potter and the Cursed Child - Parts One and Two
Government
Blood
The Final Empire
Harry Potter and the Half-Blood Prince
Royal
The Lord of the Rings
Outlander
The Shining
Interview with the Vampire
The Queen of the Damned
The Way of Kings
Light
The Hobbit
Magic
The Wise Man's Fear
Dragonfly in Amber
Quality
Fire
Animal
Old Magic
The Vampire Lestat
Dead Until Dark
Death
Harry Potter and the Goblet of Fire
Harry Potter and the Chamber of Secrets
A Dance with Dragons
Do Androids Dream of Electric Sheep?
Living Dead in Dallas
Mort
Pet Sematary
Coraline
Carrie
The Dragon Reborn
olor
Name
Harry Potter and the Order of the Phoenix
Dark
The Green Mile
The Color of Magic
State
Harry Potter and the Deathly Hallows
Religion
American Gods
The Eye of the World
'Salem's Lot
The Return of the King
ppening
A Feast for Crows
New
Space
Attraction
Guilty Pleasures
The Shadow Rising
All Together Dead
Dead in the Family
Club Dead
Time
The Fellowship of the Ring
Group
The Hero of Ages
Summer Knight
Drums of Autumn
Anansi Boys
The Graveyard Book
From Dead to Worse
Dead as a Doornail
Definitely Dead
Dead to the World
Dead and Gone

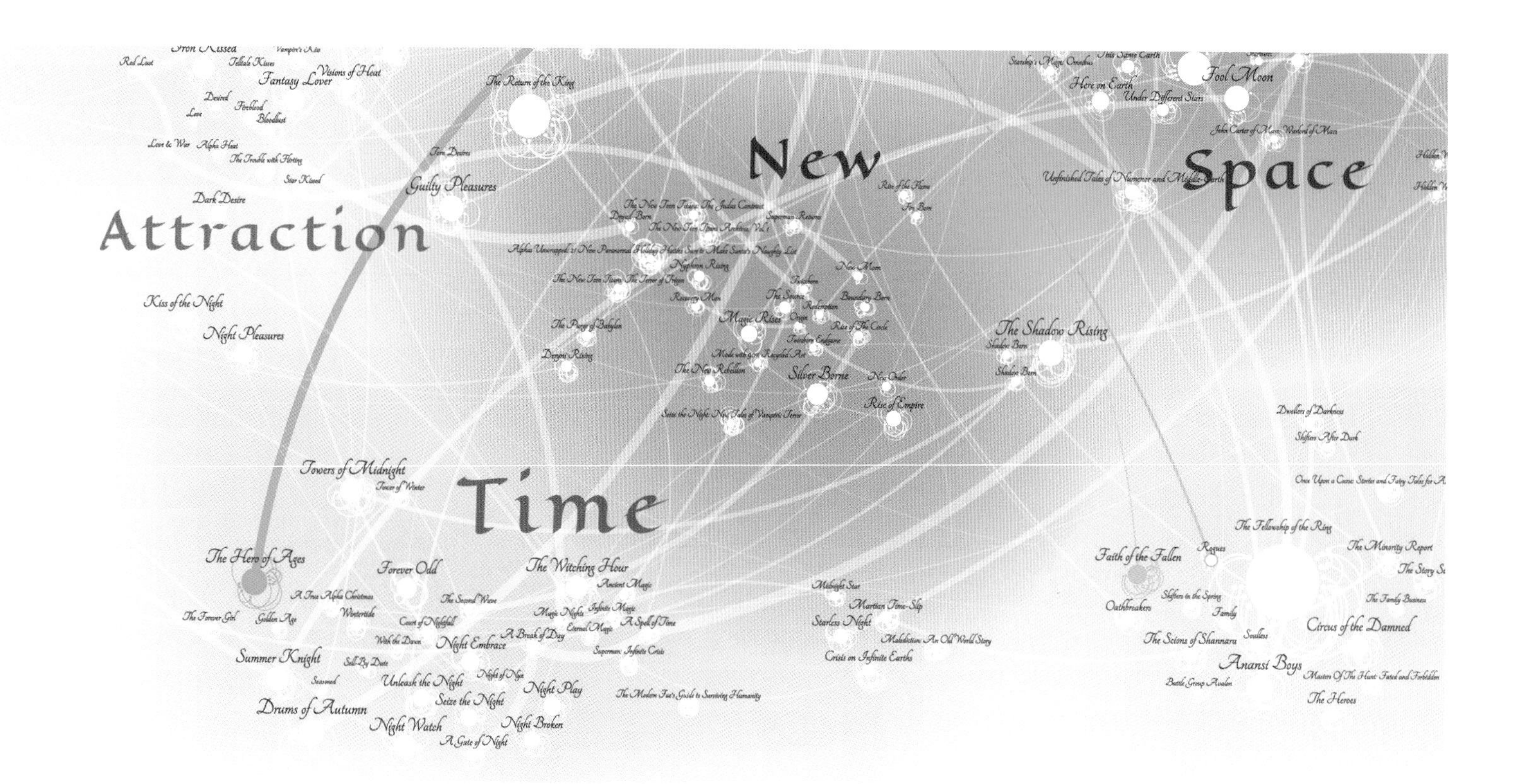

Fig.5.16

Zooming in on the lower-left section of the map that focuses on the themes of “time,” and “new.”

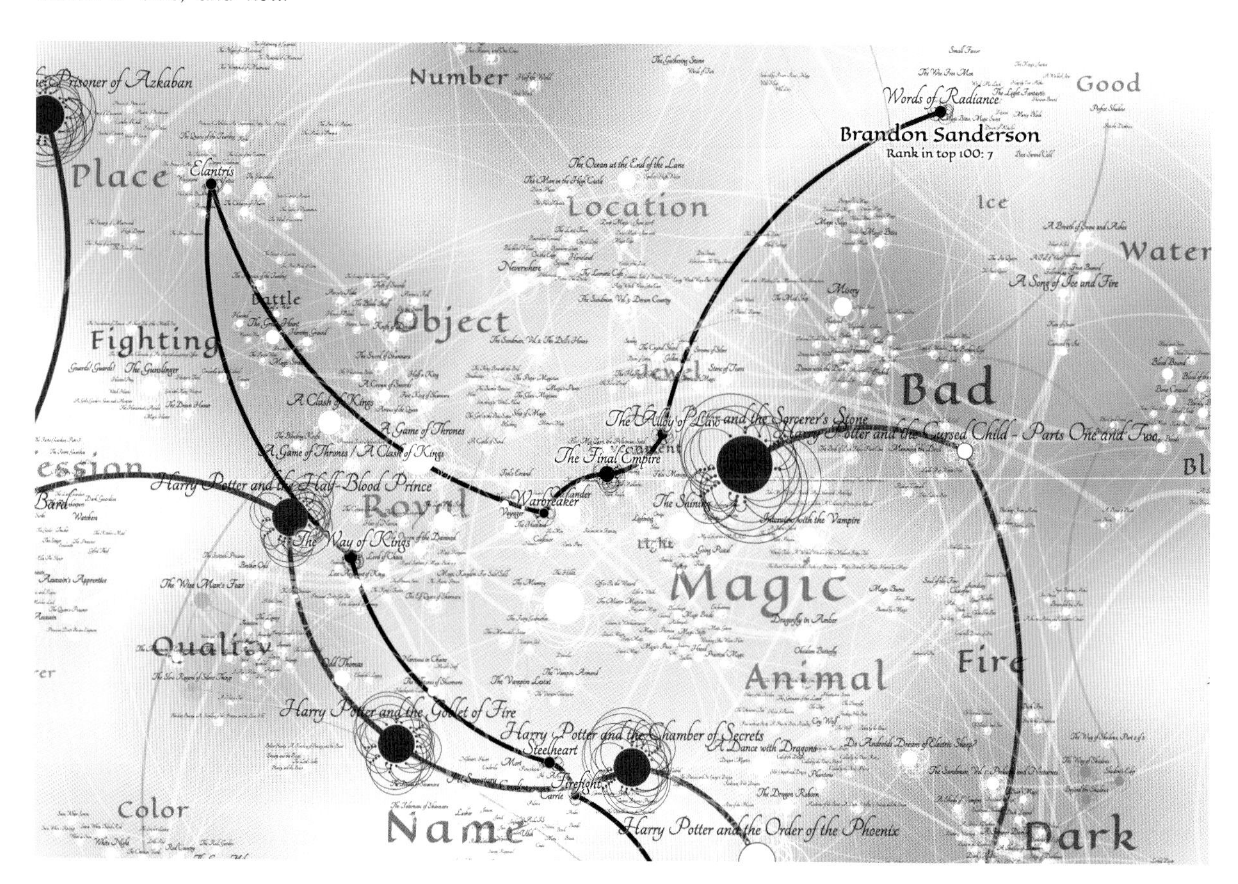

Fig.5.17

Hovering over a specific book circle will highlight the author, their rank in the top 100, and all of the other books by the same author.

Fig.5.18

The title that I made for the online version.

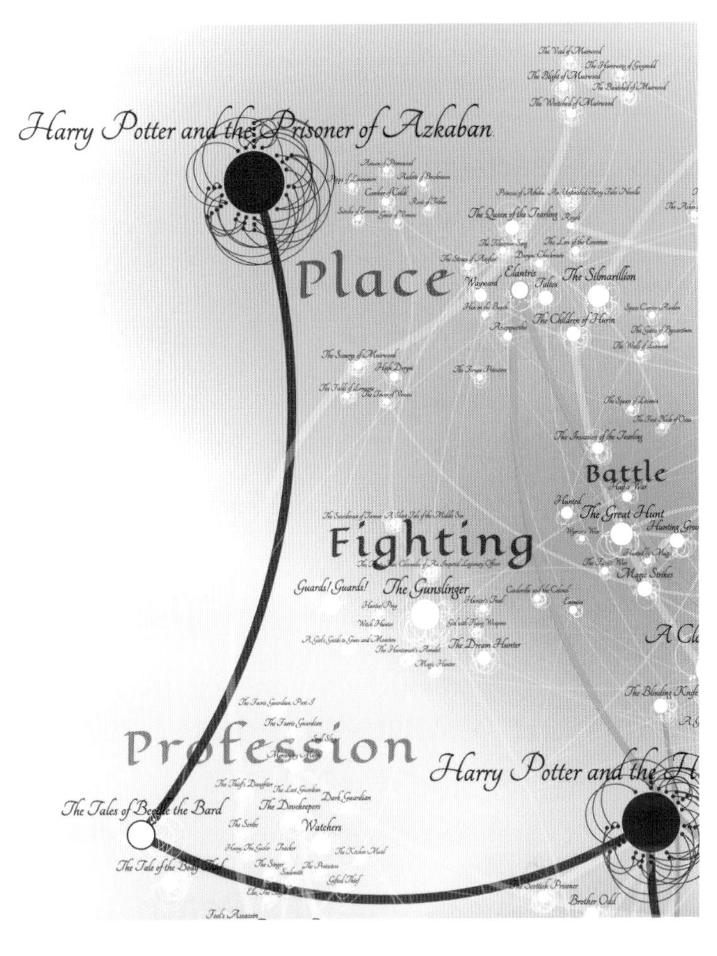

Fig.5.19

Zooming in on the upper-left section where the third Harry Potter book stands out from the other books.

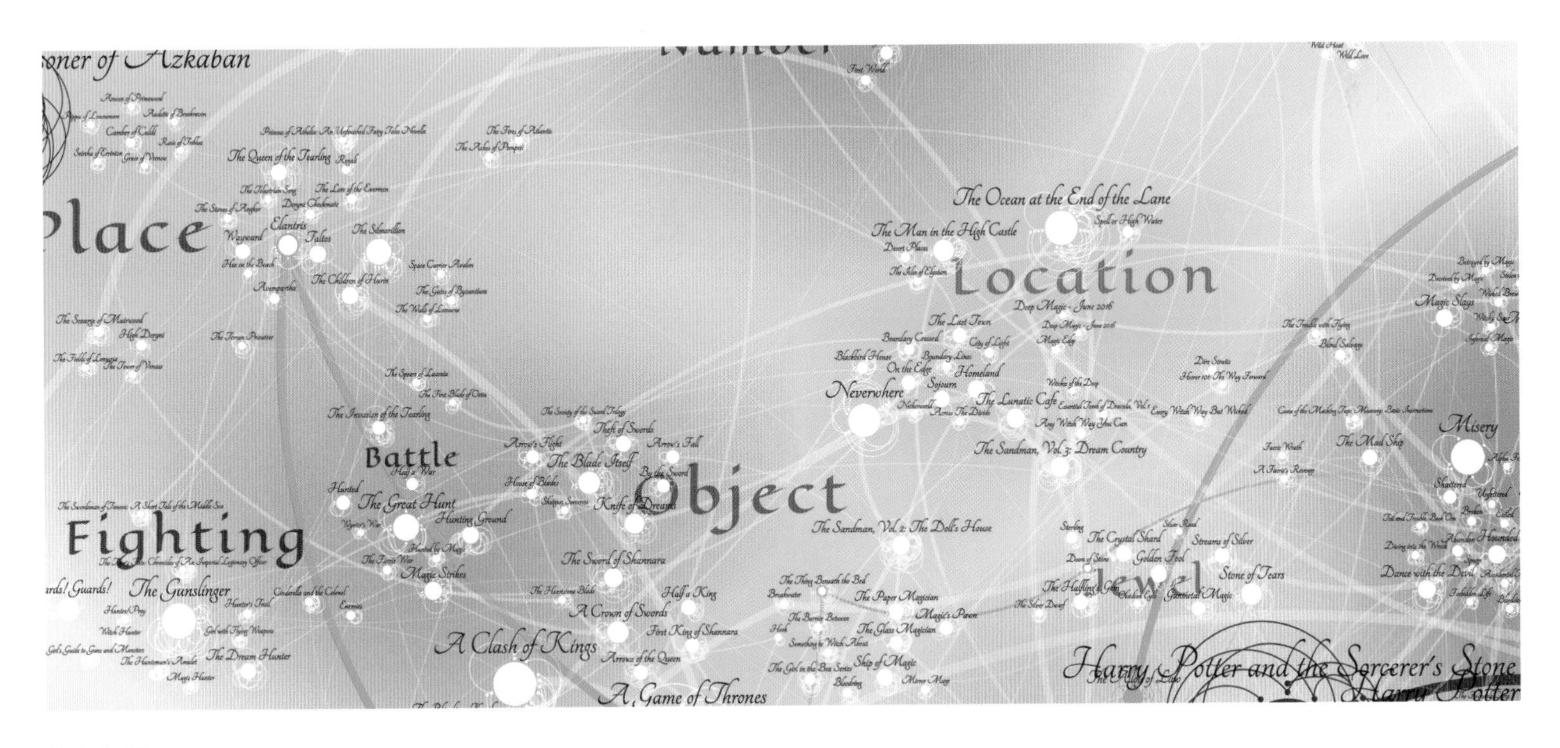

Fig.5.20

Zooming in on an upper-middle section of the map that focuses on the themes of "object," and "location."

characters, conversations, and themes below to explore them. Take advantage of the fact that some characters, conversations, or themes will disappear as you filter down; their co-appearances and co-occurrences are often times just as interesting as the songs that are left.

If you get into a bad state, reset.

↓

Ambition

a1 a2 a3 a4

Personality

p1 p2 p3

Contentment

c1 c2 c3

Miscellaneous

m1 m2 m3

Legacy

l1 l2 l3

Relationship

r1 r2

Death

d1

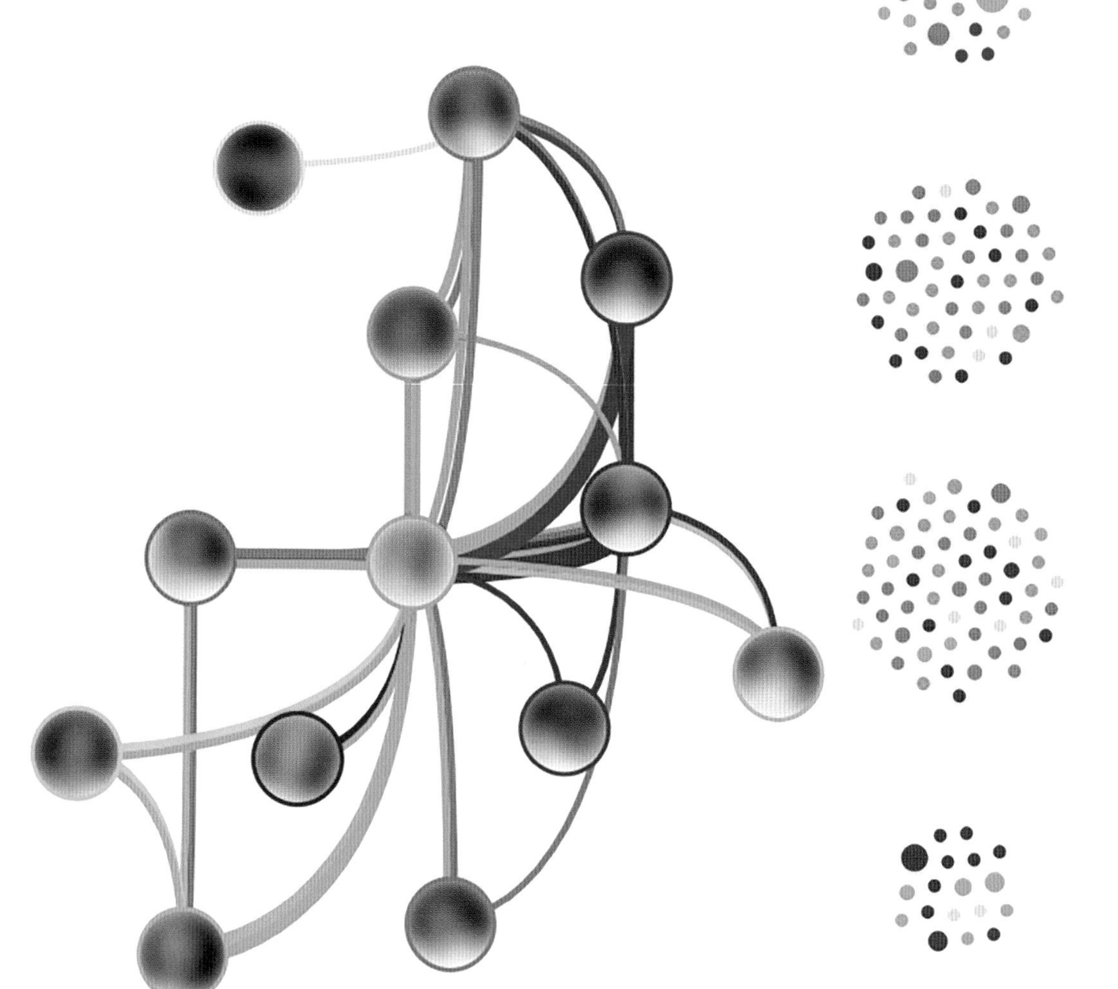

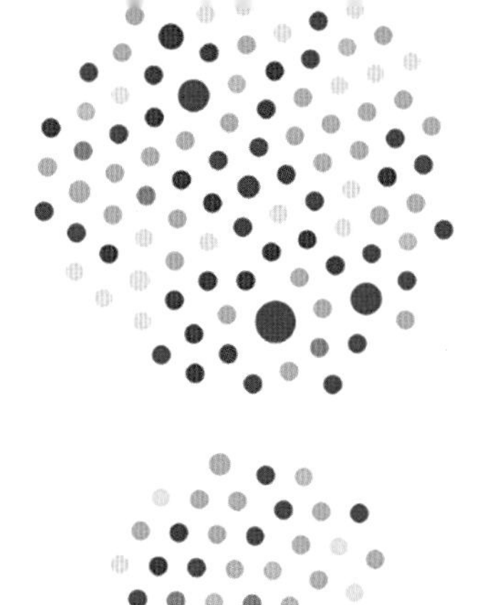

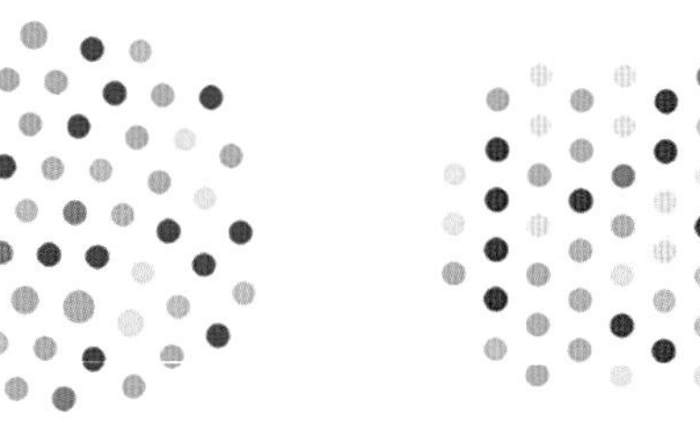

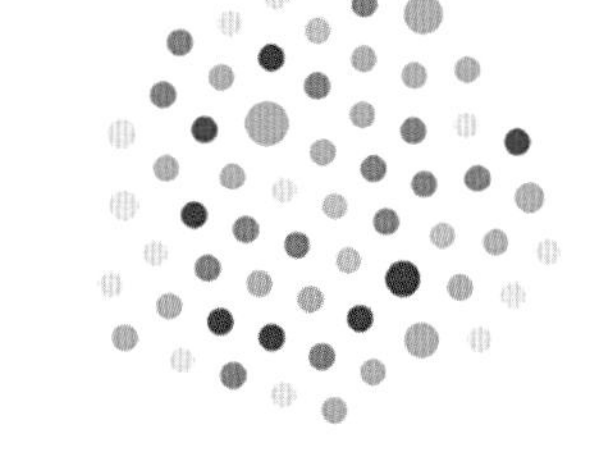

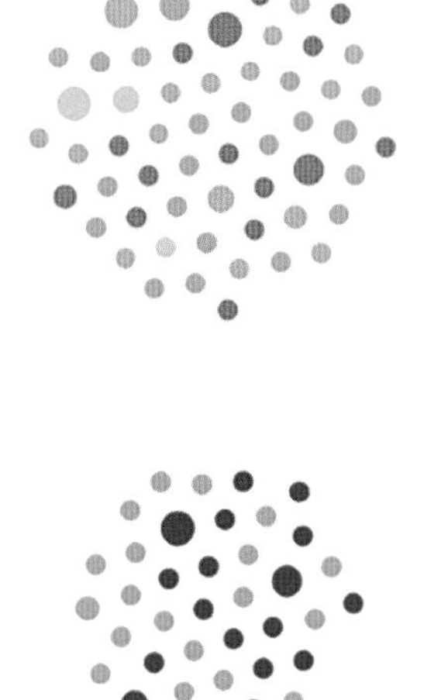

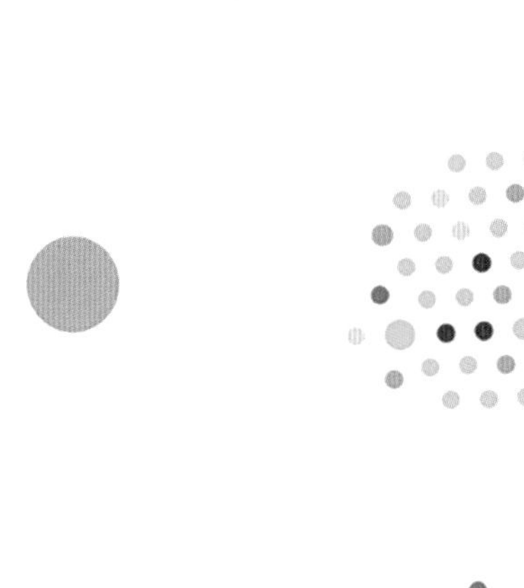

Every Line in Hamilton

SHIRLEY

In the summer of 2016, I got really, *really* obsessed with *Hamilton: An American Musical*. It was quite a unique experience because all of the show's lines and dialogues were contained in songs, so I could get the whole plot by listening to the cast recording. I had it on repeat for months, and it got to a point where I was analyzing lyrics and searching for recurring themes throughout the musical. At one point, my boyfriend (now husband) suggested I turn it into a data visualization. I was really resistant at first ("that's beyond obsessive!"), but eventually gave in ("ok, I guess I *am* that obsessive.") I had been talking to Matt Daniels from *The Pudding*—a collective of journalist-engineers that work on visual essays—about working on a story together and pitched the idea to them. I wanted to create a visual tool to analyze character relationships, recurring phrases, and how they evolved throughout the musical—and they agreed.

I had originally budgeted one month to work on the project, but it ended up taking three months on and off. It took so much time and was so all-encompassing that I didn't have the time to work on a project for the "Books" topic, and I asked Nadieh if I could turn my *Hamilton* visualization into a *Data Sketches* project. It was a musical, but I made the point that I had created the dataset using *Hamilton: The Revolution* (a detailed book about the creation of the musical, lovingly referred to as the "Hamiltome" and co-written by Lin-Manuel Miranda, the creator of *Hamilton*), and Nadieh thankfully agreed. (; ・ ∀ ・)϶

Data

Because the dataset I needed wasn't readily available, I needed to create it myself. First, I went through the "Hamiltome," which includes the lyrics for every song and notes about its inspirations, creation, and musical influences. I went through all of the lyrics and marked any repeated phrases with a corresponding number, marked the page with a post-it, and noted it in my sketchbook—this whole process took two days (Figure 5.1). I then typed up each of those recurring phrases in order to group them into broad themes.

I couldn't do a simple text search for this, because oftentimes the phrases were repeated with slight changes in wording. And instead of trying to write some sophisticated algorithm, Matt suggested it would be faster to manually collect them instead.

The next step was to manually enter all the data. Thankfully, the full set of lyrics were already available online in a fan-made Github repo, but it was a text file that wasn't formatted to support metadata. I assigned a unique ID to each character and created a CSV file (`characters.csv`) to note which character sung which lines. And since I was already at it, I also noted when the character was directing the line(s) at another character in conversation (Figure 5.2, top). Finally, I assigned a unique ID to each recurring phrase, created another CSV file (`themes.csv`), and recorded only the lines with the repeated phrases I had noted in my first run-through (Figure 5.2, bottom). This whole process took another painstaking two days.

One of the reasons it took so long is because I entered all my data in CSV format in a *text editor* because it didn't occur to me to use Excel... I've since learned my lesson.

Next, I wrote a script to join the actual lyrics with the metadata in `characters.csv` and `themes.csv` into one master JSON file. I started by assigning the metadata to each individual line, but when I went to draw each of those lines as dots on the screen, I saw that there were way too many of them (Figure 5.3, left). I realized that as the metadata were usually the same across a set of consecutive lines (sung by the same character), it would make the most sense to assign metadata to those sets of lyrics instead. When I drew the new dataset on the screen (with each dot size mapped to the the number of lines it represents), it looked much more manageable (Figure 5.3, right).

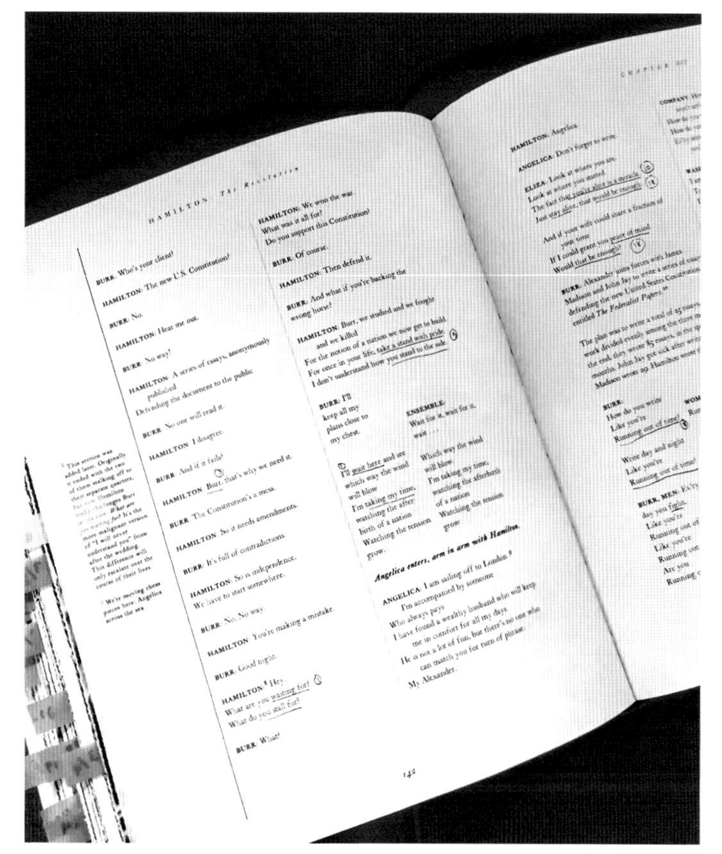

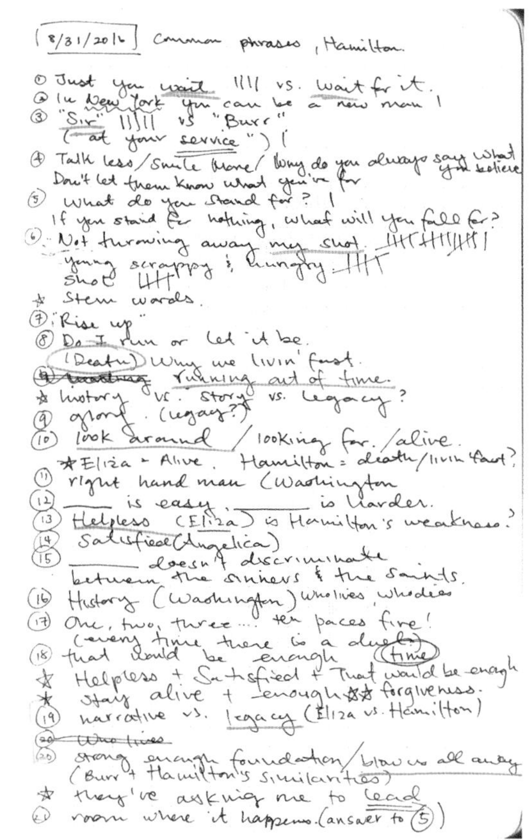

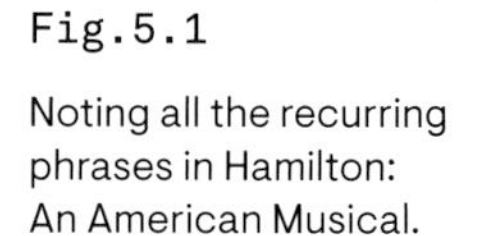

Fig.5.1

Noting all the recurring phrases in Hamilton: An American Musical.

Fig.5.2

CSV files for recording characters and conversations metadata (top) and recurring phrases (bottom).

characters.csv

1422 lines (1422 sloc) | 16.7 KB

1	characters	excluding	directed_to	lines	song
2	1			1-5	1
3	4			6-11	1
4	10			12-15	1
5	11			16-19	1
6	1			20-23	1
7	2			24-27	1

themes.csv

256 lines (256 sloc) | 2.68 KB

1	themes	notes	lines	song
2	1		27	1
3	1		54	1
4	1		76	1
5	3		2	2
6	3		4	2
7	3		5	2

Fig.5.3

Each dot as a line (left) versus as a set of consecutive lines sung by a character (right).

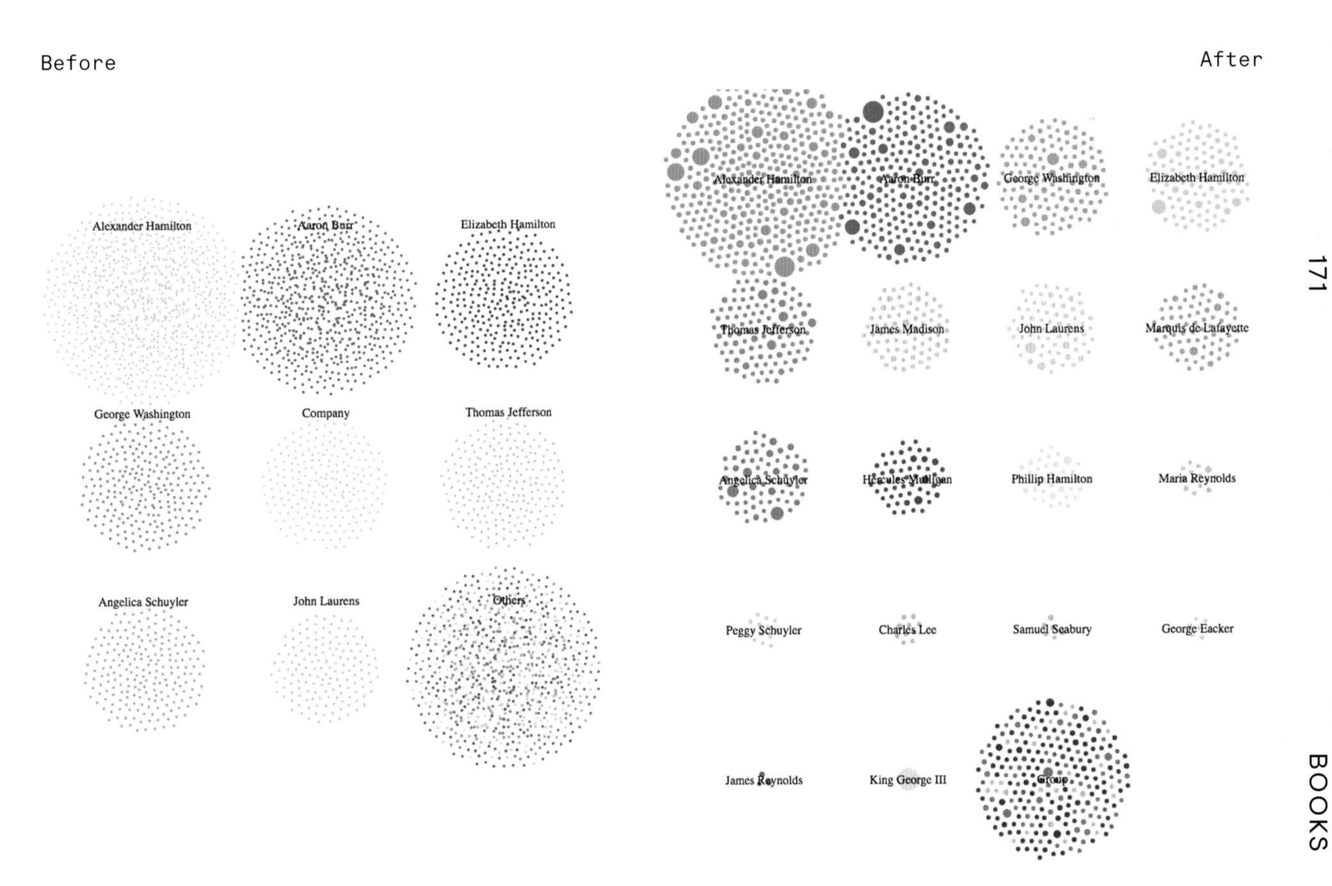

I had originally gathered the metadata in separate files because I thought it would save me time not to have to record metadata for every single line of lyrics. But as it took me a good few days to write the scripts and join the separate files anyway, it might have been faster to just start with one master CSV file where I had columns for lyrics, characters, and recurring phrases all together. I learn something new with every project. ¯_(⊙︿⊙)_/¯

Sketch & Code

I knew I wanted to visualize the lyrics to be able to see who was singing, when, and for how long. To do that, I thought of two ways of depicting lines of lyrics: as circles or as long, narrow bars (Figure 5.4). To indicate recurring phrases, I played around with the idea of a diamond that stretched across the bars. And to deal with lines where a group of characters sing together, I thought of stacking narrower bars on top of each other.

Once the initial sketches were done, I got back on my computer to see how feasible my ideas were. Positioning things always turns out more painful than I give it credit for, and after some bugs and mishaps with math and SVG paths, I ended up with the first vertical version and, eventually, the horizontal version with the themed diamonds (Figure 5.5).

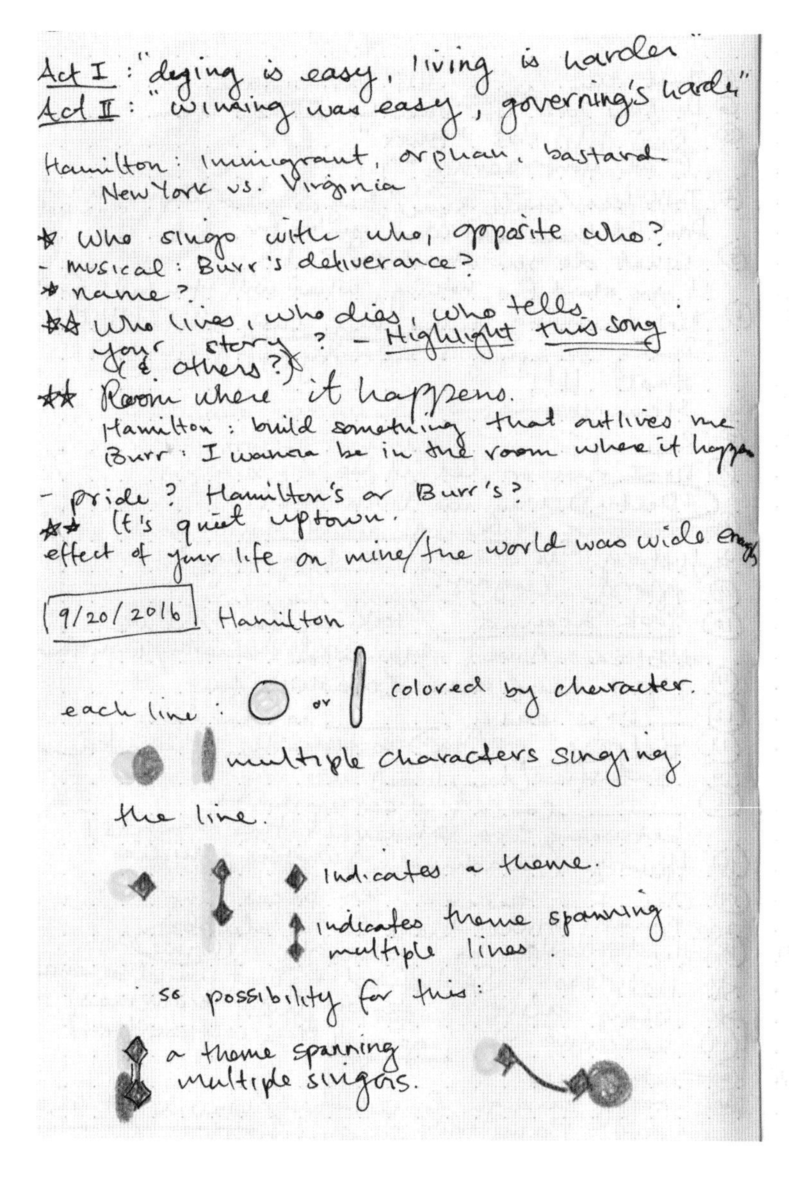

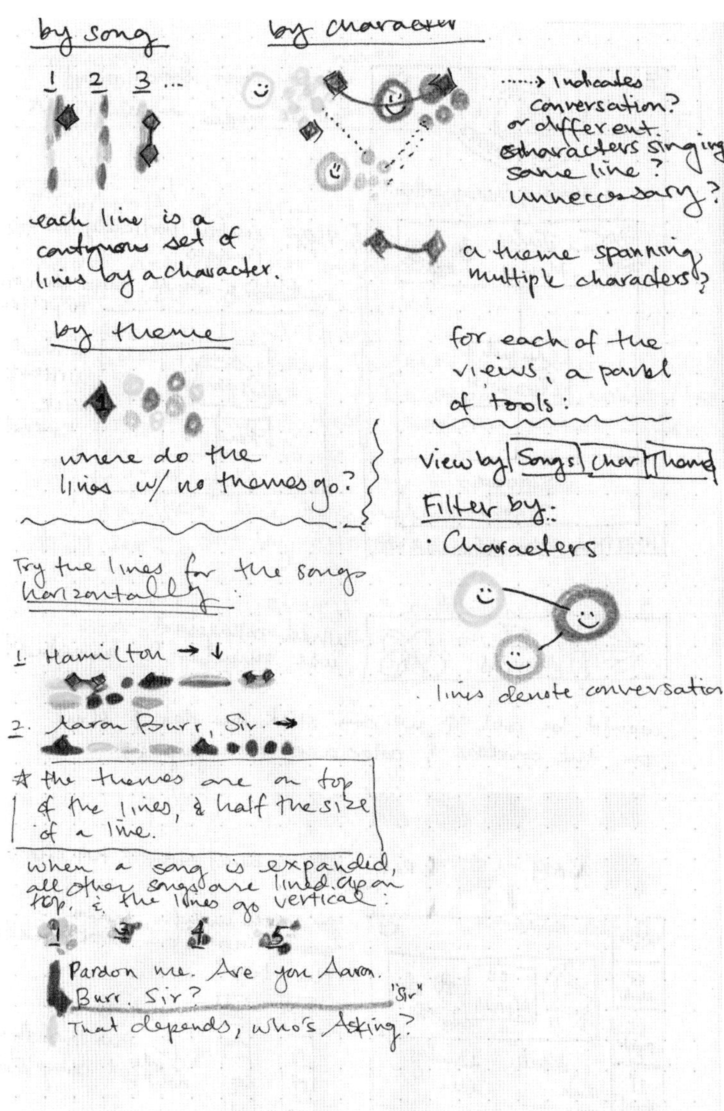

Fig.5.4

Sketches working through how to represent each line in a song, the characters and their conversations, and the recurring phrases.

Fig.5.5

First attempts at positioning all songs.

1. Vertical positioning with bug

2. Vertical positioning working as intended

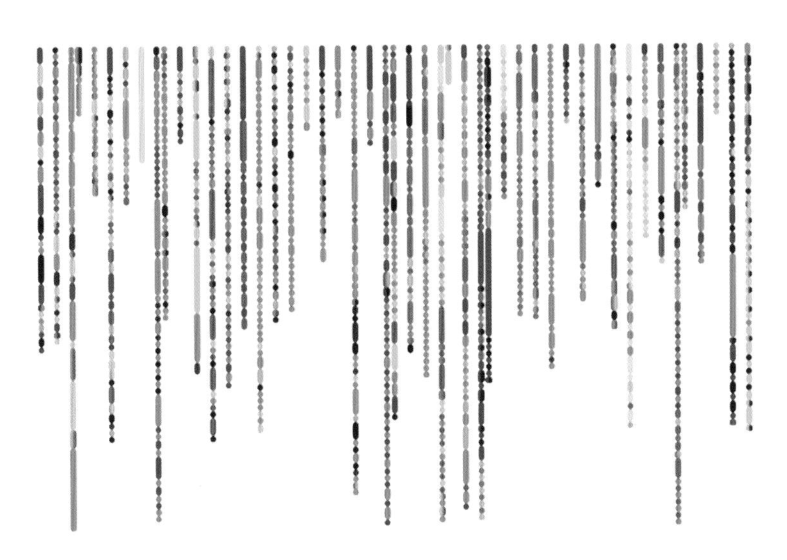

3. Horizontal positioning with theme diamonds

Now, if I thought the positioning took a while to work through (that only took a few days), implementing the filters was an absolute nightmare. I wanted to create a tool where I (and eventually the reader) could filter all the lyrics by any set of characters, conversations, and/or recurring phrases to do my analysis. I thought this would be straightforward to implement, but unfortunately, there were a lot of possible filter combinations and edge cases that I didn't expect and had to work through. I spent a few *weeks*, on and off, pacing around my bedroom and living room to work through all the logic and bugs:

I hadn't felt so frustrated and happy and alive with a piece of code in a long time.

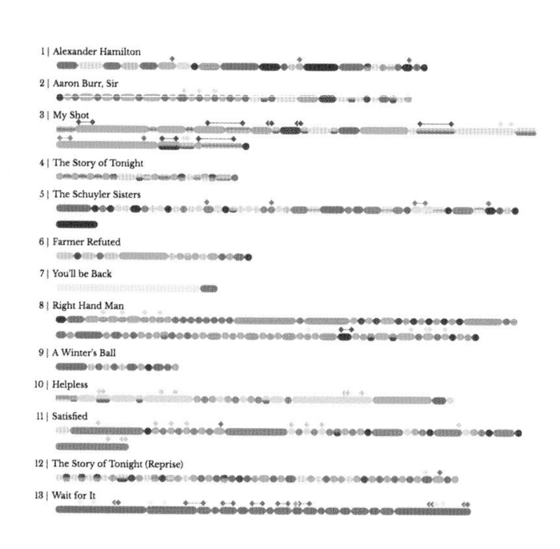

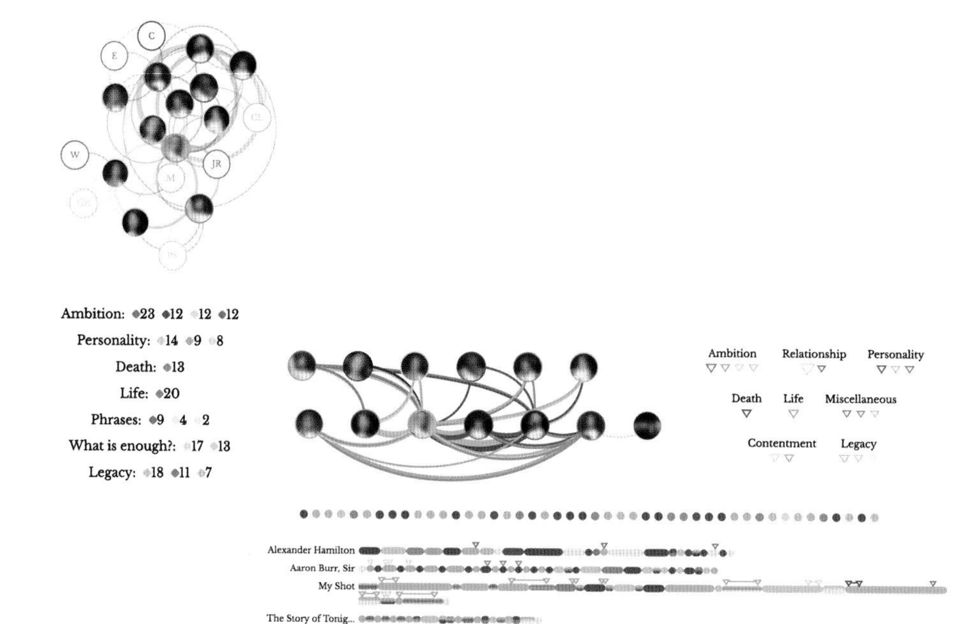

Fig.5.6

Different visual iterations of the filter tool.

After a few weeks, I was able to work out a set of logic operations that I was satisfied with:

- Filtering by characters is an AND operation; I only keep a song if *all* the selected characters have a line in it
- Filtering by conversations is an OR operation; I keep any song with even one of the selected conversations in it
- Filtering by recurring phrases is an OR operation; I keep any song with even one of the selected phrases in it
- Filtering between categories is an AND operation; I only keep a song if all of the selected characters, conversations, and phrases are included in it

I used AND operations for characters because most characters are in many songs, so if I did an OR operation, I'd end up with all the songs after selecting just a few characters (which would negate the point of filtering in the first place) (Figure 5.7). On the other hand, the opposite is true for conversations and phrases, so if I did AND operations for them, I'd have no songs left after only a few selections. One of the reasons working through the filter logic took so long is because I wanted to strike a good balance where the filter would whittle down the number of songs to a more manageable number while still being useful for analysis.

Fig.5.7

The filter tool with multiple characters selected.

Filters

(reset filters)

Ambition Relationship Personality

Death Life Miscellaneous

Contentment Legacy

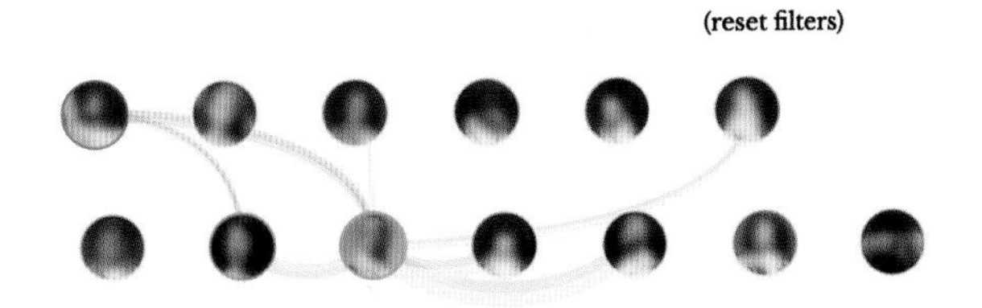

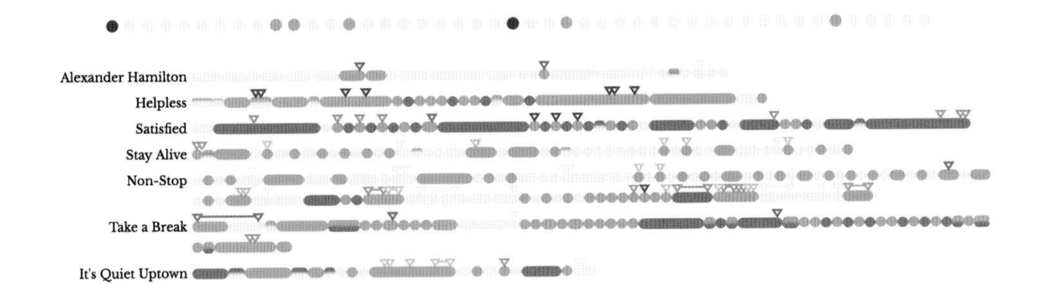

Filters

(reset filters)

Ambition Relationship Personality

Death Life Miscellaneous

Contentment Legacy

Take a Break

Schuyler Defeated

Stay Alive (Repris...

Another reason the filters took so long to work through was because certain combinations lead to "dead ends," where the filters are so specific that only a few songs remain. And because most characters, conversations, and phrases aren't in those remaining songs, selecting any one of them would lead to an empty data state. To help me work through this UI problem, I listed all of the possible states that a character, conversation, or phrase could be in when filtered and decided after a lot of consideration that I'd disable selection on any "dead ends" so that we'd never enter an empty data state. Here are the four possible states and their visual characteristics I worked out:

- It is selected (colored and 100% opacity)
- At least one of its corresponding lines are highlighted because of some other filter applied (colored and partial opacity). For example, if Eliza the character is selected, her corresponding conversations and themes should be highlighted as well.
- At least one of its corresponding lines are in the song, but not highlighted (gray and partial opacity)
- None of its corresponding lines are in the song (missing with dotted outline)

Once I was certain I had fixed all the bugs and covered all the edge cases, I was finally able to move on to the analysis and story.

At first, I filtered by the main characters "Alexander Hamilton" and "Aaron Burr" because I was curious about how their relationship evolved throughout the musical. But Taia, my Hamilton expert, convinced me that there were enough Hamilton-Burr analyses out there already, and that the characters Eliza Hamilton (née Schuyler, Alexander Hamilton's wife) and Angelica Schuyler (Eliza's sister) would be much more interesting to explore instead. I wholeheartedly agreed.

I filtered by Eliza and Angelica, then by Eliza and Alexander, Angelica and Alexander, and finally by their conversations. I looked at what phrases they sang the most and was happily surprised to find that for Eliza, it wasn't "helpless" (the title of her main song), but "look around at how lucky we are to be alive right now" and "that would be enough/what would be enough." That was the point at which I really fell in love with her character, with her optimism, and with how much she matured throughout the story. I knew then that I had to center the story around her.

With my main story figured out, I decided to start outlining and working on my rough draft (Figure 5.8). From the beginning, I wanted to appeal to a wider audience that might not be familiar with the visualizations that I'm used to. I wanted to ease them in slowly and get them used to all the different layouts available in the filter tool. And because I knew that it would be a lengthy article, I wanted to create a delightful enough experience to keep my readers scrolling. I used `D3.js`'s new *force simulation* module to position the dots and have them explode out, dance around, and zoom back together on scroll (Figure 5.9). It was a really fun effect.

I learned the importance of delight from Tony Chu's talk, "Animation, Pacing, and Exposition."[1]

But after the introduction, I didn't know how to include all of the interesting insights I found through my analysis. I had a long stretch of writer's block and went through three rounds of rough drafts, none of which I was satisfied with (and none of which anybody will ever see. (; ๑´ㅂ`๑)). I knew before I started that I would struggle with writing the most (I'm a horribly slow writer), but I reassured myself that I had the visuals covered, and even though I was slow, I wasn't a *bad* writer. How hard would it be to write and make visuals at the same time? Turns out, very, *very* hard. While I could do both tasks separately, I had never given thought to how I'd weave both the visuals and the words together.

This whole struggle gave me an even bigger respect for data journalists, who often do both.

↳ Design to Maximize for Delight

When I first started creating visualizations, I thought that every visual element had to have a purpose; there shouldn't be flashiness for flashiness's sake. But after watching Tony Chu's "Animation, Pacing, and Exposition" talk I decided to give it a try for my Hamilton project. On scroll, I made the dots dance around the screen—a frivolous addition, but it really delighted my friends when I showed them and, more importantly, kept them scrolling.

I've been a firm believer ever since that adding subtle—and sometimes flashy—touches to my visualizations can give readers a much more enjoyable experience, even if they add nothing to the visualizations' readability and understandability.

[1] "Animation, Pacing, and Exposition" by Tony Chu: https://www.youtube.com/watch?v=Z4tB6qyxHJA.

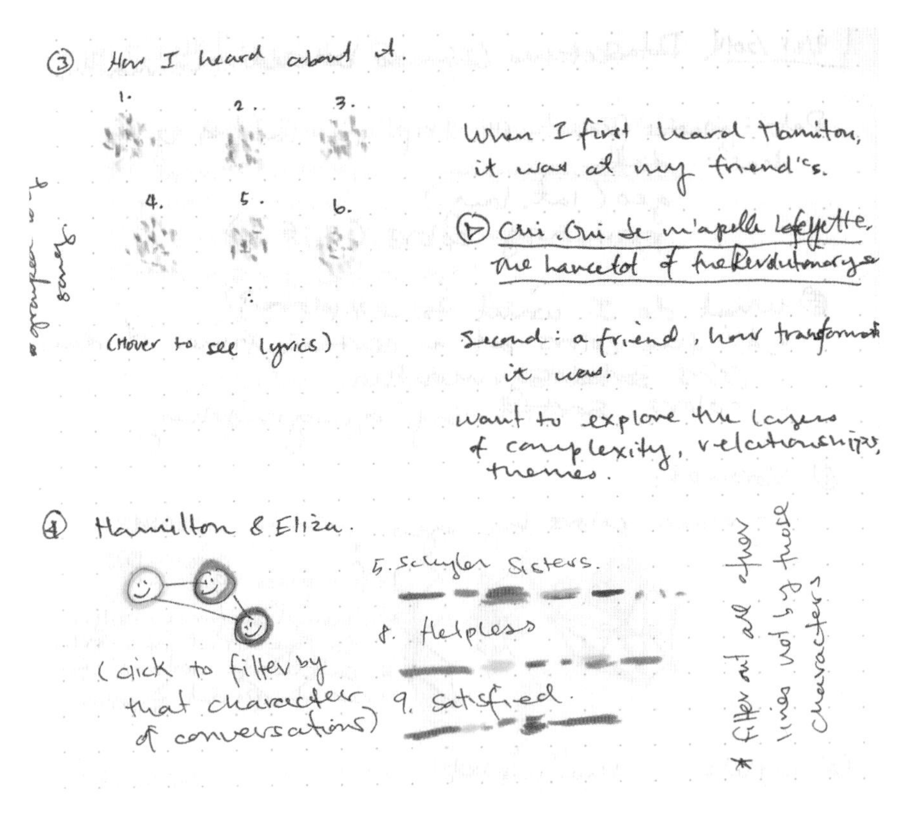

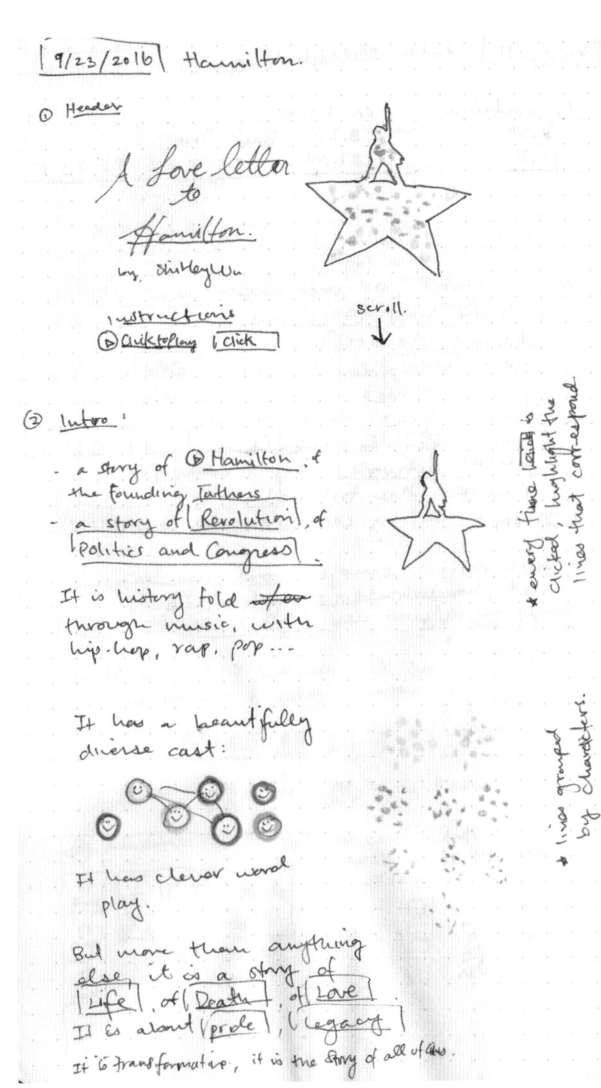

Fig.5.8

My first attempt at sketching the intro section.

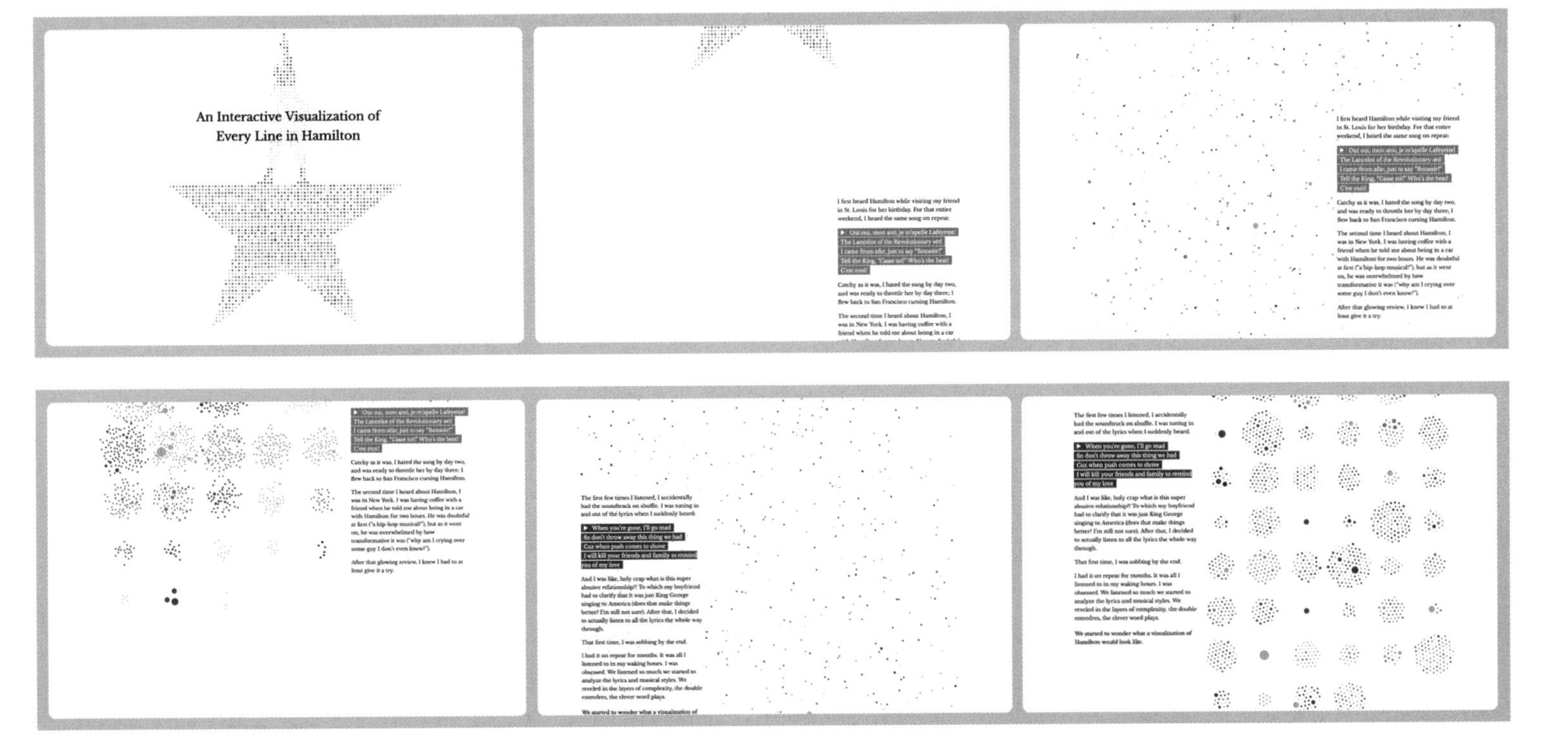

Fig.5.9

Dots exploding out and coming back together on scroll.

I decided to take a short break to work on my “Presidents & Royals” project, and came back to this one right after finishing “Presidents & Royals.” The break really helped to clear my head and when I got back to my notebook to brainstorm, I had an idea right away.

My biggest struggle was finding a way to convey recurring phrases simply and clearly. In the previous iteration, I liked that the diamonds pointed out where the phrases were, but they also made the visualization more cluttered. (The diamonds were colored by theme, which made it even more confusing since I was already using color to identify characters). I learned an important lesson from this experience not to overload visual channels.

For this iteration, I decided to take a different approach: as I couldn’t use color anymore, how else could I visually represent a categorical data type? And then it hit me: I could use symbols—and even better, since *Hamilton* was a musical, I could use *musical* notations! I decided to mimic the long arcs and labels I would often see in music sheets and used a shorthand label to denote a recurring phrase and an arc when that phrase was repeated for more than one consecutive set of lyrics:

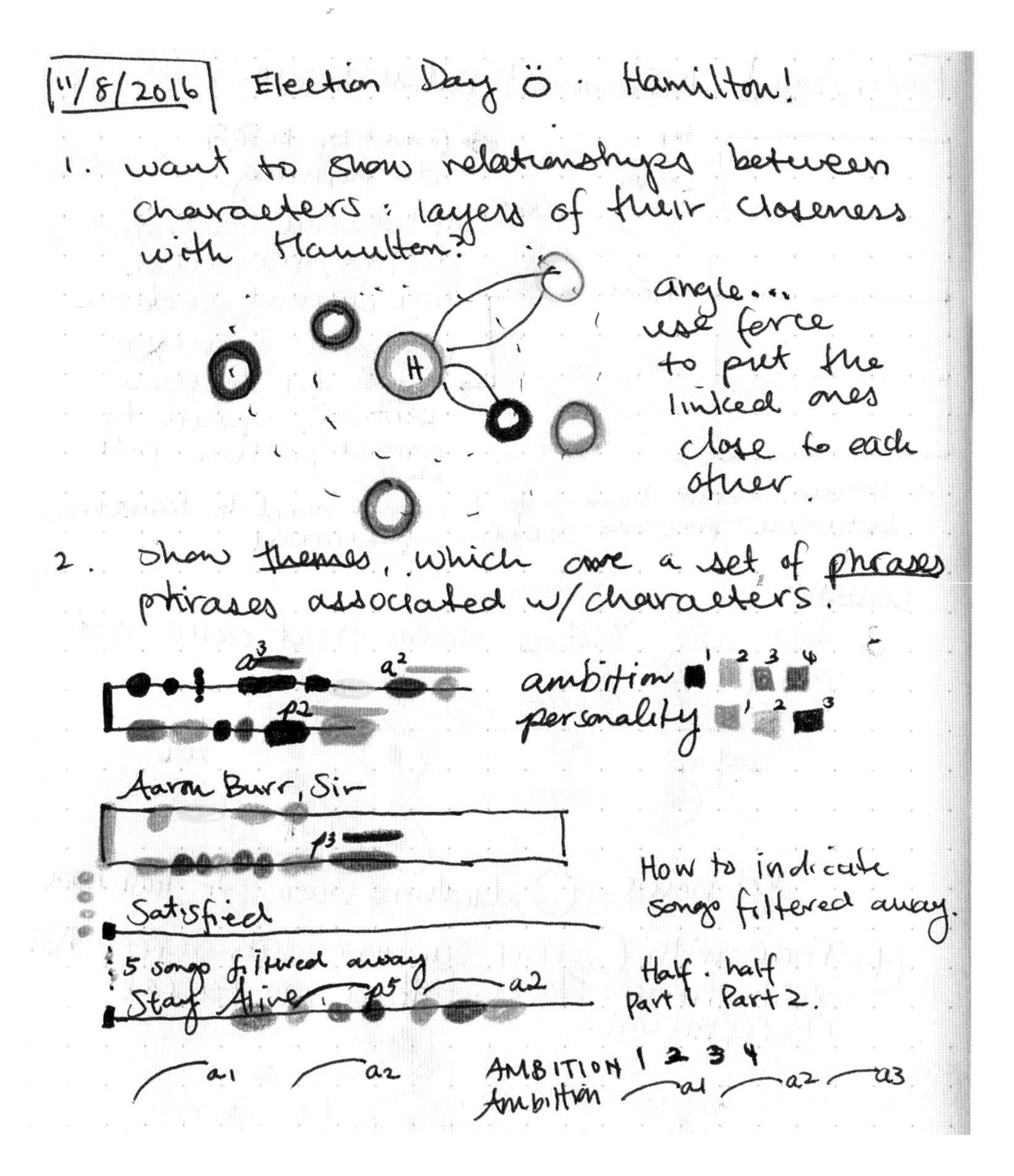

Fig.5.10

Sketch of a new design for recurring phrases, fashioned after musical notations.

I loved the effect; not only were the arcs and labels simpler and cleaner; they were also great visual metaphors. To add to the metaphor, I tried to add musical staves, which I hoped would help indicate the length of the lines (Figure 5.11, top). I liked it, but unfortunately it caused confusion when I showed it to others, and I ended up removing the staves for a much cleaner look (Figure 5.11, bottom).

With the visualizations figured out, outlining the write-up was much easier to do and I managed to finish most of my final draft on a flight to Singapore. The biggest change between my final draft and all the previous ones was that I moved away from merely reciting the numbers (like I did in my "Presidents & Royals" project). Instead, I wrote about my own love for *Hamilton* and what made me create a whole project around it. I also detailed Angelica's and Eliza's character growths that I was really excited to have (and could only have) discovered with my visual filter tool. Not only did this change in approach make it easier for me to write, I think it was also a major reason this project ended up being so well-received.

17 hours...never again...

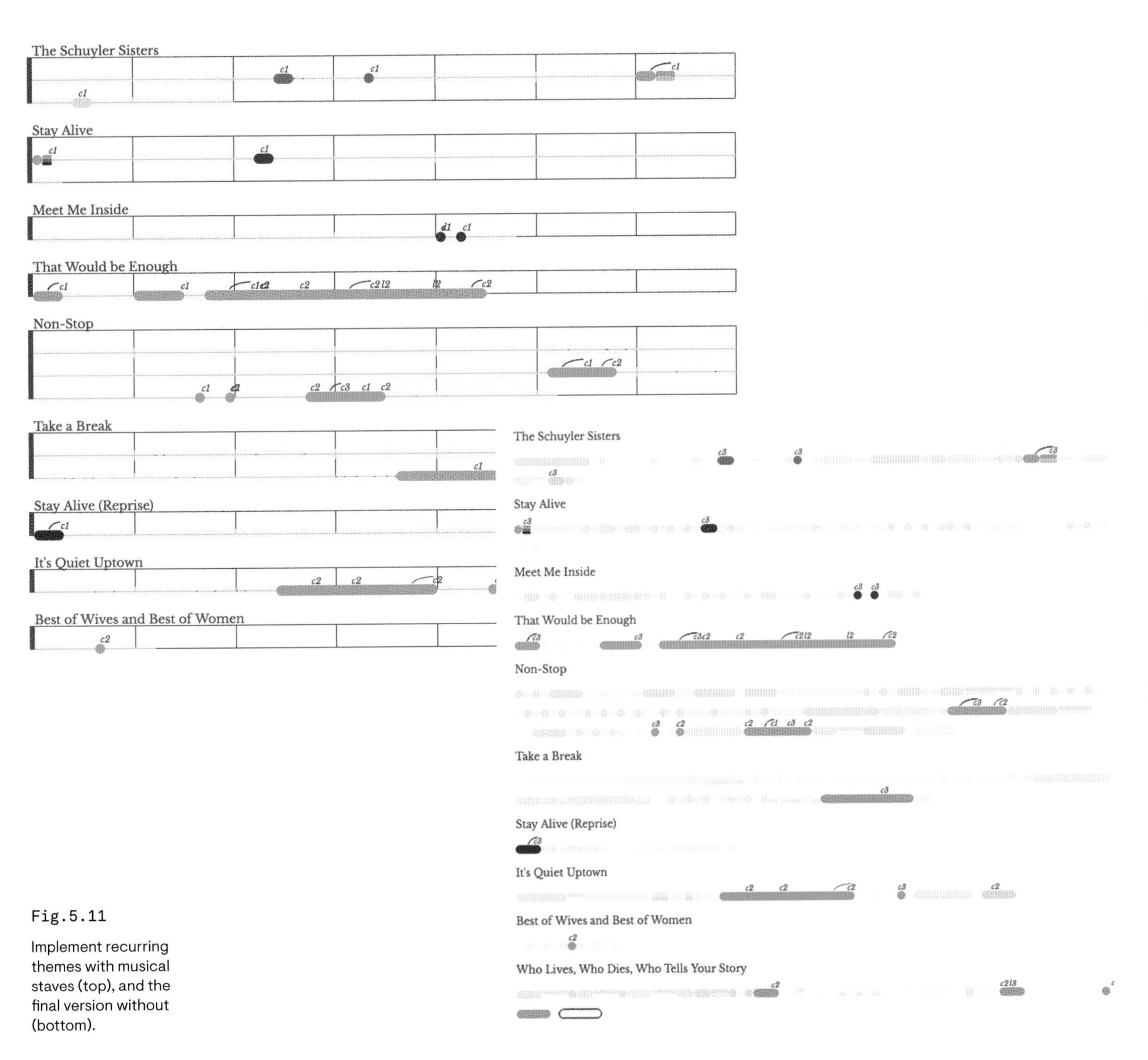

Fig.5.11

Implement recurring themes with musical staves (top), and the final version without (bottom).

Code

When I was almost to the finish line, I showed the project to friend and fellow developer Sarah Drasner for feedback on the animation. Her immediate reaction was: "Canvas! You need canvas!" At ±1,700 path elements, SVG was doing alright—as long as they didn't animate. But on scroll, you could start to see the lag.

I rendered and animated everything with canvas and noticed the visualization's performance was indeed *much* better. The next step was to add the hover interaction back in, and this was where I encountered the most frustrating bug I've seen in awhile.

With the bug, I would hover over a line and, though the canvas and hidden canvas were clearly positioned at the same *x* and *y coordinates*, the tooltip would react incorrectly. It took me hours of agony to realize that, because I scaled both of the canvases by 2x to make sure canvas displayed crisply on retina screens, the underlying hidden canvas image data *also* scaled by 2x. And as I wasn't multiplying the mouse positions by two, the tooltips that popped up were all at about half the *x* and *y* positions of where I was actually hovering.

For an explanation of hidden canvas, see the lesson "Canvas Interactions" in this chapter.

Everything was fortunately smooth going after solving this bug. I made final edits to my story, fixed some more positioning bugs with the tooltip, and spent another agonizing day making everything mobile-friendly.

Reflections

I published "An Interactive Visualization of Every Line in Hamilton" on *The Pudding* on December 13th, 2016 (my birthday!). It did better than I ever could have imagined; it went "viral" for a few days amongst *Hamilton* fans, was picked up by the musical's official Twitter and Facebook accounts, and was even quote-tweeted by Lin-Manuel Miranda (creator of the musical and the original actor for Alexander Hamilton). It was a great birthday.

One of the most important lessons I learned from this project was that though accuracy and precision are very important in a data visualization as to not mislead, I've found that people rarely remember the numbers I give them. Instead, they are much more likely to remember how my stories made them *feel*. I've kept that lesson close to heart in all the projects I've worked on since.

This project also had a huge impact on my career. I worked on it as I was just starting to freelance, and had originally thought I could finish it in the month of September 2016. By October, I had to take on two more client projects to make ends meet and ended up working on my "Olympics," "Travel," and "Presidents & Royals" projects for *Data Sketches*, as well as a visualization for d3.unconf, my own portfolio website, and the *Data Sketches* website at the same time. When the project went viral, I saw a few comments about how I must have too much free time to have created something like this, and I could only laugh because it was so far from the truth. I worked my absolute hardest every waking moment, because I was so incredibly determined to make it as a freelancer.

And for that, I'm so grateful for this project, because if launching *Data Sketches* made my name known in the data visualization community, *Hamilton* went viral enough that it put my name on non-data visualization peoples' radars. This and my "Culture" project with Google News Labs really cemented my freelance career, and I was able to be profitable within seven months of starting.

Canvas Interactions

As I mentioned in the "SVG versus Canvas" lesson, canvas is pixel-based and has no notion of individual shapes. So how do we implement user interactions like click and hover and display the corresponding data in canvas? There are a few techniques we can use to accomplish this, including *hidden canvas* and *quadtree*.

Hidden Canvas

Hidden canvas is a technique that uses `getImageData()`, a Canvas API where we can pass in a starting *x/y* coordinate and a width/height, to get back the RGBA data for all pixels within that area. To implement hidden canvas, we have to:

1. Create two canvases: a visible one with the visualization, and a hidden one to use for reverse lookup.
2. Fill each data point in the hidden canvas with a unique RGB color.
3. Store each pairing in a JavaScript object with the stringified RGB color as *key* and data point as *value*.
4. Register a `mousemove` or `click` event on the visible canvas.
5. On callback, get the mouse position relative to container and pass it in to `getImageData()` with 1px width and height.
6. Use the returned RGB color to look up a corresponding data point.

Hidden canvas can be implemented with vanilla JavaScript and is the most performant for large datasets. Unfortunately, it's also the most time-consuming to implement. (╥﹏╥)

Quadtree

Quadtree is an efficient data structure for looking up points in two-dimensional space. `D3.js` has a great implementation that I love to use:

1. Initialize `d3.quadtree()` with dataset and an *x/y* accessor.
2. Register a `mousemove` or `click` event on the canvas element.
3. On callback, get the mouse position relative to container and pass it into `quadtree.find()` to get the corresponding data point.

For most canvas interactions, I prefer using quadtree because it's *so* much easier to implement. The only caveat is that for very large datasets, the data structure can sometimes take up a lot more memory and affect performance. In those cases, I tend to revert back to the hidden canvas approach.

Every Line in Hamilton

↳ shirleywu.studio/projects/hamilton

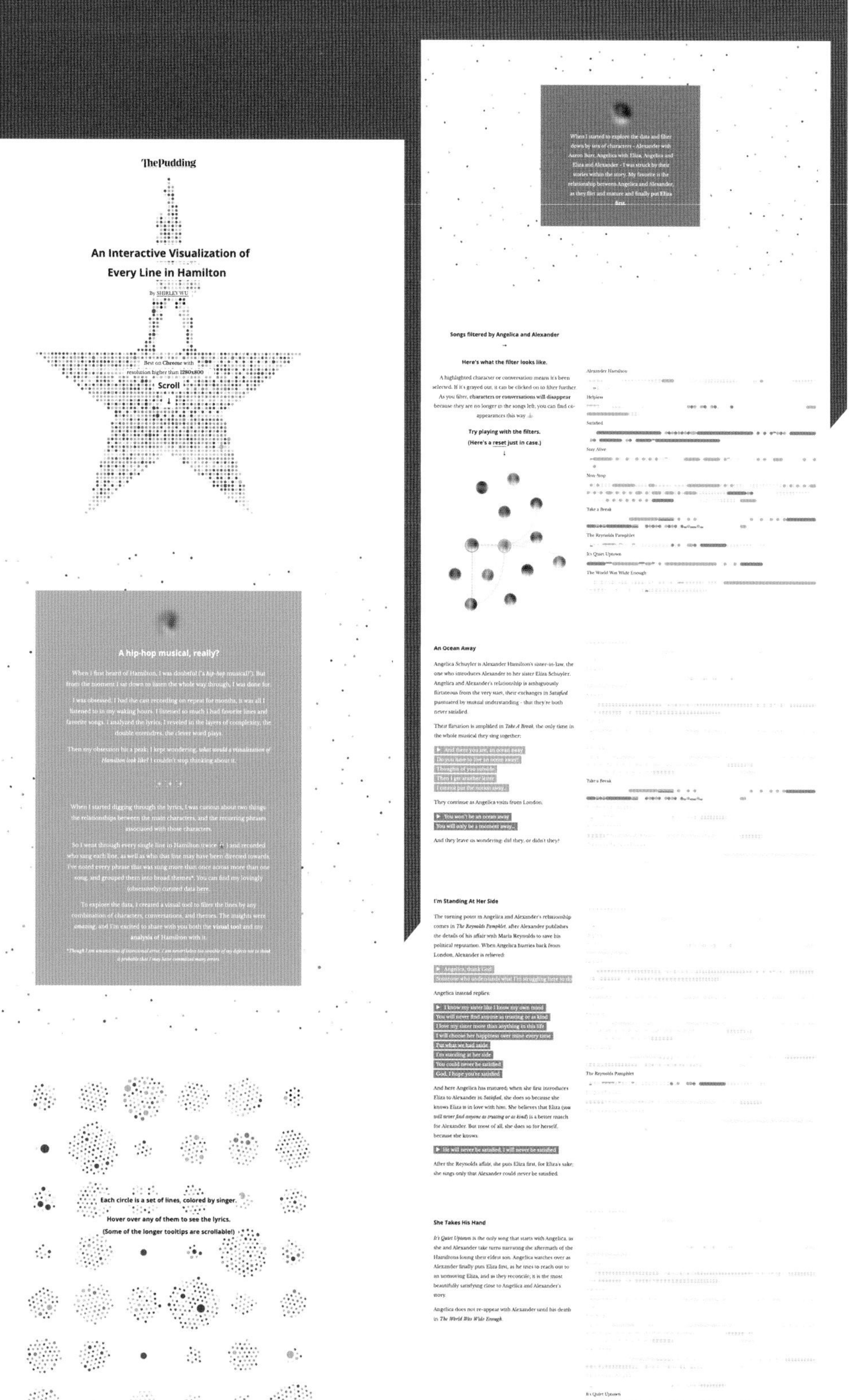

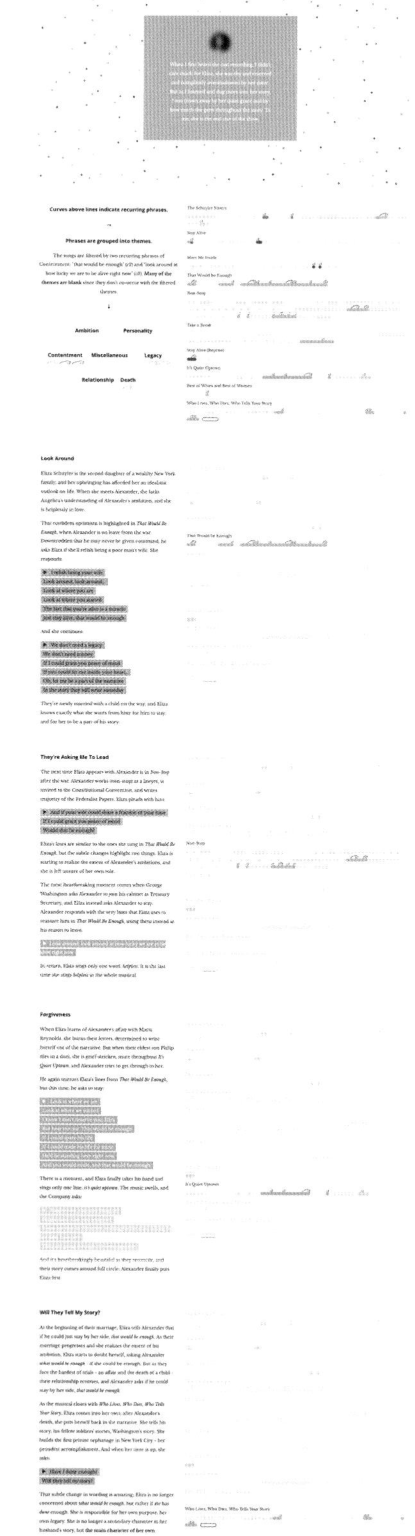

Fig.5.12

The final scrollytelling piece (without exploratory tool)—I'm still so proud of it (๑•̀ㅂ•́)و✧

Fig.5.13

The final exploratory tool where the reader can filter by characters, relationships, and themes, and dig into the remaining songs to do their own analysis.

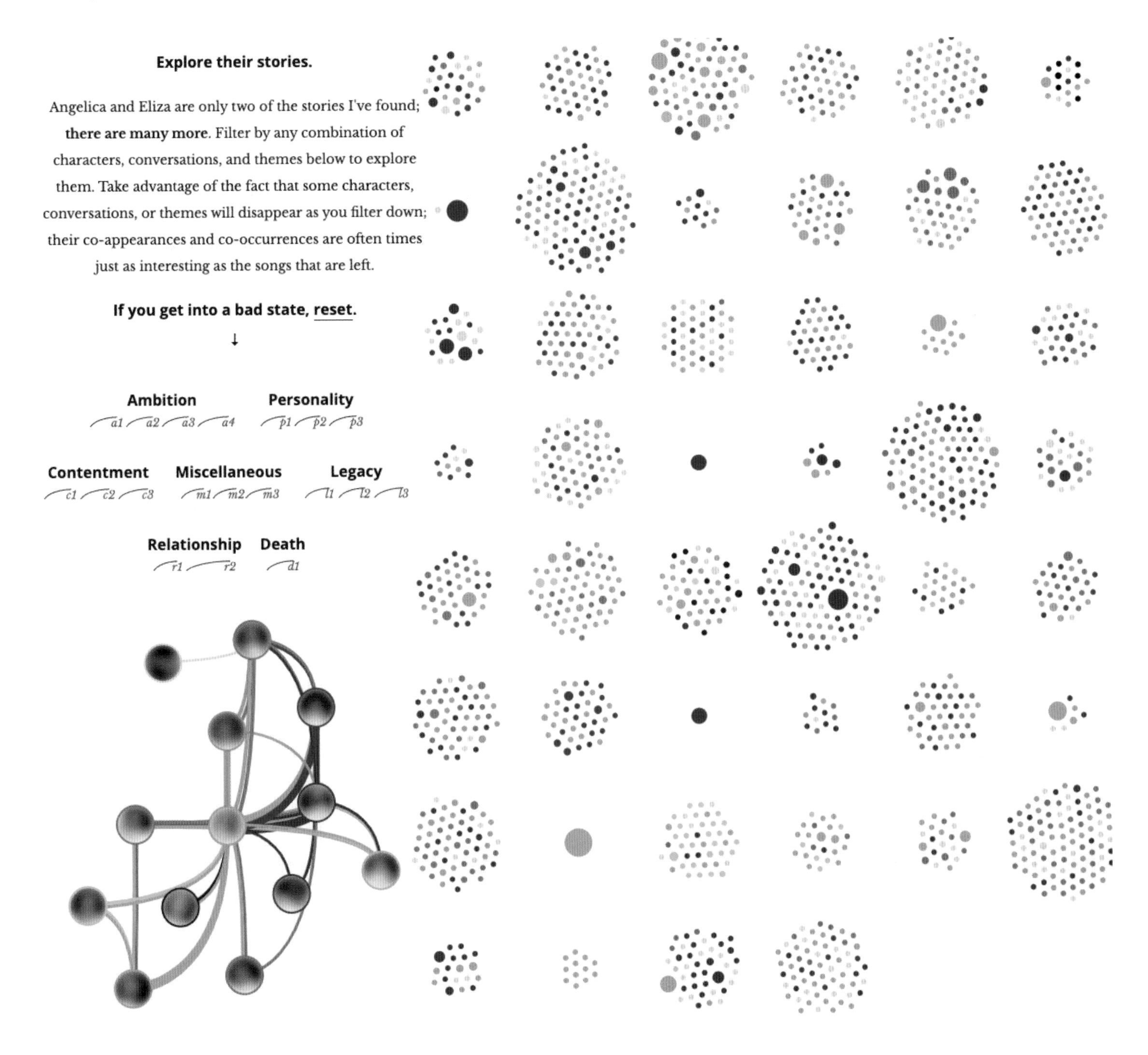

Explore their stories.
Angelica and Eliza are only two of the stories I've found; there are many more. Filter by any combination of characters, conversations, and themes below to explore them. Take advantage of the fact that some characters, conversations, or themes will disappear as you filter down; their co-appearances and co-occurrences are often times just as interesting as the songs that are left.
If you get into a bad state, reset.
↓
Ambition
a1 a2 a3 a4
Personality
p1 p2 p3
Contentment
c1 c2 c3
Miscellaneous
m1 m2 m3
Legacy
l1 l2 l3
Relationship
r1 r2
Death
d1

Thank you to **Matt+Pudding** for letting me work on this labor of love, **to Taia**, my ultimate Hamilton expert, for all the feedback and encouragement, and **to Alex**, my Eliza, who teaches me to *look around* and reminds me of *what would be enough*.

Made with 💕 by Shirley Wu for Pudding and Datasketch|es.

Songs filtered by Angelica and Alexander

→

Here's what the filter looks like.

A highlighted character or conversation means it's been selected. If it's grayed out, it can be clicked on to filter further. As you filter, **characters or conversations will disappear** because they are no longer in the songs left; you can find co-appearances this way 👍

Try playing with the filters.

(Here's a reset just in case.)

↓

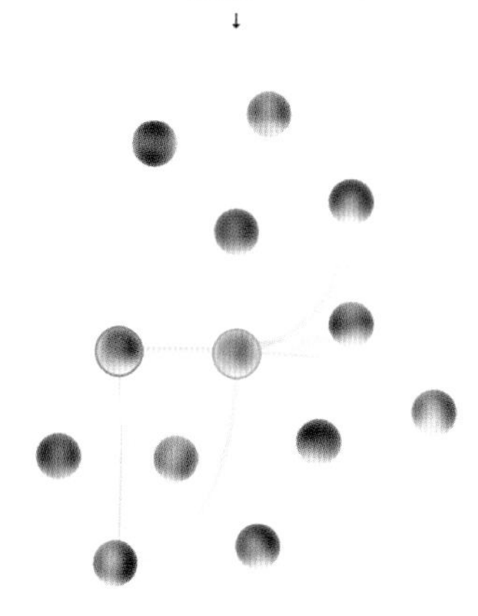

Fig.5.14 & 15

I'm really proud of the analysis I did, so I've included both Angelica and Eliza's story. One of my favorite little details (unfortunately not pictured) is that for every section of the stories I highlight the corresponding song and fade everything else in the visualization out.

An Ocean Away

Angelica Schuyler is Alexander Hamilton's sister-in-law, the one who introduces Alexander to her sister Eliza Schuyler. Angelica and Alexander's relationship is ambiguously flirtateous from the very start, their exchanges in *Satisfied* puntuated by mutual understanding - that they're both never satisfied.

Their flirtation is amplified in *Take A Break*, the only time in the whole musical they sing together:

▶ And there you are, an ocean away
Do you have to live an ocean away?
Thoughts of you subside
Then I get another letter
I cannot put the notion away...

They continue as Angelica visits from London:

▶ You won't be an ocean away
You will only be a moment away...

And they leave us wondering: did they, or didn't they?

I'm Standing At Her Side

The turning point in Angelica and Alexander's relationship comes in *The Reynolds Pamphlet*, after Alexander publishes the details of his affair with Maria Reynolds to save his political reputation. When Angelica hurries back from London, Alexander is relieved:

▶ Angelica, thank God
Someone who understands what I'm struggling here to do

Angelica instead replies:

▶ I know my sister like I know my own mind
You will never find anyone as trusting or as kind
I love my sister more than anything in this life
I will choose her happiness over mine every time
Put what we had aside
I'm standing at her side
You could never be satisfied
God, I hope you're satisfied

And here Angelica has matured; when she first introduces Eliza to Alexander in *Satisfied*, she does so because she knows Eliza is in love with him. She believes that Eliza (*you will never find anyone as trusting or as kind*) is a better match for Alexander. But most of all, she does so for herself, because she knows:

▶ He will never be satisfied, I will never be satisfied

After the Reynolds affair, she puts Eliza first, for Eliza's sake; she sings only that Alexander could never be satisfied.

She Takes His Hand

It's Quiet Uptown is the only song that starts with Angelica, as she and Alexander take turns narrating the aftermath of the Hamiltons losing their eldest son. Angelica watches over as Alexander finally puts Eliza first, as he tries to reach out to an unmoving Eliza, and as they reconcile; it is the most beautifully satisfying close to Angelica and Alexander's story.

Angelica does not re-appear with Alexander until his death in *The World Was Wide Enough*.

Curves above lines indicate recurring phrases.
→
Phrases are grouped into themes.
The songs are filtered by two recurring phrases of Contentment: "that would be enough" (c2) and "look around at how lucky we are to be alive right now" (c3). Many of the themes are blank since they don't co-occur with the filtered themes.
↓
Ambition
Personality
Contentment
Miscellaneous
Legacy
Relationship
Death
The Schuyler Sisters
Stay Alive
Meet Me Inside
That Would be Enough
Non-Stop
Take a Break
Stay Alive (Reprise)
It's Quiet Uptown
Best of Wives and Best of Women
Who Lives, Who Dies, Who Tells Your Story

Look Around

Eliza Schuyler is the second daughter of a wealthy New York family, and her upbringing has afforded her an idealistic outlook on life. When she meets Alexander, she lacks Angelica's understanding of Alexander's ambition, and she is helplessly in love.

That confident optimism is highlighted in *That Would Be Enough*, when Alexander is on leave from the war. Downtrodden that he may never be given command, he asks Eliza if she'll relish being a poor man's wife. She responds:

▶ I relish being your wife
Look around, look around...
Look at where you are
Look at where you started
The fact that you're alive is a miracle
Just stay alive, that would be enough

And she continues:

▶ We don't need a legacy
We don't need money
If I could grant you peace of mind
If you could let me inside your heart...
Oh, let me be a part of the narrative
In the story they will write someday

They're newly married with a child on the way, and Eliza knows exactly what she wants from him: for him to stay, and for her to be a part of his story.

They're Asking Me To Lead

The next time Eliza appears with Alexander is in *Non-Stop* after the war. Alexander works (non-stop) as a lawyer, is invited to the Constitutional Convention, and writes majority of the Federalist Papers. Eliza pleads with him:

▶ And if your wife could share a fraction of your time
If I could grant you peace of mind
Would that be enough?

Eliza's lines are similar to the ones she sung in *That Would Be Enough*, but the subtle changes highlight two things: Eliza is starting to realize the extent of Alexander's ambitions, and she is left unsure of her own role.

The most heartbreaking moment comes when George Washington asks Alexander to join his cabinet as Treasury Secretary, and Eliza instead asks Alexander to stay. Alexander responds with the very lines that Eliza uses to reassure him in *That Would Be Enough*, using them instead as his reason to leave:

▶ Look around, look around at how lucky we are to be alive right now

In return, Eliza sings only one word: *helpless*. It is the last time she sings *helpless* in the whole musical.

Forgiveness

When Eliza learns of Alexander's affair with Maria Reynolds, she burns their letters, determined to write herself out of the narrative. But when their eldest son Philip dies in a duel, she is grief-stricken, mute throughout *It's Quiet Uptown*, and Alexander tries to get through to her.

He again mirrors Eliza's lines from *That Would Be Enough*, but this time, he asks to stay:

▶ Look at where we are
Look at where we started
I know I don't deserve you, Eliza
But hear me out. That would be enough
If I could spare his life
If I could trade his life for mine
He'd be standing here right now
And you would smile, and that would be enough

There is a moment, and Eliza finally takes his hand and sings only one line: ***it's quiet uptown***. The music swells, and the Company asks:

▶ Forgiveness. Can you imagine?
Forgiveness. Can you imagine?
If you see him in the street, walking by her side, talking by her side, have pity
They are going through the unimaginable

And it's heartbreakingly beautiful as they reconcile, and their story comes around full circle; Alexander finally puts Eliza first.

Will They Tell My Story?

At the beginning of their marriage, Eliza tells Alexander that if he could just stay by her side, ***that would be enough***. As their marriage progresses and she realizes the extent of his ambition, Eliza starts to doubt herself, asking Alexander ***what would be enough*** - if she could be enough. But as they face the hardest of trials - an affair and the death of a child - their relationship reverses, and Alexander asks if he could stay by her side, ***that would be enough***.

As the musical closes with ***Who Lives, Who Dies, Who Tells Your Story***, Eliza comes into her own; after Alexander's death, she puts herself back in the narrative. She tells his story, his fellow soldiers' stories, Washington's story. She builds the first private orphanage in New York City - her proudest accomplishment. And when her time is up, she asks:

▶ Have *I* done enough?
Will they tell *my* story?

That subtle change in wording is amazing; Eliza is no longer concerned about ***what would be enough***, but rather if ***she*** has ***done*** enough. She is responsible for her own purpose, her own legacy. She is no longer a secondary character in her husband's story, but **the main character of her own**.

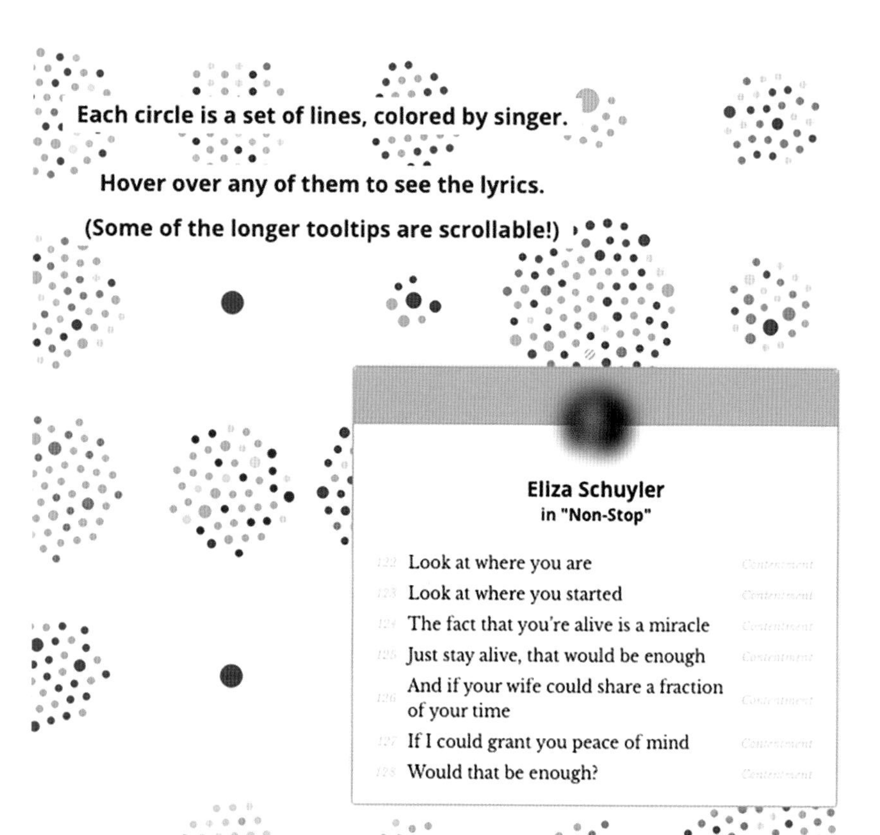
Each circle is a set of lines, colored by singer.
Hover over any of them to see the lyrics.
(Some of the longer tooltips are scrollable!)
Eliza Schuyler
in "Non-Stop"
Look at where you are
Look at where you started
The fact that you're alive is a miracle
Just stay alive, that would be enough
And if your wife could share a fraction of your time
If I could grant you peace of mind
Would that be enough?

Fig.5.16

Hovering on a dot shows the singer and the set of lyrics it represents.

Quiet Uptown, and Alexander tries to get through to her.

He again mirrors Eliza's lines from *That Would Be Enough*, but this time, he asks to stay:

Look at where we are
Look at where we started
I know I don't deserve you, Eliza
But hear me out. That would be enough
If I could spare his life
If I could trade his life for mine
He'd be standing here right now
And you would smile, and that would be enough

There is a moment, and Eliza finally takes his hand and sings only one line: ***it's quiet uptown***. The music swells, and the Company asks:

Forgiveness. Can you imagine?
Forgiveness. Can you imagine?

Fig.5.17

Probably my favorite little detail: clicking on the lyrics in the text actually plays the corresponding portion of the song and starts a progress bar that animates along with it.

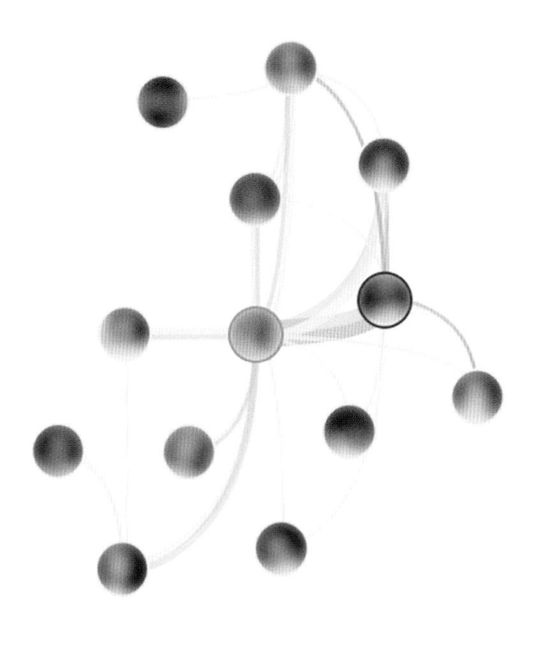

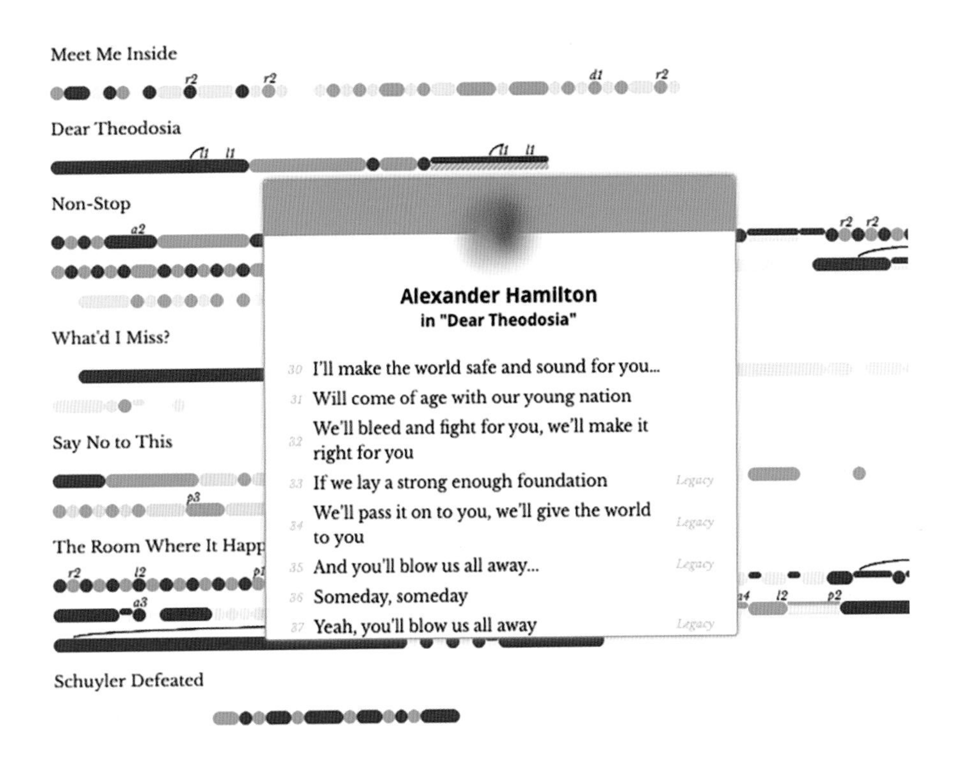

Meet Me Inside
Dear Theodosia
Non-Stop
What'd I Miss?
Say No to This
The Room Where It Happ
Schuyler Defeated
Alexander Hamilton
in "Dear Theodosia"
30 I'll make the world safe and sound for you...
31 Will come of age with our young nation
32 We'll bleed and fight for you, we'll make it right for you
33 If we lay a strong enough foundation
34 We'll pass it on to you, we'll give the world to you
35 And you'll blow us all away...
36 Someday, someday
37 Yeah, you'll blow us all away
Legacy

Fig.5.18

This is my favorite story that I don't share in my finished project: a set of filters that show the only lyrics that Hamilton and Burr ever sing together are in "Dear Theodosia". They disagree everywhere else in the musical, but in this one moment, they sing together in agreement that their children are their most precious legacies.

Fig.5.19

My favorite bug from this project courtesy of canvas!

Lin-Manuel Miranda
@Lin_Manuel
Following
Hello. Boy did @sxywu make us a gift. Hey @VAMNit check it.
Shirley Wu @sxywu
Excited to finally share "An Interactive Vis of Every Line in #Hamilton"! Filled with analysis & colorful #dataviz
polygraph.cool/hamilton/
GIF
RETWEETS
399
LIKES
3,107
12:21 AM - 14 Dec 2016
32
399
3.1K

Fig.5.20

Such a great birthday present ヾ(≧∪≦*)ノ〃!!

MU

SIC

Newly discovered

Although already released in 1972, *Starman* from David Bowie is the highest new song in the list. It never appeared in the previous 17 editions of the Top 2000 and entered in 2016 on position 270.

Prince

Another legend who passed away in 2016 (on April 21st). It seems that new people discovered his works, with all 9 songs that were in 2015's list rising significantly and 8 more songs joining in 2016.

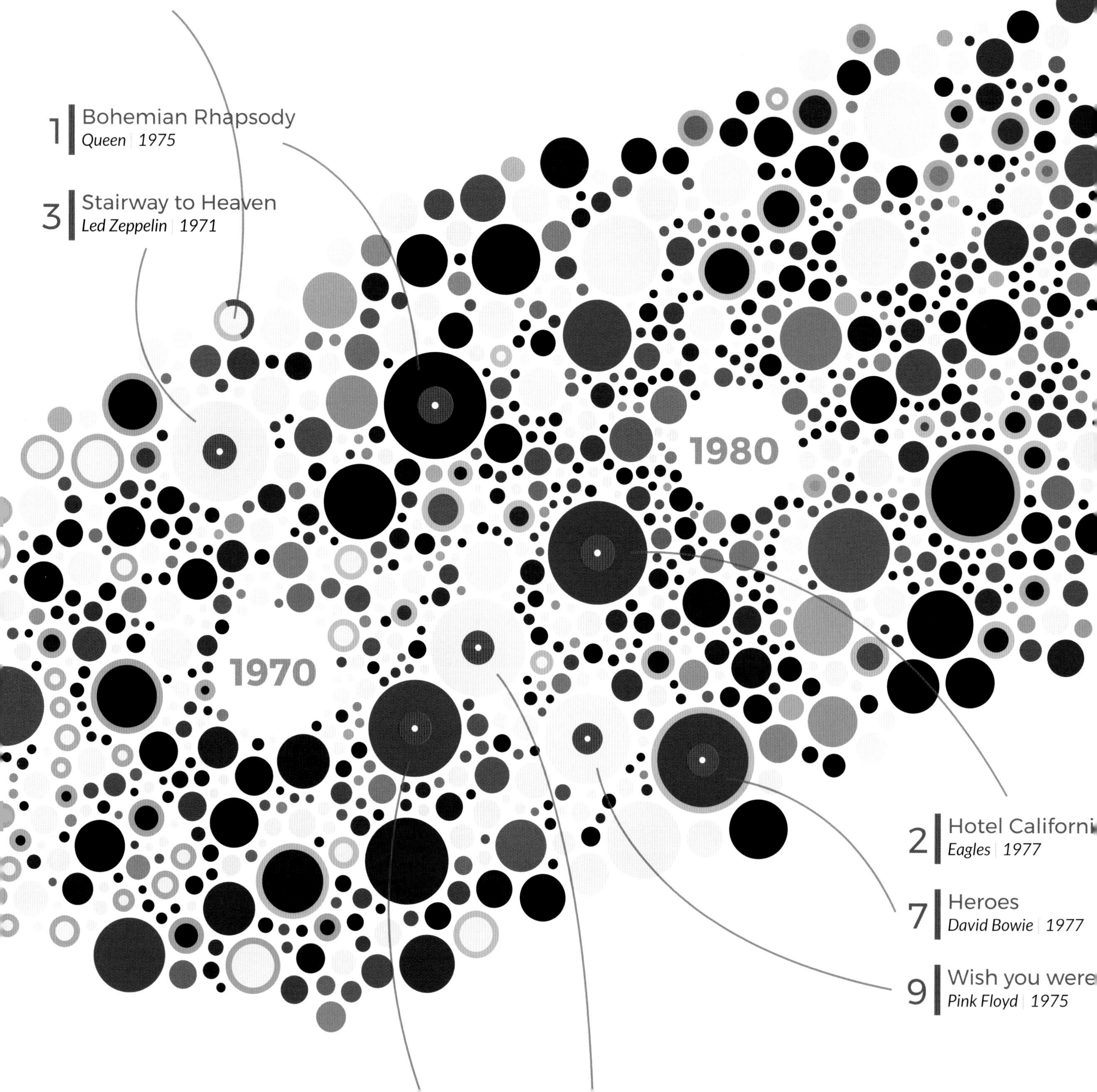

The Top 2000 ❤ the 70s & 80

NADIEH

When you say "Music in December" to somebody from the Netherlands, there's a very likely chance that they'll think of the Top 2000. The Top 2000 is an annual list of 2,000 songs, chosen by listeners, that airs on Dutch Radio NPO2 and is played between Christmas and New Year's Eve. I actually played with this data before in 2014, when I was still very new to `D3.js` (Figure 6.1).

As artists sometimes revisit a past artwork to see how their style has evolved (which I always love to see), I thought it would be fitting to try and do the same myself. So, two years after my first attempt, I decided to look at the Top 2000 songs again and visualize which decade was the most popular in terms of the years the songs were released.

Not to worry if you're not Dutch; roughly 90% of the songs in the Top 2000 are sung in English (with Queen usually in first place) so the songs should seem familiar.

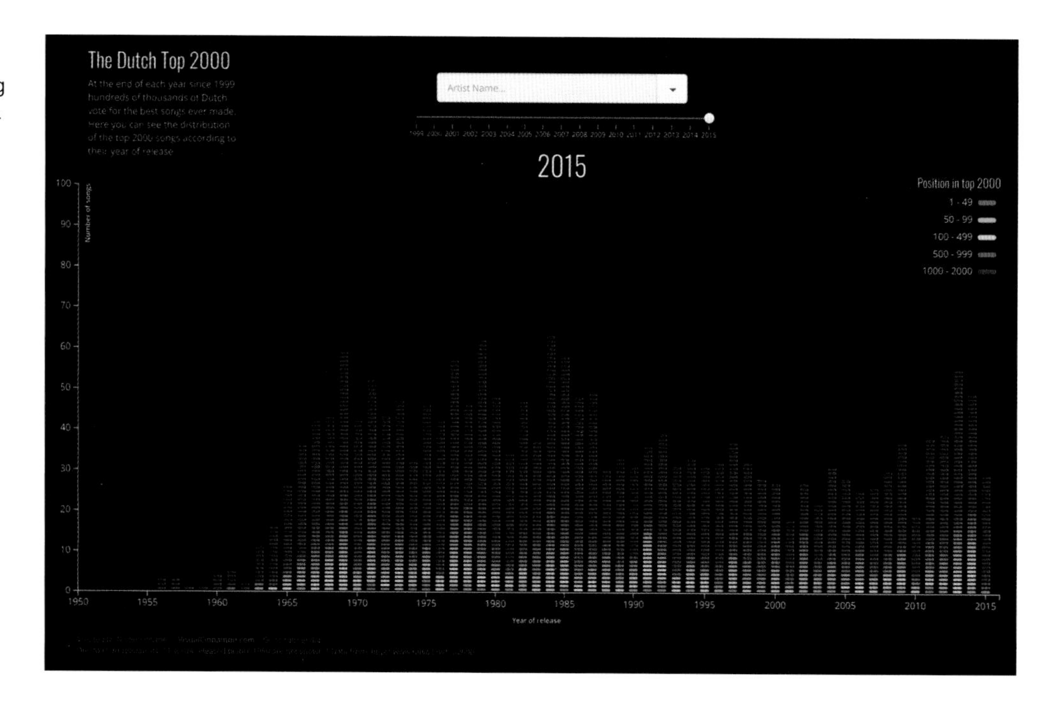

Fig.6.1

The result of visualizing the Top 2000 data in a 2014 personal project made with `D3.js`.

Data

Thankfully, the Top 2000 website publishes an Excel file of the 2,000 songs for that year, containing the song title, artist name, and year of release. On December 19, 2016, that file was released. However, I wanted another important variable—the highest rank ever reached in the normal weekly charts—because adding this to the visual would provide more context on the songs. There are a few of these in the Netherlands, and I eventually went with the Top 40 chart because this had been going non-stop since 1965 and because the Top 40 website seemed scrapeable. Next, I wrote a small scraper in R that would go through the ±50 years of music chart lists and save the artist's name, song title, song URL (as the unique ID), and chart position. This data was then aggregated to make it unique per song. I also saved some extra information per song, such as the highest position ever reached and number of weeks in the Top 40.

The Top 40 chart URLs had a year/week logic to them: www.top40.nl/top40/2015/week-42

The tricky part was to match the artists and songs from the Top 40 list I created to those in the Top 2000 list. I first tried a merge on an exact match of artist and title. That matched about 60% of the songs between both lists. Not bad, but I had actually expected more matches, thinking that artist and song names were rather fixed.

Browsing through the remaining songs, I noticed that one of the two lists would sometimes use more words than the other, such as "John Lennon-Plastic Ono Band" versus just "John Lennon." Therefore, I also searched for partial matches between the lists, as long as all the words of one song and artist were contained in the other list. That helped match 10% more.

Next came the fuzzy part. Sometimes words were apparently written slightly different, such as "Don't Stop '*Til* You Get Enough" versus "Don't Stop '*Till* You Get Enough." Using R's `stringdist` package, I applied the Full Damerau-Levenshtein distance to compare titles and artists. However, I was quite strict; only two changes were allowed on both the title and artist to create a match (otherwise, the song "Bad" from U2 could be turned into any other three-letter song title or 2–3 letter artist name). Sadly, that only gave me 2.5% more matches, and I manually checked all the matched songs after each step to correct a handful of wrong matches.

The Full Damerau-Levenshtein distance counts the number of deletions, insertions, substitutions, and transposition of adjacent characters necessary to turn string "a" into string "b."

I also tried something with the "Tips of the Week" list to check against, searching for songs that were tipped but that never made the Top 40, which gave me a few more matches. For the remaining songs I manually went through each list searching for artists or song titles with variations in how they were spelled, such as "Andrea Bocelli & Sarah Brightman" in the Top 2000 list versus "Sarah Brightman & Andrea Bocelli" in the Top 40. For the remaining 380 songs, I wasn't able to find exactly how many actually appeared in the Top 40, but after all the data processing I did along the way, I'd guess it's less than 10%.

Data

The idea for visualizing this particular dataset had been in the works for some time. During the spring, I attended a very interesting data visualization workshop given by Juan Velasco on "Information Graphics for Print and Online." Part of the workshop was to come up with an idea for an infographic. And although my small team of three people came up with 40 possible ideas, we were all intrigued by the Top 2000 songs.

We decided to have the most recent list of 2,000 songs take center stage and visualized them in a "beeswarm" manner that grouped them around their year of release. Each circle (i.e., a song) would be sized according to their highest position in the Top 40 and colored according to their rank in the Top 2000. Some of these songs would then be highlighted with color and annotations, such as "highest newcomer in the Top 2000 list."

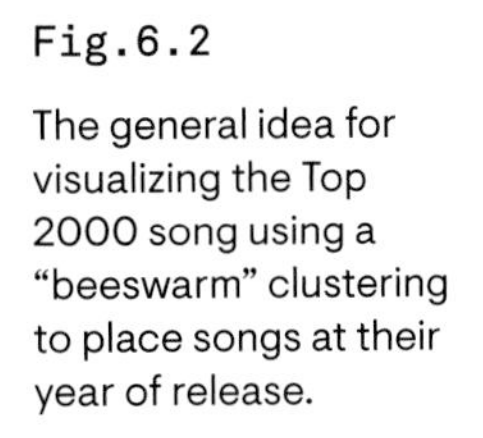

Fig.6.2

The general idea for visualizing the Top 2000 song using a "beeswarm" clustering to place songs at their year of release.

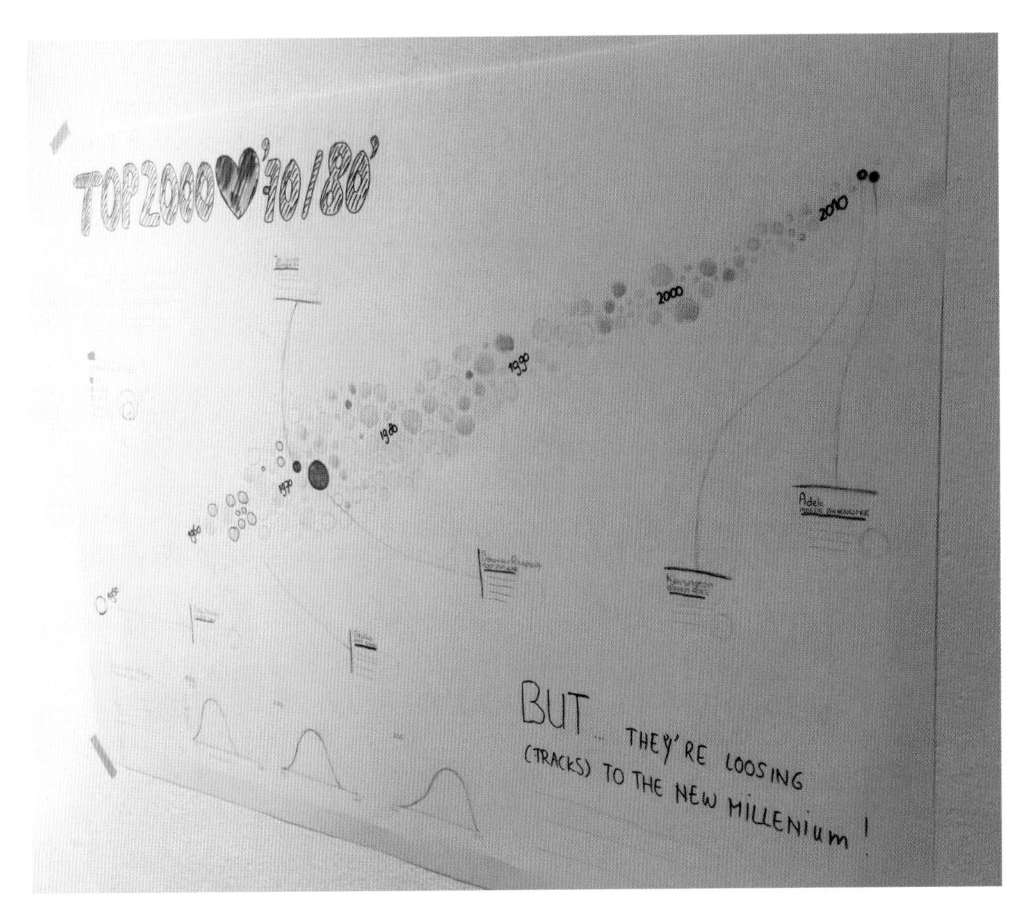

↳ Add Context Using Remaining Visual Channels

Even if getting the main insights from your data across to your audience is of the utmost importance, try to keep an open mind by adding extra details to create additional context about the information that you want to convey. This can create a more visually pleasing result, while also giving the truly interested reader even more ways to dive into and understand the information.

A way for me to think about adding extra details is to think about which "visual channels" are still free after I have the main chart standing; visual channels being those components of a data visualization that can be used to encode data, such as position, color, and size. For example, with the Top 2000 infographic during the workshop, our team knew that we wanted to use a beeswarm clustering to place all the song circles near their year of release. This would define the main visual's shape and also answer the original "Which decade is most popular in terms of song release year?" question. And while size and color are pretty common visual encodings to use with data, there are so many more visual channels that make it more interesting!

In terms of *remaining* visual channels, we chose to use a colored stroke to highlight the "interesting" songs (such as the highest riser, newcomer, or the Pokémon song), which we also annotated with text.

Finally, in the bottom section we decided to place some mini-charts that highlighted the distribution of the songs (arranged by release year) that were featured in the 1999, 2008, and 2016 editions of the Top 2000. These would highlight the fact that the bulk of the 2,000 songs from the 1999 edition of the Top 2000 were released in the 70s, but that this has slowly been moving towards newer decades for every new edition of the Top 2000.

On the second day of the workshop we also made a mobile version of the concept. This time we thought of creating a long scrollable beeswarm visual where you could theoretically listen to bits of each song and see extra information (Figure 6.3).

Code

This time I finally made something primarily static: a poster. Nevertheless, I still had to use `D3.js` to build the beeswarm centerpiece. In this case I needed a *force* along the horizontal x-axis that would cluster the songs based on their year of release, starting from the 1950s at the left all the way to 2016 at the other end. It took me several iterations to figure out the right balance of settings before it filled the region nicely around the horizontal axis, without the songs being moved away too far from their actual release year.

This is very similar to what I talked about in my "Royal Constellations" project when I used the birth date to pull the network apart along one axis.

Fig.6.3

For the mobile version we converted the poster to a very thin and long beeswarm where you could listen to small snippets of each song.

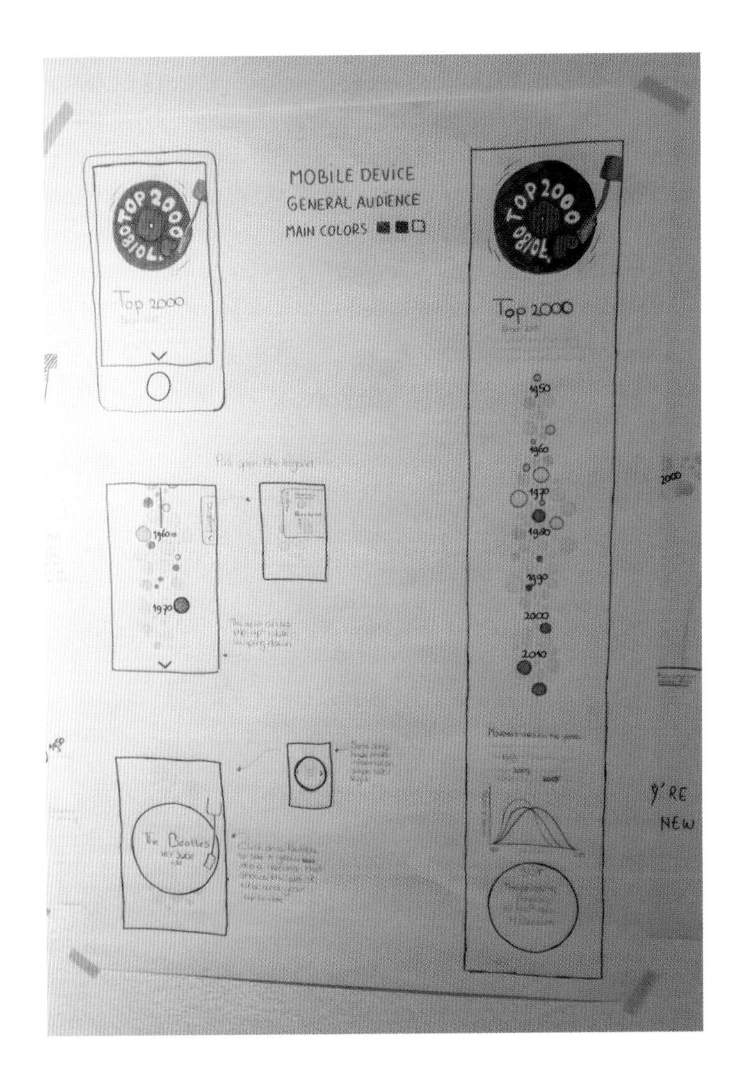

In my first attempts, the circles were *sized* according to the highest position they had reached in the Top 40 and *colored* according to their position in the Top 2000. This is what we'd come up with during the workshop. But this gave some difficulty in songs that never appeared in the Top 40; I still needed a size for these. I therefore made the "unmatched" all the same size, but that resulted in many light grey circles of about the same size and it didn't look appealing.

Fig.6.4

Using Top 40 information for circle size and Top 2000 ranking for color didn't create an appealing image.

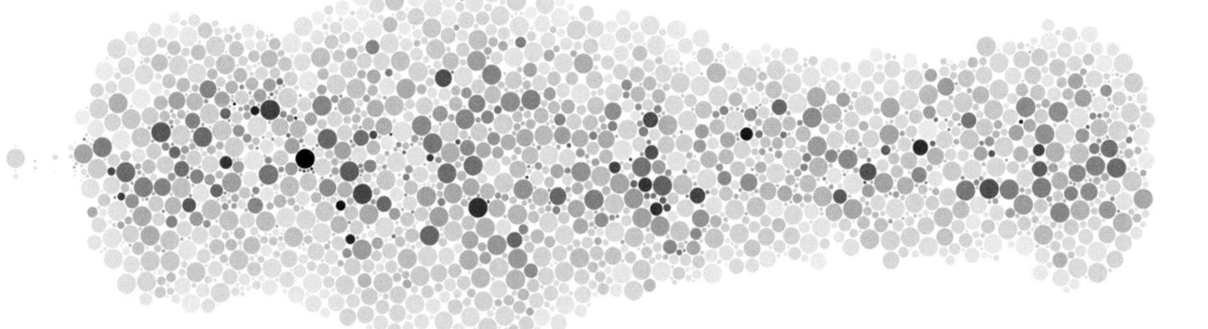

In hindsight, this choice of size and color wasn't optimal in another way as well. My visual was about the Top 2000, but I was using circle size for the "nice-to-have" variable of the Top 40 position. It made a lot more sense to switch the two; the Top 2000 rank determined the circle size and the Top 40 rank became a color. Now, the biggest circles—which were easier to locate—actually represented the highest ranking songs. And I could use a very light grey (a color more often used for "missing" data) for songs that never appeared in the Top 40. Plus, it immediately gave a great visual improvement and only needed a few tweaks to the code.

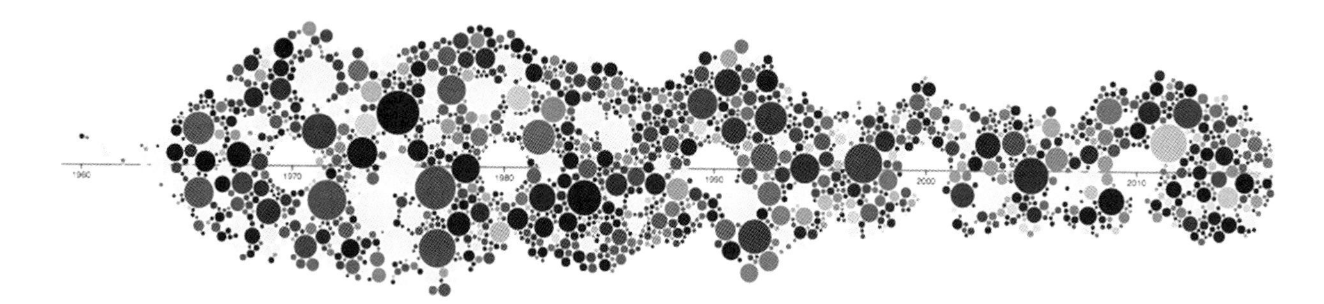

Fig.6.5

Switching the scales for the Top 2000 and Top 40 rankings immediately made the visual more visually appealing and effective.

I then started to mark out the circles (songs) that I wanted to annotate later. Already during the workshop we decided to keep the visual very black and white, inspired by the intense blackness of vinyl records. Using red, the color of the Top 2000 logo, to mark songs that had something interesting about them and blue for the artist/band with most songs in the list. I also highlighted all the songs by David Bowie and Prince, who passed away in 2016, by adding yellow and purple strokes around their songs, respectively.

Since the top 10 songs from the list were the biggest circles, I thought it would look nice to mark these as small vinyl records to make them stand out even more.

The "vinyls" are nothing more than a very small white circle on top of a small red circle.

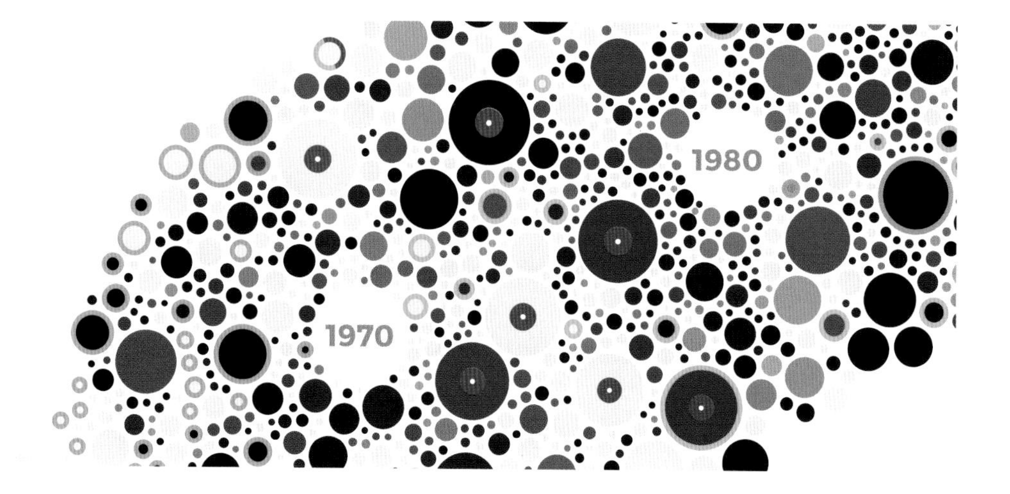

Fig.6.6

Using colored strokes to mark certain artists, and turning the top 10 songs into tiny vinyl records with a red and white circle on top.

↳ Design to Maximize for Delight

Those top 10 songs didn't have to look like tiny vinyl records to make them stand out. However, by adding a touch that fits with the topic that's visualized, it made the total visual just a little bit more fun to look at.

Outside Strokes with SVG

Although possible in certain vector drawing programs, such as Adobe Illustrator, you cannot do an *outside* stroke on SVG elements, such as circles or rectangles, in the browser. Thus when you stroke an element, the width of that stroke is centered on the outline of the element. However, for data visualizations (and especially for smaller circles) it's quite important that part of the circle's area isn't "hidden" behind a stroke.

Thankfully, an outside stroke can easily be mimicked; plot a circle in the color that you have in mind for the stroke. The radius of this circle should be just as big as your "actual" circle *plus* how wide you want the stroke to be. Next plot your actual circle on top and it will look like the background circle is an outside stroke (Figure 6.7).

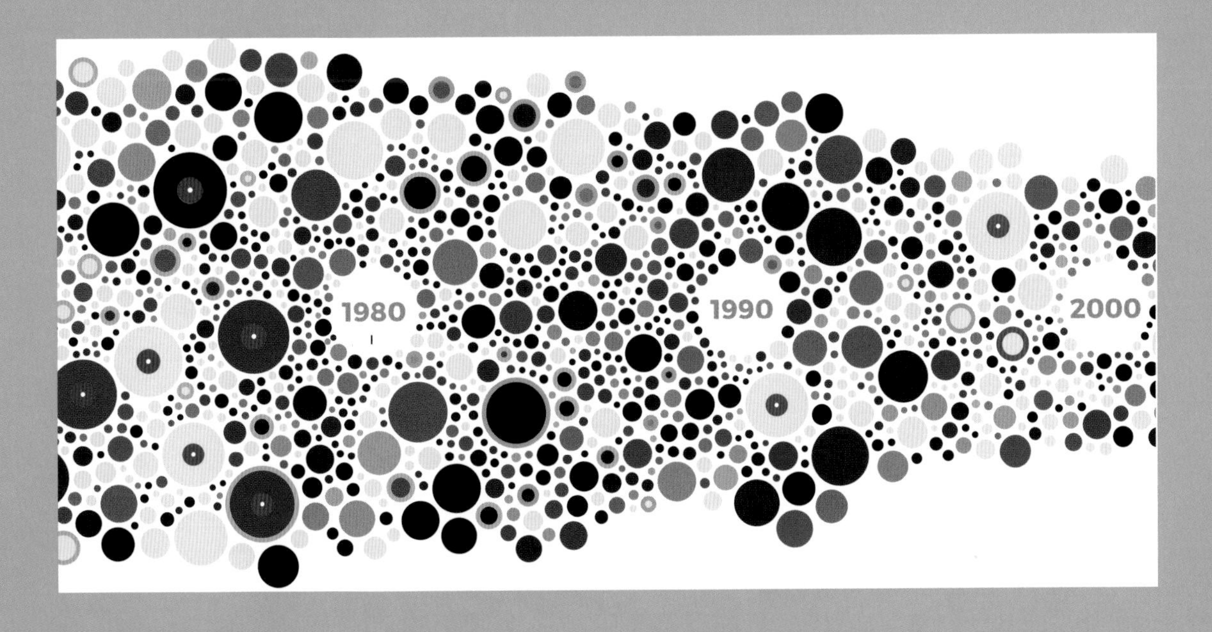

Fig.6.7

The colored strokes are colored circles behind the grey circles that are just a little bigger.

With those relatively simple elements done and being sure I wouldn't change anything anymore, I copied the SVG element of the beeswarm from my browser and pasted it into Adobe Illustrator. There I started adjusting it to look like the poster that our little group had made during the workshop (from Figure 6.2). Such as turning the beeswarm 25 degrees, just for the effect of making it look a bit more interesting, and placing annotations around it. For the red "notable" songs I used the data itself together with the Top 2000 website to search for some interesting facts, like Justin Timberlake having the highest ranking song from 2016. I placed these texts using an underlying grid to keep things nicely aligned in columns and rows (Figure 6.8). After finishing the beeswarm/top part of the infographic, I capitulated on keeping this visual totally static and made a small interactive version online just to be able to hover over each circle and see which song it is (Figure 6.9).

You can either use the free "NYT SVG Crowbar" tool, or literally copy the SVG element from the Chrome devTools into Illustrator.

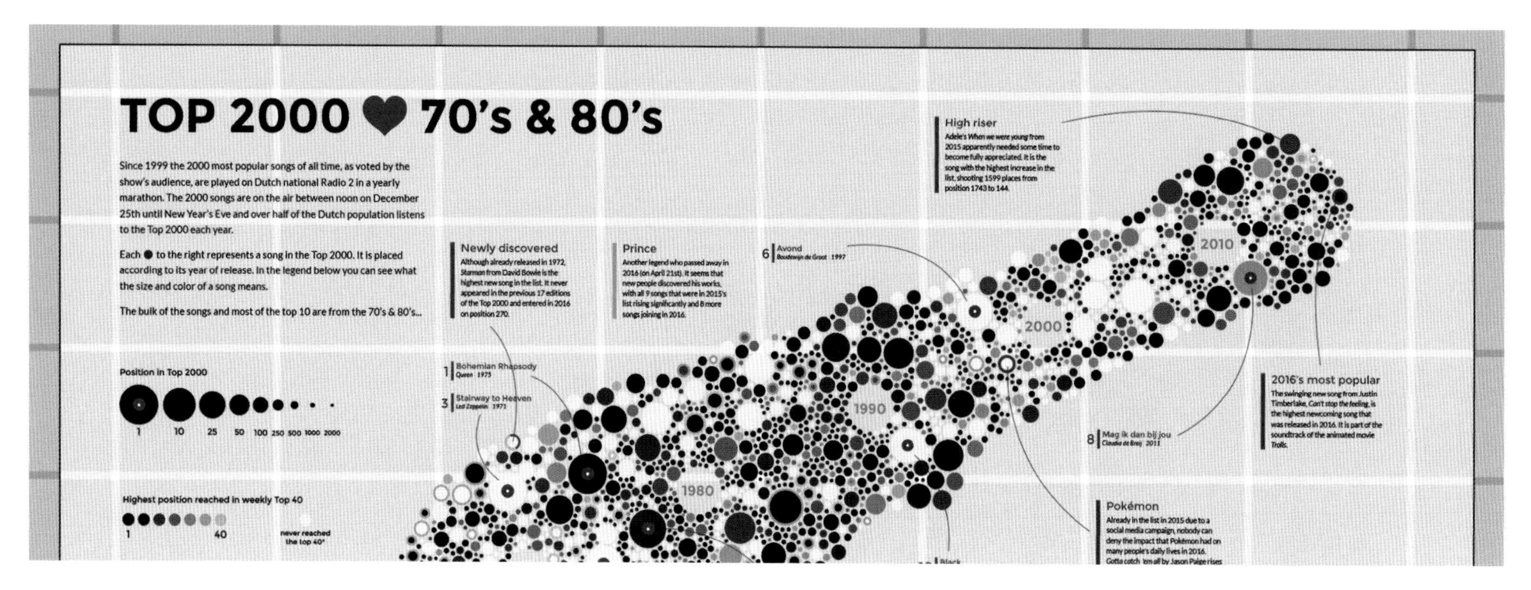

Fig.6.8

The underlying grid that I used in Illustrator to lay out all the text.

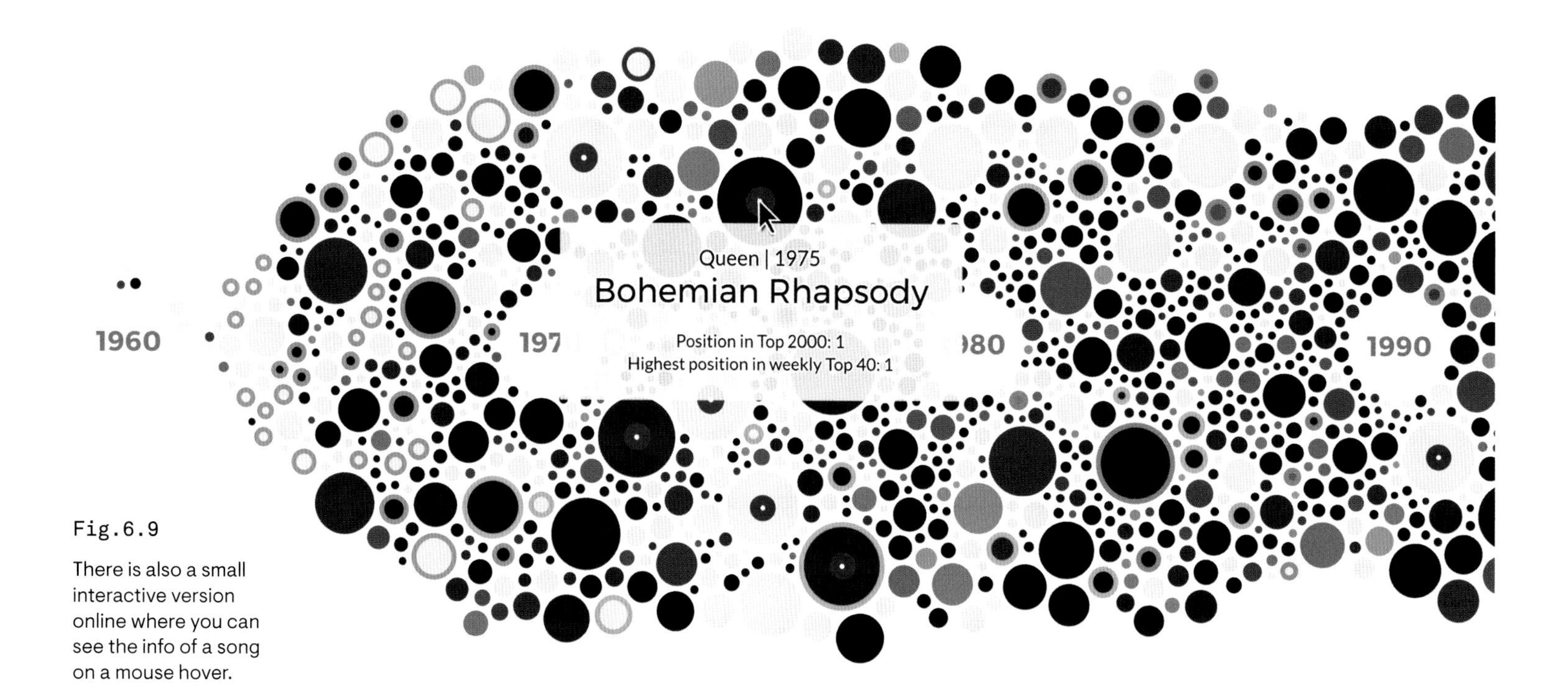

Fig.6.9

There is also a small interactive version online where you can see the info of a song on a mouse hover.

Finally, I wanted to incorporate the fact that the distribution of the songs across release year has been changing towards the 90s and 2000s. I had mentioned this fact to my team during the workshop, and we'd placed three simple line charts in the lower left of our design to highlight it (Figure 6.2). But now that I wanted to *actually* create the charts, it wasn't quite clear what visual form would convey the idea best. I already had the full history of every Top 2000 since the first one aired in 1999 from my previous visual on the topic from two years ago. I appended the 2016 data and started making some simple plots in R using `ggplot2`. That it should probably be a histogram or something similar was clear to me from the start, but should I smooth it down? How many years to show? Should they overlap or be displayed as "small multiples"? (Figure 6.10).

In the end I chose to go with a small multiple histogram of four editions picked from the past 18 years, but overplotted with a smoothed density curve to make the general shape more easily comparable between the four charts.

In Figure 6.11 you can see what I took straight from R. I played with the color to also encode the height. Eventually, however, I made them all the same grey on the poster, since I didn't want the histograms to draw too much attention.

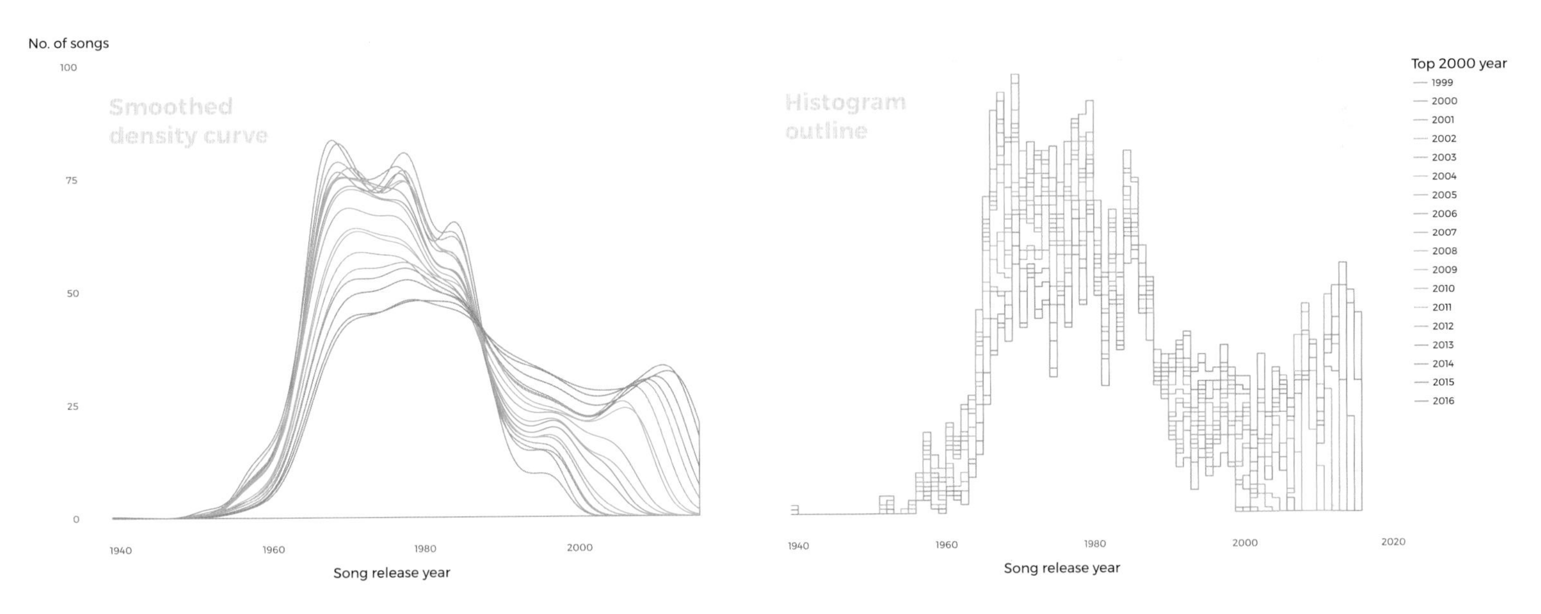

Fig.6.10

Comparing the trend of song release year across all 18 editions of the Top 2000.

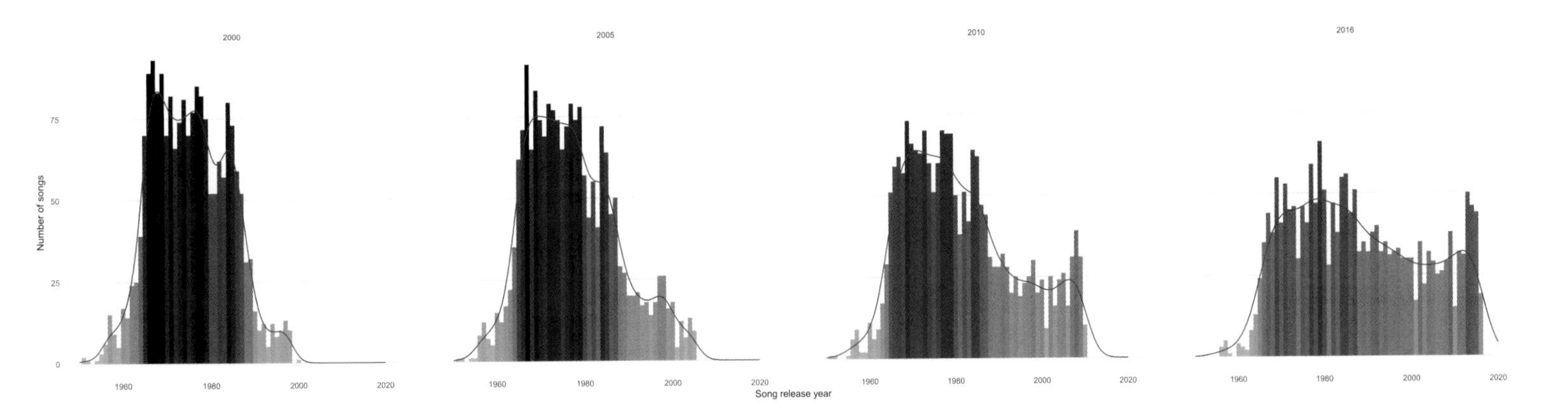

Fig.6.11

Four histograms from four different years of the Top 2000 showing the distribution of song release year and its steady move towards more recent decades.

↳ Combine Tools

It's perfectly fine if you create your visual with several different tools. Just like people, finding a unicorn tool that can do everything you need "the best way every time" doesn't exist. Personally, I always start with R; I use it to read in the data, do some initial statistics, clean and transform the data, and make (lots of) simple plots to get a sense of the data. However, although R's `ggplot2` package can do a lot to visualize data, I need more creative control for the final result, and it often has to be interactive and on the web. I find that creating my visuals in JavaScript with the help of `D3.js` provides what I need.

For static visuals, I always end with Adobe Illustrator, or other vector-based tools, to add some final touches (such as legends and annotations—things that can take a lot of time to add with code). Of course, not all projects lend themselves to using your tools of choice, but just keep in mind that you don't need to always create a visual from scratch with *one tool*. Try to learn more tools and programs and acquaint yourself with the strengths and weaknesses of each.

Reflections

One of the main reasons why I decided to create a poster instead of an interactive visual came down to other time commitments. Making a static visualization is always *much* faster for me to create, even if it's partly based on something that I started in `D3.js`. For static visuals you just don't have to consider browser bugs, performance, responsive design, *and* interactivity!

Preparing and working on the data scraping and cleaning took about 20 hours of time, the ideation and sketching about three, and the coding/creating approximately 20–30 hours. Finishing a static visual after creating so many interactive ones always reminds me how much I like making printable static visuals.

A funny thing happened though. I generally share my (personal) projects only on Twitter, maybe Instagram if I'm in a good mood. But I decided to also post this infographic on my LinkedIn account, due to most of my connections there being Dutch at the time. And somehow, it got picked up beyond my direct connections and went a little viral in the Netherlands. Totally unexpected! I thought this visual was too complex for that. Majorly exciting to experience though! A few hundred thousand views and thousands of likes later, I even had some businesses reach out to me to create a data visualization for them. Which was *perfect* timing, since I was officially a freelancer from the moment the Top 2000 finished airing the number 1 song at the very start of 2017.

The Top 2000 ❤ the 70s & 80

↳ Top2000.VisualCinnamon.com

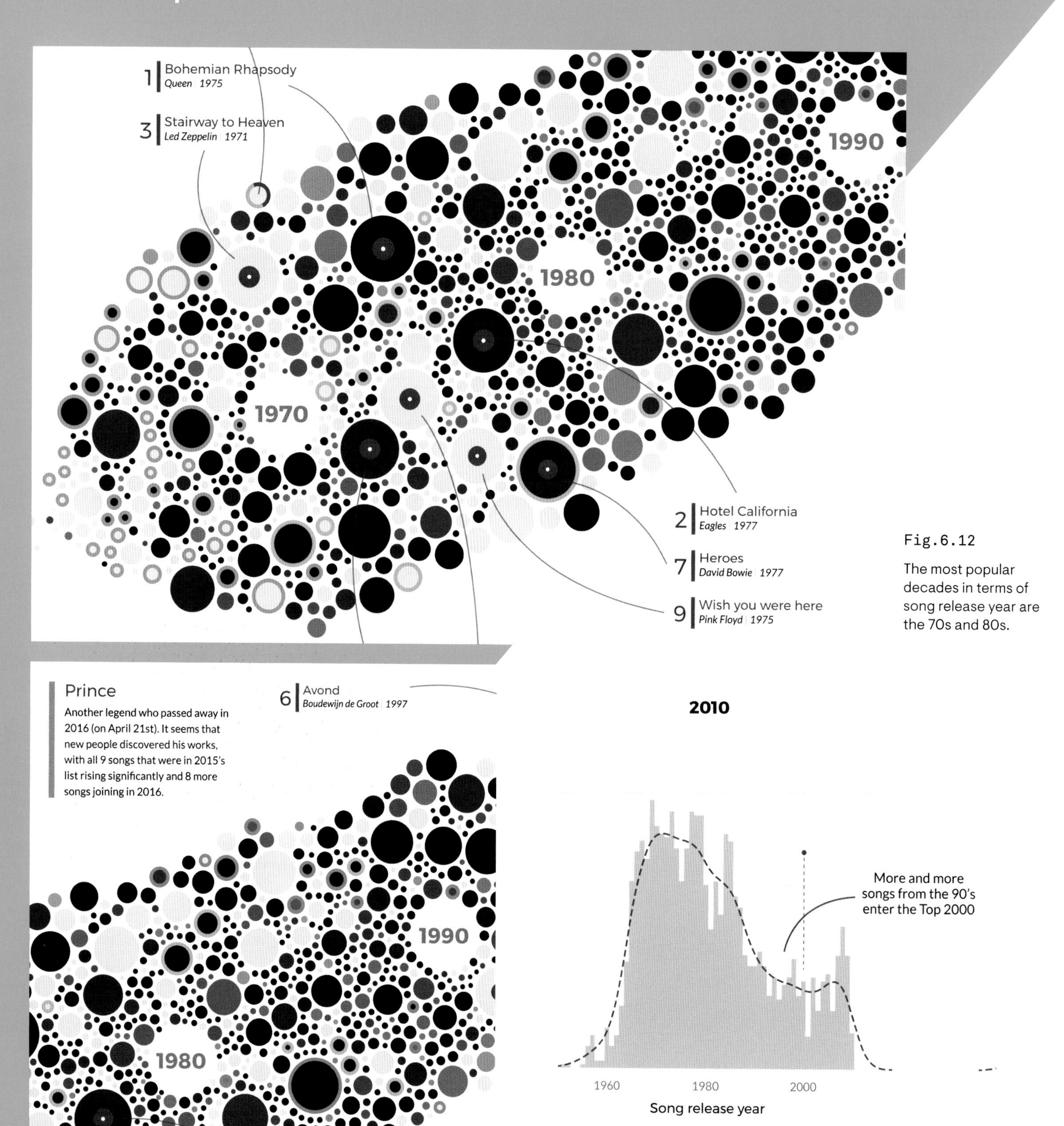

Fig.6.12

The most popular decades in terms of song release year are the 70s and 80s.

Fig.6.13

Highlighting all of Prince's songs in the list with a purple stroke.

Fig.6.14

With each new Top 2000, more songs from the 90s enter the list, even though the first edition was aired in 1999.

Fig.6.15

TOP 2000 ❤ 70's & 80's

Since 1999 the 2000 most popular songs of all time, as voted by the show's audience, are played on Dutch national Radio 2 in a yearly marathon. The 2000 songs are on the air between noon on December 25th until New Year's Eve and over half of the Dutch population listens to the Top 2000 each year.

Each ● to the right represents a song in the Top 2000. It is placed according to its year of release. In the legend below you can see what the size and color of a song means.

The bulk of the songs and most of the top 10 are from the 70's & 80's...

Position in Top 2000

1 10 25 50 100 250 500 1000 2000

Highest position reached in weekly Top 40

1 40 never reached the top 40*

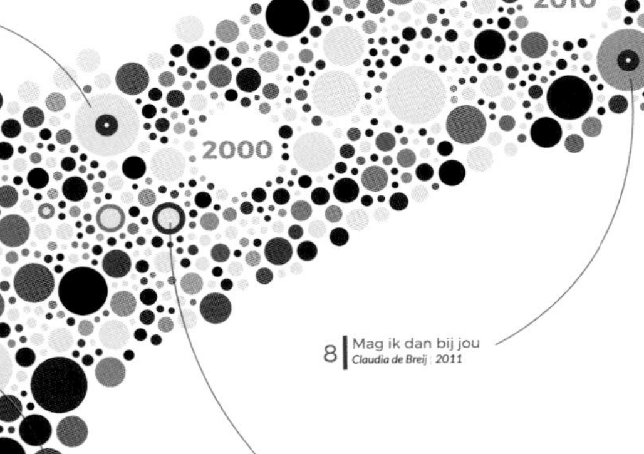

Golden oldie
The oldest song in the list, Billie Holiday's *Strange Fruit*, is from 1939. It's 17 years older than the second-oldest song. If it will make the 2017 edition remains to be seen, it's barely in now, on position 1989.

Newly discovered
Although already released in 1972, *Starman* from David Bowie is the highest new song in the list. It never appeared in the previous 17 editions of the Top 2000 and entered in 2016 on position 270.

Prince
Another legend who passed away in 2016 (on April 21st). It seems that new people discovered his works, with all 9 songs that were in 2015's list rising significantly and 8 more songs joining in 2016.

1 | Bohemian Rhapsody — *Queen* | *1975*

3 | Stairway to Heaven — *Led Zeppelin* | *1971*

6 | Avond — *Boudewijn de Groot* | *1997*

High riser
Adele's *When we were young* from 2015 apparently needed some time to become fully appreciated. It is the song with the highest increase in the list, shooting 1599 places from position 1743 to 144.

2016's most popular
The swinging new song from Justin Timberlake, *Can't stop the feeling*, is the highest newcoming song that was released in 2016. It is part of the soundtrack of the animated movie *Trolls*.

8 | Mag ik dan bij jou — *Claudia de Breij* | *2011*

Pokémon
Already in the list in 2015 due to a social media campaign, nobody can deny the impact that Pokémon had on many people's daily lives in 2016. *Gotta catch 'em all* by Jason Paige rises 1434 spots to position 232!

10 | Black — *Pearl Jam* | *1991*

2 | Hotel California — *Eagles* | *1977*

7 | Heroes — *David Bowie* | *1977*

9 | Wish you were here — *Pink Floyd* | *1975*

David Bowie
Passing away only days after the release of his new album *Blackstar* on January 10th 2016. His legend remains strong with 26 songs in the Top 2000. His most popular song *Heroes* jumps from 34 to position 7.

The Beatles
No other artist or band has more songs in the Top 2000 as the Beatles. With 38 songs they are responsible for 14% of all titles before 1970. Nonetheless, only 5 years ago they still had 50 songs in the list.

4 | Piano Man — *Billy Joel* | *1974*

5 | Child in Time — *Deep Purple* | *1972*

Spread across release years of the 2000 songs

For 4 editions of the Top 2000

The charts on the right represent all 2000 songs from 3 past editions of the Top 2000 (held in 2000, 2005, 2010) and the most recent 2016 edition.

The songs are stacked according to their year of release. The higher a rectangle, the more songs that are in the Top 2000 list from that release year.

The black dotted line represents a smoothed curve over all 2000 songs. This makes the comparison between the 4 charts easier.

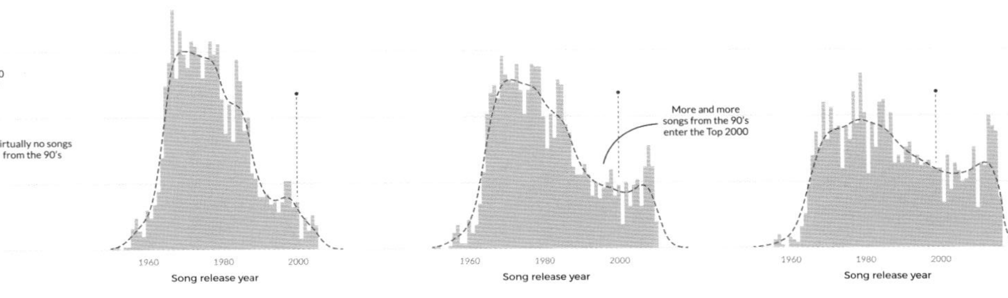

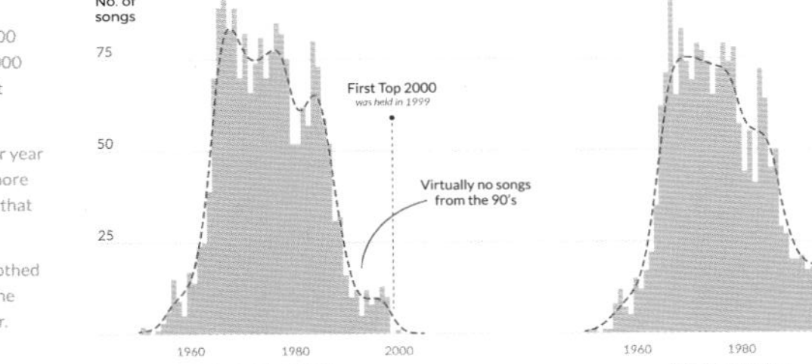

But they're losing tracks to the new Millenium

It makes sense that the Top 2000 will be more spread out for each new edition, since there are more songs to choose from. However, if we compare the distributions of the Top 2000 songs over 4 editions, we see that, especially, the 90's has been gaining a lot of popularity.

Even though all songs from the 90's were out in the 2000 edition, only a few songs from that decade were chosen. Whereas in the 2016 edition the number of songs from the 90's has risen significantly. This could be due to a new generation who has grown up during the 90's taking over from those who voted in the early 2000's (who apperantly didn't appreciate the new music).

Created by Nadieh Bremer | VisualCinnamon.com for the December edition of data sketch|es

Visit tinyurl.com/2016top2000 for the interactive visual and see the name & title of each song

Data | Top 2000 list from Radio 2 | Top 40 info from Mediamarkt's Top 40

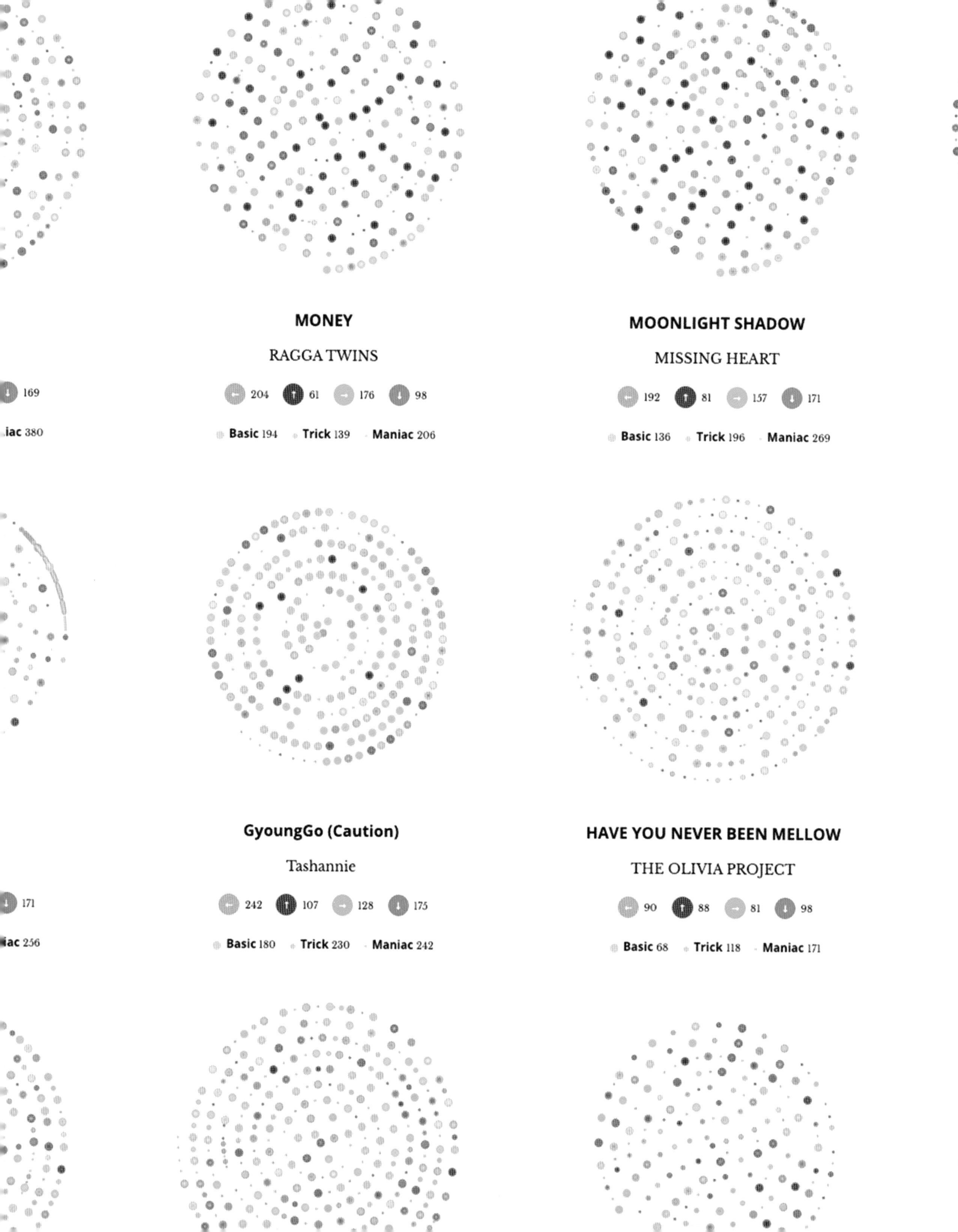

169
iac 380
MONEY
RAGGA TWINS
204 61 176 98
Basic 194 Trick 139 Maniac 206
MOONLIGHT SHADOW
MISSING HEART
192 81 157 171
Basic 136 Trick 196 Maniac 269
171
iac 256
GyoungGo (Caution)
Tashannie
242 107 128 175
Basic 180 Trick 230 Maniac 242
HAVE YOU NEVER BEEN MELLOW
THE OLIVIA PROJECT
90 88 81 98
Basic 68 Trick 118 Maniac 171

Data-Driven Revolutions

SHIRLEY

When Nadieh and I first agreed on the topic "Music," I felt quite lost. I didn't know what I wanted to do, except to perhaps explore something related to K-Pop. But K-Pop was just too broad of a topic and I didn't know it well enough (anymore) to find a good angle. I was lamenting this lack of an angle with a friend when I suddenly remembered the game *Dance Dance Revolution* (DDR).

DDR originally started as an arcade game in Japan and eventually was released as video games for home consoles. Basically, the player would stand on a mat or a platform that served as the "controller" and step on any of the four arrows on the mat to "press" the controller buttons. A combination of arrows would appear on the TV screen in front of them, and the players would have to step on (or stomp on) the same arrow(s) on the mat, timing their steps with the arrows scrolling up the screen. It was a popular game across the world and a huge part of my teenage life.

I first came across DDR in 2001 at a friend's house. Since there were no arcades with DDR anywhere near where I lived, I begged my parents to buy me the game. (This was easier said than done; there were absolutely *no* video games in our house at the time, so a DDR set meant not only buying the game itself, but also the PS2 and the mat that came along with it.) It took me two years to convince my parents, and the summer I got it, I was on it every day for hours. I was the type that played the same song on the same difficulty over and over until I mastered it, and I played it regularly until I left for college five years later.

Data

With hundreds (maybe thousands) of songs, each with at least three difficulty levels, and each level comprised of songs with different beats per minute (BPMs) and "Groove Radar" values ("Stream," "Voltage," "Air," "Freeze," and "Chaos"—essentially the characteristics of the step choreographies), there was *a lot* of data. But never in my wildest dreams could I have imagined that I would find *all the steps from 645 songs*. DDR Freak is a website dedicated to DDR and provides (among many other things) "step charts": PNG images with every step of a song laid out, so that fans could practice the song without having to pay for time on the machine:

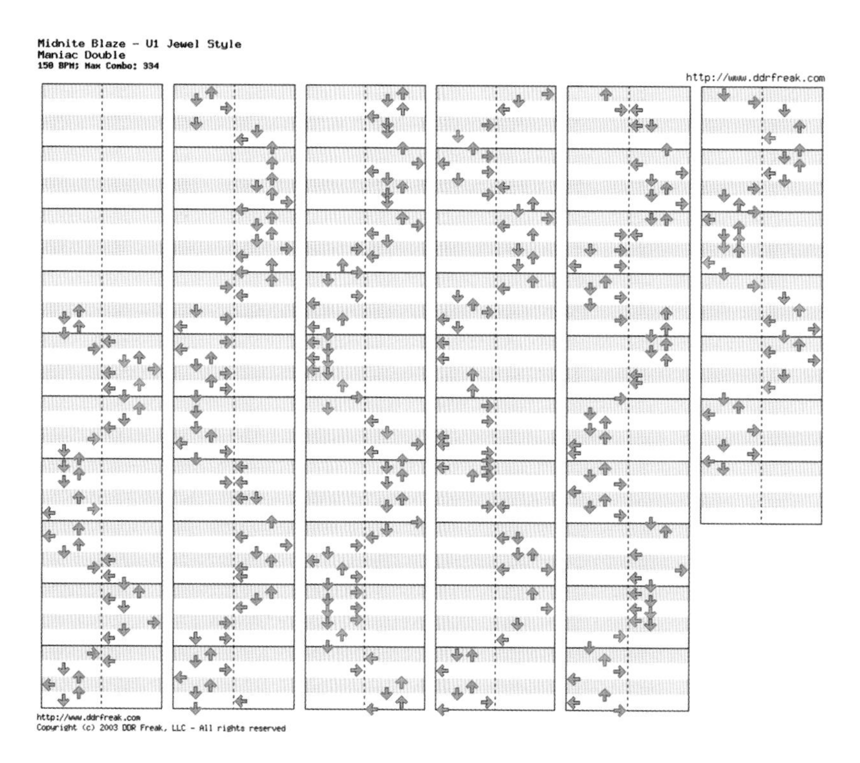

Fig.6.1

Step chart for the DDR song "Midnite Blaze."

As soon as I saw the step chart images, I started talking to my Computer Vision friend about how we could potentially get the data out of that image. He started brainstorming but also encouraged me to email DDR Freak and ask if they had the raw data. I was skeptical (the site had been inactive for at least five years) but I found an email for Jason, the founder of DDR Freak, and sent him a message without expecting a response.

Amazingly, Jason responded *18 minutes later* with a zip file of all the songs he had on hand. How awesome is that?? (Thank you so much, Jason!)

The data was saved in several different data formats, including .dwi (or Dance With Intensity, a freeware version of DDR) and .stp (DDR Freak's own proprietary format). I found an explanation for the .dwi file format online, and learned that 0 meant no arrow, 2 represented the down arrow, 4 the left arrow, 6 the right arrow, and 8 the up arrow. Each number defaulted to 1/8 of a beat, but everything within parentheses indicated 1/16 steps, brackets indicated 1/24 steps, and so on.

With that information, I was able to reverse-engineer the .stp files (which followed the same rule but was much more reader-friendly) and wrote a `Node.js` script to parse the file and store the steps, along with their song name and difficulty level, in a JSON file.

Sketch

Because I was traveling for the majority of the project timeframe, I wanted to do something relatively simple. I had seen TeamLab's *Crystal Universe*—an immersive installation where lights explode or linger or dance all around the room based on the input we, the participants, give it—a few weeks earlier, and was incredibly inspired by its beauty (Figure 6.2).

I wanted to create something similar, where each step is a light colored by its arrow direction and animated based on its position in the song. I started by mapping the steps for each song's two modes (single and double) and three difficulty levels (basic, trick, maniac) on top of each other (Figure 6.3).

To distinguish different modes and difficulties from the next row of music, I went back to the music sheet and staves idea from my *Hamilton* project, but this is where I got stuck. The musical staves again didn't do me much good, and the steps were spread so far apart from one line to the next that I couldn't see the whole song or any interesting patterns at a glance. I wasn't happy with the aesthetic of this detailed song view, and I also couldn't figure out the rest of the interface and how this song view would fit into it.

I've since learned about "data art," where data is used to drive the aesthetics, but doesn't necessarily have to give any patterns or insights.

Fig.6.2

A photo of TeamLab's Crystal Universe, taken by me at the Pace Gallery in Palo Alto, December 2016.

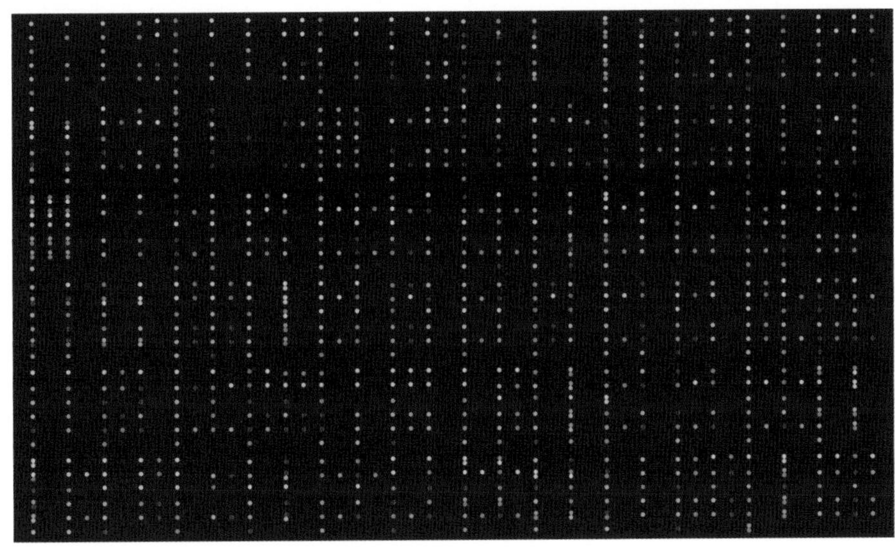

Fig.6.3

First attempt at replicating Crystal Universe with a DDR song.

While still stuck, I went to sleep with two goals for the interface in mind:

1. I wanted each song to be compact so that I could quickly scan across multiple songs and compare them.
2. I also wanted to use the visual metaphor of a song being "continuous" instead of being broken up into multiple rows like the previous version.

I woke up the next day convinced that spirals were the answer to my problems. Spirals, like circles, are very compact, and a spiral is basically one long continuous line that I could map my steps to.

I giddily set about looking into the math behind spirals and was super happy to find that for an Archimedean spiral (a spiral where the distance between each spiral branch is the same) the radius was equivalent to the angle. So I thought I could loop through each beat in a song while increasing the angle and radius by some constant number.

Until I realized that, if I used that formula, I would end up with a spiral where the beats on the inner spirals would be closer together than those on the outer spirals, because as the radius increases, the arc length for a constant angle also increases (Figure 6.4). This would spread the beats further apart as the radius increases, but what I actually wanted was for the beats to be placed equidistantly along the length of the spiral. When I Googled again, I realized that the problem wasn't as simple as I thought. The StackExchange answer I found had formulas with integrals and sigmas and symbols I haven't had to think about in a decade. To make matters even harder, the answer seemed to skip steps in its derivation. I spent an hour trying to figure out the steps between equations in order to understand how the final equation was derived. No luck.

Fig.6.4

Drawing a dot while increasing the angle and radius by some constant number (left), or drawing a dot whenever the radius was equivalent to the angle (right)—neither of which was what I wanted visually.

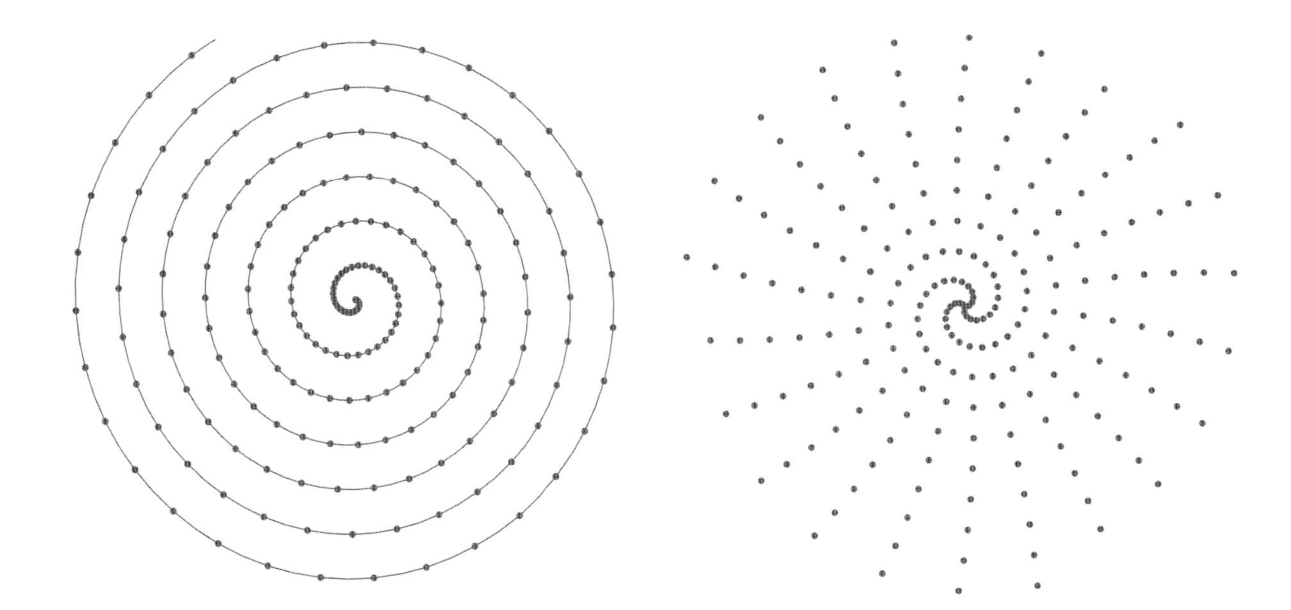

Then two miraculous things happened:

1. I posted my plight on Twitter, and Andrew Knauft came back with his own step-by-step derivation. Unfortunately, the derivation was still too complex for me to figure out (at least, not quickly enough for this project). ~(>_<。) \
2. In a chance encounter, I struck up a conversation with front-end developer Isaac Kelly at a bar in San Francisco. After a few email exchanges where we tried to solve the problem together, he brute-forced the answer.

We've since run into each other at some art+technology conferences. He even wrote a very funny blog post about the whole experience.[1]

Because I was already falling behind on the project deadline and running out of time, I took Issac's brute-force approach instead of trying to figure out the StackExchange formula. Issac's approach was very programmatic (which definitely helped me understand it quicker) and was dependent on two constant numbers: a really small number to increase the *angle* by (I ended up using 0.02), and the desired arc *length* between two beats. The program repeatedly increased the *angle* by that small number until it hit the desired arc *length*, and that was where the beat would be placed. It then remembered the new *angle* where the beat was placed and repeated the process. I adapted this code to draw the steps (represented as a colored circle) to the beats they belonged, and terminated the `while` loop only when the song ended and I ran out of steps to draw:

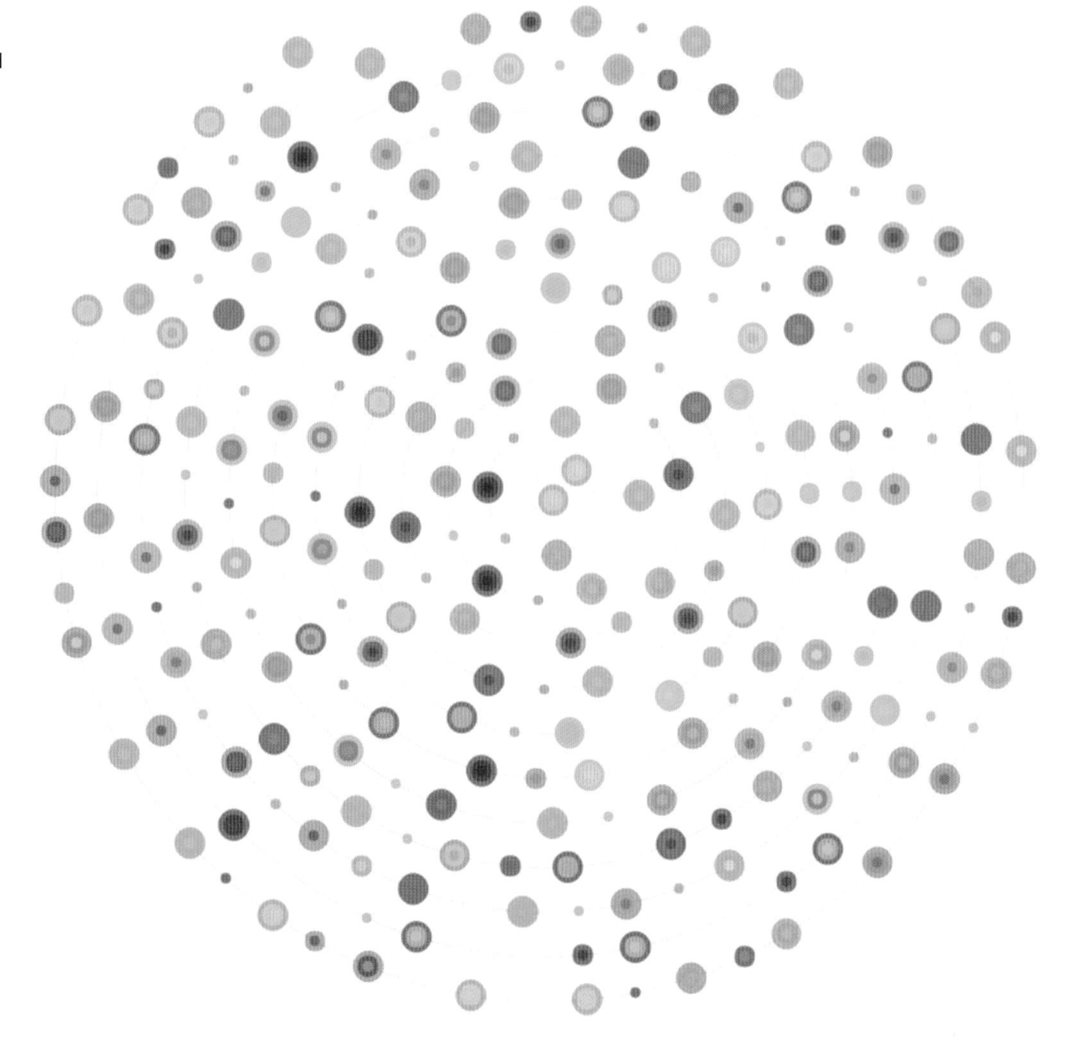

Fig.6.5

A DDR song visualized as a spiral.

[1] Issac Kelly, "Mark Equidistant Points on a Spiral": https://www.issackelly.com/blog/2017/01/20/mark-equidistant-points-on-a-spiral

I then mapped arrow direction to color, and though I didn't have any particular reasons behind each color assignment, I did want them to be distinct from each other: left is orange, up is red, right is green, and down is blue. I also mapped difficulty to circle radius: I made it so that the higher the difficulty, the smaller the circle would be.

This was meant to be a visual pun on the fact that it's harder to see all the steps in more difficult songs. (o≧∀≦o)

I drew all of these in HTML5 canvas as there were so many steps and so many songs, and I applied an overlay blend mode on the circles. For each song, I also added a legend explaining the colors and sizes and an interaction to filter by arrow direction and difficulty (Figure 6.6). I was hoping that the filter would help me see some patterns, but unfortunately because there were so many steps and so many songs, I was too overwhelmed and wasn't able to find any.

Although the result was aesthetically pleasing, it was also very misleading since the overlay blend mode changed the colors of the overlapping circles.

For my final step, I added an expand button for each song that scaled the spiral by 1.5x and increased its viewport by 2x when clicked. This allowed the visitor to both view each spiral in more detail and to see some of the longer songs whose spirals were cut off.

Reflections

Overall, I'm happy with where I ended up given where I started. With the spiral, I can compare multiple songs, and both the big spirals (long songs) and smaller spirals stand out. But for the sake of compactness, I sacrificed being able to see at-a-glance patterns within a song. One of the aspects that make DDR so fun is that a song will have many sets of steps that repeat throughout; with this visualization, it's hard to find those. I'm not sure if, given enough time, I would have been able to come up with a visualization that could show both the trends *across* songs as well as the patterns *within* songs.

I could also see that some of my favorite DDR songs had a lot of steps closer together. No wonder I was so out of breath whenever I played them!

I was also unsatisfied with the performance. Even though I used HTML5 canvas for every song, there were so many songs that the web page rendered very slowly. Given more time, I would have created an overview page with a simplified version of the visualization and only showed the song in detail once selected.

Either way, my favorite part about this project wasn't actually the process or what I ended up with; it was instead the kindness of strangers willing to give a helping hand, proving that (some) people are *awesome*.

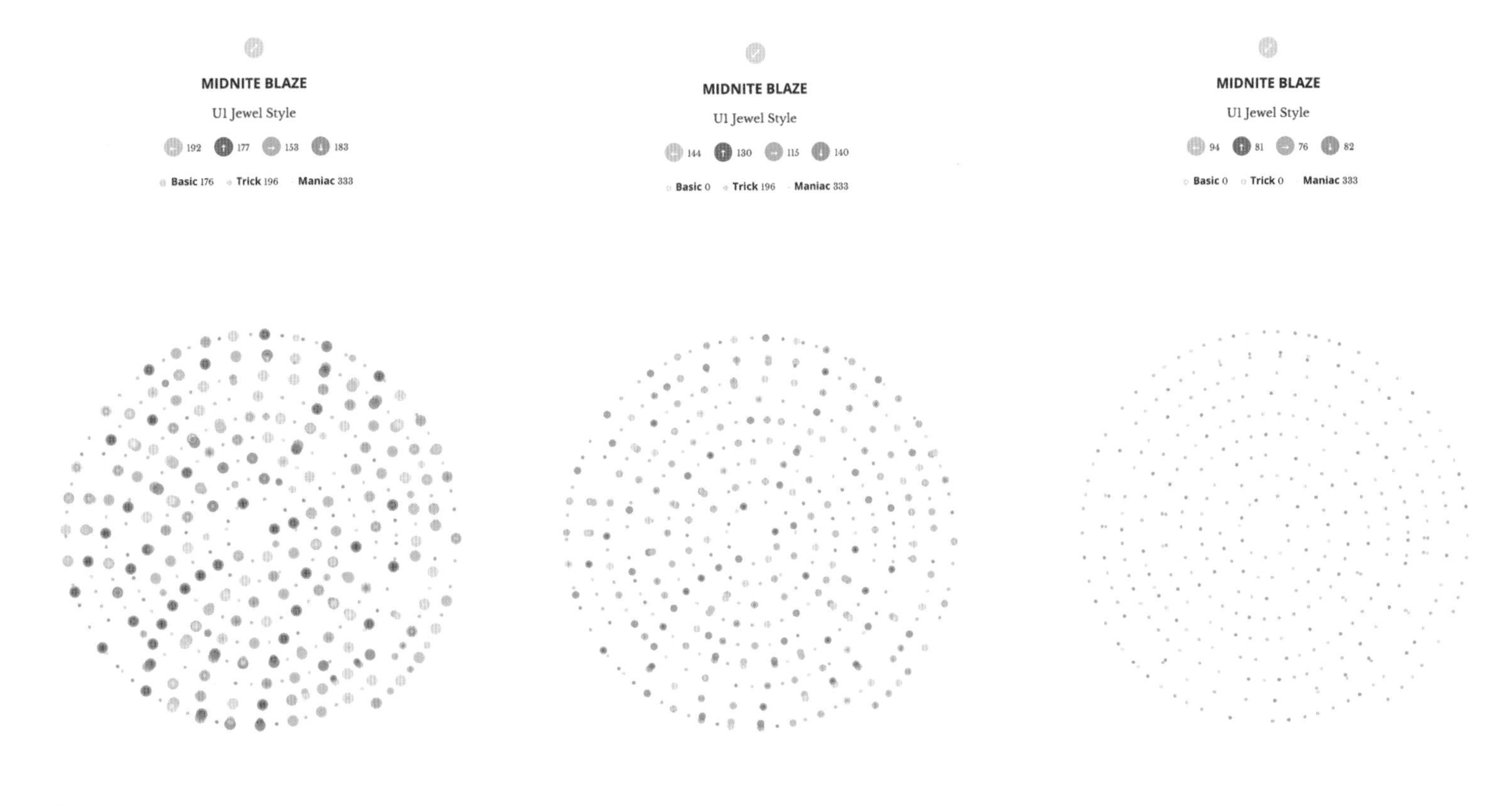

Fig.6.6

"Midnight Blaze" with all steps (left), without the "basic" steps (middle), and without the "trick" steps (right).

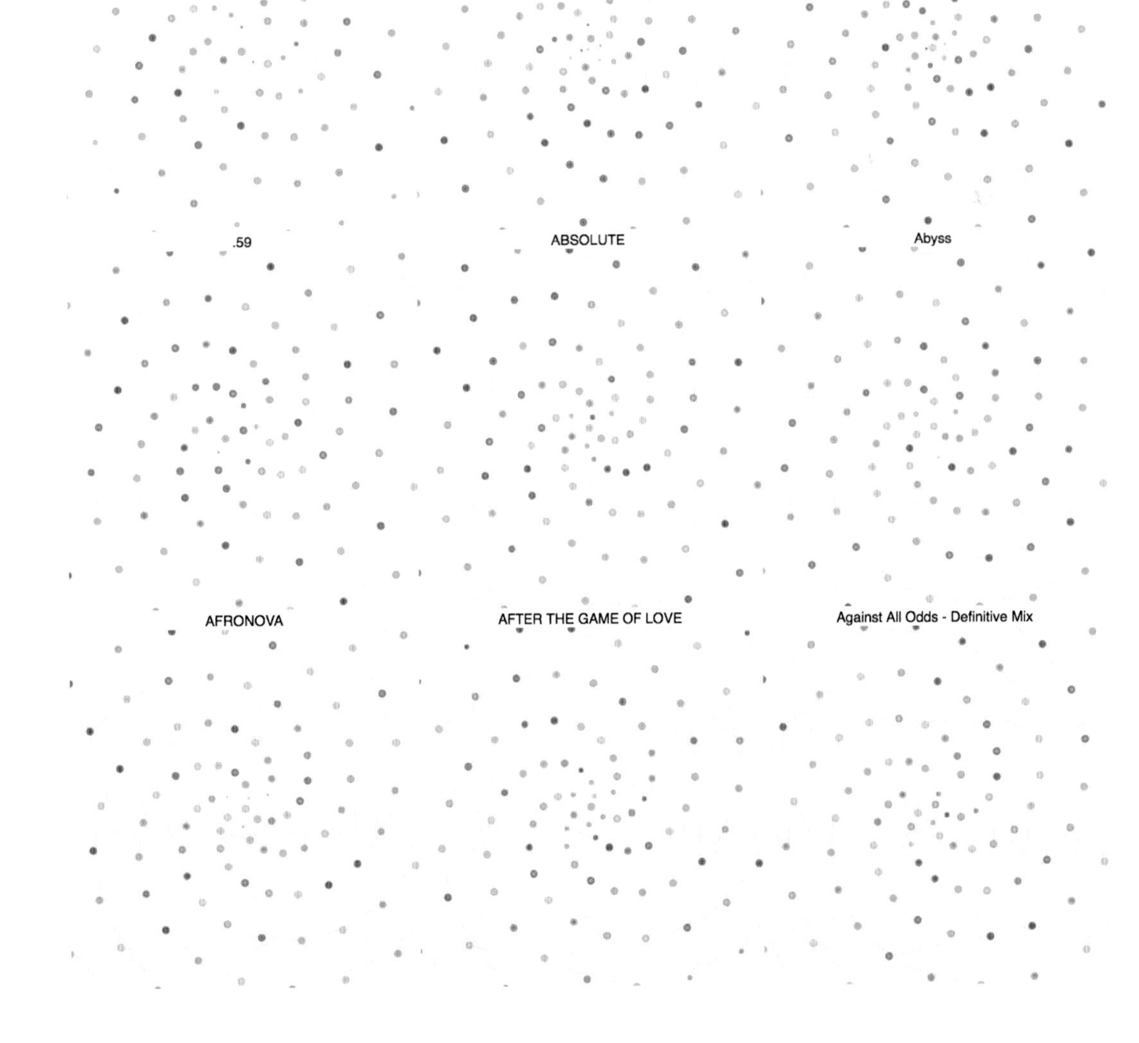

Fig.6.7

My favorite bug from this project, where the spirals ended up looking like roses instead!

Data-Driven Revolutions

↳ shirleywu.studio/projects/ddr

DATA DRIVEN REVOLUTIONS

BY SHIRLEY WU

I came across DDR in middle school, and was immediately intrigued. It was the first time I had begged my parents for a video game, and the only time my parents bought me one. I played all through high school with my friends and my little sister, and it is a part of some of my fondest memories. So when I found DDR Freak while looking for DDR data, I was beyond ecstatic; when I found their step charts, I knew I had to visualize them.

At 250 songs, each with different modes (Single, Double, 6Panel, etc.) and difficulty levels (Basic, Trick, Maniac), that's a lot of data; I had to be selective with what I showed. Thus, each song only shows steps from Single mode, and only shows both arrows (jumps) when expanded. I've also compressed each song as much as possible, so that you can compare multiple songs at a time. I hope that you find fascinating patterns, and if at first you don't, try filtering by a difficulty level.

Made with data from DDR Freak (thank you Jason!) and ♥ for December data sketch|es.

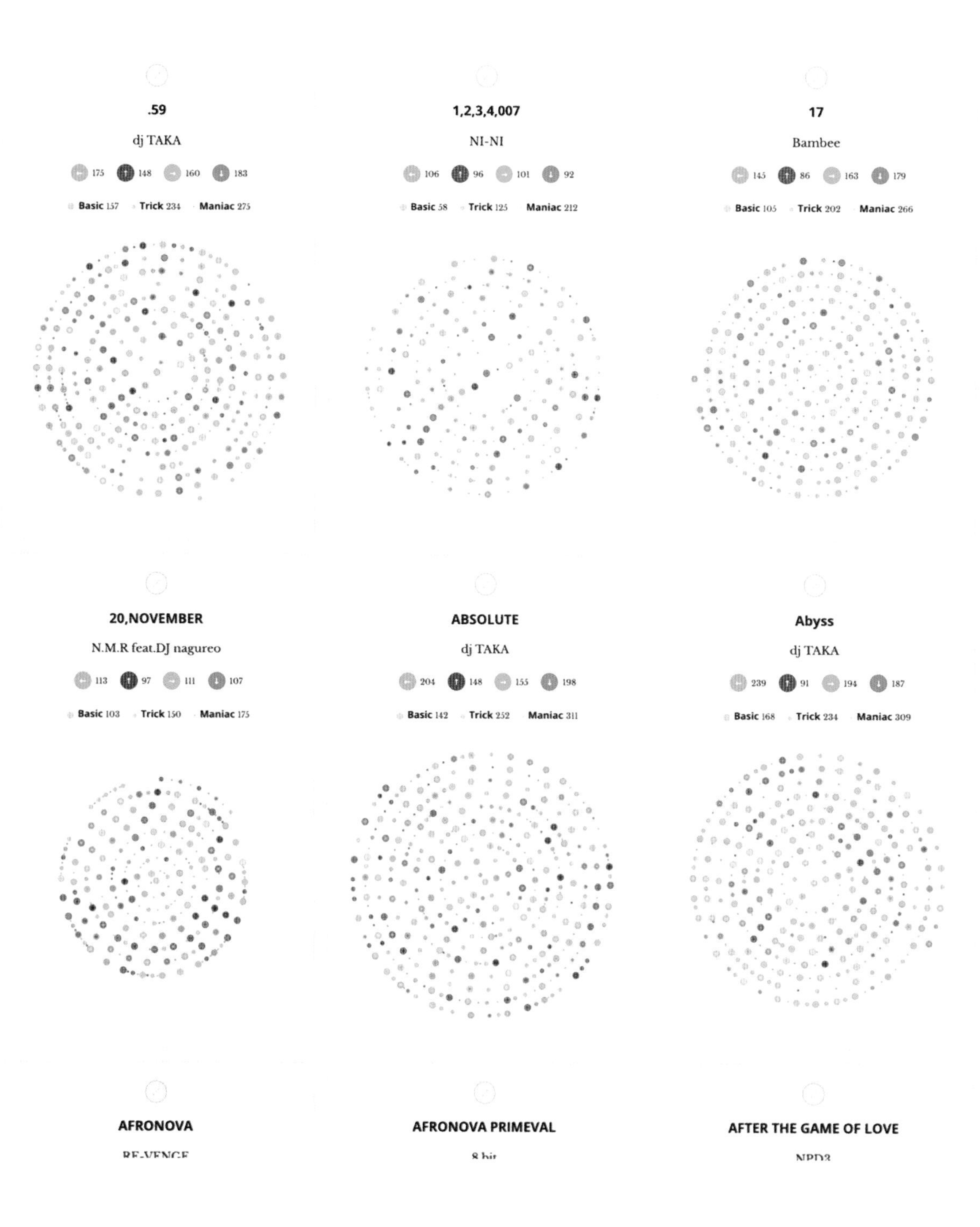

Fig.6.8

The final visualization with six out of the 645 available DDR songs.

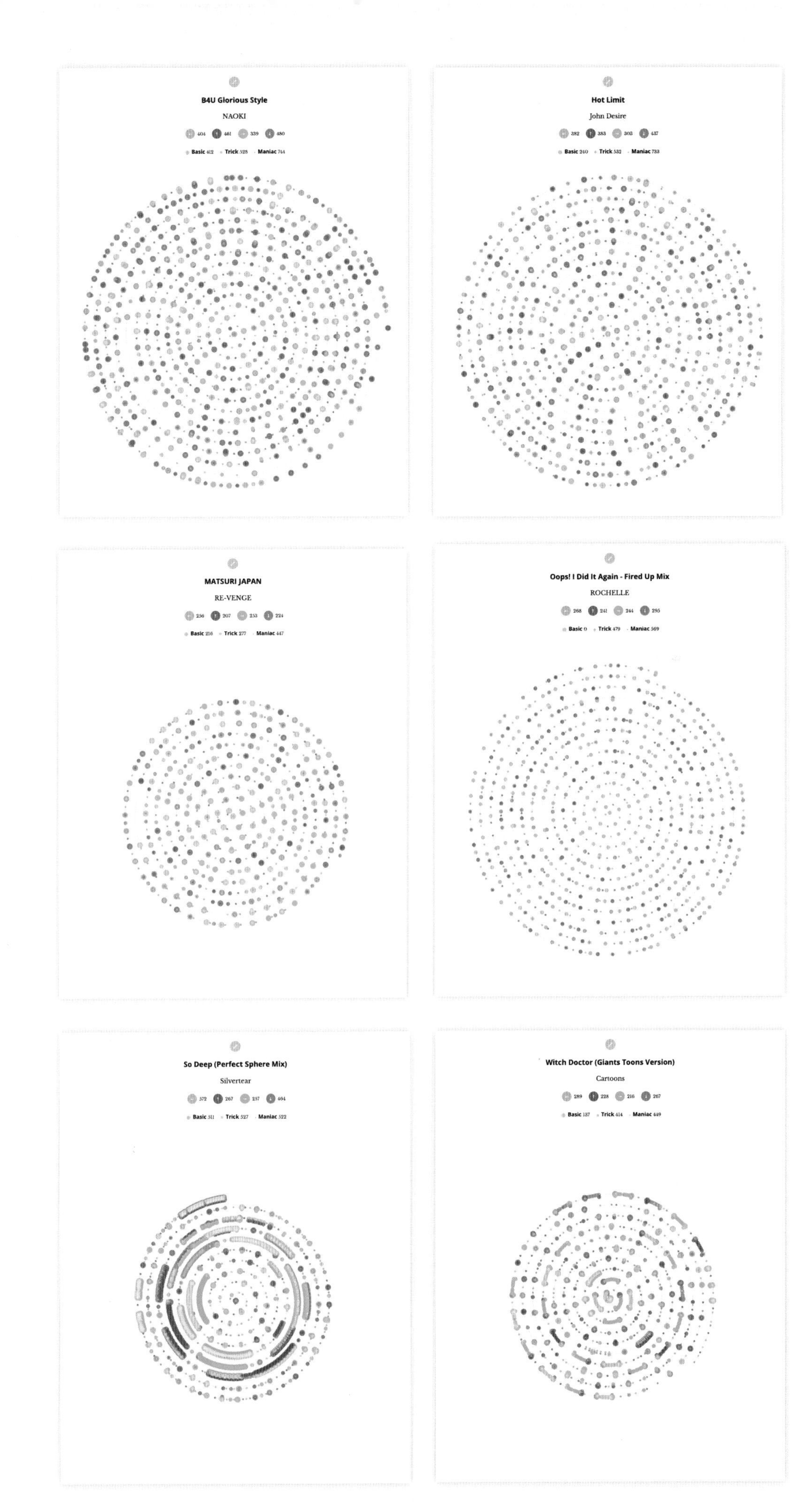

Fig.6.9

Some of the most complicated (or pretty!) looking DDR songs.

NOSTA

ALGIA

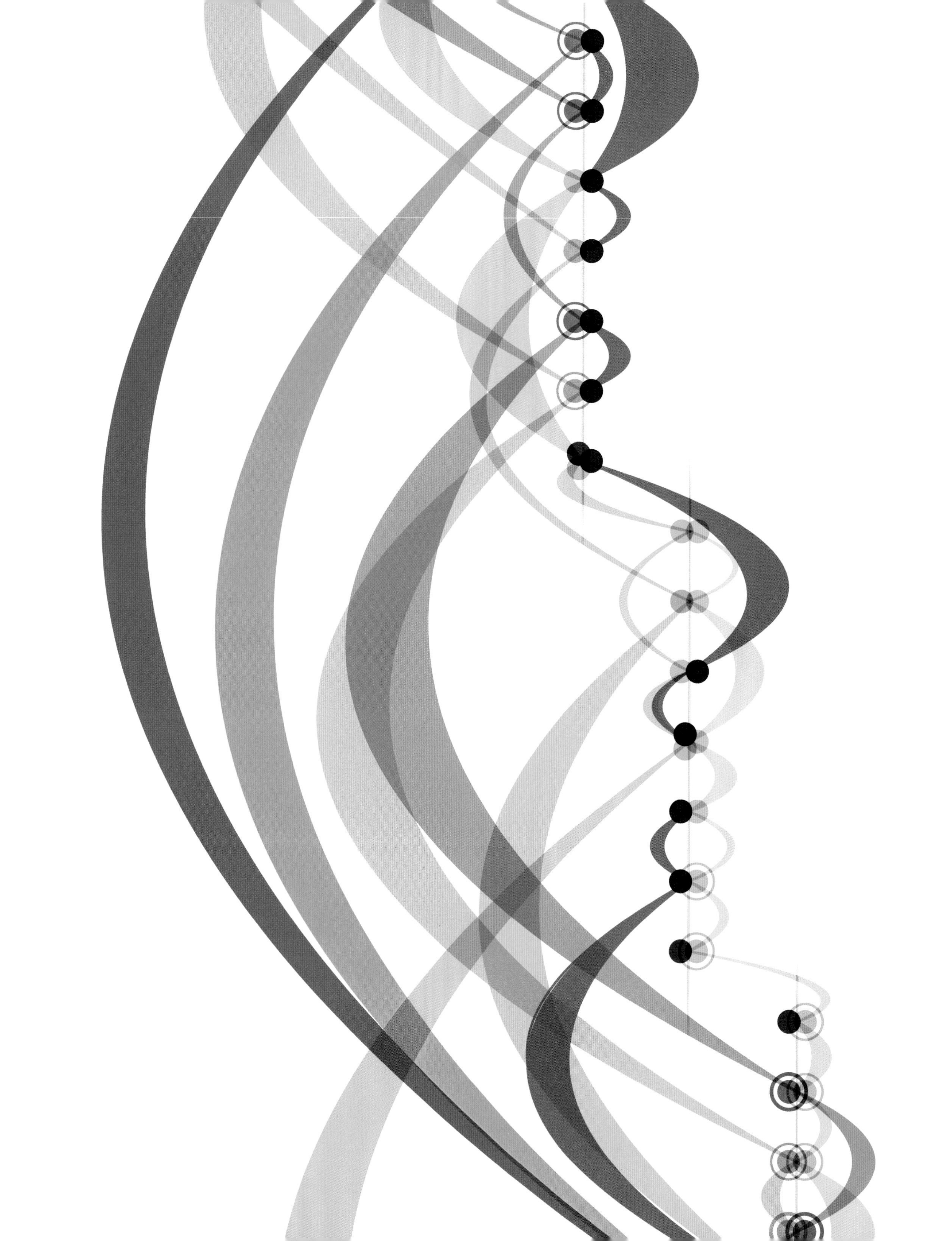

All the Fights in Dragon Ball Z

NADIEH

I had several choices for the topic of this project's "Nostalgia" theme—including video games, manga, and anime—and I didn't know which one would provide the most interesting dataset. I therefore scoured the web for all the topics I had in mind. I loved playing *The Legend of Zelda* on the Game Boy, especially "Minish Cap" and the "Oracle of Ages/Seasons" combo. There was something about the 2D bird's eye view of Game Boy's *Zelda* games that worked exceptionally well for me, whereas in 3D games I just kept falling off bridges and running into corners... Unfortunately I couldn't find any interesting data. I looked at the original *Cardcaptor Sakura* (CCS) manga but also didn't find enough data that I could use to create something elaborate. Neopets is where I first encountered HTML & CSS at 14, even though I didn't continue using any web languages until I saw the magic of `D3.js` some 12 years later. However, I found out my account was hacked, my pet stolen, and I wasn't in the mood to investigate that further at the time.

Eventually I turned my focus to the last subject I had in mind: *Dragon Ball Z* (DBZ). This anime was my very first introduction to the Japanese animation scene, airing on Cartoon Network in the Netherlands when I was about 13 or 14. Although I quickly turned solely to manga in general, I stayed loyal to the DBZ anime and loved watching it all the way to the Fusion Saga. (I eventually stopped watching DBZ because I was too annoyed by the characters Majin Buu and Gotenks 눈_눈)

Data

While randomly going through the extensive Dragon Ball Wiki,[1] I learned that *Dragon Ball Z* is very much alive despite originally airing about 20 years ago. I also learned that *Dragon Ball Kai*—a filler-free version of DBZ—aired in the United States not too long ago, as did *Dragon Ball Super*, a new series taking place right after DBZ chronologically. I came across the page to each DBZ saga (you can think of a saga as being a season or story arc) and saw a list of all the major fights that took place as well as the characters involved, in order of appearance. This dataset immediately stuck out to me. Because DBZ is an anime that revolves around fighting, it felt so appropriate to map out all of the fights that happened during the run of the show, including who fought whom, who were the bad guys, and what transformations were active (such as "Super Saiyan," or SSJ for short).

No filler for DBZ basically means 30% less episodes!

I'm afraid that due to copyright reasons I'm unable to show any images of Dragon Ball Z in this book.

So I manually went through all ±17 sagas that make up the 291 episodes and copy-pasted the fights with the characters into Excel. With the help of some Excel functions (such as `LEFT`, `RIGHT`, `LEN`, and `FIND`), I split up the fights into the separate characters involved, and after two more hours of manual cleaning, I had a nice list of ±200 fights.

I use Excel from time to time to do highly manual, one-off, data preparation.

↳ Data Can Be Found in Many Different Ways

Be open to the weird ways in which data can be found online. For this project there was no nice-looking, already formatted spreadsheet that I could download and use for my visual. Instead, it came down to manually copying text lists from a website into Excel. I'm glad that other people had already put in the effort to create this list of fights on the DBZ Wiki!

Next, I needed to arrange the data in a slightly different order. Originally each fight was on one row, but I needed *each character per fight* in one row. Because I'm more comfortable in R than in Excel, I created this additional dataset with a few lines of code in R; but I think it could also have been done in Excel. So generally, all the data was gathered and prepared without *truly* needing to write a single line of code!

Another "data gathering" step that came up later in the process was finding appropriate animated GIFs to highlight the most special moments and fights, plus nice shots of the main characters for the legend. I went through *hundreds* of GIFs and images, trying to find some decent stuff.

Seriously, it took me hours going through all the GIFs and images, but it was a lot of fun too.

[1] Dragon Ball Wiki: https://dragonball.fandom.com/wiki/Main_Page

↳ Design to Maximize for Delight

I didn't need the animated GIFs highlighting some of my favorite moments in DBZ to show the fight data. However, I felt that they made a very big difference in how much *fun* it would be to look through the whole visual, as it draws the viewer in more. These images also show a sneak peek of DBZ to those that have never seen or heard of it before.

It's important to keep your viewer engaged, especially with more complex visuals. Small touches—such as animated GIFs, images, weird design styles and flourishes, and interesting legends—can help with that. But don't go too far; these should be small additions to your visual should not take center stage!

Sketch

After finding the DBZ fight info on the Wiki, I came up with the concept to place each fight on a line, one after another, where each saga in the series would have its own column. Lines would be drawn through the fights of each character, as if that person was flying from fight to fight. Figure 7.1 a shows my very first sketch of the idea, while Figure 7.1 b shows another sketch where I started looking into the details a bit more. As I already knew that I would have to create the formulas for the "swooshing" SVG paths myself, I tried to see where to place the anchor points of the SVG *Quadratic Bézier Curves*[2] (which makes the lines swoosh outward).

The ability of strong characters to fly in DBZ is an important part of the show.

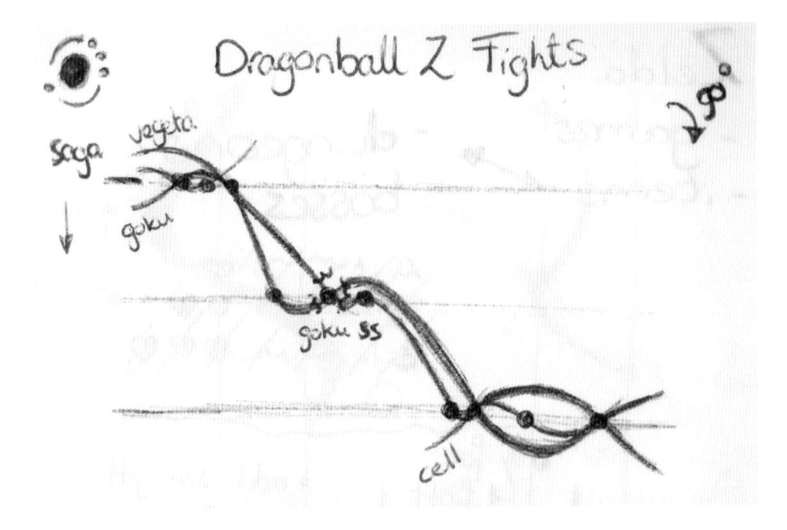

a

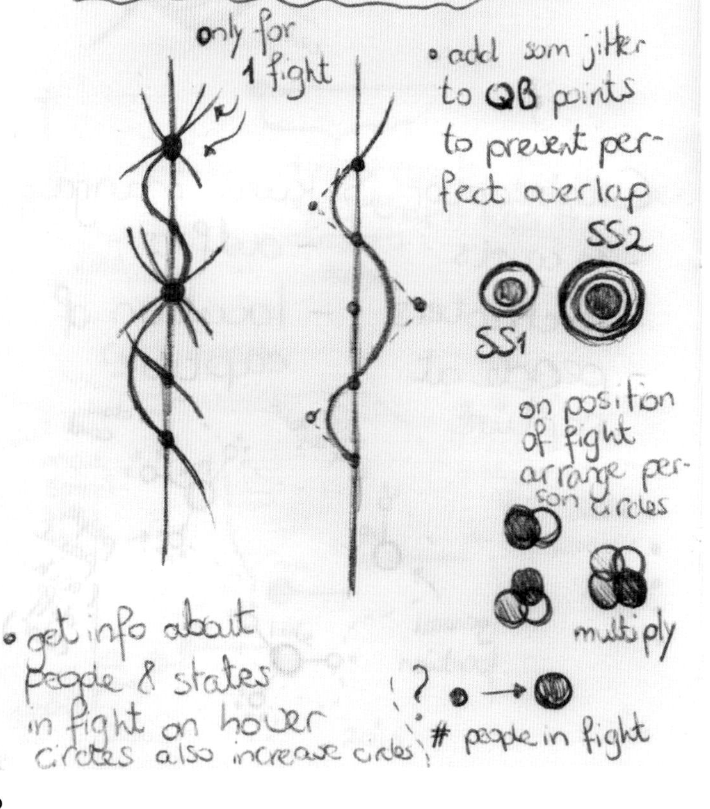

b

Fig.7.1
(a & b)

Sketches showing the general layout of the fights across sagas and connecting characters between fights.

[2] SVG Quadratic Bézier Curve Example: http://blogs.sitepointstatic.com/examples/tech/svg-curves/quadratic-curve.html

A while later, having already programmed the basic layout of the fights, I realized that I didn't like the lines to have only one level of thickness. It was making the overall visual look a little plain, and it didn't convey the more dynamic nature of the fights. So I set out to create a shape that would look like a line of varying levels of thickness, as if pressure was being used on a pen that draws the lines. I created another simple sketch (Figure 7.2) that had the shape that I was looking for. Only after I drew my desired end state of a shape was I able to figure out the SVG path formula that would be needed to recreate it.

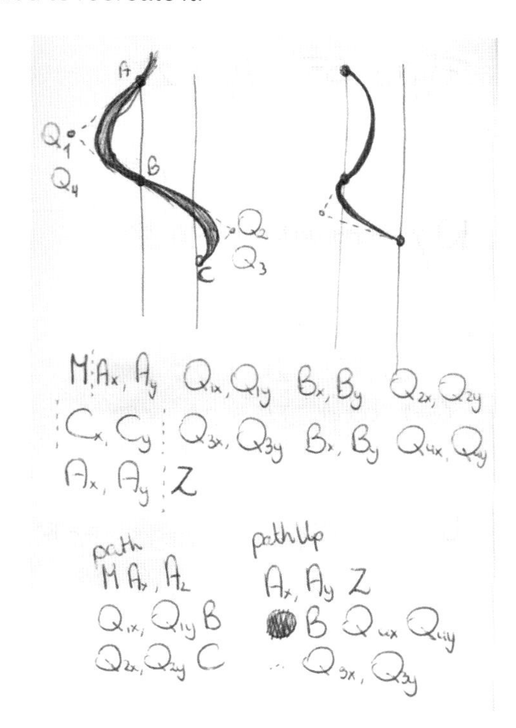

Fig.7.2

Figuring out how to create a shape of varying thickness in SVG path coordinates.

One part that I was constantly struggling with, and have struggled with practically every project, is the overall layout. What fonts would fit this design? How should I place the intro text and any legends? How can I make the page as a whole look interesting? I sketched out a general idea for the overall layout (Figure 7.3) and I have to admit, as usual, the end result wasn't quite as visually pleasing as I would have hoped, but I was reasonably OK with it this time.

For this project the sketches were *very* basic. It was really about getting the abstract shapes on paper. Any form of design was completed while I was already programming with the actual data.

a

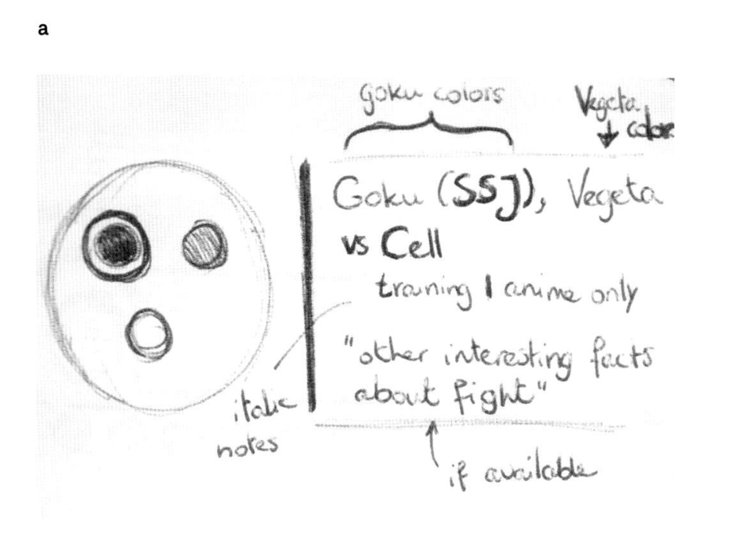

b

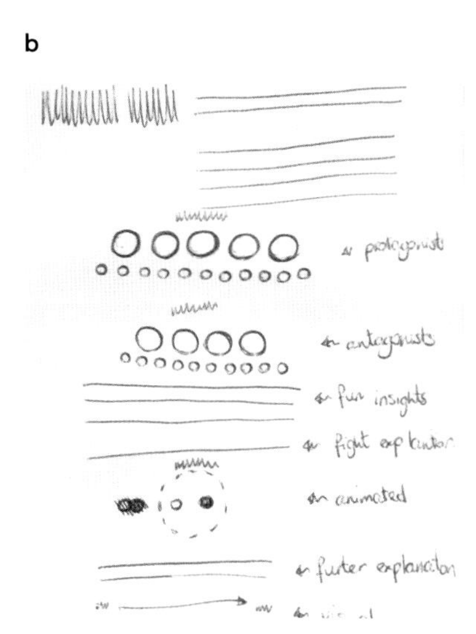

Fig.7.3 (a & b)

Figuring out the layout of the tooltip. Drawing the layout of the intro section.

Code

As per my sketches, I wanted to represent every fighter in a battle with a circle and then have these circles partly overlapping. Each saga would be placed in its own column to make the distinction between different story arcs more apparent. With some simple geometry that was achieved relatively quickly (Figure 7.4 a). I then colored the circles differently for each character, taking inspiration from the character's appearance. As another level of detail, I added extra rings around the main circle to reveal possible "transformation levels," such as Super Saiyan (SSJ). As the number of unique transformations throughout the anime was manageable, I wrote a bunch of `if-elseif-else` statements to set the number of outer rings based on the specific transformation's name (Figure 7.4 b).

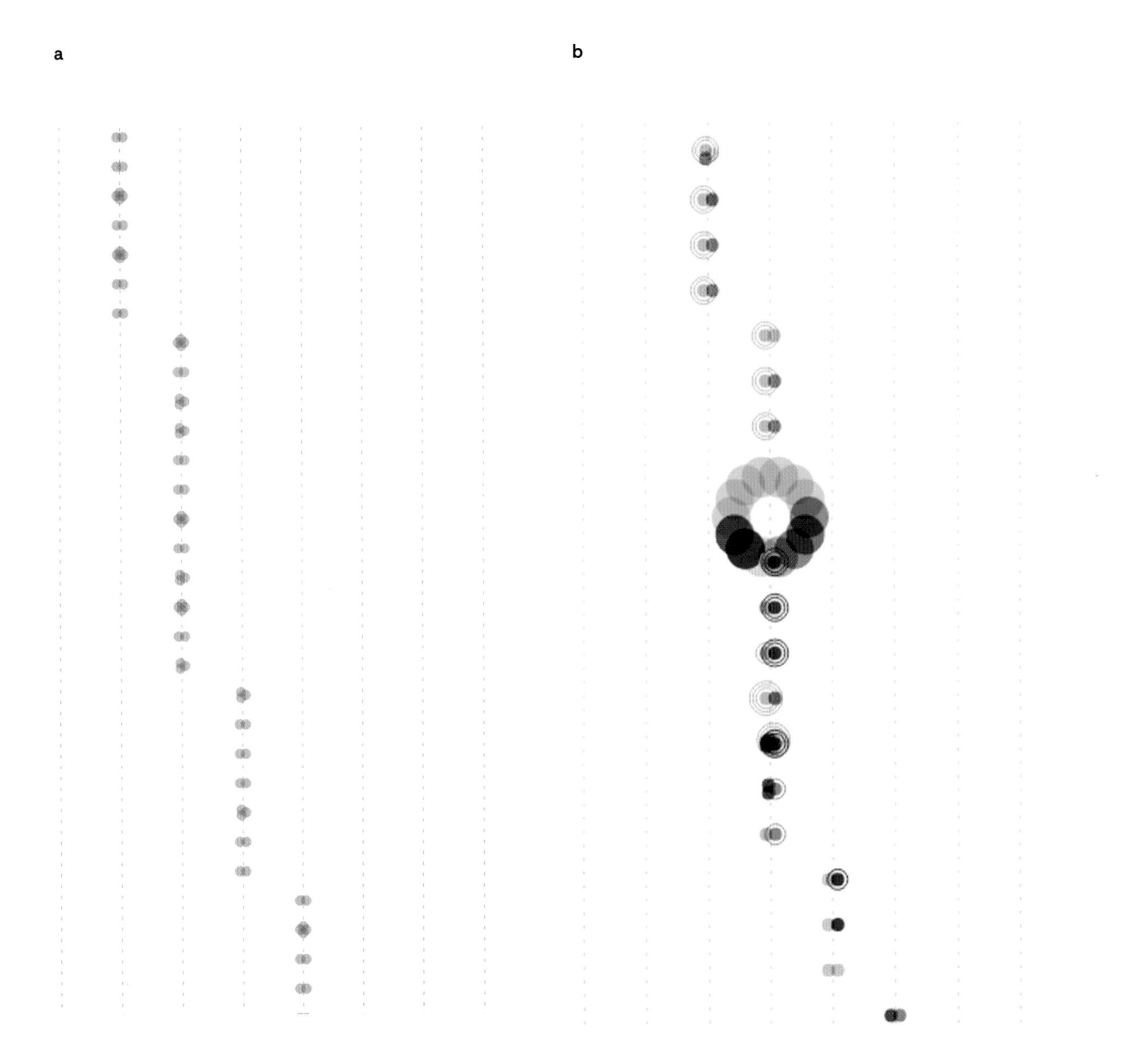

Fig.7.4
(a & b)

The basic outline of the visual was achieved in relatively little time.

Next came the custom SVG paths that connected all the fights from one character, using SVG Quadratic Bézier Curves to create a "swoosh" effect that made it look more visually pleasing. Of course, I didn't start by creating the complete path; I built things up incrementally, adding more and more of the required complexity bit by bit (Figure 7.5). I started with straight lines from fight to fight, adding curves to the lines, cutting the lines between sagas for the minor characters to reduce clutter, and so on.

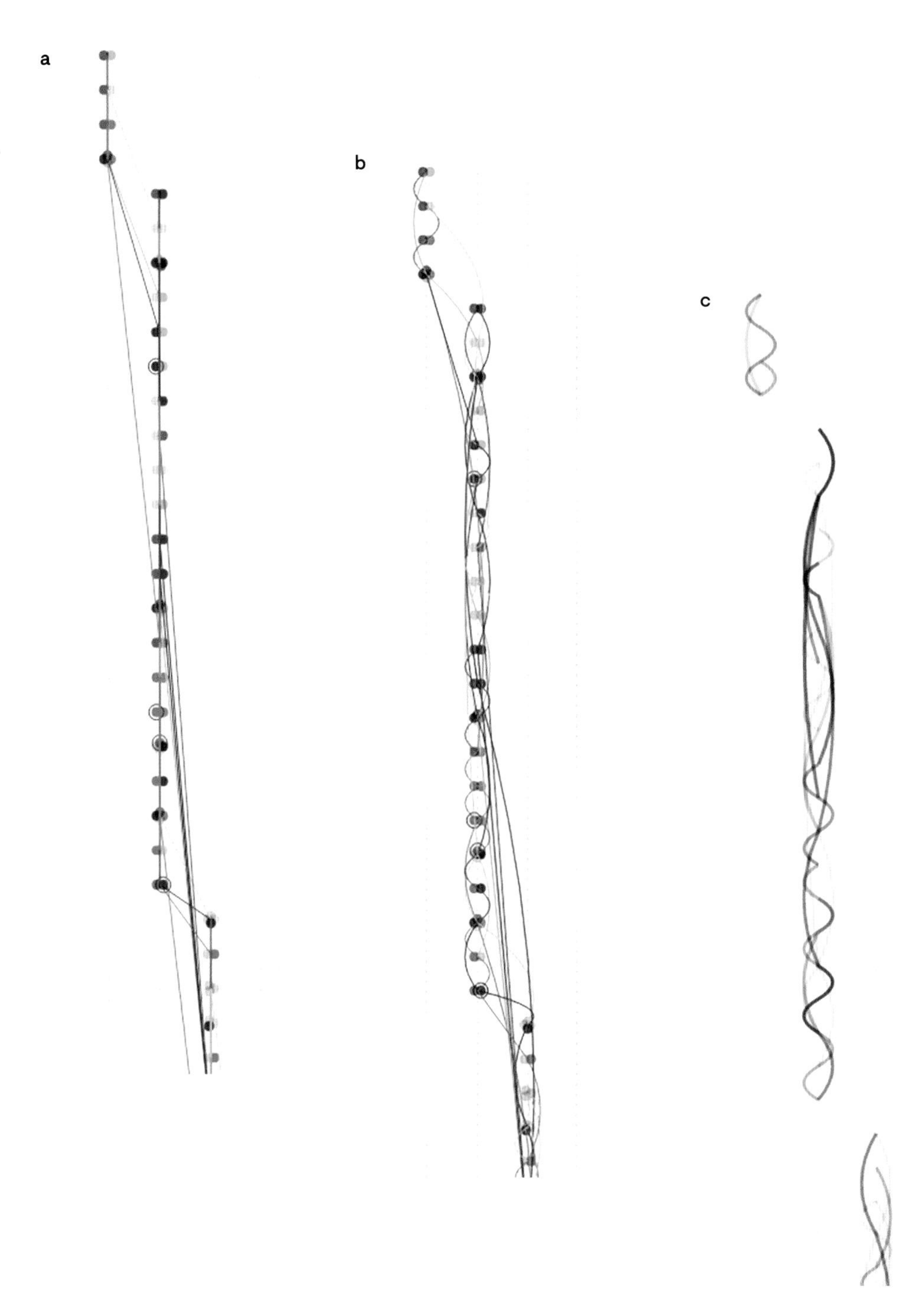

Fig.7.5 (a,b,c)

Slowly building up the complexity of the lines.

That did make all the swooshes look rather similar, creating a bit of a boring result and the occasional overlap of lines. I therefore made the swooshes move farther outward if the distance between two consecutive fights was larger, and added a bit of randomness in the amount of distance on top of that so no swoosh would overlap (Figure 7.6).

I liked the end result more with the added randomness, but *in my mind* the swooshing lines had different widths, from thin at the fight circles to broad during the bend in between fights. I therefore turned the line into a shape by adding another piece to the SVG path that would run back from bottom to top, creating a closed loop with slightly different values for the amount of swoosh, which I could fill with a color (Figure 7.7).

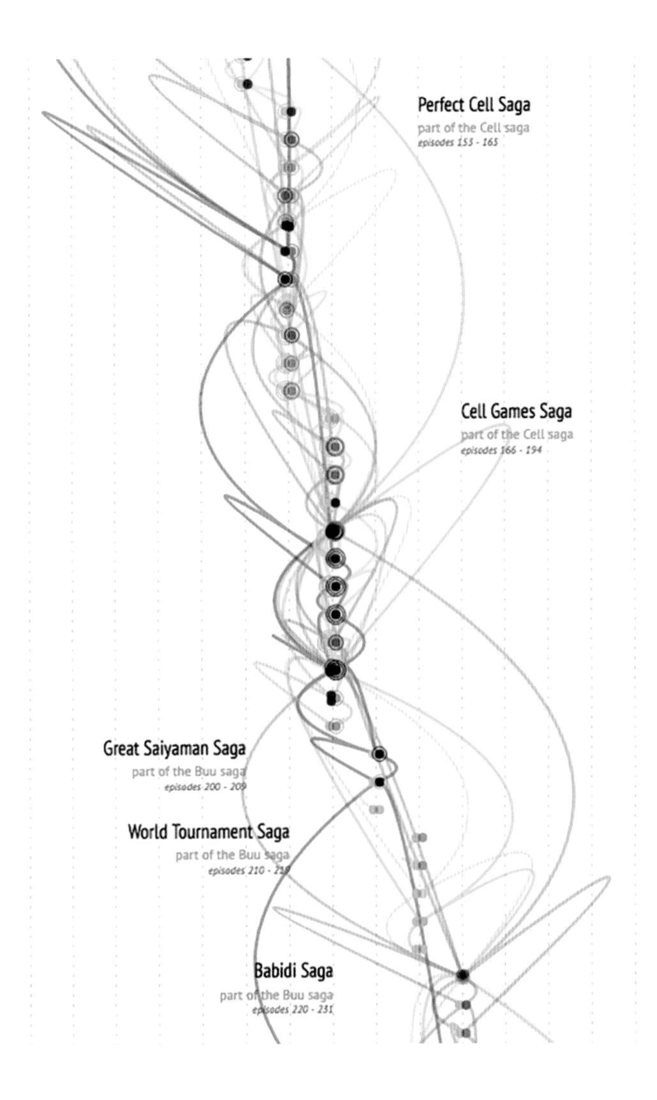

Fig.7.6

I obviously had errors in my line math here...

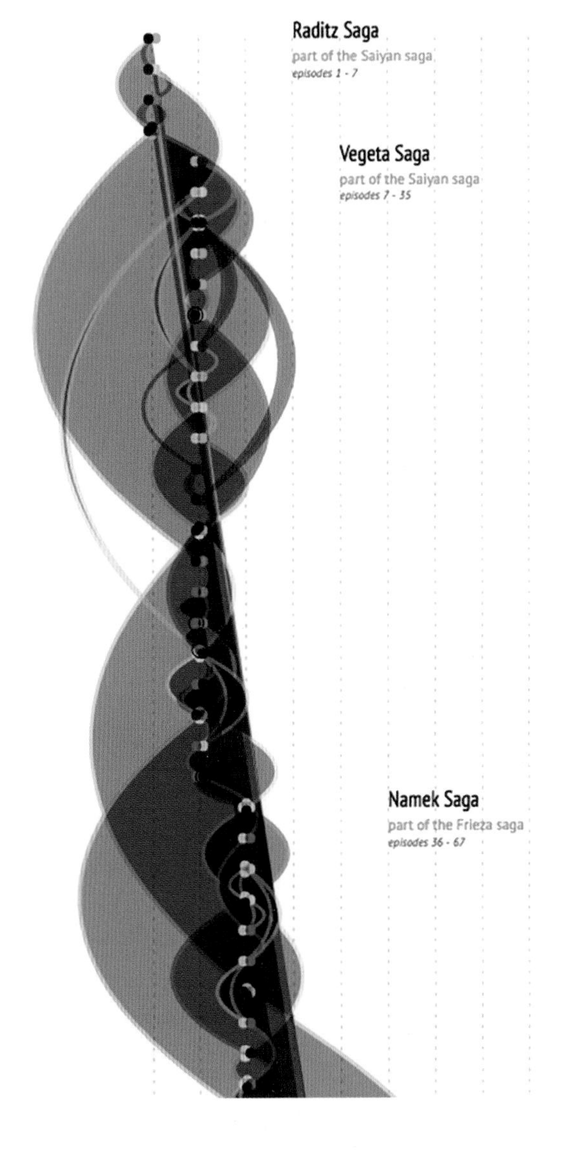

Fig.7.7

Filling my new shapes showed that I wasn't doing the path calculations totally correct.

Thinking about math and hand-crafted SVG paths yet again was an interesting exercise. And thankfully, it gave me the result that I was looking for. Figure 7.8 shows the comparison of a simple stroked line versus the varying thickness shape. Personally, I find the latter looking a lot more visually appealing.

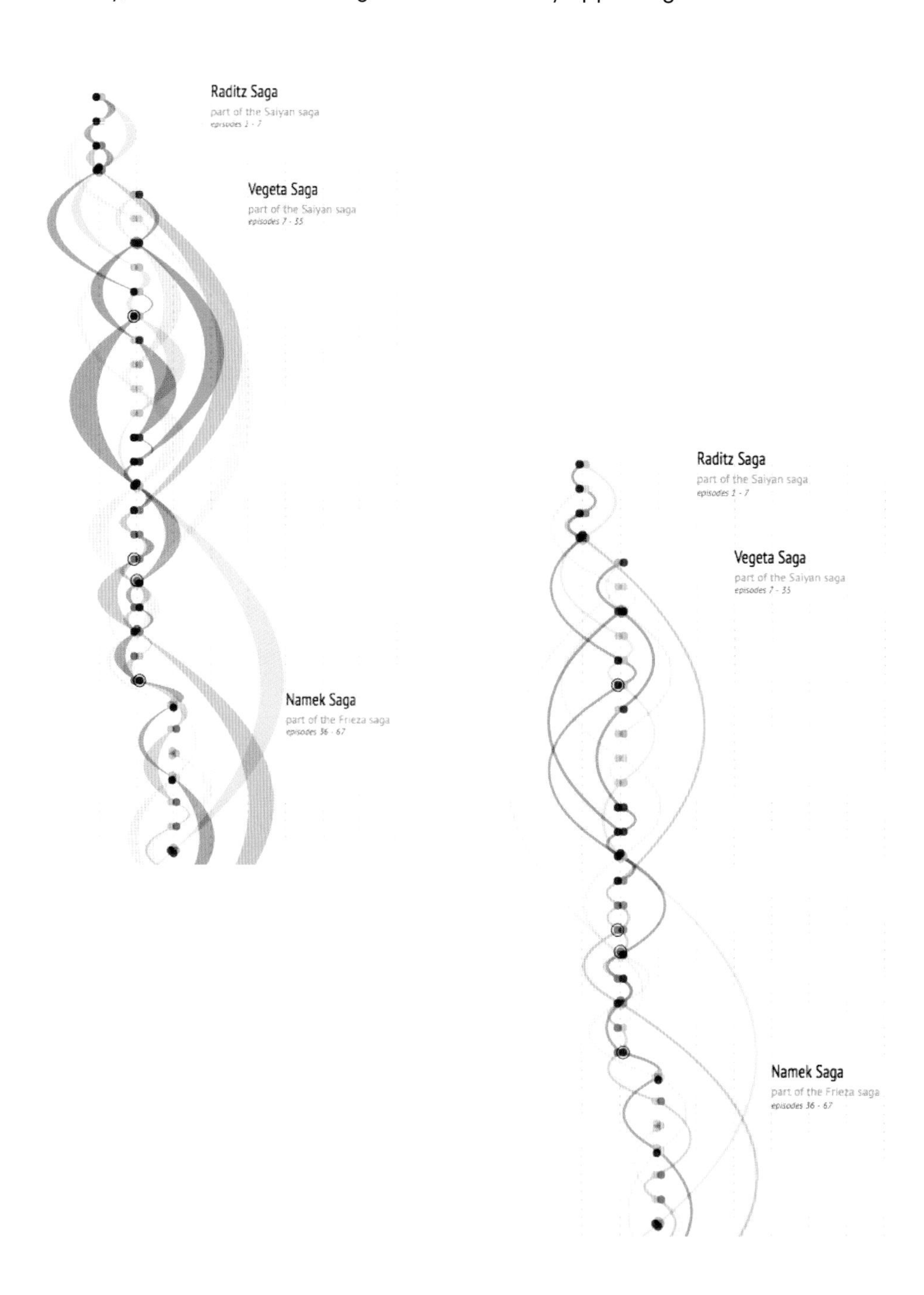

Fig.7.8 (a & b)

Comparing before and after, going from line to varying thickness path.

With about 200 fights to include, this visual is quite long; in fact, it's about six-and-a-half times as long as it is wide. This will never fit on *any* screen. When I showed the visual to Shirley, she said that she wanted to know what section she was looking at with respect to the whole and suggested I add a mini map with a full overview that would always be visible. I thought it was a great idea, but I also dreaded building it, due to its complexity to program.

Nevertheless, one a Saturday in January I sat behind my laptop for hours and slowly built up the mini map (Figure 7.9) ᕦ(ò_óˇ)ᕤ.

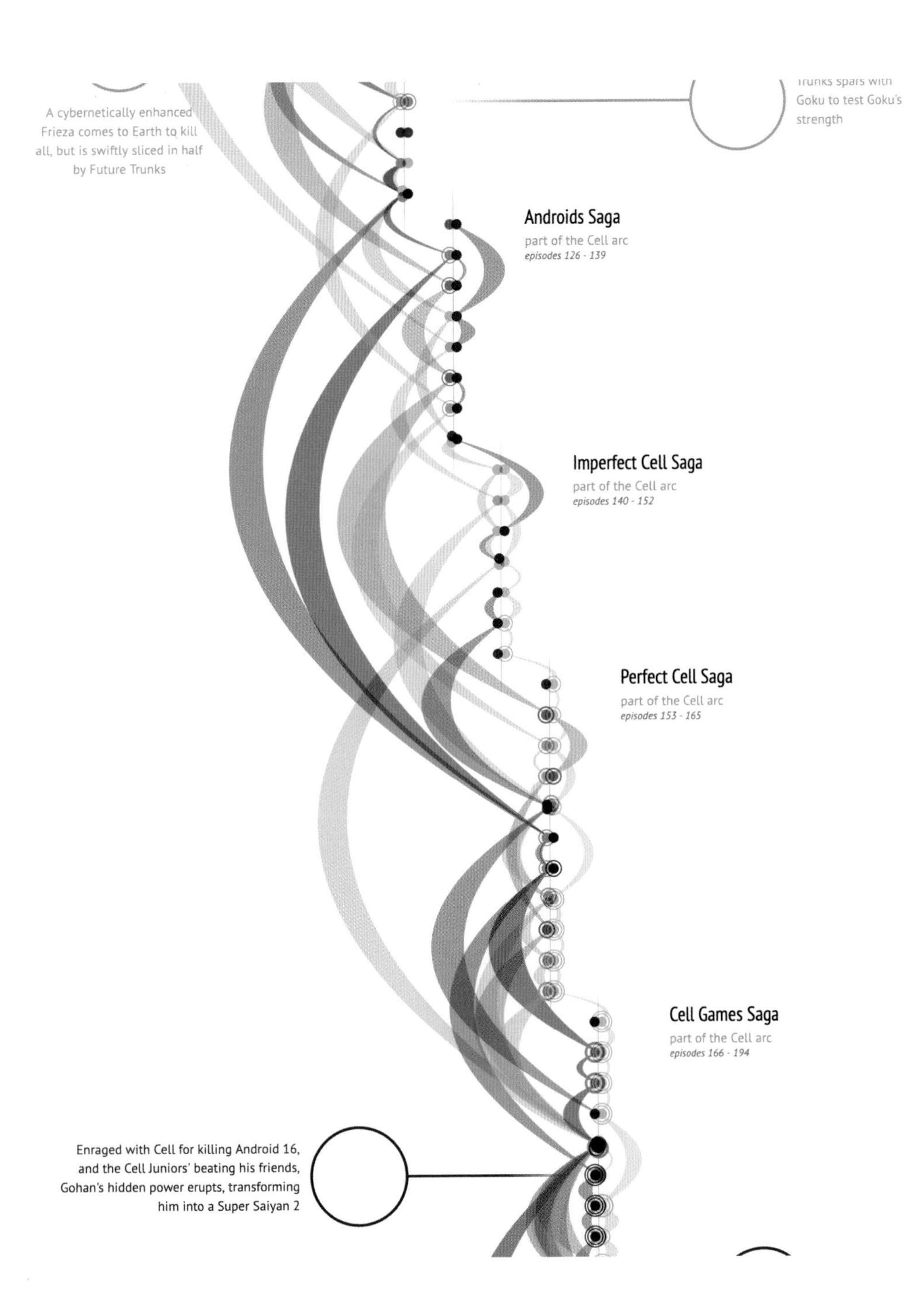
Trunks spars with
Goku to test Goku's
strength
A cybernetically enhanced
Frieza comes to Earth to kill
all, but is swiftly sliced in half
by Future Trunks
Androids Saga
part of the Cell arc
episodes 126 - 139
Imperfect Cell Saga
part of the Cell arc
episodes 140 - 152
Perfect Cell Saga
part of the Cell arc
episodes 153 - 165
Cell Games Saga
part of the Cell arc
episodes 166 - 194
Enraged with Cell for killing Android 16,
and the Cell Juniors' beating his friends,
Gohan's hidden power erupts, transforming
him into a Super Saiyan 2

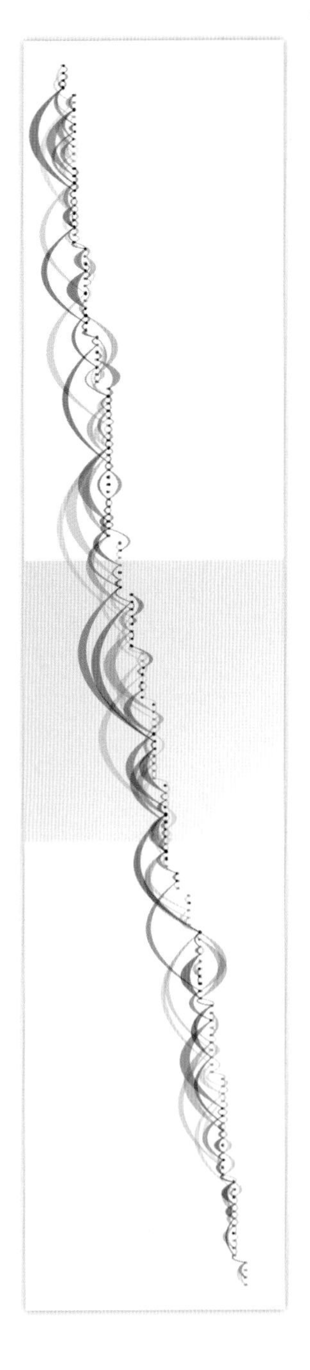

Fig.7.9

Adding a mini map of the total visual on the right of the page, showing the viewer where they are in relation to the whole.

I also got another great suggestion from a friend to make it easier to distinguish the good guys from the bad guys. In the new version all the good guys have their lines swooshing on the left, whereas the bad guys are on the right. This made it possible to see for example that the character Vegeta (in blue) starts out as a main antagonist, but, after moving around a bit, *mostly* turns to the good side.

Apart from the female Android 18 soundly defeating some of the good guys during a few fights, it's basically only guys that are fighting.

And those were just the static parts of the visual! I also added a transition effect that occurs when you hover over a fight (Figure 7.10). A white circle that appears as a background, with everything getting bigger and the circles moving outward. The only downside is that it's quite performance heavy, because so many paths and circles change opacity. (In retrospect, this is another perfect example of a visual that I should've made in canvas instead of SVG due to the number of changing elements, but I didn't have enough canvas knowledge at the time to pull it off.)

Maybe Vegeta didn't turn into a "good guy" for the right reasons at first, until he sacrificed himself in the Majin Buu saga; *cue tears.*

Finally, I added a lot of small details that only DBZ fans would likely notice. For example, I turned the animated GIFs into links that led to the *exact* point in YouTube videos of the episode. I also replaced the GIFs with static images for those viewing in slower-performing browsers or on mobile. I annotated every fight manually to explain why it was an important fight, or what occurred. I can say that I gave it *my all*, and in general it's the attention to detail that can make a true difference overall! (๑•̀ㅂ•́)و✧

Pretty proud of those links! (^.~)

Fig.7.10 (a & b)

The transition that makes the fight more visible when you hover over it.

a

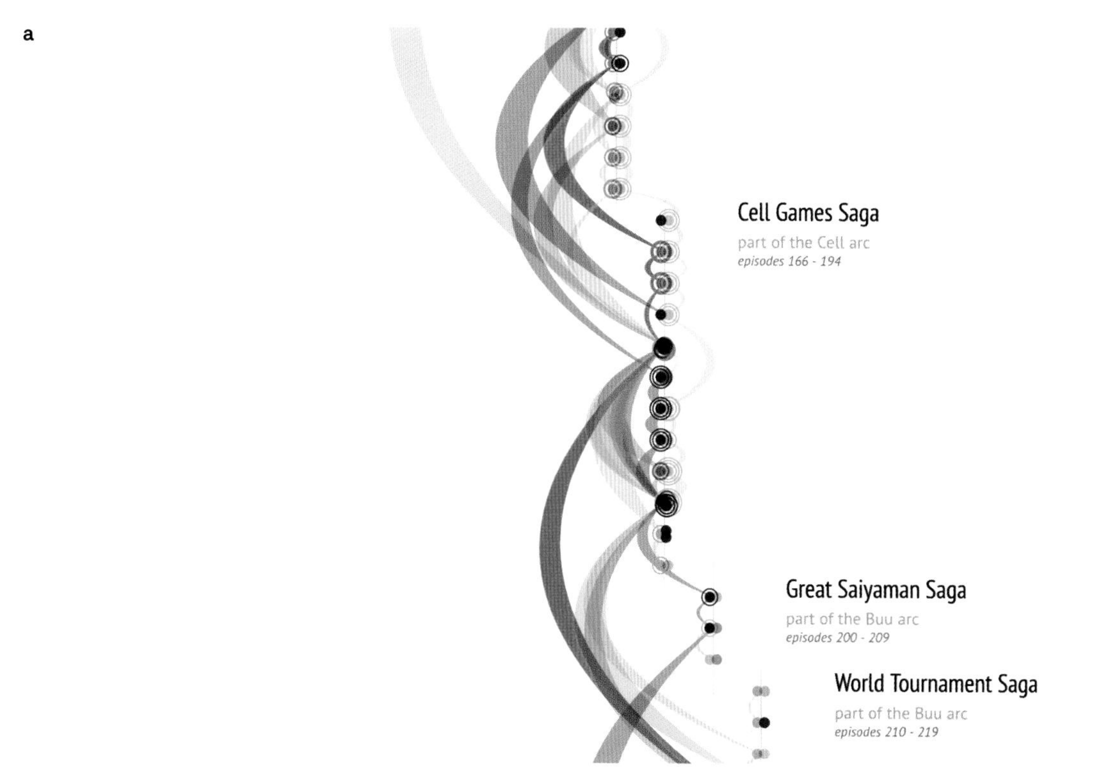

b

Reflections

When I started this project I also went into freelancing full time, and this project was the first one where I truly tracked my hours with a tool. Compared to the first six projects, this one felt like it took much more time. The final verdict came in at—*drumroll*—75 hours of time spent on this visual. (╥﹏╥) Not something that is sustainable to keep up next to my new full time job...

However, it was great fun to dive into a topic I hadn't really thought about for 15 years. It gave me a lot of enthusiasm to try and turn it into something worth looking at and informative. And although I usually shy away from the negativity found on Reddit, I had some interesting and fun discussions on the insights that we figured out about *Dragon Ball Z* after I was notified that someone shared my project in a DBZ subreddit.

Plus, I started watching *Dragon Ball Super*, which is a real nostalgia trip every episode.

> This was exactly the kind of reaction I was hoping for: making true fans see something they'd never realized.

All the Fights in Dragon Ball Z

↳ DragonBallZ.VisualCinnamon.com

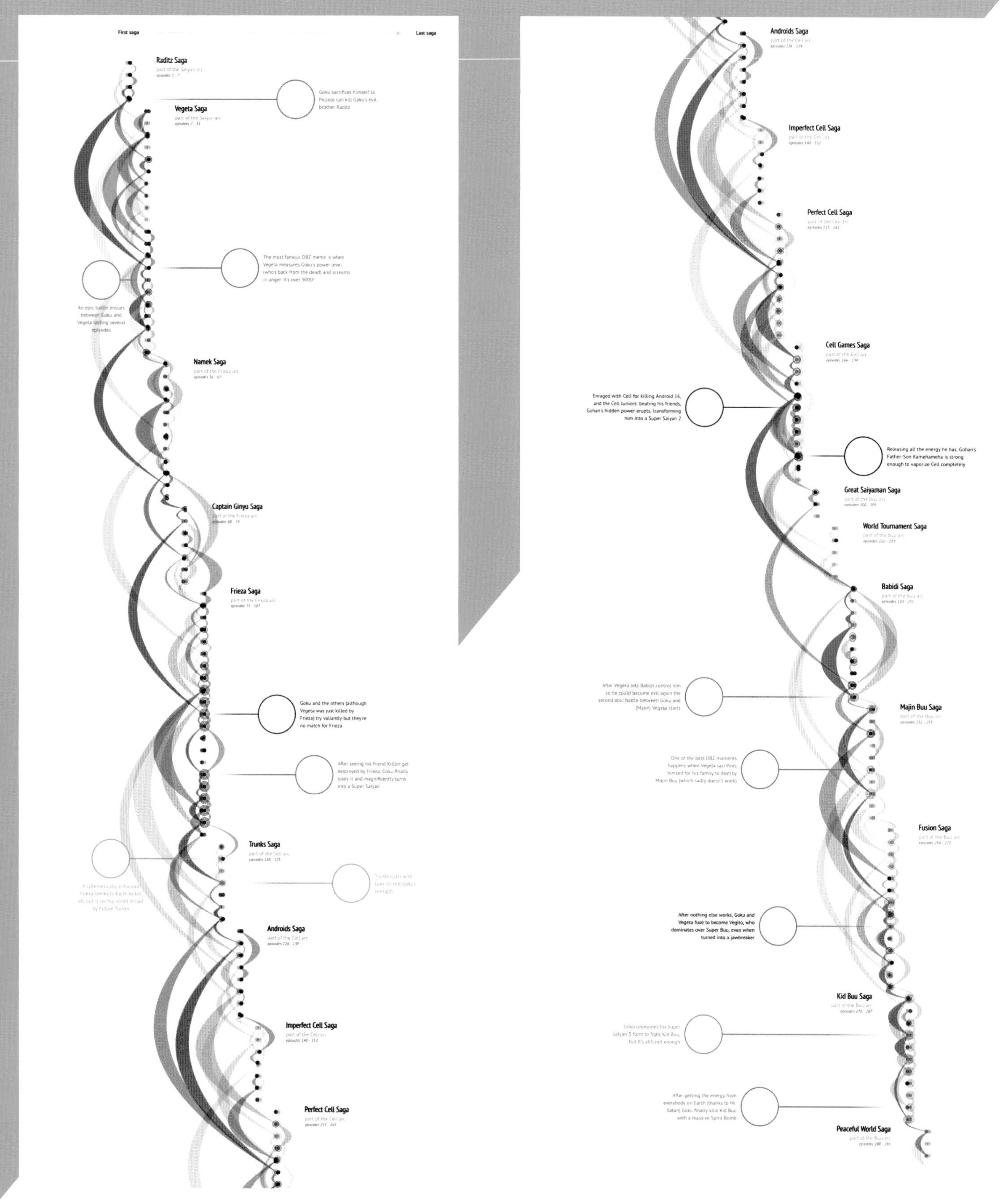

First saga
Last saga
Raditz Saga
part of the Saiyan arc
Goku sacrifices himself so Piccolo can kill Goku's evil brother Raditz
Vegeta Saga
part of the Saiyan arc
The most famous DBZ meme is when Vegeta measures Goku's power level (who's back from the dead) and screams in anger 'It's over 9000!'
An epic battle ensues between Goku and Vegeta lasting several episodes
Namek Saga
part of the Frieza arc
Captain Ginyu Saga
part of the Frieza arc
Frieza Saga
part of the Frieza arc
Goku and the others (although Vegeta was just killed by Frieza) try valiantly but they're no match for Frieza
After seeing his friend Krillin get destroyed by Frieza, Goku finally loses it and magnificently turns into a Super Saiyan
Trunks Saga
part of the Cell arc
A cybernetically enhanced Frieza comes to Earth to kill all, but is swiftly sliced in half by Future Trunks
Trunks spars with Goku to test Goku's strength
Androids Saga
part of the Cell arc
Imperfect Cell Saga
part of the Cell arc
Perfect Cell Saga
part of the Cell arc
Androids Saga
Imperfect Cell Saga
Perfect Cell Saga
Cell Games Saga
part of the Cell arc
Enraged with Cell for killing Android 16, and the Cell Juniors' beating his friends, Gohan's hidden power erupts, transforming him into a Super Saiyan 2
Releasing all the energy he has, Gohan's Father-Son Kamehameha is strong enough to vaporize Cell completely
Great Saiyaman Saga
part of the Buu arc
World Tournament Saga
part of the Buu arc
Babidi Saga
part of the Buu arc
After Vegeta lets Babidi control him so he could become evil again the second epic battle between Goku and (Majin) Vegeta starts
Majin Buu Saga
part of the Buu arc
One of the best DBZ moments happens when Vegeta sacrifices himself for his family to destroy Majin Buu (which sadly doesn't work)
Fusion Saga
part of the Buu arc
After nothing else works, Goku and Vegeta fuse to become Vegito, who dominates over Super Buu, even when turned into a jawbreaker
Kid Buu Saga
part of the Buu arc
Goku unleashes his Super Saiyan 3 form to fight Kid Buu, but it's still not enough
After getting the energy from everybody on Earth (thanks to Mr Satan) Goku finally kills Kid Buu with a massive Spirit Bomb
Peaceful World Saga
part of the Buu arc

Fig.7.11

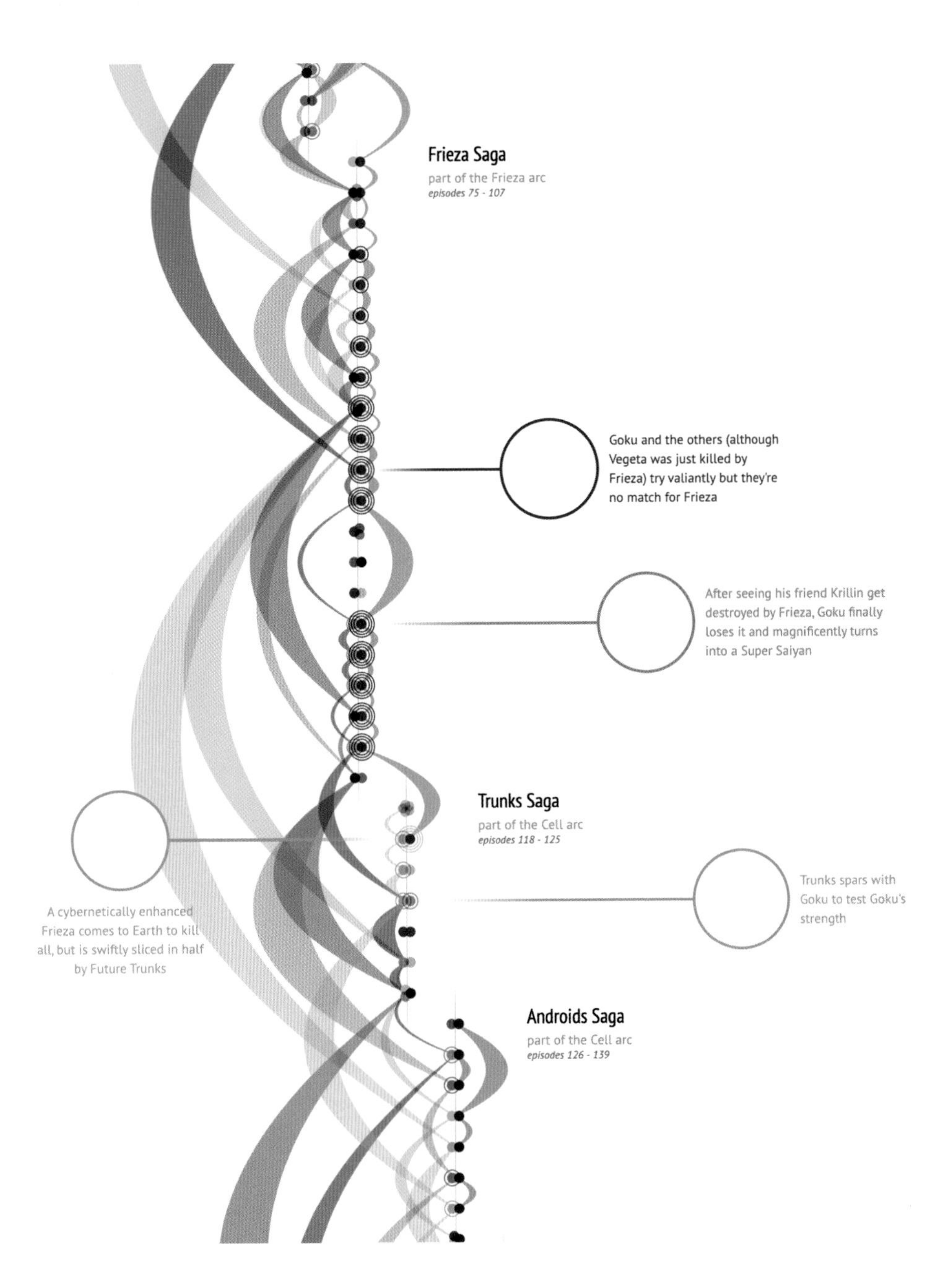
Frieza Saga
part of the Frieza arc
episodes 75 - 107
Goku and the others (although Vegeta was just killed by Frieza) try valiantly but they're no match for Frieza
After seeing his friend Krillin get destroyed by Frieza, Goku finally loses it and magnificently turns into a Super Saiyan
Trunks Saga
part of the Cell arc
episodes 118 - 125
A cybernetically enhanced Frieza comes to Earth to kill all, but is swiftly sliced in half by Future Trunks
Trunks spars with Goku to test Goku's strength
Androids Saga
part of the Cell arc
episodes 126 - 139

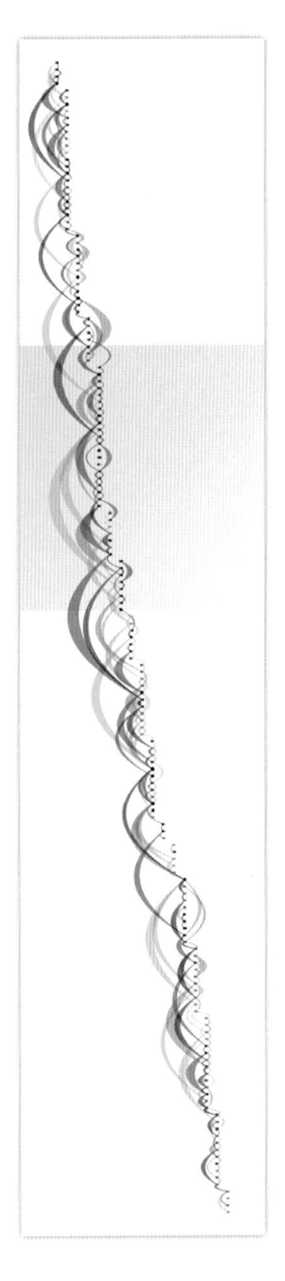

Fig.7.12

Showing the fights from one of the best sagas in the anime, the Frieza Saga, with epic final fights between (Super Saiyan) Goku (orange line along the left), and Frieza (lilac line arcing along the right side).

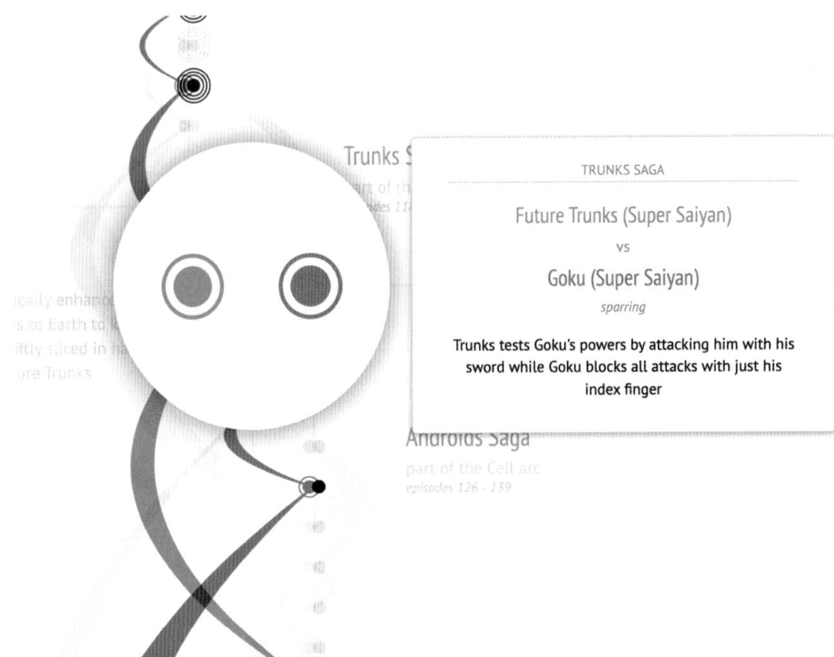
TRUNKS SAGA
Future Trunks (Super Saiyan)
vs
Goku (Super Saiyan)
sparring
Trunks tests Goku's powers by attacking him with his sword while Goku blocks all attacks with just his index finger
Androids Saga
part of the Cell arc
episodes 126 - 139

Fig.7.13

Hovering over a specific fight makes it bigger with all the circles separated. A tooltip appears that names the characters involved and shows possible other interesting facts about the fight.

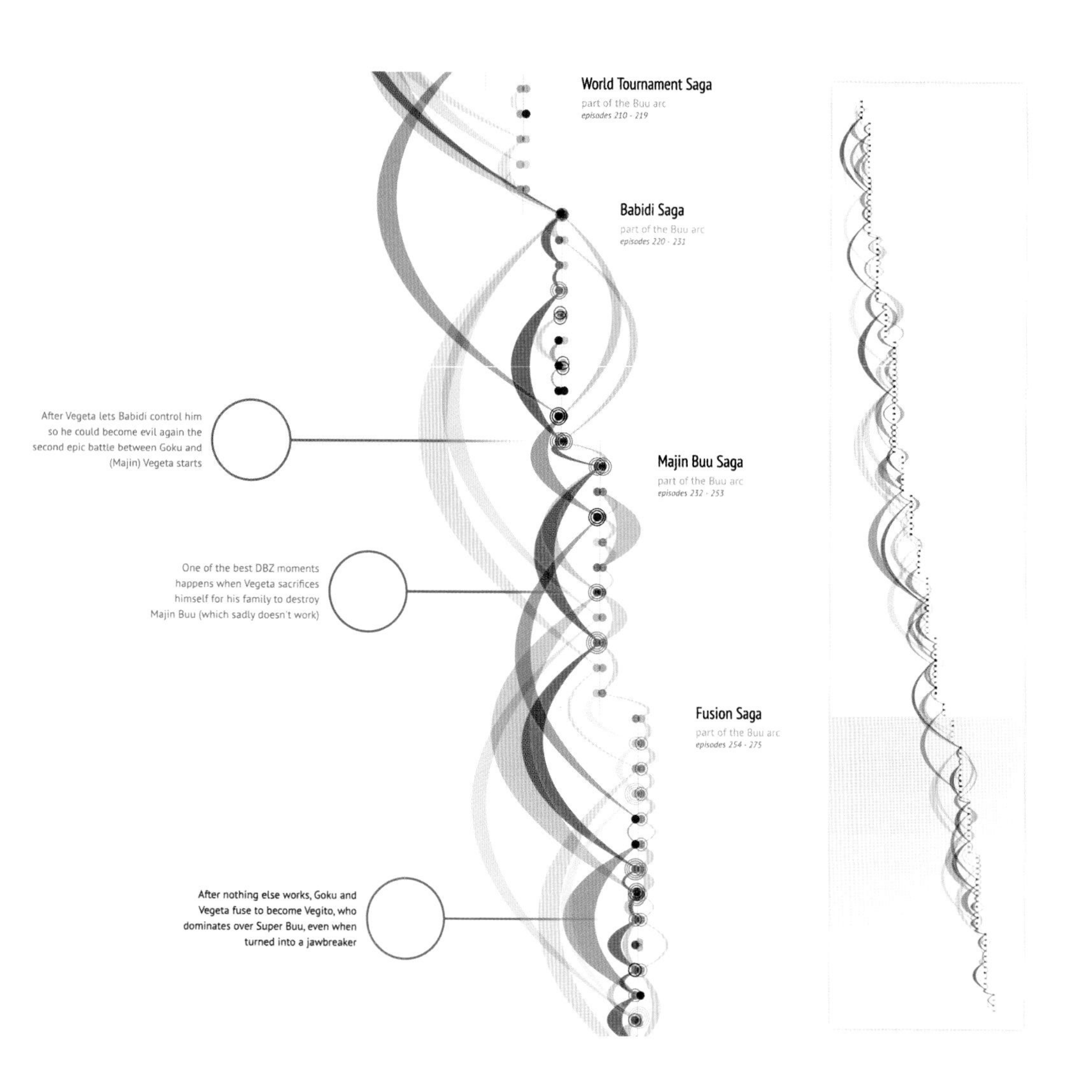

part of the Buu arc
episodes 232 - 253
One of the best DBZ moments happens when Vegeta sacrifices himself for his family to destroy Majin Buu (which sadly doesn't work)

part of the Buu arc
episodes 254 - 275
After nothing else works, Goku and Vegeta fuse to become Vegito, who dominates over Super Buu, even when turned into a jawbreaker

Fig.7.14

Strange things happened during the later sagas, with characters fusing together to become (temporarily) a new, and stronger, character.

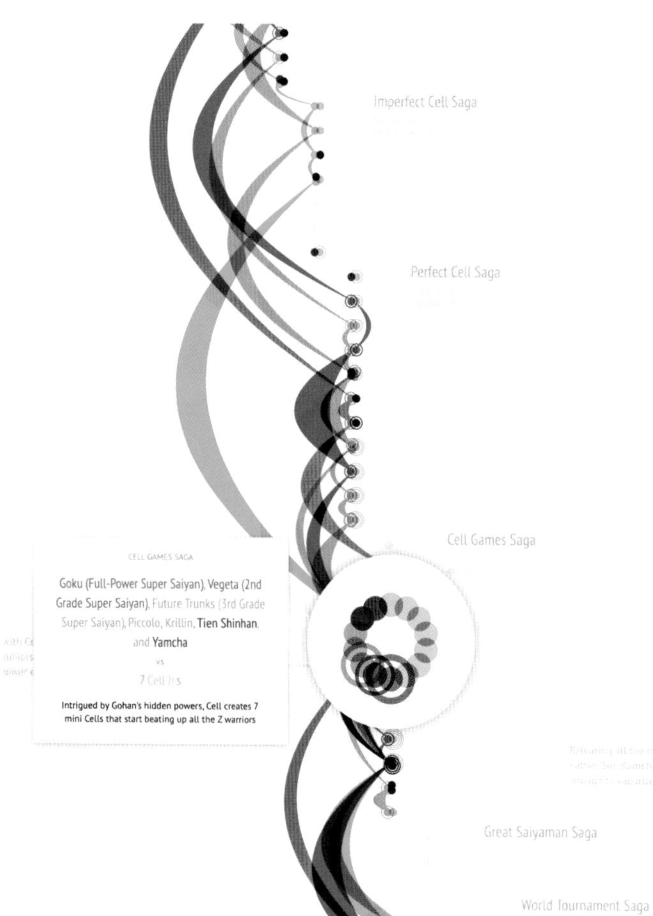

Imperfect Cell Saga
Perfect Cell Saga
Cell Games Saga
CELL GAMES SAGA
Goku (Full-Power Super Saiyan), Vegeta (2nd Grade Super Saiyan), Future Trunks (3rd Grade Super Saiyan), Piccolo, Krillin, Tien Shinhan, and Yamcha
vs
7 Cell Jrs
Intrigued by Gohan's hidden powers, Cell creates 7 mini Cells that start beating up all the Z warriors
Great Saiyaman Saga
World Tournament Saga
part of the Buu arc
episodes 210 - 219
Babidi Saga
part of the Buu arc
episodes 220 - 231
After Vegeta lets Babidi control him so he could become evil again the second epic battle between Goku and (Majin) Vegeta starts

Fig.7.15

Hovering over the biggest fight of the series, with almost all the good guys fighting against 7 mini versions of the main antagonist.

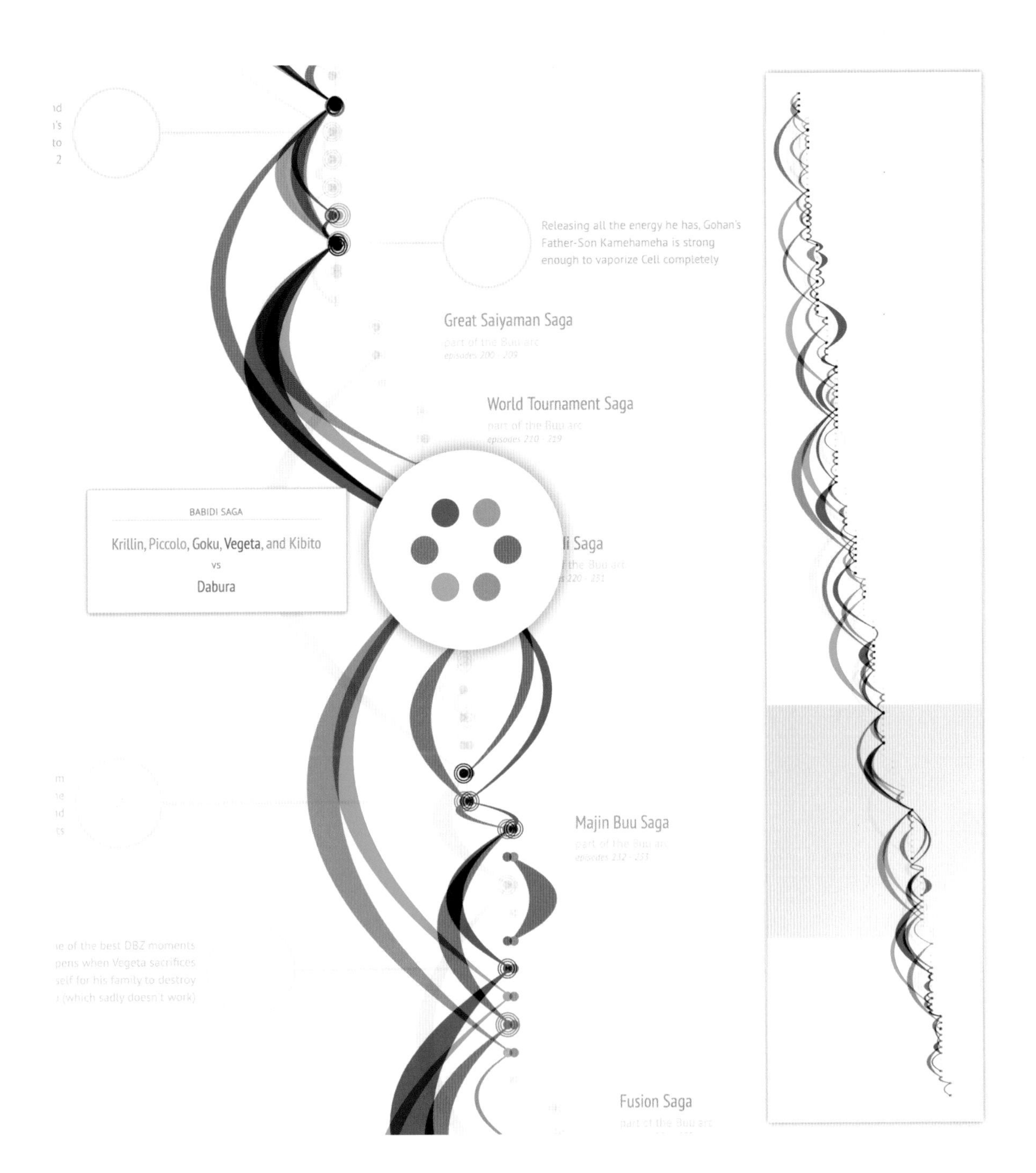

Fig.7.16

Another big fight with many good guys trying to beat a new antagonist

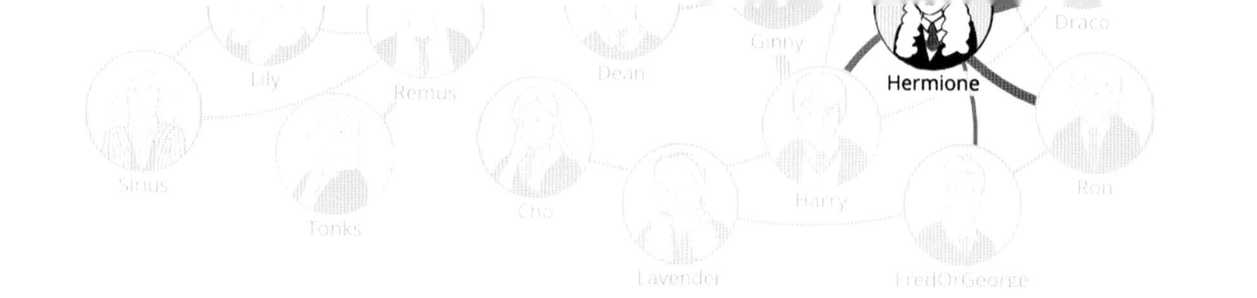

♥♥

Hermione

❶ st film ⑤ th book

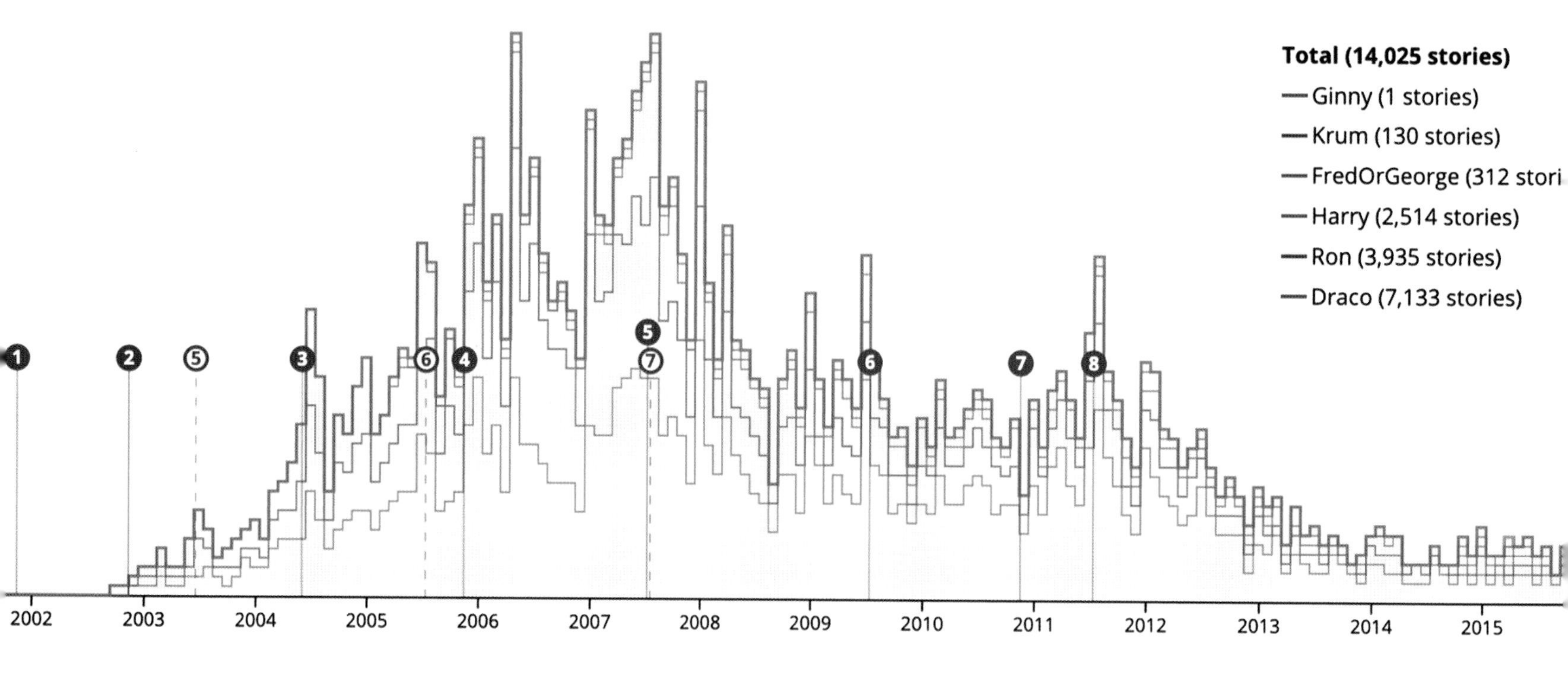

Hermione ♥ Draco

1 4 19 100 518 2.7k reviews

❶ ❷ ⑤ ❸ ⑥ ❹ ❺ ⑦ ❻ ❼ ❽

Total (7,133 stories)

*hover to see stories

The Most Popular of Them All

SHIRLEY

Alright, I'll be honest; I conned Nadieh into doing "Nostalgia" just so I could finally do a project exploring the *Harry Potter* universe. The Harry Potter book series (and *Toy Story*) will always have a special place in my heart: when the first movie came out, I was eleven years old and eagerly waiting for my Hogwarts letter. I'm the same age as actors Emma Watson and Daniel Radcliffe, who played Hermione Granger and Harry Potter, respectively. When the end credits rolled for the final movie in 2011, I had the distinct feeling that my childhood had ended.

Having said all that (sentimental stuff), I still went through a few different potential angles before settling on my final idea. At first, I wanted to do something on *Harry Potter and the Cursed Child*, but realized there were probably copyright issues with the script. I then bought a Marauder's Map hoping to visualize that somehow, but got stuck trying to figure out what kind of data I could extract from it.

Then one day, I suddenly remembered my original "Books" idea: to use Harry Potter fanfiction as a proxy for fan reactions to the movies and books.

Data

To create the dataset, I decided to scrape HarryPotterFanfiction.com because of its rich and Harry Potter specific metadata: in addition to categorizing by genres and ratings like many other fanfiction sites, it also organizes content by eras ("Hogwarts," "post-Hogwarts," "Marauders," "New Generation," etc.) and character pairings.

Like my "Movies" project, I used the Node.js *http* module to request for each page, but I also decided to use `html2json` to turn the page responses into JSON. I did this because with the "Movies" page responses I only needed to grab the ID of each movie, so I only had to write one regular expression. But for this project, I wanted to grab a lot of metadata and I didn't want to write regular expressions for all of them. With the HTML response converted to JSON, I only had to write a small function to loop through the children objects for each story and find the metadata I wanted (though I still had to do some work cleaning up the raw strings).

Because the website would only give me 25 stories per page, I had to loop through and request approximately 3,300 pages. I was worried that the server would time me out for making so many requests at once, but because I made those requests synchronously (asking for the next page only after the current page had responded and I had completely parsed through it), there were just enough seconds between each page request that I never had any issues.

I usually try to find the simplest and quickest way to gather a static dataset because I know I don't have to worry about performance, but this time that laziness actually paid off (๑•̀ㅂ•́)و✧

When I had all the data, I wanted to get a sense of it by logging some statistics to Terminal, the command line interface for Mac. I looked primarily at the publication years (Harry Potter fanfiction spiked in 2007 when the final book and fifth movie came out, but has been dwindling significantly ever since), and the most popular pairings (unsurprisingly, Harry/Ginny followed by James/Lily). I also looked at which characters appeared the most often in these stories (Harry was the most popular, followed by Hermione in second, the author's made-up "Original Character" in third, and Ron in fourth—which really says something about Ron's lack of popularity).

I had originally tried to log statistics to Terminal as a quicker way to analyze data (as opposed to building visualizations from scratch to test my hypotheses, like I did with my previous projects), but found that it didn't work out too well for me. My knowledge was limited to logging the min/max and some medians, and though they gave me a basic understanding of the data, they were very superficial and didn't help me find the deeper, more interesting insights I was seeking. And though I knew there were many more sophisticated approaches out there, I didn't know when or how to use them, so I decided to move onto the sketch section.

Fig.7.1

Logging some statistics to Terminal.

```
[ [ 'undefined', 18569 ],
  [ 'Other Pairing', 18416 ],
  [ 'Harry/Ginny', 15200 ],
  [ 'Ron/Hermione', 14171 ],
  [ 'James/Lily', 12057 ],
  [ 'Draco/Hermione', 8826 ],
  [ 'Sirius/OC', 6993 ],
  [ 'OC/OC', 5491 ],
  [ 'Remus/OC', 4322 ],
  [ 'Draco/OC', 3866 ],
```

```
[ [ 'Hogwarts', 23177 ],
  [ 'undefined', 14964 ],
  [ 'Marauders', 13299 ],
  [ 'Post-Hogwarts', 12914 ],
  [ 'Next Generation', 10270 ],
  [ 'Other', 4137 ],
  [ 'Pre-Hogwarts', 1879 ],
  [ 'Founders', 517 ],
  [ 'The Cursed Child', 7 ],
  [ 'Fantastic Beasts', 5 ] ]
```

Sketch

I decided to start my brainstorming by writing down all the metadata I had. From there, I identified genres, eras, and pairings as potentially interesting metadata that might correlate with a story's number of reviews, which I used to measure its popularity and "success." I also jotted down some questions for the data: which pairings were the most popular, and which other pairings did they usually appear with? Who did a character get paired with the most?

My first idea was to create a timeline of stories for each character and denote pairings for that character on top of the timeline. But it was around that time that *The Pudding*, a digital publication dedicated to visual and data-driven journalism, came out with a piece called "The Shape of Slavery"[1] (Figure 7.2). I liked how they used both the size and color of the dots on a map to encode two related dimensions (the size of Black population and their percent of total population). I was also visually inspired by the sequential color scale they used, which ranged from a deep blue to magenta to orange to yellow on a navy background.

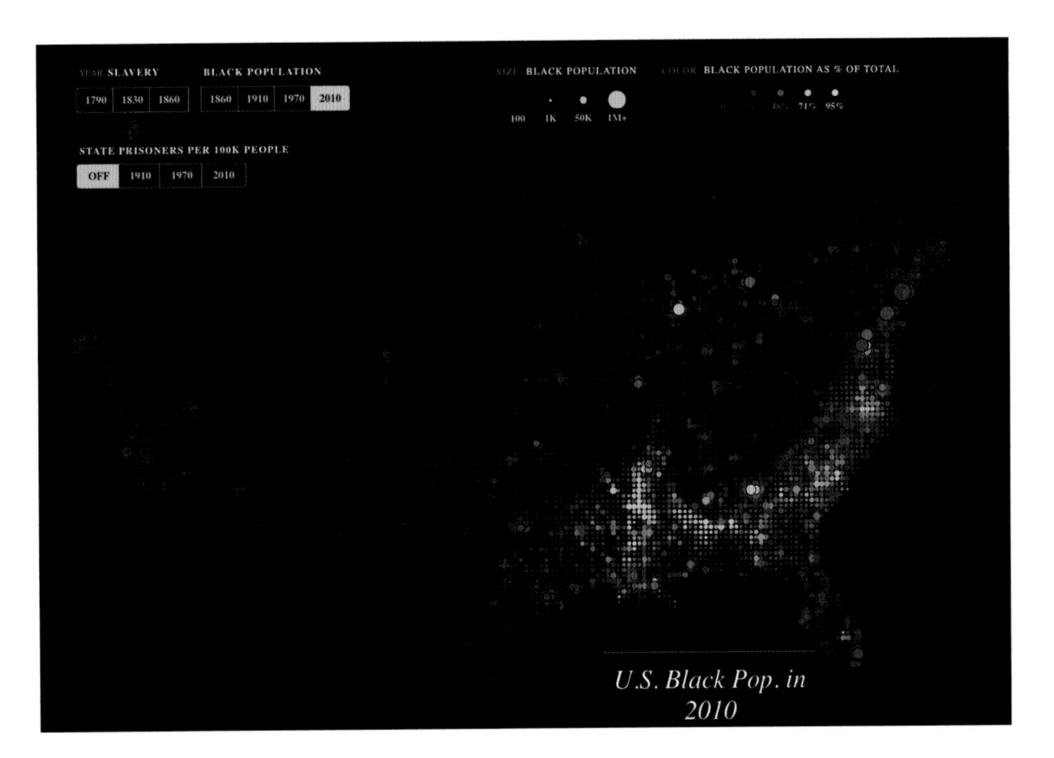

Fig.7.2

"The Shape of Slavery," a visual essay from *The Pudding*.

So, inspired by their visualization, I decided to give the dots a try. Since there were roughly 80,000 stories on HarryPotterFanfiction.com, I decided to have each dot represent up to 100 stories, color the dots by the average number of reviews (and thus the popularity) for those 100 stories, and place those dots over a timeline. I was keen on using a timeline because in my data exploration I noticed that there were definite spikes and declines in the number of stories across the years, and I knew a timeline would give good context. To place the dots, I grouped stories by the month of their publication (which determined their x-position), then stacked each set of 100 stories on top of each other. It ended up looking like a histogram made up of dots (Figure 7.3).

[1] "The Shape of Slavery" data visualization by The Pudding: https://pudding.cool/2017/01/shape-of-slavery/

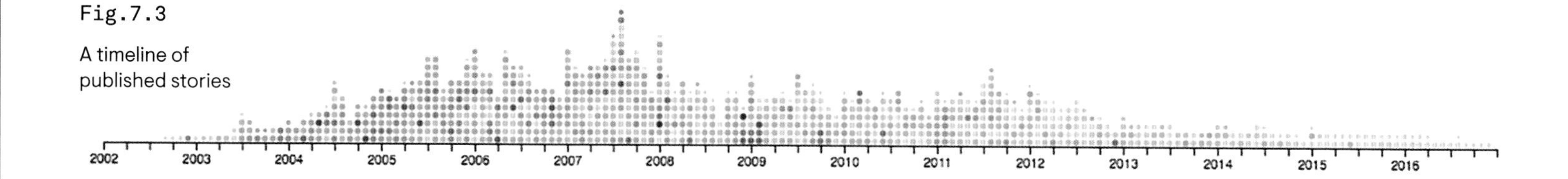

Fig.7.3

A timeline of published stories

To get a sense of character pairings, I also sketched a matrix of the top five most popular characters along with their partners and placed all the stories at the intersection of those pairings (Figure 7.4, right).

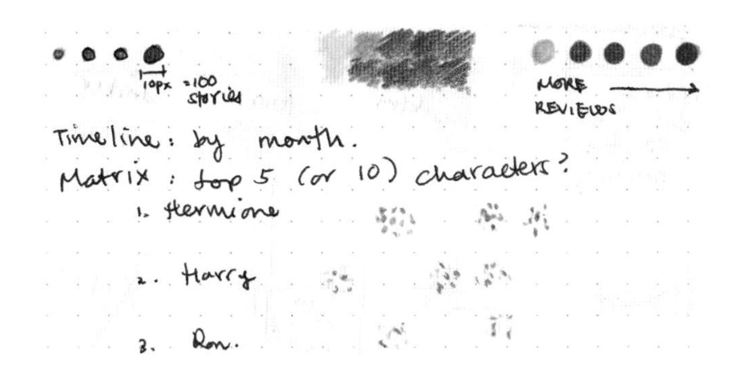

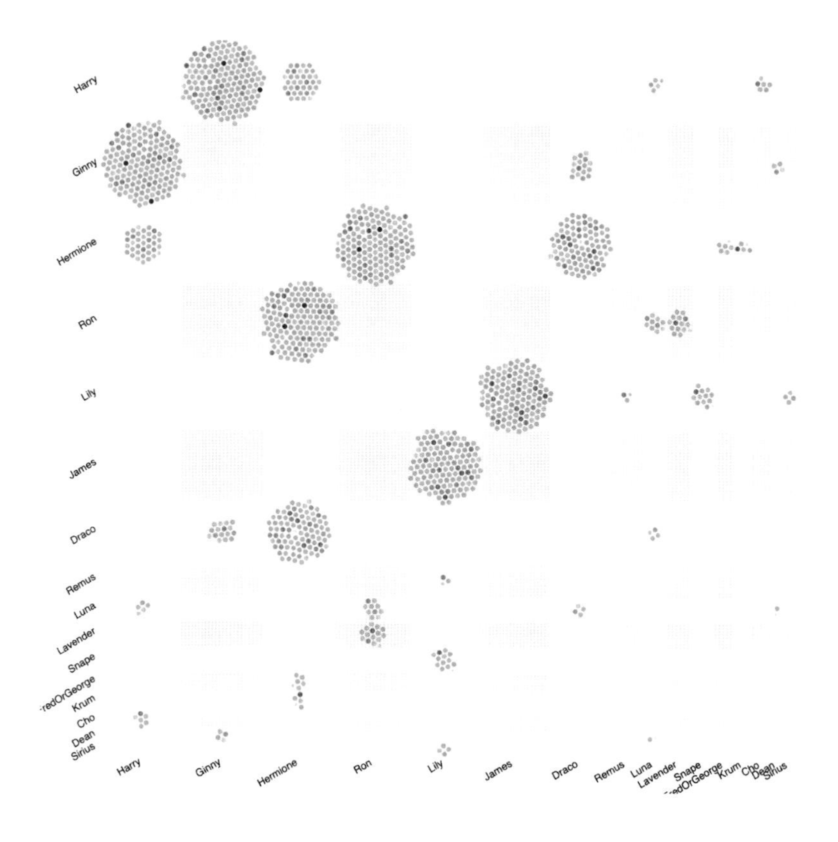

Fig.7.4

Sometimes my sketches are really simple, just jotting down some ideas (left), and the matrix idea implemented (right).

I quite liked the timeline. I could immediately see overall trends, including when the most stories were published and when those stories had the most reviews, which really showed when Harry Potter and the fandom peaked. I had mixed feelings about the matrix, though; on one hand, I liked being able to see who a character was paired with and in how many stories. On the other hand, I didn't like how much space the matrix took and how it mirrored itself across the diagonal and thus had redundant information (a drawback of matrices).

Looking back now, I do like how I varied the size of the row/columns for less popular characters so that the matrix didn't take up as much room.

I ultimately decided to keep the timeline, but removed the matrix. With that decision, I wasn't sure how to move forward with the visuals I had and what angle I wanted to approach the data from. It was around that (uncertain) time that I hung out with my friends RJ Andrews and Catherine Madden for an afternoon, and two beautiful things came out of that day.

First, while showing them screenshots of my progress, RJ noticed my timeline of Draco/Hermione was stacked on top of Ron/Hermione (Figure 7.5). He excitedly pointed out that one pairing was canon (Ron/Hermione, official in the books) and the other was non-canon (Draco/Hermione, a fan pairing), yet the non-canon had more stories. And thus we were able to narrow in on an interesting angle: how do canon and non-canon relationships compare in popularity? I re-thought all of my visuals around that question.

Fig.7.5

Timeline of Hermione/ Draco and Hermione/ Ron pairings.

Draco/Hermione (795)

2002 2003 2004 2005 2006 2007 2008 2009 2010 2011 2012 2013 2014 2015 2016

Ron/Hermione (478)

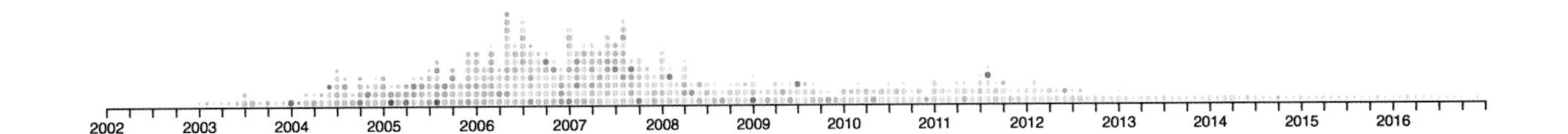

One of the main things I decided to do was visually separate canon versus non-canon pairings using color: I chose pink for canon and purple for non-canon. I chose warmer colors since, to me, they represented relationships more, and I always loved the look of pink and purple together. Since I wanted to map those colors to the number of reviews, I called `chroma.js`'s `scale` function with an array of colors to create a gradient that traveled through all of them.

`chroma.js` is an amazingly handy library for working with color.

And in that same afternoon, Catherine offered to draw the Harry Potter characters I needed in my visualization (Figure 7.7). Her illustrations really added another dimension of aesthetic and polish to the project, and I still can't thank Catherine enough for these beautiful icons. (*´▽`*)/

Fig.7.6

Sketch of idea comparing canon and non-canon pairings.

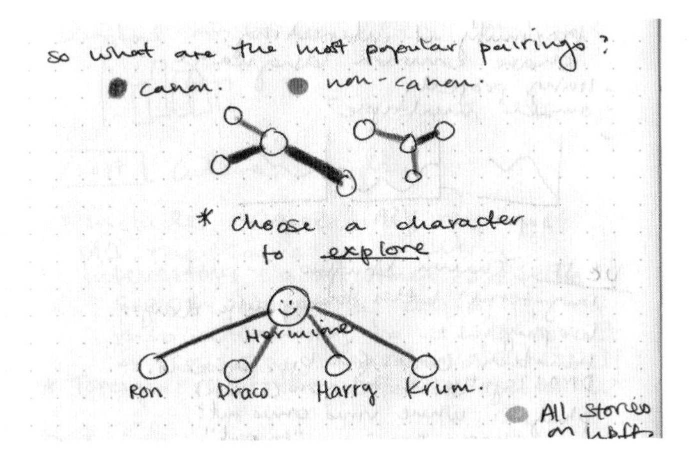

Fig.7.7

Catherine's gorgeous illustrations of Harry Potter characters.

Code

For this project, the hard part was really in the ideation; the coding was pretty straightforward. Once I sketched out the layout I liked, I got back to my timeline and added book release dates, film dates, and GIFs for context:

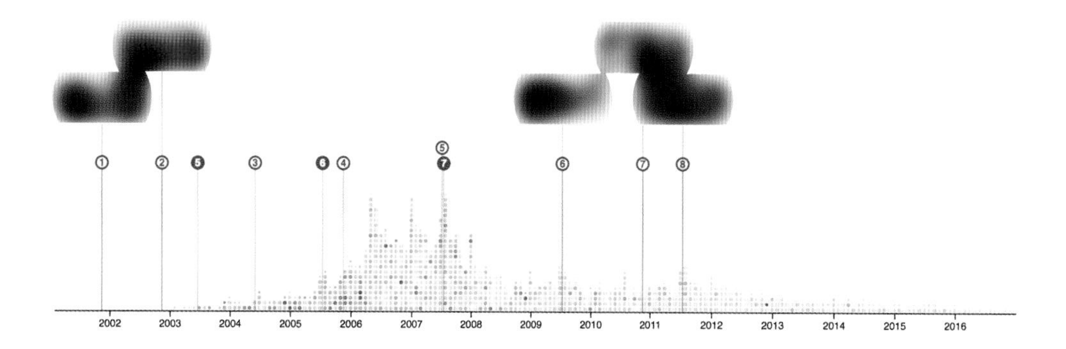

`Fig.7.8`

Timeline with book dates (filled circle), film dates (white circle), and GIFs for context (blurred for copyright reasons).

And immediately, some interesting things jumped out: fanfiction really spiked in number between the publication of the sixth and seventh books (2005 and 2007 respectively) and also during May 2006 (I still haven't figured out why that month is significant). There were also spikes before and after a book or movie release and spikes during the winter holidays (presumably because people had more time to relax and write stories).

Next, I started looking into how I wanted to depict multiple pairings for a character. I started by overlapping a line chart for each pairing on top of each other. Unfortunately, the lines were very thin and hard to distinguish from each other, so I tried filling them to a baseline to make them appear more solid and hopefully easier to distinguish (Figure 7.9). I used `D3.js`'s *area* module to accomplish this.

`d3.area()` works by passing it a set of points, and it draws a line between all of the points and fills the area below the lines down to the baseline.

But again, the area charts, while pretty, were still hard to read at a glance. I then tried stacking the area charts, which was only slightly easier to read; the pairings with the least stories blended into the others for most of the months. To try and mitigate this, I played around with a few of the curve options in `D3.js` and I ultimately decided on `d3.curveStep`, which had its downsides (when the lines are too close together, it's hard to tell where one ends and the next starts), but I still liked it the most because of how clean it looked (Figure 7.10).

By default, `D3.js` connects an array of points with just a straight line, but we can give it additional options to draw the line with different types of curves, which may help distinguish the lines that are closer together

Once the pairings timeline was done, I turned my attention to visualizing the genres for each pairing. I was curious to see if specific pairings leaned more heavily towards specific genres, and it turned out that they did. For example, Ron/Hermione stories tended to lean more towards drama, fluff, and humor, while top Draco/Hermione genres were drama and angst, with two or three times more in the horror/dark genre than all of Hermione's other pairings. James/Lily's top genre was humor, followed closely by drama, and Harry/Ginny's top genres were drama and action/adventure.

To show this, I initially considered displaying the cumulative numbers of stories for each genre, colored and placed according to how "dark and serious" or "light and fluffy" it was. I was also interested in seeing if specific genres rose or fell because of occurrences in the books or films, so I initially had plans to do a timeline along with a horizontal spectrum of the darkness/lightness of genres. I ended up scrapping that idea though, because I couldn't quite figure out how to quantify the darkness/lightness of a genre on a continuous spectrum (how much lighter was humor than drama, for example), and kept only the timeline (Figure 7.11).

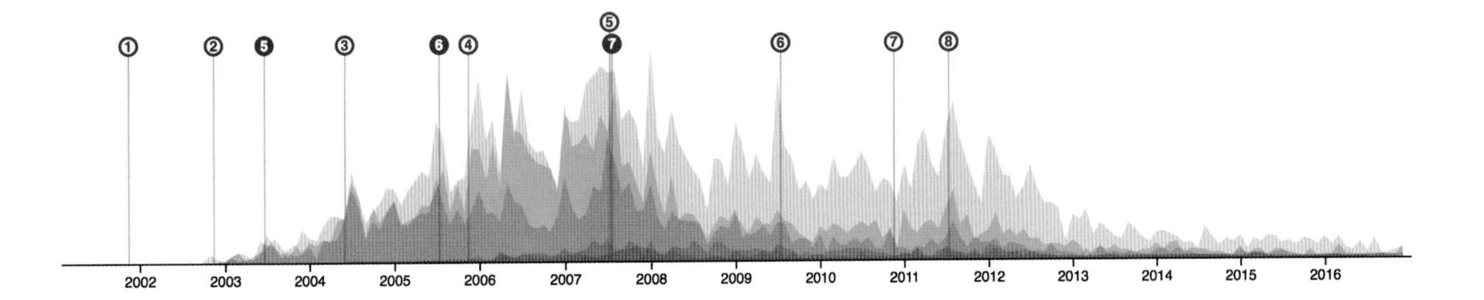

Fig.7.9

Timeline with pairings as area charts overlapping each other.

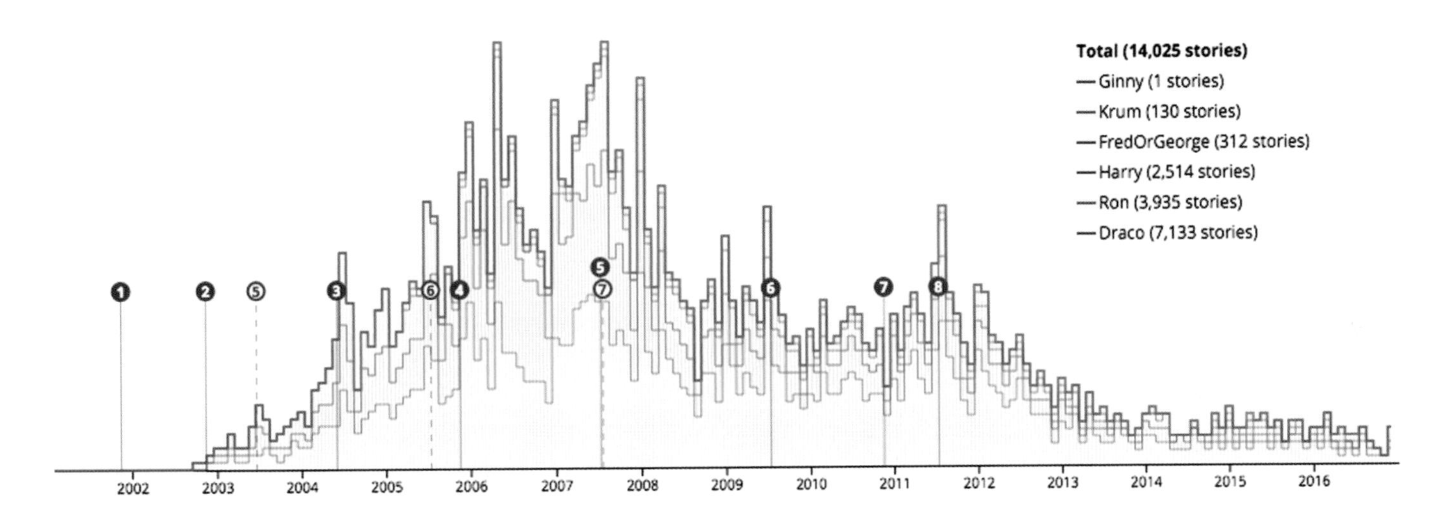

Fig.7.10

Timeline with pairings as area charts stacked on top of each other.

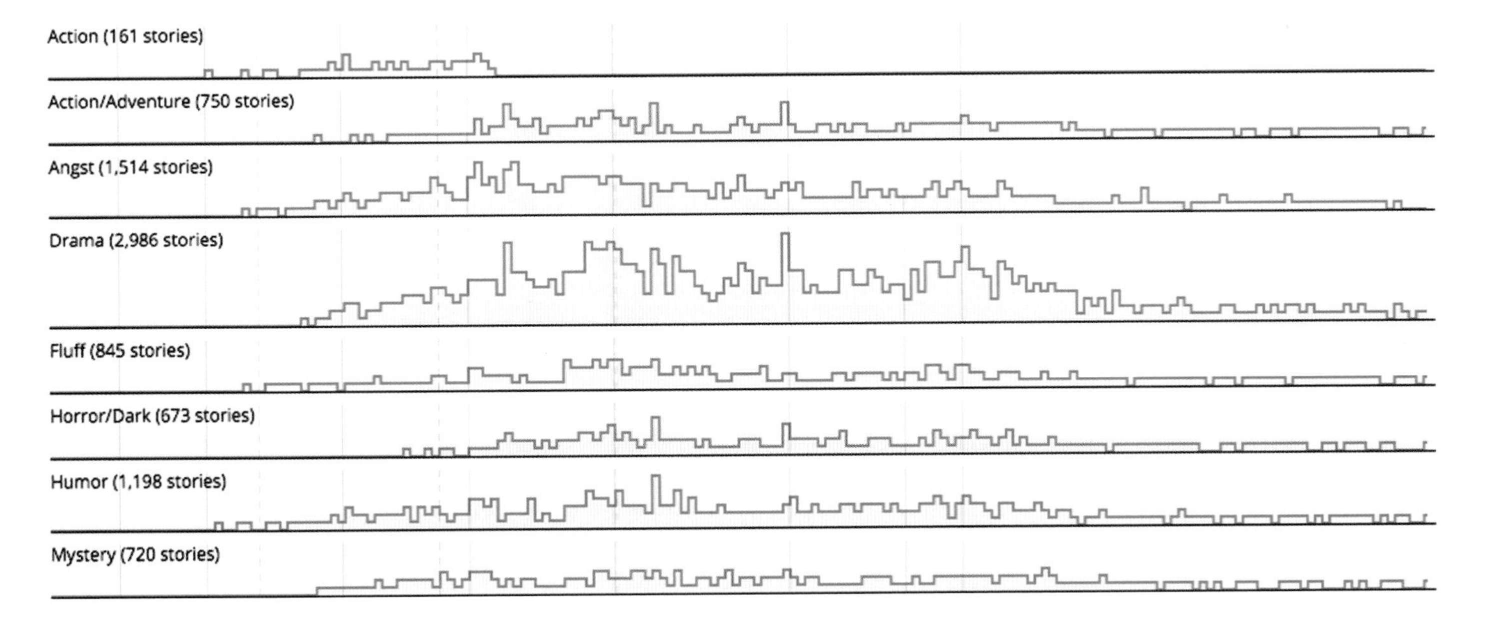

Fig.7.11

Genres organized into timelines.

I lined the genre timelines up with the pairing timeline and used a canvas element to color each block by number of story reviews (I converted my original dots into blocks to fit in the step area chart I ended up creating). I added a legend above the timelines to map the color gradient back to the number of reviews. I also added a hover interaction for each block of stories to show the titles, authors, and number of reviews for the stories within that block:

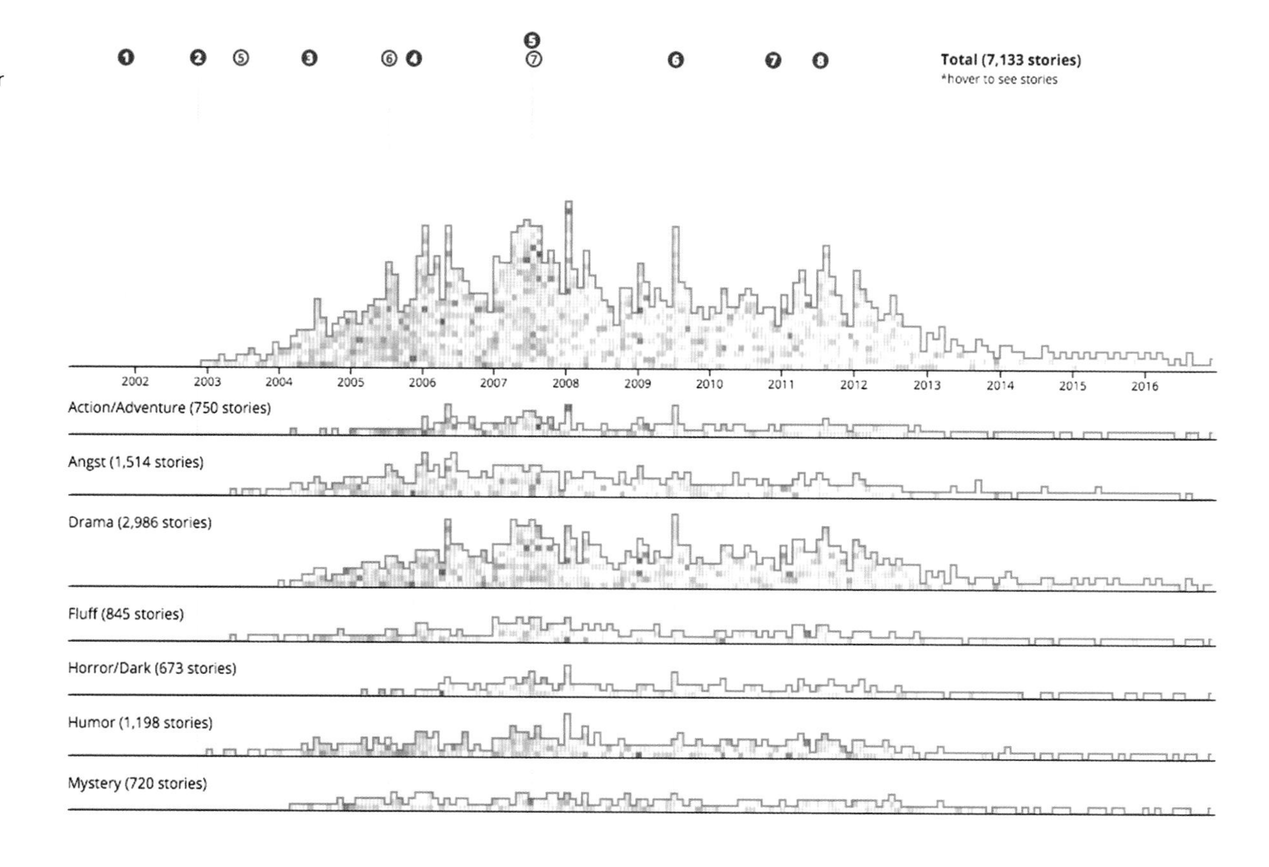

Fig.7.12

Final visualization for each pairing.

Finally, I put in the graph of characters linked by their pairings and had it double as navigation for selecting and exploring each character (Figure 7.13). It defaults to Hermione, who has the most romance stories with the most suitors. (o≧∀≦)o

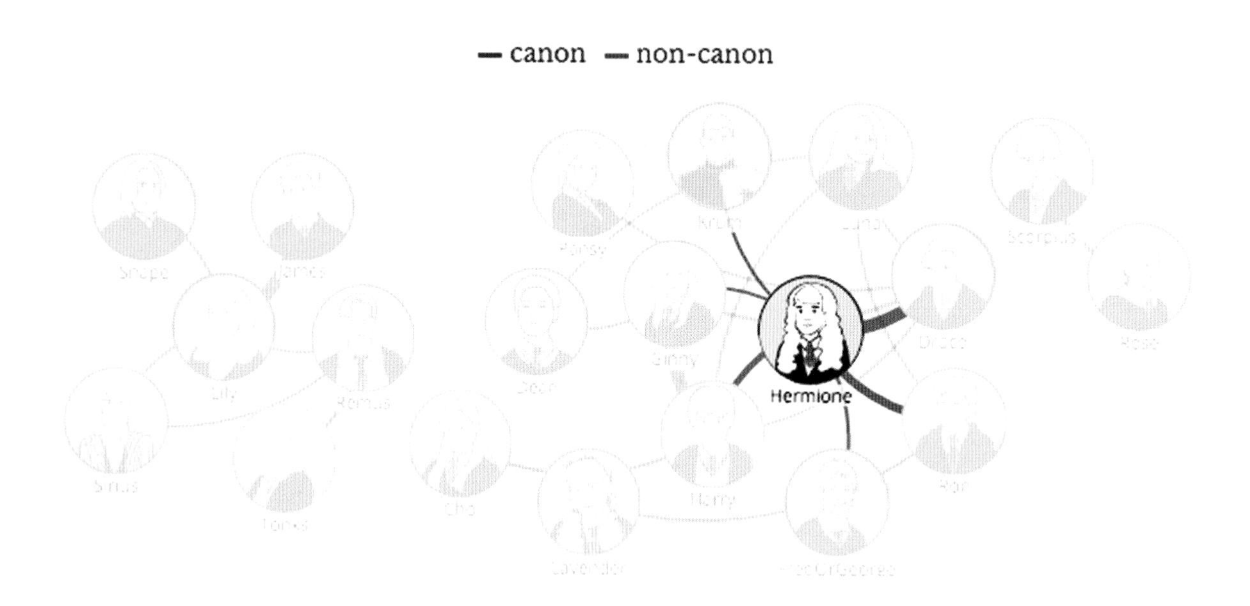

Fig.7.13

Graph of the characters, linked by their relationships.

Reflections

I really like the overall look of the project (Catherine's character illustrations really added to the aesthetics of the visualization), but I have mixed feelings about the fact that every visual is some form of a timeline. I really wonder if I should have changed up the visuals to break up the monotony. I'm also dissatisfied that though I was able to eventually see some interesting trends between genres and pairings, that I didn't have the time to annotate those findings.

Despite all that, I'm happiest about the fact that I managed to do this whole piece with gradients of only two colors: pink and purple. Generally, I'm quite over-reliant on color in my visuals and wanted to try something more subtle in comparison, and I feel I was quite successful on that front!

The Most Popular of Them All

↳ shirleywu.studio/projects/hpff

The Most Popular of Them All: a look at fanfiction's favorite ships

BY SHIRLEY WU

There are many aspects to the *Harry Potter* franchise: books, movies, a new play, studio tours and theme parks. But when it comes to gauging fan reaction, there's probably no better place than fanfiction; with ~760,000 stories on fanfiction.net, ~110,000 on Archive of Our Own, and ~80,000 on Harry Potter Fanfiction - that's a lot of reactions.

For this visualization, I looked into Harry Potter Fanfiction because of its more manageable numbers, meaningful metadata, and (I'll be real) relatively family-friendly content. First interesting find: the number of stories peaked in **2007** at **12,613** stories, held steady at around 7,000 until **2011**, and have been rapidly declining since. Second interesting (though not unexpected) find: at **52,908** stories, **Romance** is the most popular genre - almost twice as popular as the next biggest genre, **Drama** (**29,429**).

So of course, I was curious about all the ships: who does the fandom want to see with whom? The top two - **Lily** and **James**, and **Ginny** and **Harry** - are definitely canon, but the third - **Hermione** and **Draco** - is non-canon. (I have this theory that Dramione blew up after Hermione punched Draco in the third movie, but alas, the data is inconclusive.) The most popular leading lady is undoubtedly **Hermione**, with a whopping **six** suitors.

Explore Hermione's stories, or select another character to see theirs:

— canon — non-canon

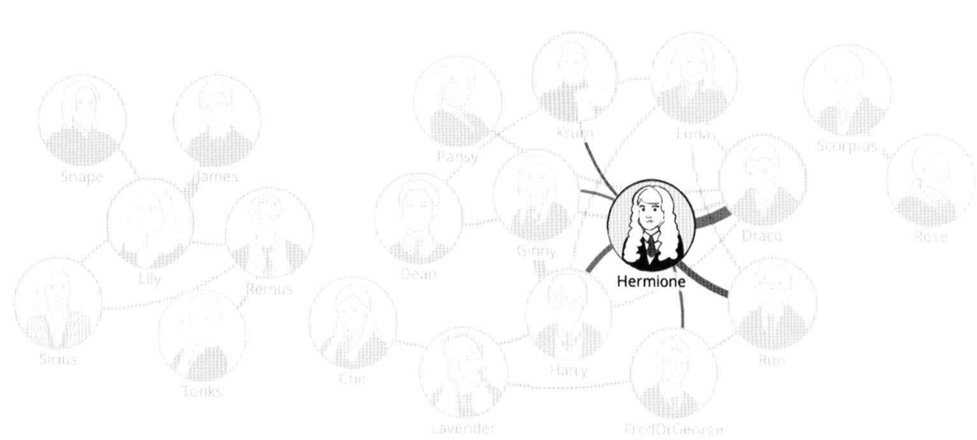

Fig.7.14

Selecting a character in the graph switches the visualization to center around that character instead.

Hermione
❶ st film ⑤ th book

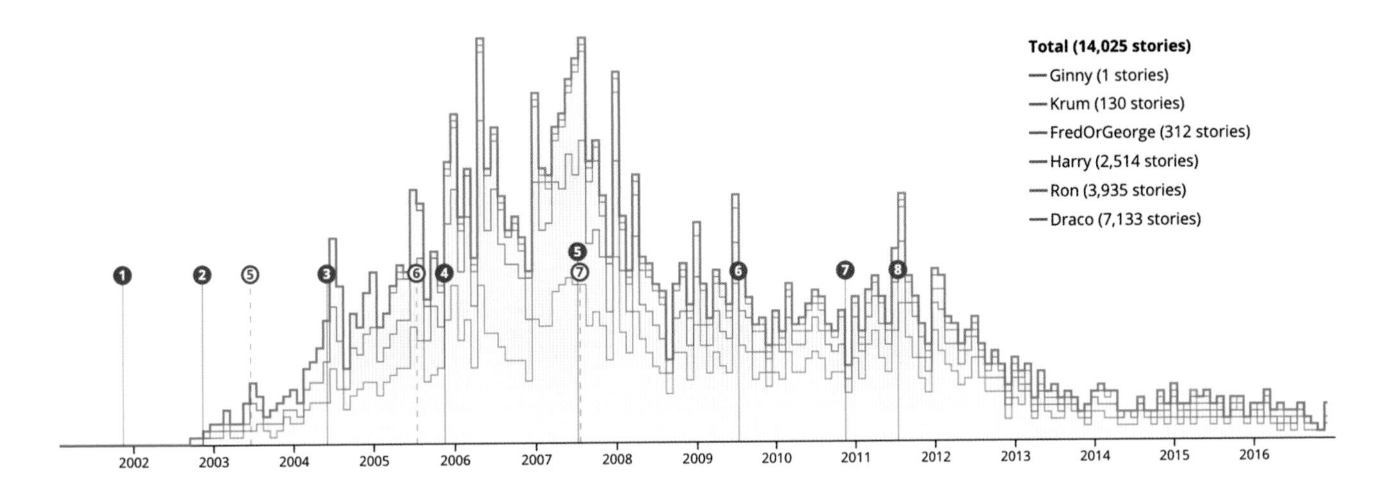

Total (14,025 stories)
—Ginny (1 stories)
—Krum (130 stories)
—FredOrGeorge (312 stories)
—Harry (2,514 stories)
—Ron (3,935 stories)
—Draco (7,133 stories)
2002
2003
2004
2005
2006
2007
2008
2009
2010
2011
2012
2013
2014
2015
2016

Hermione ♥ Draco

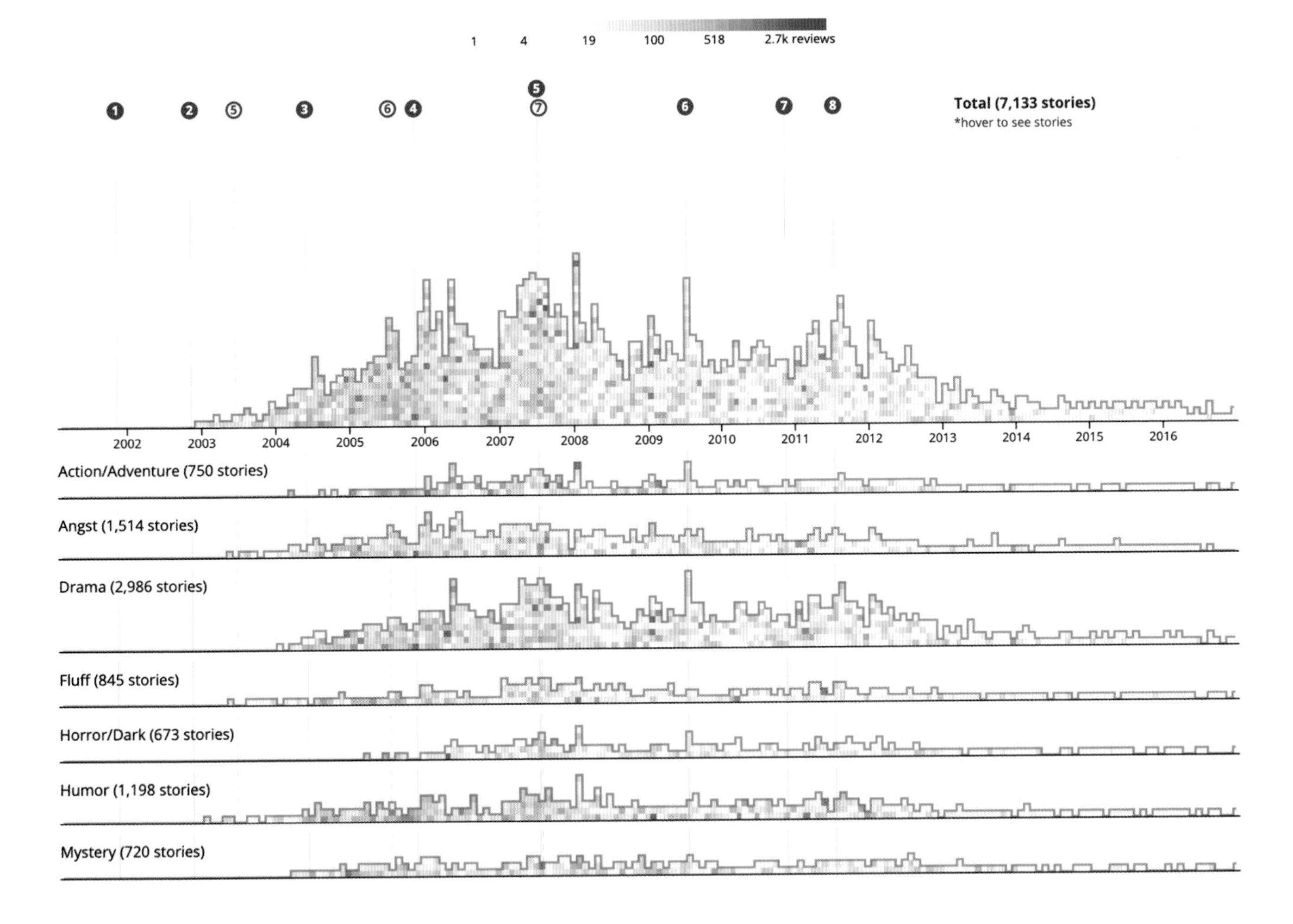

1
4
19
100
518
2.7k reviews
Total (7,133 stories)
*hover to see stories
2002
2003
2004
2005
2006
2007
2008
2009
2010
2011
2012
2013
2014
2015
2016
Action/Adventure (750 stories)
Angst (1,514 stories)
Drama (2,986 stories)
Fluff (845 stories)
Horror/Dark (673 stories)
Humor (1,198 stories)
Mystery (720 stories)

Hermione ♥ Ron

Total (3,935 stories)
*hover to see stories

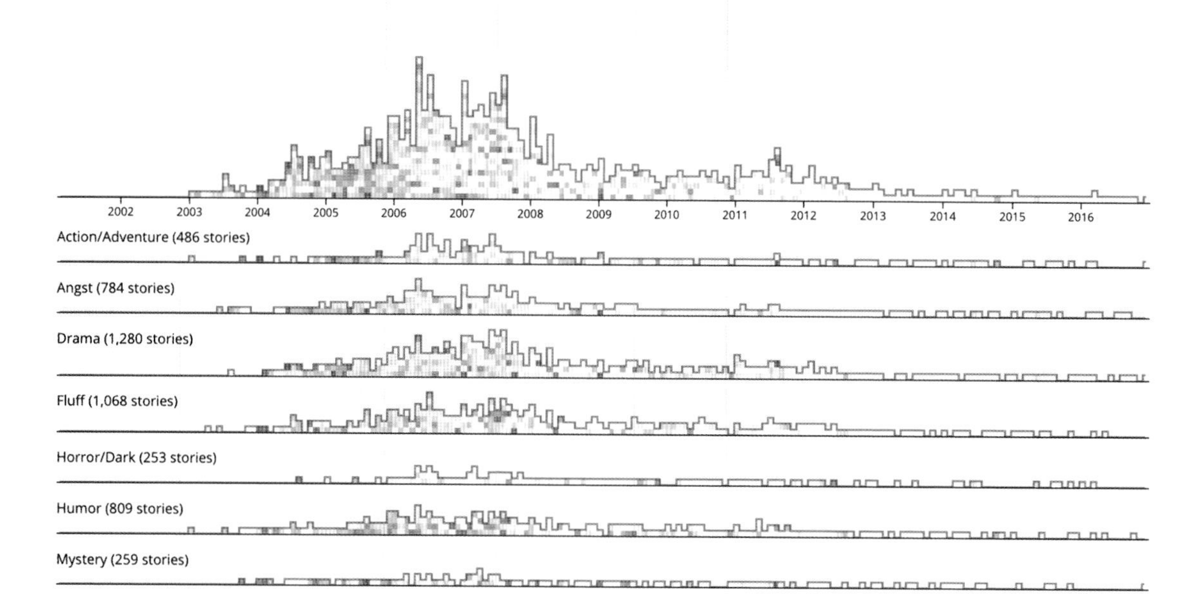

Hermione ♥ Harry

1 4 19 100 518 2.7k reviews

Total (2,514 stories)
*hover to see stories

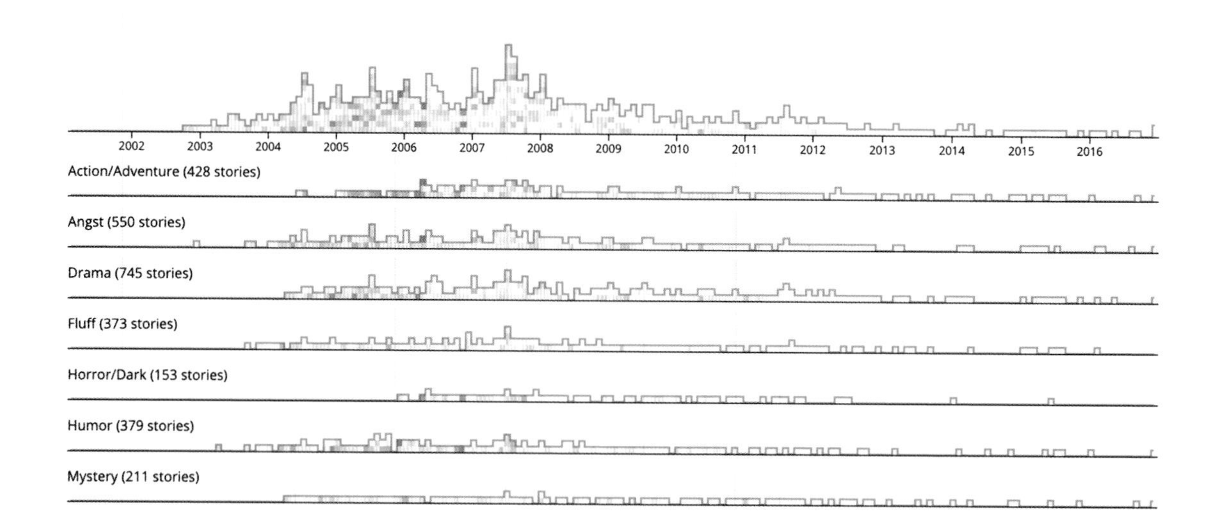

Hermione ♥ Ginny

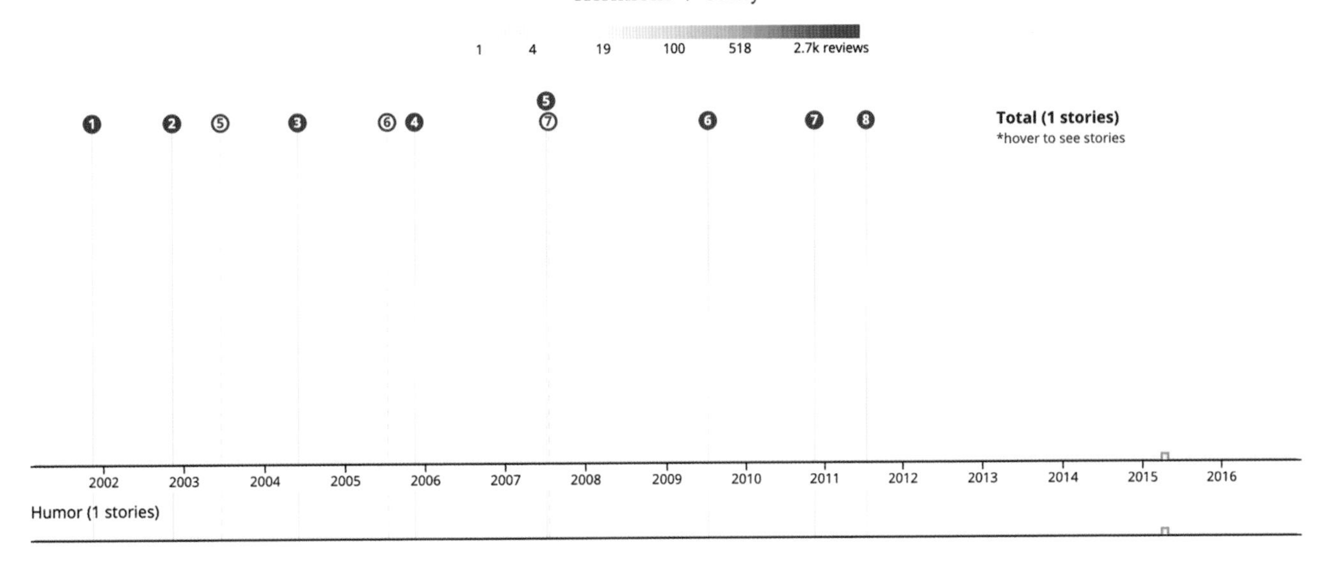

A tribute

My family immigrated to America when I was 10 years old, and I spoke no English. I studied every day, motivated by the need to understand what was going on around me. My dictionary-reading paid off and half a year later, I was deemed "fluent enough". I was able to read most grade-school books by then, but the Harry Potter series eluded me; I had checked out *Sorcerer's Stone* many times, but found it too difficult every time.

When *Goblet of Fire* came out that summer, my parents came back from Costco with the hardcover. Money was still tight in our household back then, and I had never owned such a beautiful book; I was touched and determined to get through the whole book. When I did, I was ecstatic. I binged the first three books, and was proud when I joined my first Harry Potter conversation.

I have only good memories associated with Harry Potter, from that first hardcover (my parents made it a tradition to buy me every hardcover since), to the 3am screenings with my college floormates, to the full-theatre standing ovation as the credits rolled on the final film. It's given me a whole other world to daydream about, and a sense of belonging in this one.

This year is the 20th anniversary of *Philosopher's Stone*.
With all my love, thank you JKR ♥

many many thank you's to Catherine Madden for
the beautiful illustrations of every character (and of me)

made with ♥ for January: datasketch|es
a monthly collaboration between Nadieh Bremer and Shirley Wu

NAT

URE

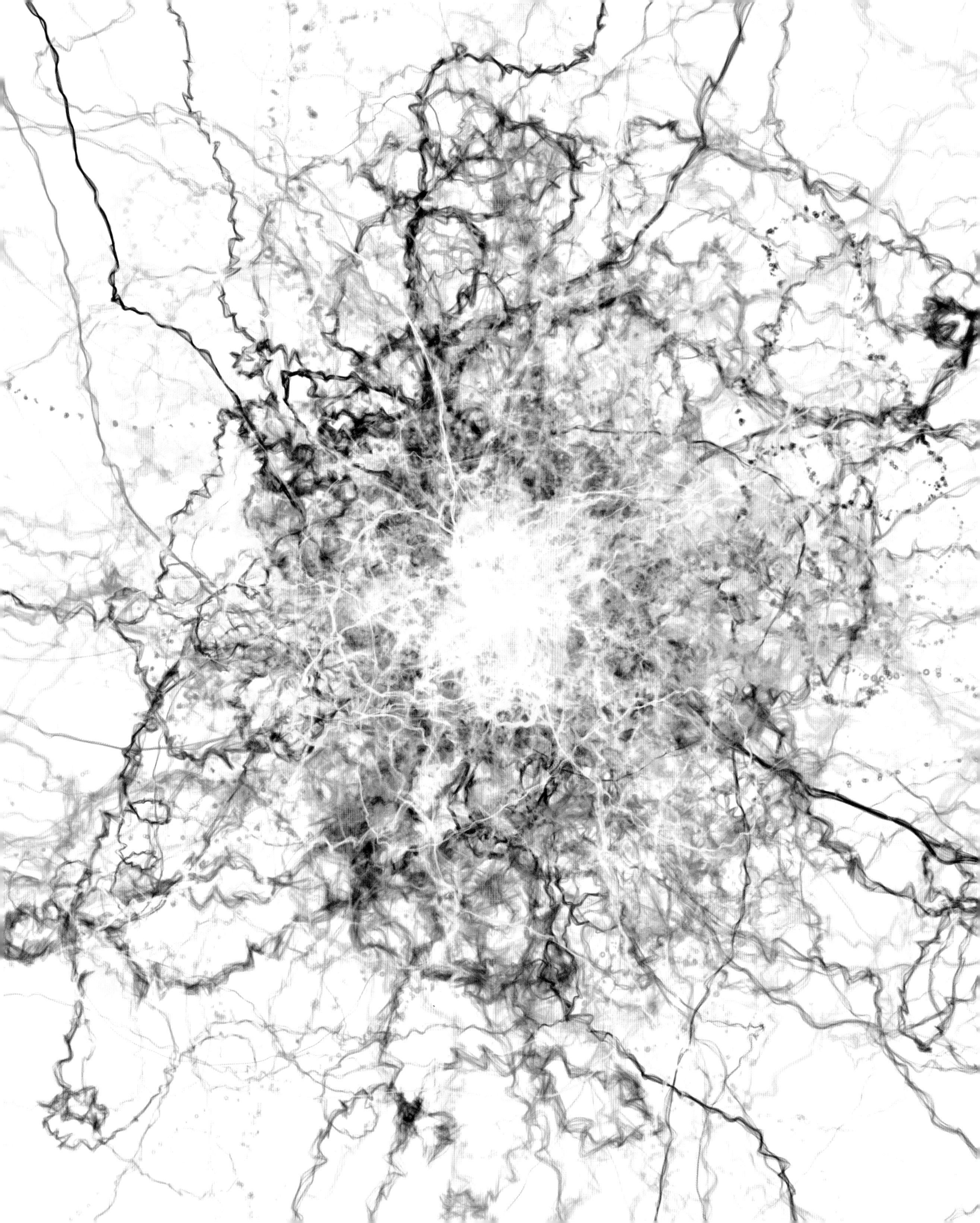

Marble Butterflies

NADIEH

Nature—what a lovely theme. I thought about what defines nature for me; it makes me think of evolution and the random mutations that lead to adaptations to the environment. This in turn reminded me of generative art and the semi-randomness which often guides those pieces of art. I've always loved generative art and this topic seemed like a perfect opportunity to give it a try. After spending a lot of time on my previous visual, I set a goal to finish this one within 20 hours to force myself to deliver within a reasonable number of hours.

Data

To start, I needed a subject to guide the generative art concept. The first things I was drawn to were flowers, but I couldn't find a *single* dataset about flowers that contained information such as the main color, average number of petals, blooming period, etc. I only came across images of flowers to be used for image processing training, or taxonomy lists (in other words, lists that classify flowers into specific groups, families, and species). I therefore turned my attention to butterflies, for which I also have a fascination. However, I encountered the same problem here; I only came across taxonomy lists.

Don't ask my boyfriend how many times he's had to wait for me on vacations while I was trying to perfectly capture a butterfly.

The best dataset that I was able to find, I *kid you not*, came from the website "Gardens with Wings."[1] It had a section that divided butterflies by main colors, wing shape, size, and species (Figure 8.1). I felt that scraping the site was more trouble than the manual approach, so I just copy-pasted information that I needed from each page into an Excel file, resulting in a list of 86 different butterflies.

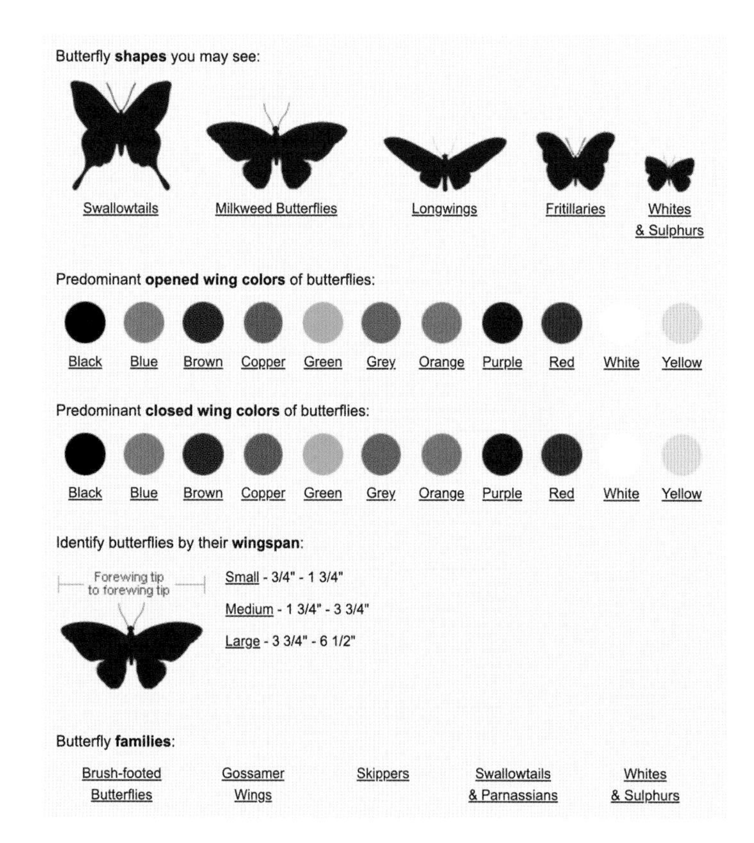

Fig.8.1

The different options you have to search for butterflies on the website "Gardens with Wings." Website design, usability, and UX – Fred Miller; website data and project management – Patty Bigner.

↳ Data Can Be Found in Many Different Ways

Just like my previous project about *Dragon Ball Z*, this one also came down to gathering the data totally manually after I had given up hope to find a more "proper" already structured dataset.

[1] Gardens with Wings: http://gardenswithwings.com/

Sketch

Once I figured out the "butterfly and generative art" angle, it seemed most fun to me to mimic a butterfly's path across the screen. They always seem to fly quite randomly and change direction quite suddenly (and that was it, really, the full idea! (•ω•)).

At first I thought about using the different path options to build up the butterfly's route; for example, the Skipper butterfly's path would travel in straight lines, and Swallowtails would be arcs. I tried sketching out this idea, but then quickly switched to coding it instead. No sketch could capture the delicateness of the lines I had in mind. I wrote down several questions or ideas I wanted to keep in mind while coding, such as "Should a line (e.g. butterfly) quit after a certain time?"

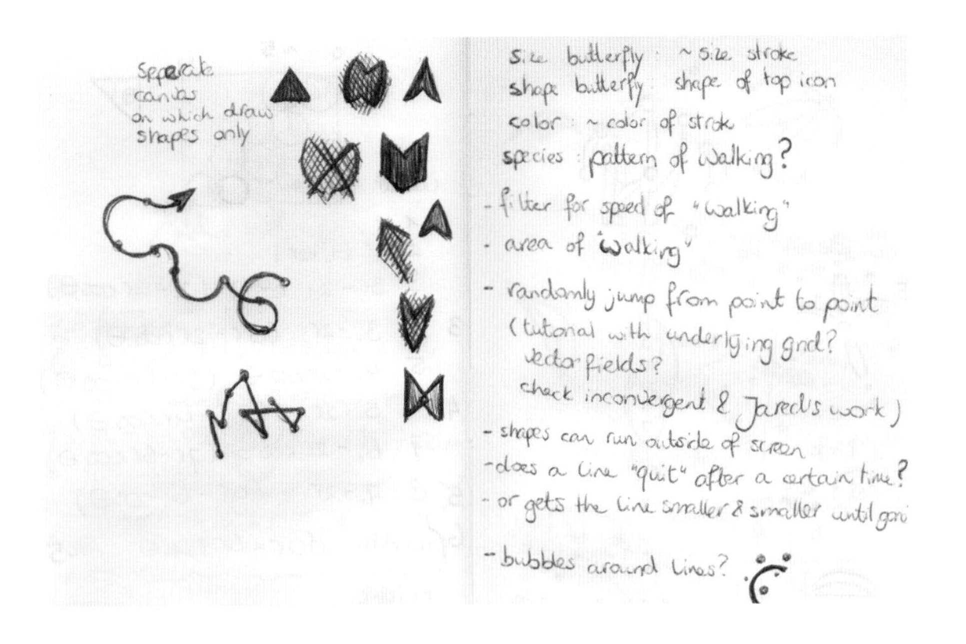

Fig.8.2

This "sketch" with questions to keep in mind is what I used to figure out how to draw the butterfly lines.

Code

I knew there would be many paths going all across the screen at once with no interactivity. So for this project, I finally did it right and decided to start building the visual with canvas instead of D3.js. For my very first attempt I just wanted the simplest thing on the screen, which in my mind was representing the butterflies with random straight paths, switching to a different direction every so often (Figure 8.3).

For a short explanation of canvas, please see "Technology & Tools" at the beginning of the book.

Well, that looked absolutely dreadful! (⊙_⊙) This taught me that making random things look beautiful is *difficult* (and I developed even more respect for people like generative artists Jared Tarbell and Anders Hoff).

I started from a generative-like example[2] I created about a year before for my butterflies, which in itself was based on a demo from Sketch.js that shows a stream of circles bursting from a line moving across the shape of an infinity symbol (∞).

[2] Color blending—infinity showcase: http://bl.ocks.org/nbremer/cbf61944aeb3204d3e4986ea645afc2b

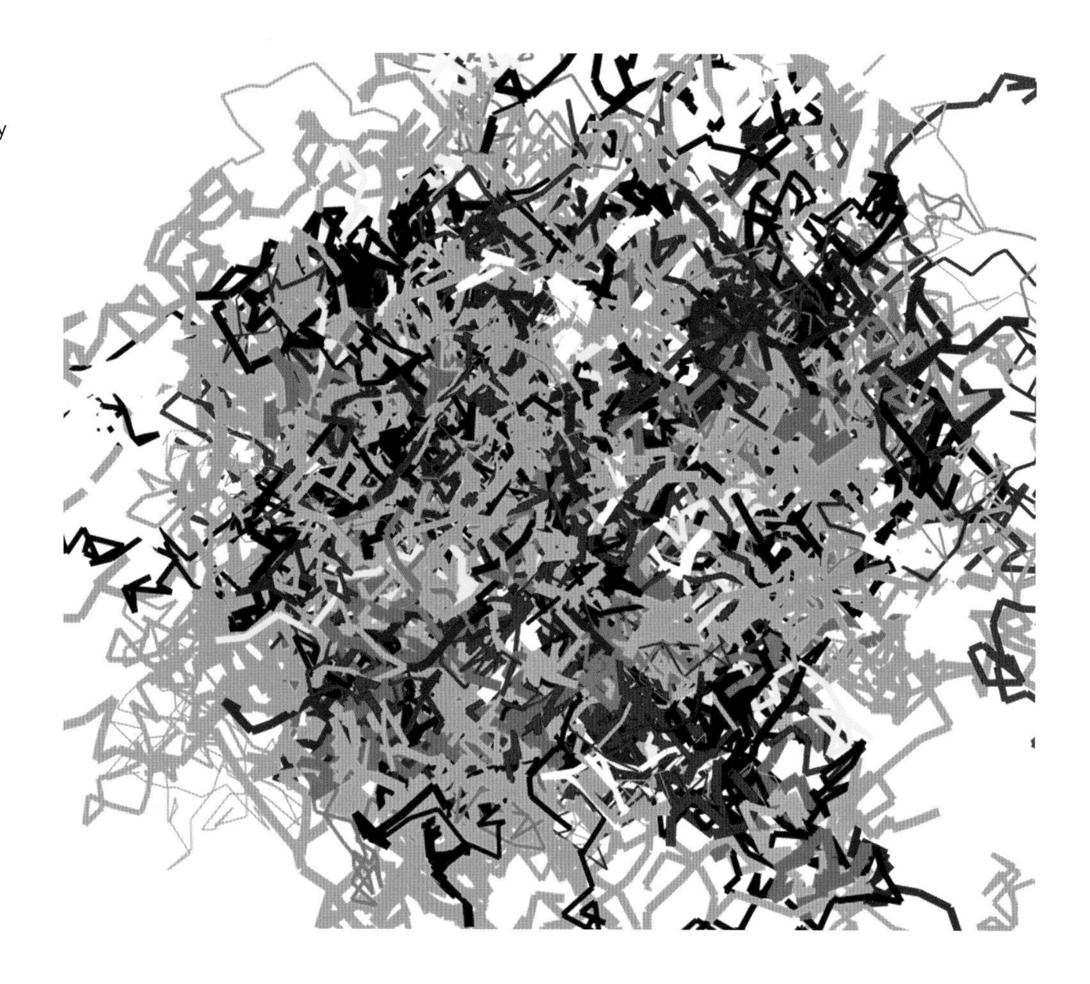

Fig.8.3

Very first result on my screen, using only straight lines.

The horrible screenshot in Figure 8.3 showed me that I would probably need some smoothing of the curves. And that's when I found a tutorial on Anders Hoff's site[3] about creating nicely curved lines through a set of points and then slightly jittering them. It looked absolutely *amazing*, and I immediately knew I wanted to try something similar.

Thankfully, I could start from the code of a Stack Overflow question that draws a so-called curved spline through a set of points, in essence smoothing out the original line and taking off any edges. In Figure 8.4 you can see some of my earliest tests, first just getting a smooth curve and next starting to gently jitter the path of the butterflies on each redraw to get a more random looking path. Once I also dropped the opacity a lot and got to the smoky bottom line in Figure 8.4, I *finally* felt that this could turn out ok after all. (ﾉ◕ヮ◕)ﾉ*:･ﾟ✧

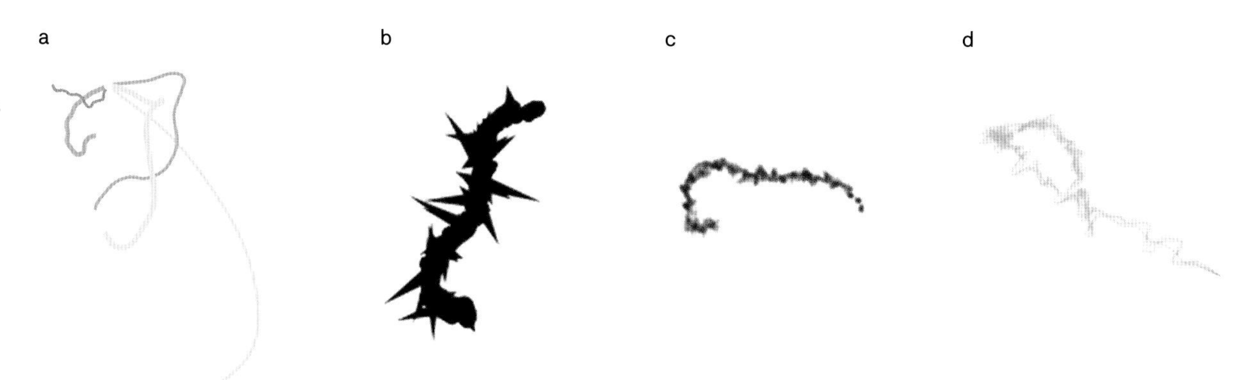

Fig.8.4 (a,b,c,d)

Slowly working towards random, smoky looking butterfly paths.

[1] Anders Hoff "Sand Spline": https://inconvergent.net/generative/sand-spline/

For ease, I let all my paths begin from the center and move outward. I started to add some rules based on my actual butterfly data. Each frame, a few butterflies are created, randomly picked from the dataset of ±50 butterflies. The color of the butterfly path is defined by the main color of the butterfly. However, to get more diversity I used the `Tinycolor.js` library to pick randomly between five very similar colors. The opacity and thickness of the path, and the way a butterfly flutters across the screen, is governed by the butterfly's wingspan, which I split into "small," "medium," and "large" categories. Naturally, larger butterflies have thicker paths.

I filtered out brown-colored butterflies because I didn't like the color on the screen.

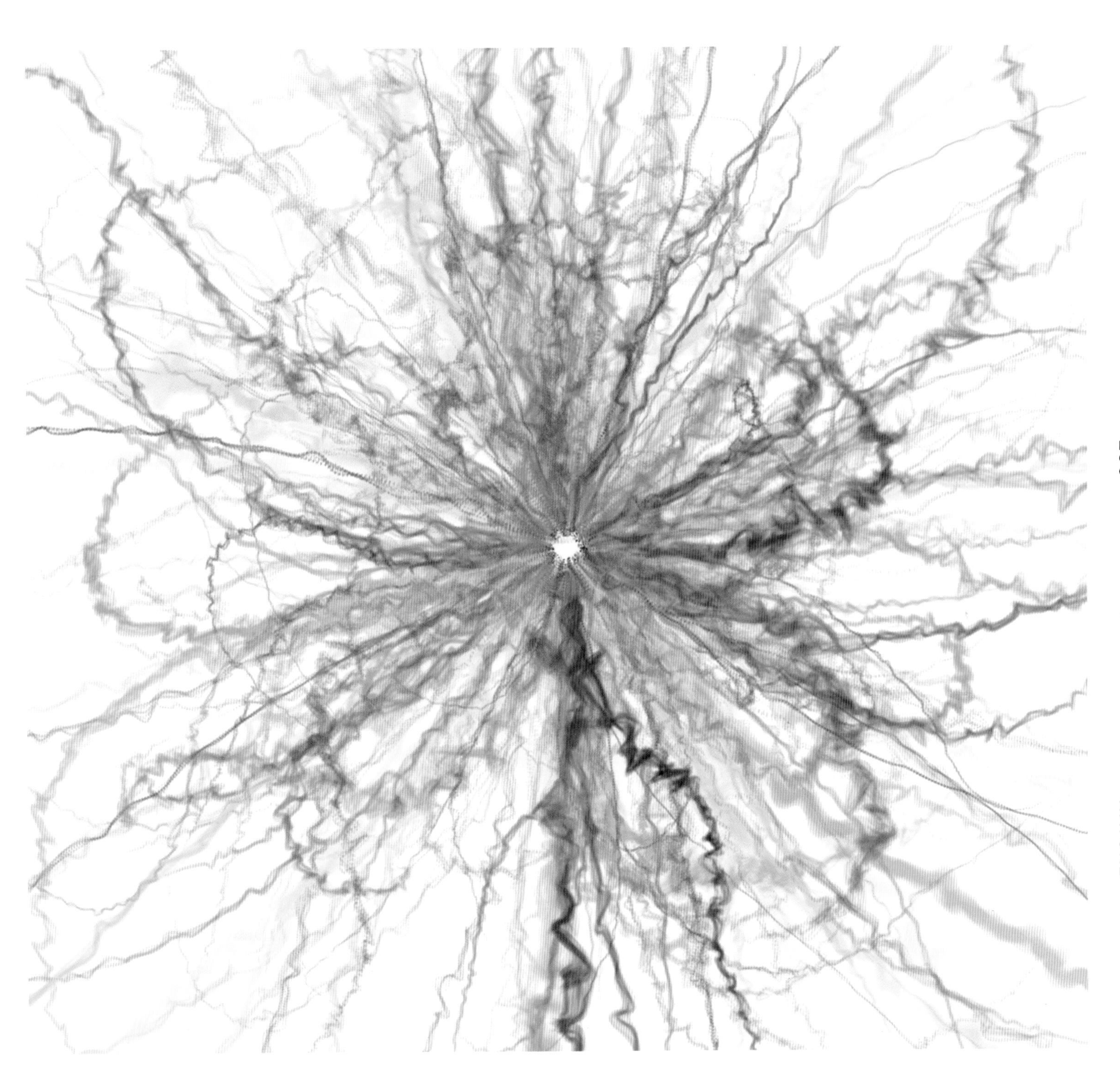

Fig.8.5

Butterfly-based random paths starting from the center that get jittered on each redraw.

I then turned to the paths themselves. In my initial setup, the paths were quite straight. A butterfly would dash over the screen never to be seen again. To better mimic a butterfly, I started to make the paths more curved. However, it took some effort to make the effect look natural. Once a butterfly curved to one side, it would have to flutter for a while before curving in a different direction, versus quickly jumping from one direction to the next.

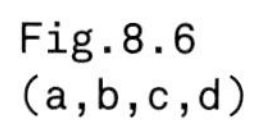

Fig.8.6 (a,b,c,d)

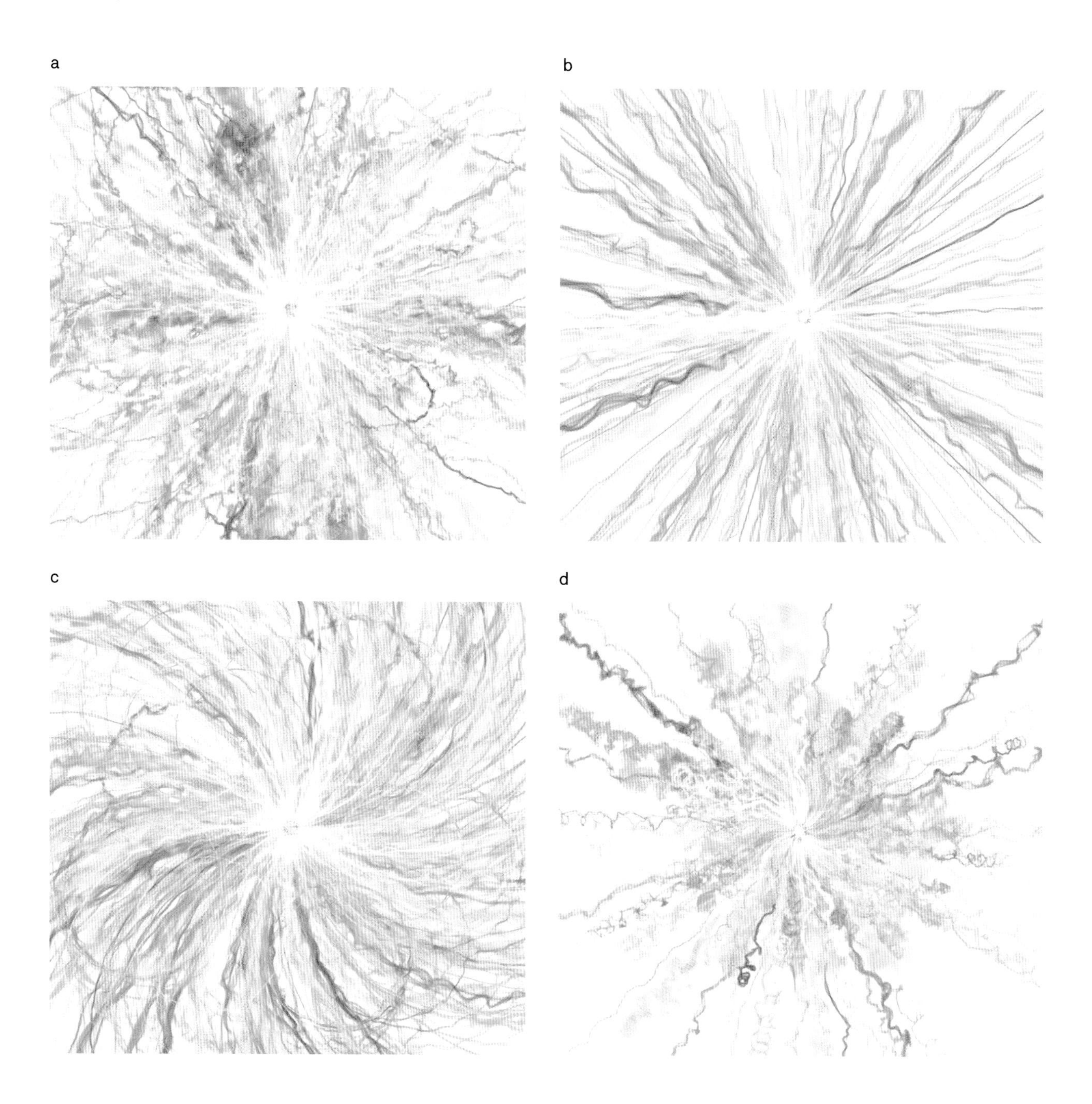

Drawing paths only in solid lines seemed a bit boring, so I started playing around with different types of paths. The smallest butterflies got a dotted line. One particular species, the Skippers, was drawn with larger circles across the path instead of a solid line. I did have to experiment a lot to figure out some decent settings for the random number generators, because jittering a set of circles around would eventually result in a very thick blobby line.

Things were starting to look better and more diverse now. I played around with different bubble effects and I quite like the "scatter" approach in Figure 8.8.

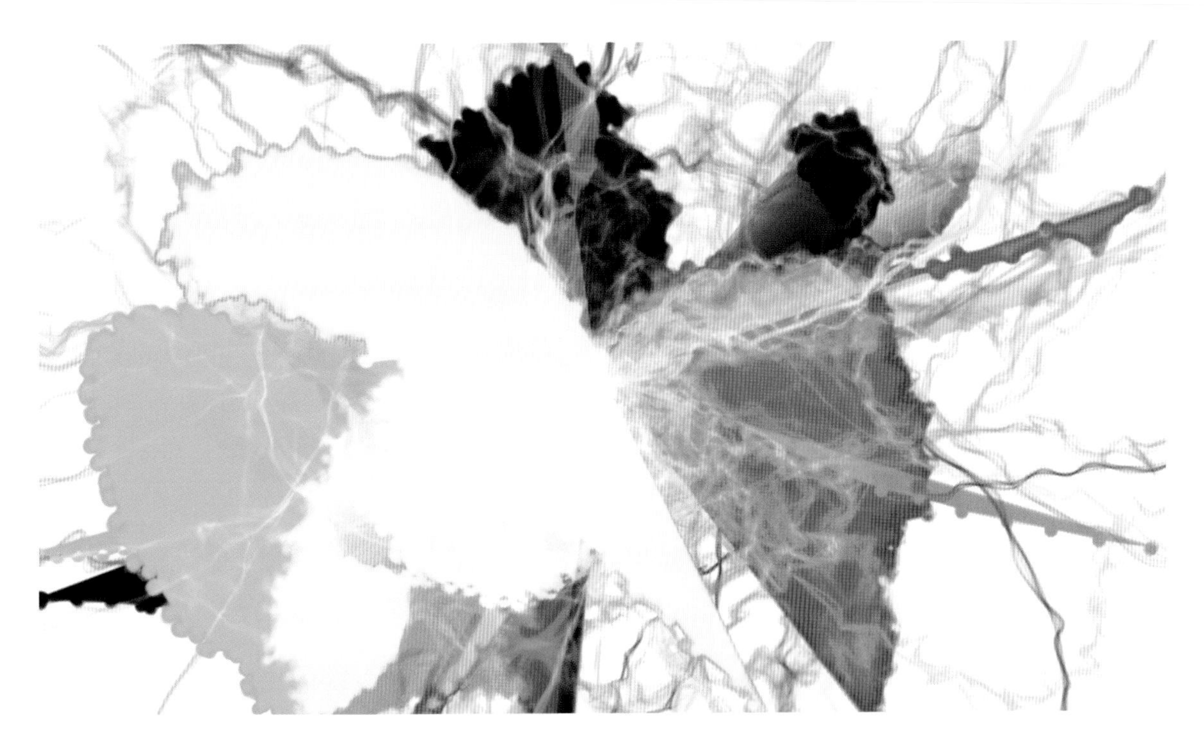

Fig.8.7

Clearly, I had a bug in my code when creating the new circle based paths!

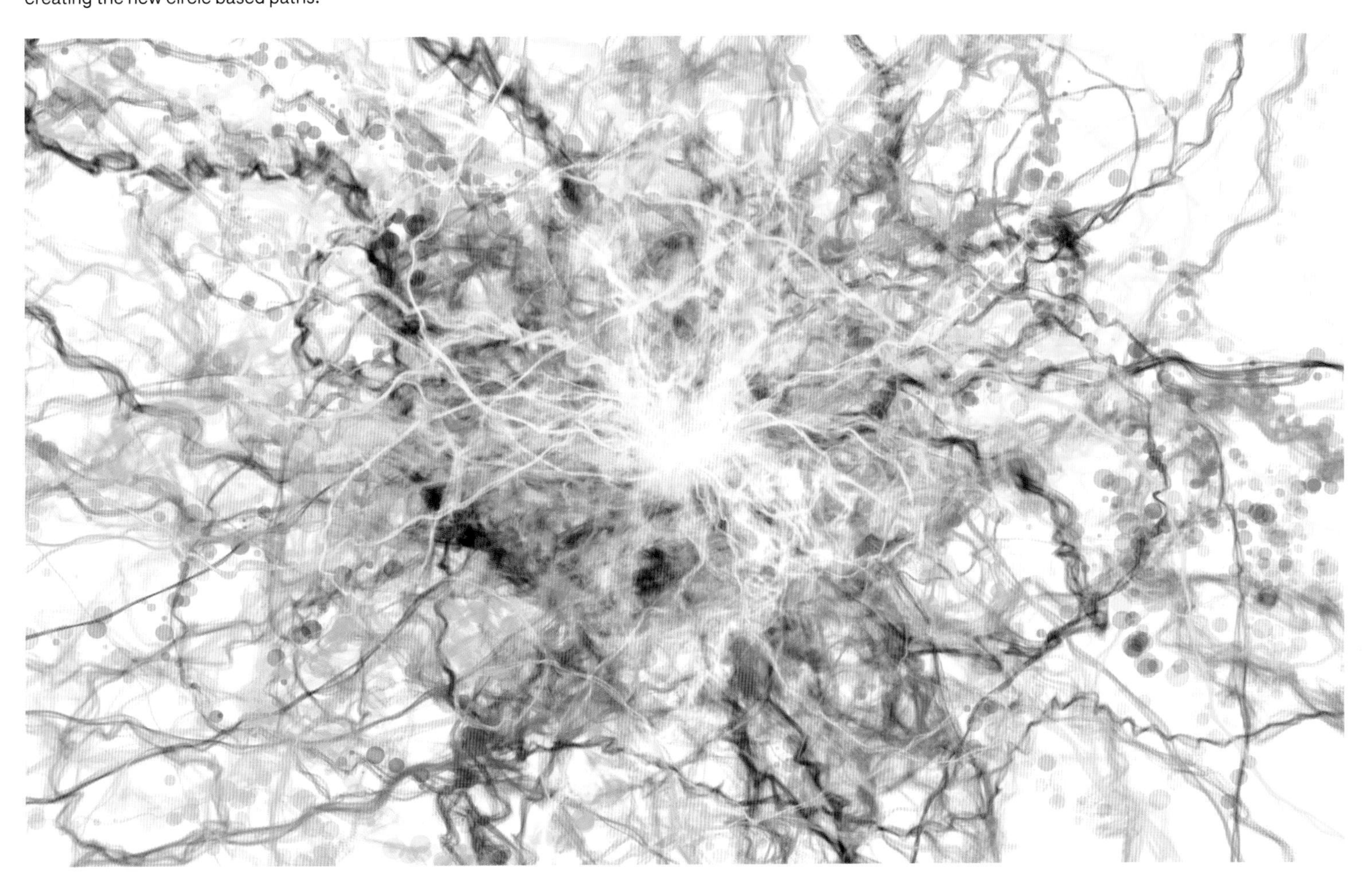

Fig.8.8

One of my favorite results, although this effect wouldn't remain looking good when the code was running for more than a few seconds.

↳ Design with Code

Although there's always an element of "designing with code" in my projects, this one, like the "Royal Constellations" project, came down to designing *the whole visual* with code. My sketch only illustrated my idea vaguely; I just couldn't quite capture what I had in mind. Figure 8.3 doesn't indicate that this would ever turn into something beautiful. But by keeping an eye open for inspiration from others, and continuing to iterate and play with the code, I ended up with a visual that I was pleased with.

For fun I added the hexagon and circle of our *Data Sketches'* logo in the center and played around with placing a butterfly-like shape in the middle, but it (naturally...) took more time to figure out than expected 눈_눈 (about 3–4 hours).

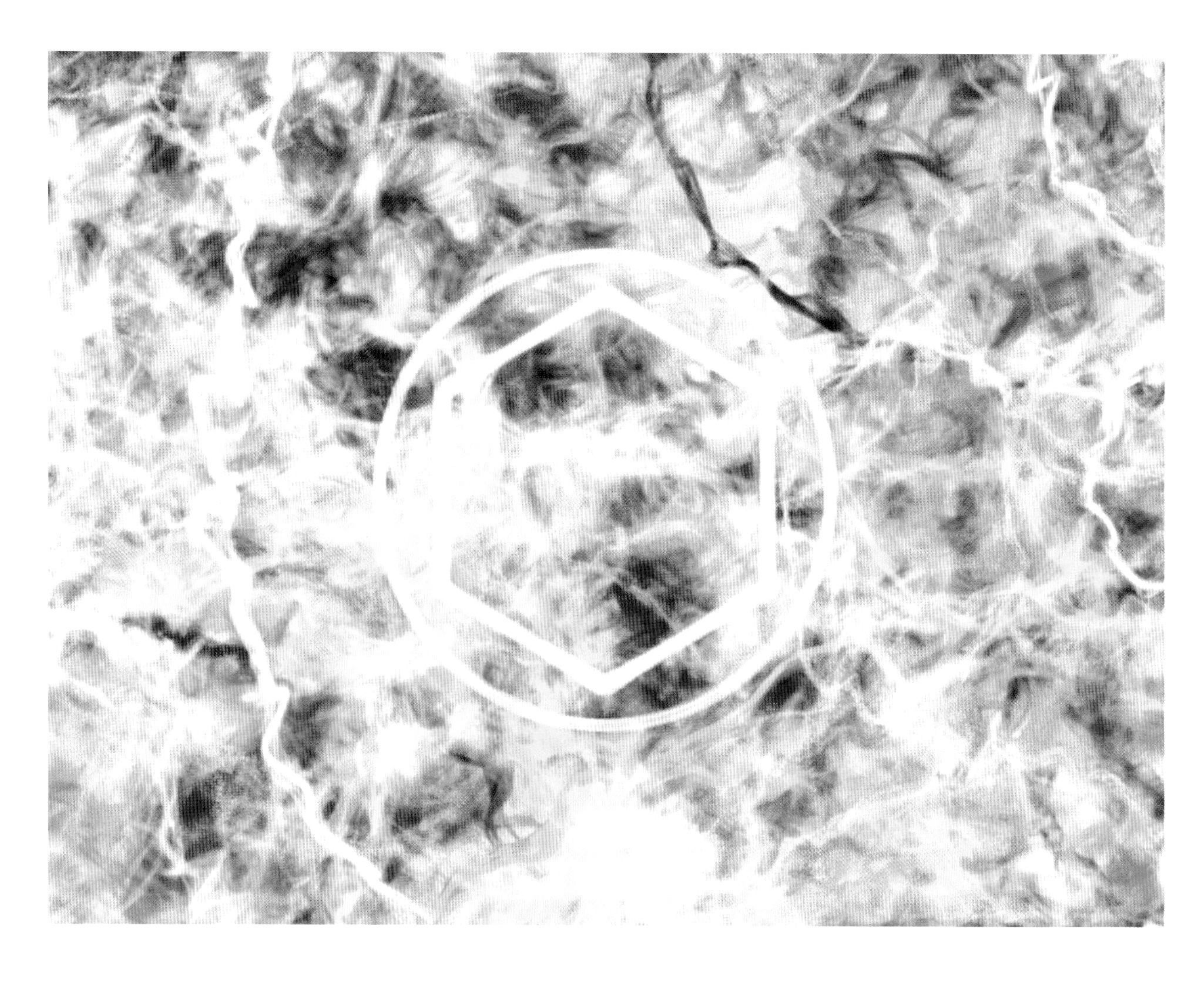

Fig.8.9

Having the paths start from outside the screen and adding a hexagon and circle (from the *Data Sketches'* logo) in the center.

Using the `openType.js` library I could recreate a font as a canvas path to spell out the name of the piece; "Marble Butterflies" (Figure 8.10). And then I could also wiggle the letters around to get a less clean look to fit better with the overall style.

I called it "Marble Butterflies" because the smoky lines across the screen remind me of patterns in marble.

After what seemed like endless tweaking of the path's settings to get better looking results, I checked out my timing app and saw that I'd reached 19:45 hours for this project, so it was time to finish and "ship it" to reach my goal of staying under 20 hours.

These write-ups also take an additional few hours!

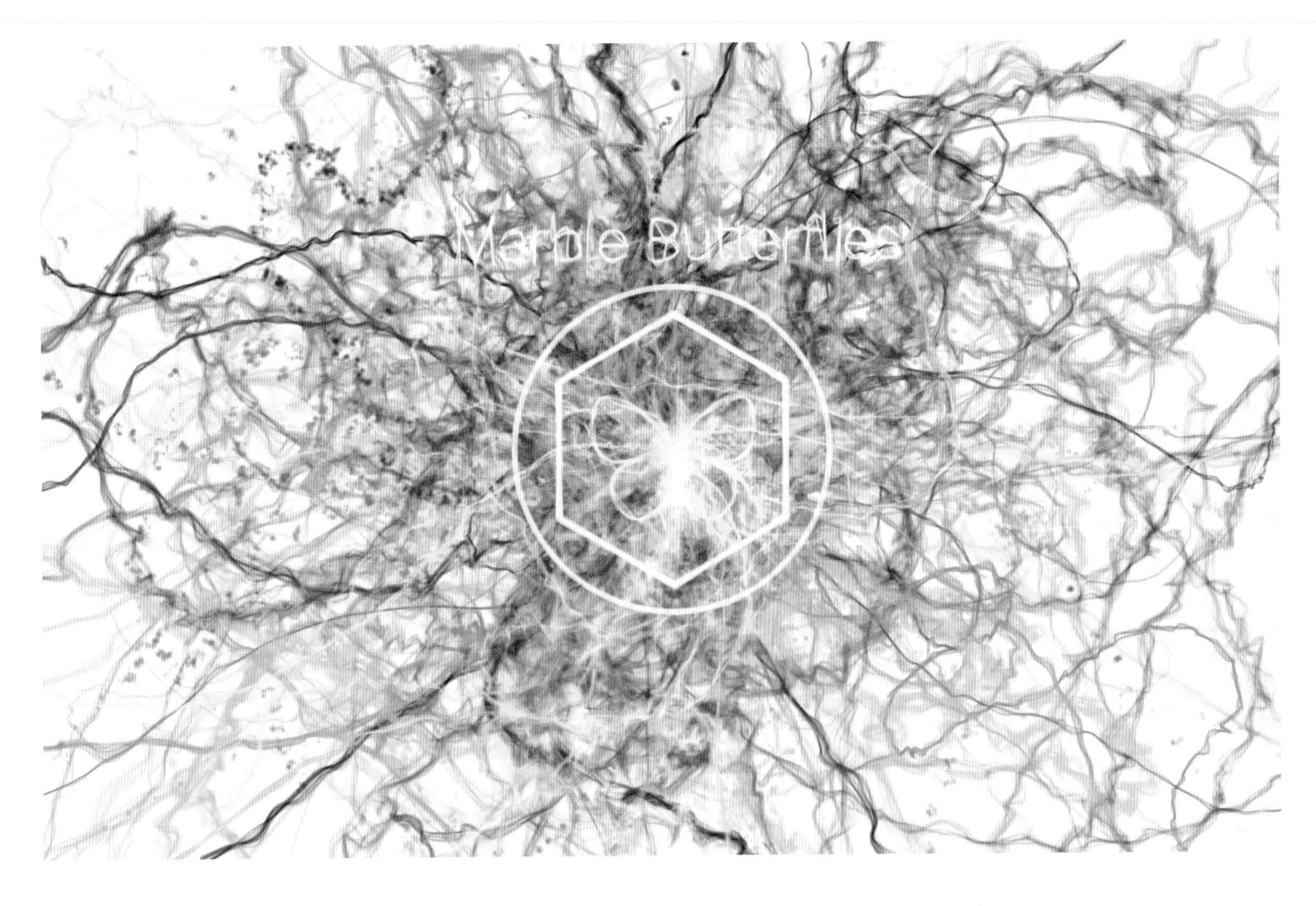

Fig.8.10

Adding a central butterfly shape and the title of the project in wiggled letters as a final touch.

One thing I really like about this piece is that due to the randomness, every version that anyone opens will be a completely unique piece of data art!

Reflections

It was quite fun *not* to have to create a visual that needed to be "understood" for a change. I was free to just experiment and kept the parts that looked nice. In the future I would definitely like to continue investigating the more artful or generative side of dataviz.

Marble Butterflies

↳ MarbleButterflies.VisualCinnamon.com

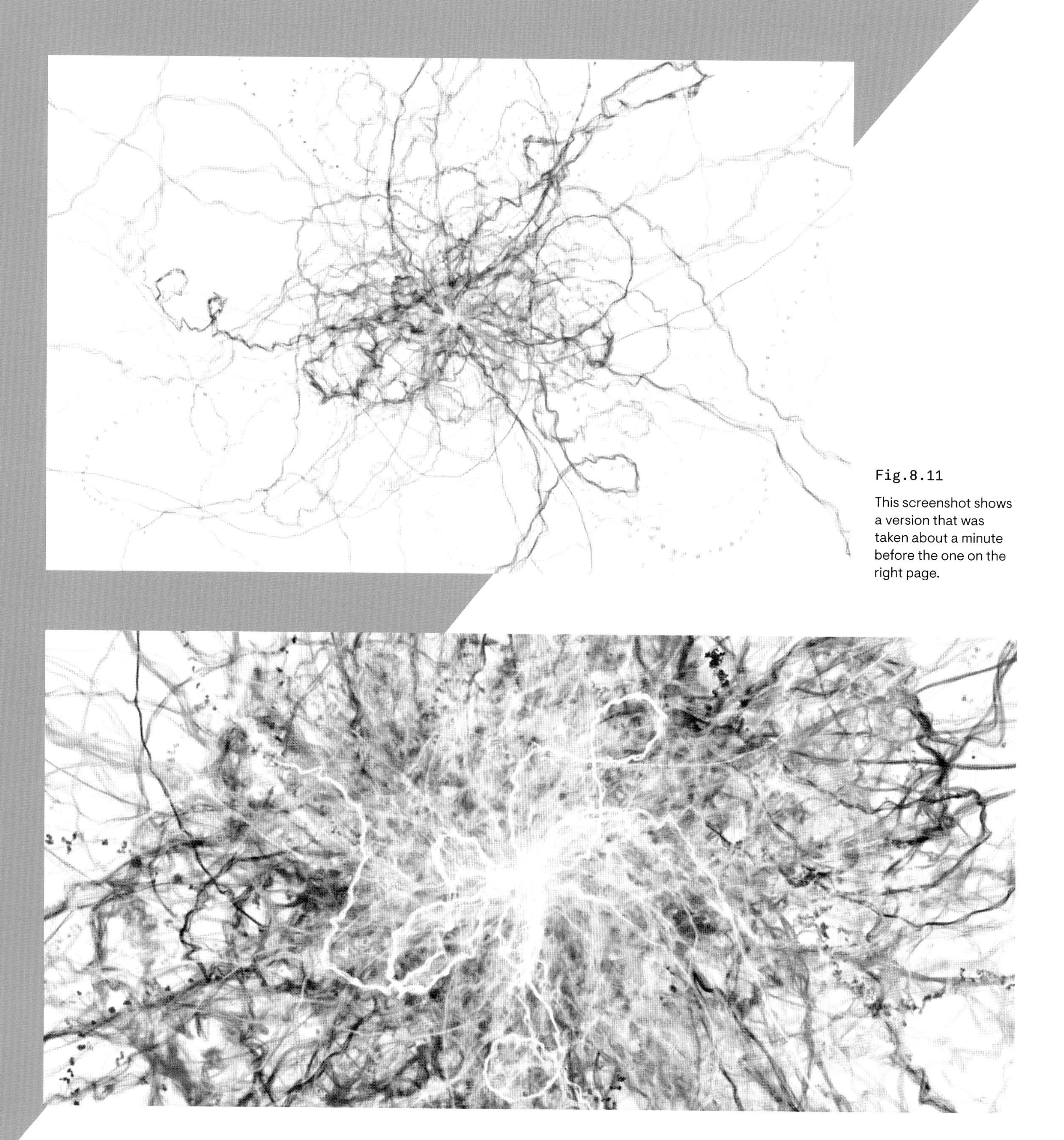

Fig.8.11

This screenshot shows a version that was taken about a minute before the one on the right page.

Fig.8.12

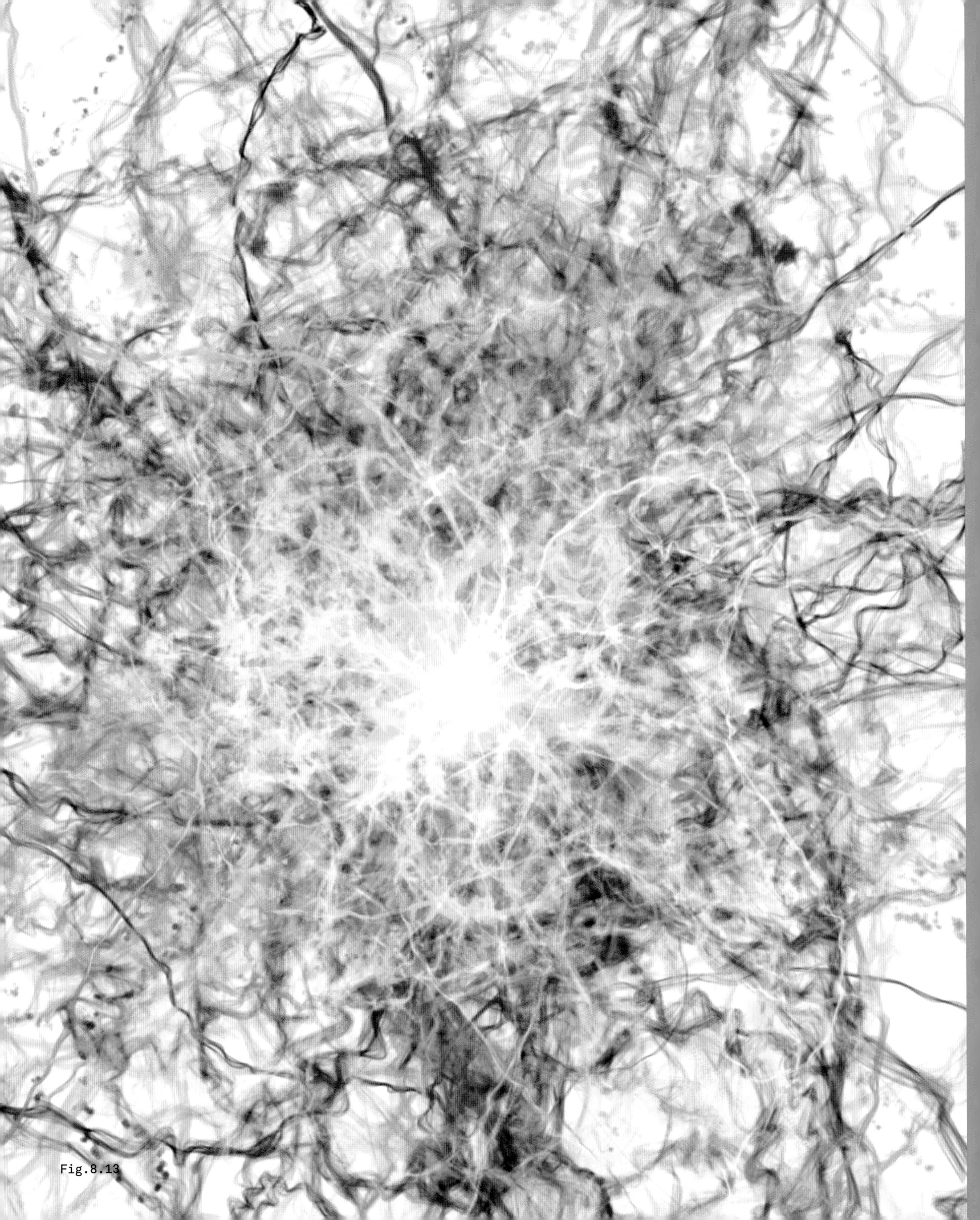

Fig.8.13

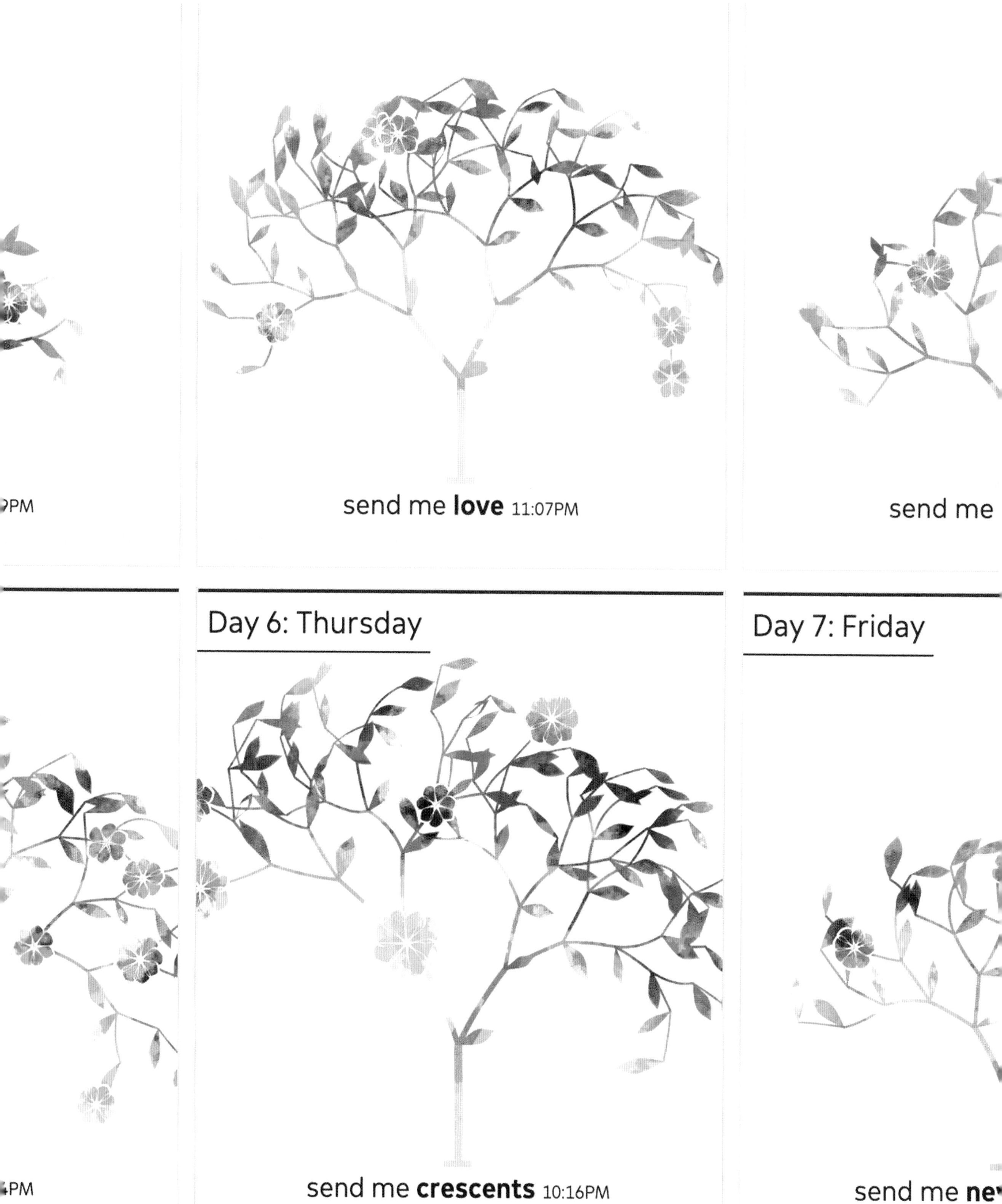
send me love 11:07PM
send me
Day 6: Thursday
Day 7: Friday
send me crescents 10:16PM
send me

Send Me Love

SHIRLEY

When we first decided on the topic of "Nature," I wanted to do something with the *BBC's Planet Earth 2*. I even started gathering data for the first episode, jotting down which animals appeared when, but it didn't go any further than that. (I really wasn't excited enough about the idea for the amount of manual data entry it required.)

More than a year later, I had finished a piece for the San Francisco Museum of Modern Art (SFMOMA) that I thought could maybe double for *Data Sketches*. I have to admit it's still sort of cheating; the data isn't about nature, it's about texts sent to SFMOMA. But I looked at individuals' texts and their interactions with SFMOMA, so it could fit under *human* nature (*≧▽≦). I'm very thankful that Nadieh was okay with me bending our rules a little bit (again).

Data

In June 2017, SFMOMA launched "Send Me SFMOMA," a service where people could text requests like "send me love," or "send me art," and SFMOMA would respond with an artwork in their collection that best matched the request. They received more than five million texts from hundreds of thousands of people. A few months later, they contacted me to do something with that data, and I was over the moon with excitement.

When I first started looking at the data, I did some simple explorations: what the attributes were, what keywords were most popular, and what artworks came back the most often (Figure 8.1). I also knew that I wanted the end result to be a painting, because I'm a cheesy individual and I wanted to make art out of art. (*´艸`*)

The idea I ended up pitching to SFMOMA was to trace an individual's "journey" throughout the month, what "positive" or "negative" thing they might have asked for, and what kind of artwork they got back. I wanted to see if I could experience what that person experienced and feel what that person felt.

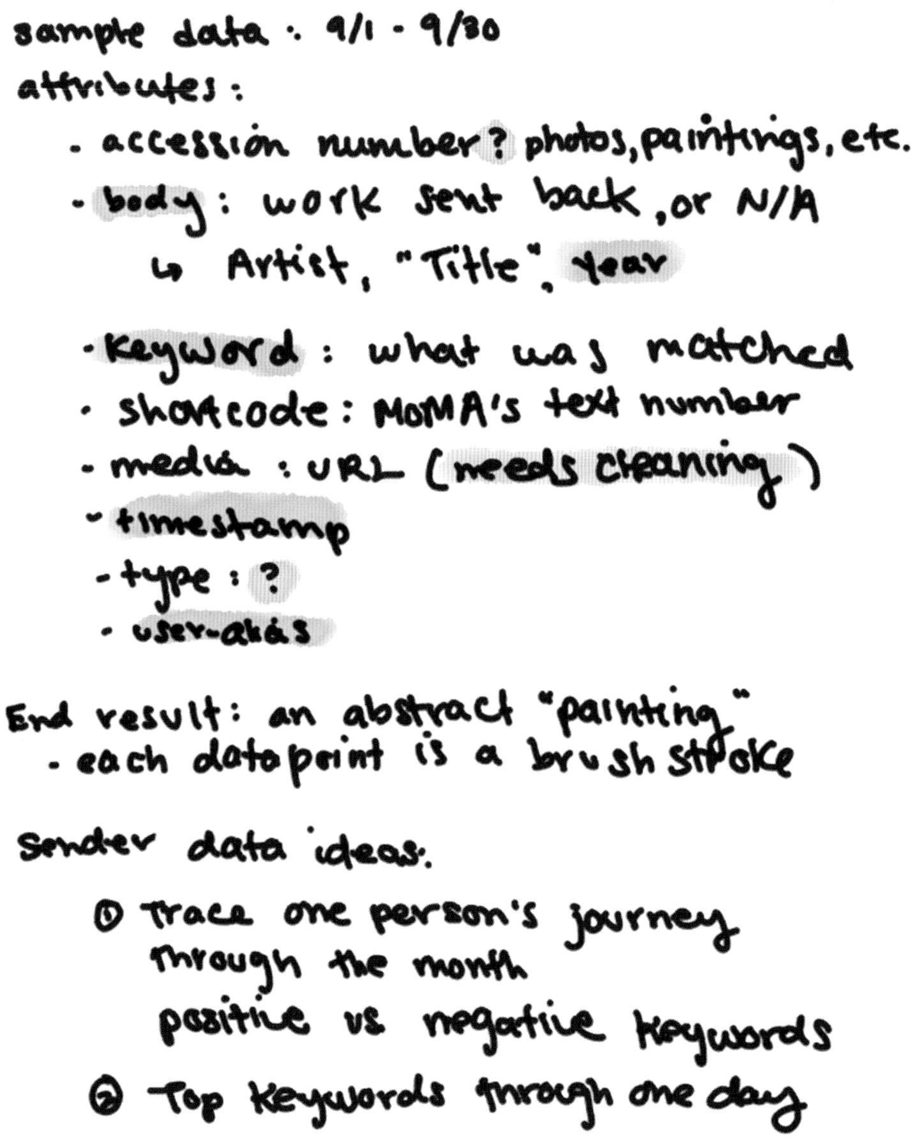

Fig.8.1

Notes from my initial data exploration.

sample data: 9/1 - 9/30
attributes:
- accession number? photos, paintings, etc.
- body: work sent back, or N/A
  ↳ Artist, "Title", year
- keyword: what was matched
- shortcode: MOMA's text number
- media: URL (needs cleaning)
- timestamp
- type: ?
- user-alias

End result: an abstract "painting"
- each datapoint is a brush stroke

Sender data ideas:
① Trace one person's journey through the month
positive vs negative keywords
② Top keywords through one day

As far as data collection, preparation, and cleaning went, this was probably one of the most straightforward projects I've done for *Data Sketches* (and I'm really grateful for that). For the data collection, SFMOMA gave me a huge file of all the texts they ever received and another huge 1TB file of all the texts they ever sent back. From that second file, I wrote a script to only retrieve texts and information from the first 100 participants. I decided to focus on the first 100 participants because it otherwise would have been a lot to analyze, and I hypothesized that the project's early adopters would have the most interesting interactions.

I ran the keyword in each text through `sentiment`, a Node.js package that gives a score for how positive or negative a word is, and my cousin Max (a junior in college studying Computer Science) wrote a script to get additional metadata for each artwork—including date, title, artists, country of origin, color palette, and Shannon Entropy (how "chaotic" the artwork is)—from the SFMOMA API.

Sketch

The very first thing I did once I had the data was to try and implement my own version of Tyler Hobb's generative watercolor effect.[1] It didn't turn out quite well:

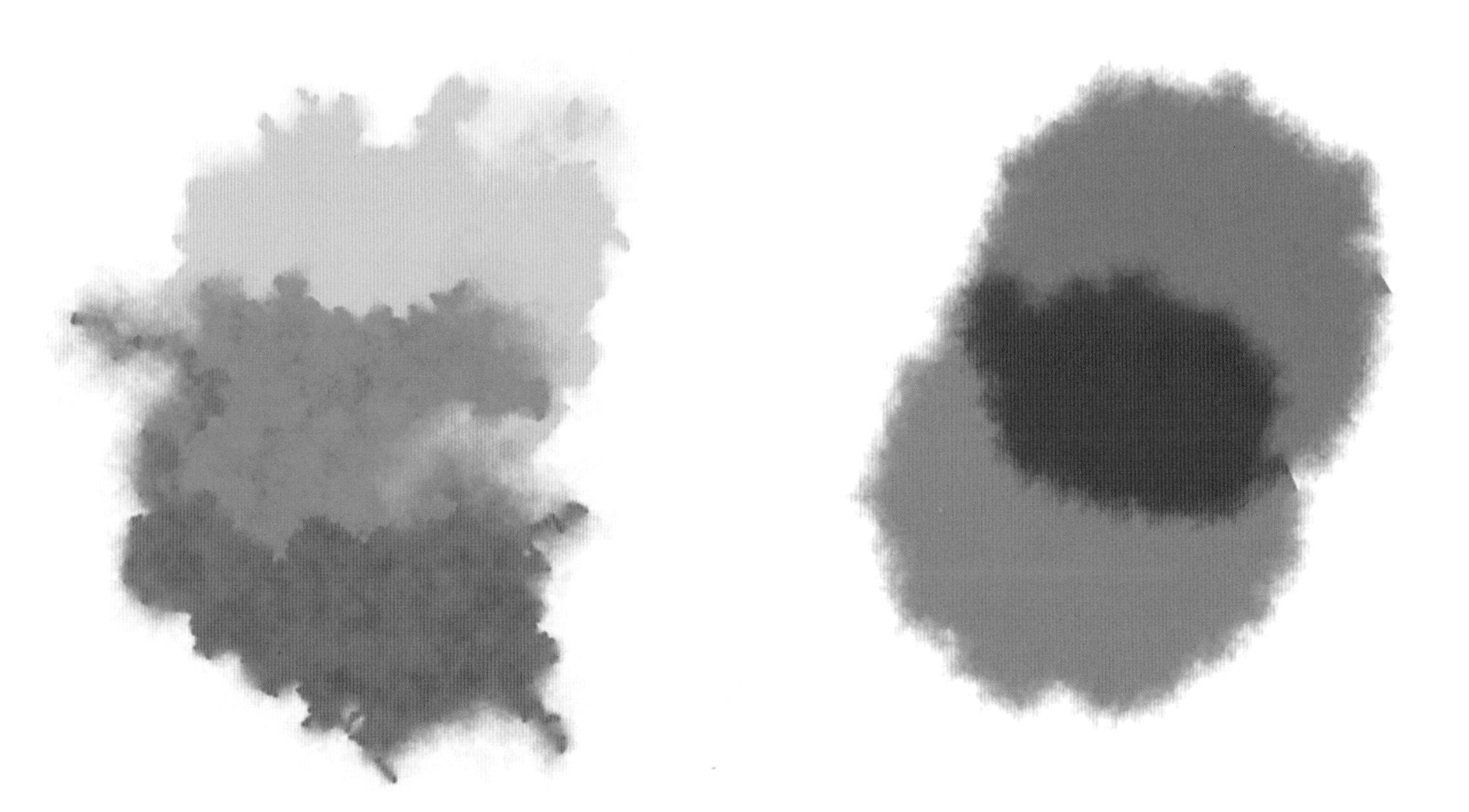

Fig.8.2

Tyler's implementation (left) versus my own attempt (right), a real case of expectation versus reality.
(¯口¯;)

[1] https://tylerxhobbs.com/essays/2017/a-generative-approach-to-simulating-watercolor-paints

But thankfully when I tweeted about this, my friend (and developer) Taylor Baldwin came to the rescue with a link to his own implementation,[2] and later on Tyler chimed in with a fix.[3] I was able to get my version looking more like Tyler's (Figure 8.3), but once I plugged in the actual data with each "paint stroke" colored by the artwork SFMOMA texted back, it was a mess again (Figure 8.4).

Around the time I started working on this project, I had moved to Tokyo for three months. Japan is a country that reveres nature and the change in seasons; when I arrived in March, flowers were everywhere. I remember walking past an ad of illustrated flowers in a subway station and doing a double take because I just loved the whimsy and simple beauty of it. I knew then and there that I wanted to work flowers and nature into the project.

As soon as I got back in front of my computer, I jotted down my ideas for how to incorporate flowers into the visualization. At the same time, I started looking at the SFMOMA API to see what kind of data I could get back about each artwork.

Fig.8.3

Programmatically generated "watercolor" strokes.

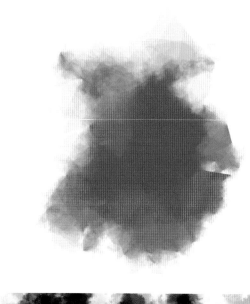

Fig.8.4

Programmatically generated "watercolor" effects with data plugged in.

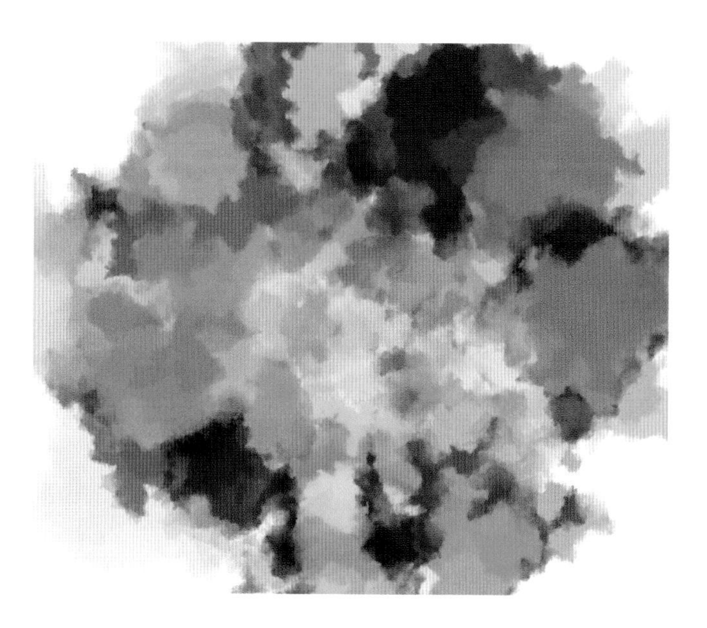

[2] https://github.com/rolyatmax/watercolor-canvas
[3] https://twitter.com/tylerxhobbs/status/959178572375232517

I started working on the flower petals a few days later, taking what I learned from the "Movies" project about SVG curve commands, as well as newly learned functions like p5.js's `randomGaussian` (randomness centered around a given number) and `noise` (an implementation of Perlin noise that simulates a more natural looking randomness) to get a more natural flower petal shape (Figure 8.5). I was especially proud of the hand-drawn effect I achieved, where the lines of the petal vary in width as if they were drawn with a thin brush (Figure 8.6). I was able to do this by first drawing the petal shape with an SVG path, calling `getPointAtLength` to get the *x/y* coordinates every few pixels along the path, and drawing a circle in canvas for every *x/y* coordinate I got back. The most important part, though, was that as I drew the circles, I varied their *radius* with Perlin noise so that they increased and decreased in size in a very natural looking way (if I had used randomness instead, it would have looked very messy and blotchy). I was inspired to try this approach after watching a series of explainers on Perlin noise by Dan Shiffman,[4] where I learned how it was implemented.

p5.js is a JavaScript library for creating generative art and interactive experiences on the web.

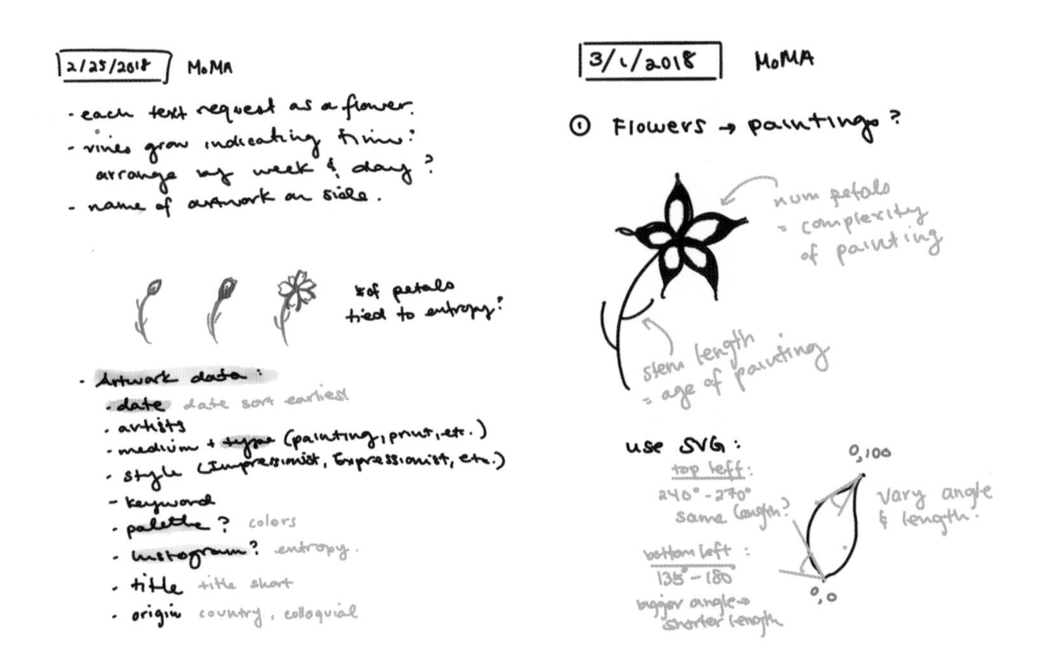

Fig.8.5

Notes on what metadata to map to a flower (left) and how to code more natural looking flowers (right).

Fig.8.6

Iterations on natural flower shapes, with SVG to color the petals and canvas to produce the hand-drawn effect. I'm extremely proud of the hand-drawn effect; I love it so much.

[4] Dan Shiffman, Introduction - Perlin Noise and p5.js Tutorial: https://www.youtube.com/watch?v=Qf4dIN99e2w

Code

This was the project that proved that even if the data and sketch parts went smoothly and I knew exactly what I wanted to do, the actual execution (code) could still take a really, really long time.

For my very first code attempt, I drew each text message as a flower placed around a spiral (I was able to repurpose the code from my "Music" project (*•̀ᴗ•́*)و̑̑), colored with the most predominant color in the artwork SFMOMA sent back, and animated by the time the text message was sent (Figure 8.7). Not only was it aesthetically unappealing, it was so computationally heavy that my Macbook Pro (with 16GB RAM!) was choking up. I immediately moved the watercolor calculation into a web worker so they wouldn't clog up the UI.

Much to my sadness, I also decided to get rid of the hand-drawn effect because it was too computationally expensive to loop through *hundreds* of points along each flower petal when I was trying to draw hundreds of flowers, and the result was too subtle to appreciate on such small flowers anyway.

Fig.8.7

First attempt at drawing each text message as flowers.

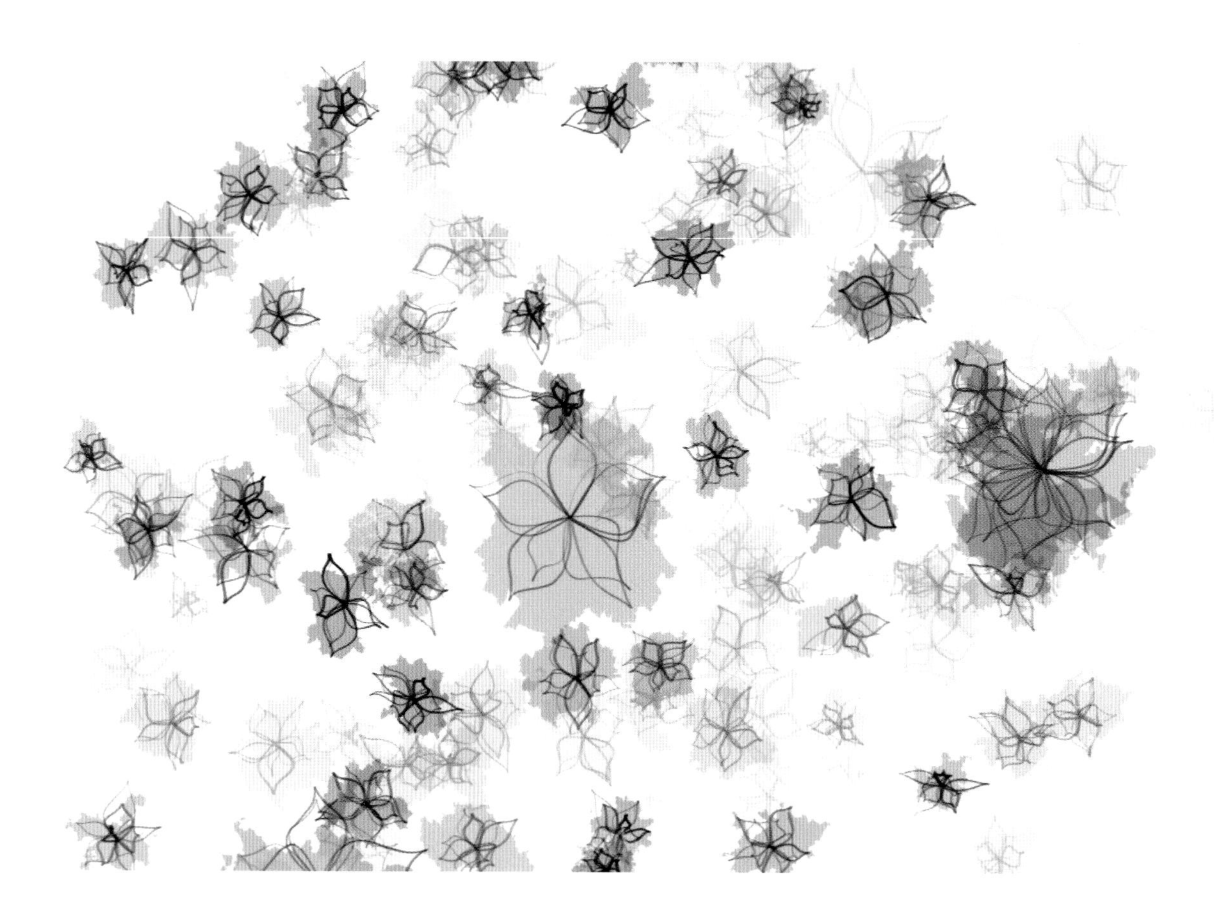

I didn't know how to deal with the aesthetics, until I went to the joint graduation exhibition of five art universities at the National Art Center in Tokyo and saw a gorgeous piece of a pregnant woman with flowers surrounding her. The flowers were cut out of the paper, showing the watercolor underneath.

Web Workers

In data visualization, the performance bottleneck is usually in rendering to the DOM (Document Object Model), but once in a while the data calculation will be heavy enough to tie up the main CPU thread. And as JavaScript is single-threaded, any time-consuming calculations will cause the page to become unresponsive and prevent user interaction until they've finished processing, which is horrible for user experience. Most of the time, we can move the super-heavy work (like calculating node-link positions with force simulations) offline and load it in a file with the calculations already done. But sometimes the file will be too large or we need to do the calculations on the fly, and that's when web workers are super helpful.

For rendering bottlenecks, consider limiting the number of DOM operations or using canvas instead of SVG; see the lesson "SVG vs. Canvas" on page 82 of my "Olympics" chapter for details.

Web workers are background threads we can spin up to communicate back and forth with the main thread, where regular JavaScript code is executed and user interactions are handled. When we need to do some heavy computation, we can send the data to the worker with `postMessage` and the main thread will be free to keep executing other code or to listen for user events. The worker will receive the data in an `onmessage` callback, execute the code, and pass the calculated data back to the main thread to handle.

It's important to keep in mind, though, that web workers don't necessarily execute the code any faster (as they're still running on the CPU); they just don't freeze up the web page. They also don't have access to the DOM, so we can't manipulate SVG elements or draw directly to a canvas element from within a web worker.

I was immediately taken by the cutout effect, and I knew I wanted to recreate it (Figure 8.8). I was able to achieve the effect by using three layers of DOM elements:

1. A canvas element with the watercolor effects
2. A second canvas element that I filled fully white first, then set the blend mode (`globalCompositeOperation`) to `destination-out` so that any subsequent flowers I drew subtracted from the white background like cutouts
3. A final SVG element on top that I used to draw the outlines for each flower petal

I liked the look of this cutout effect very much, but soon realized that having a flower for every text message was just too much, so I decided instead to only draw a flower for each positive text message that was sent. I tried circles and squares for the neutral and negative text messages, and eventually ended up with leaves (Figure 8.8).

Even with the flowers and leaves figured out, I was still unhappy with the layout of the flowers, especially because text messages that were sent in close procession translated to flowers that overlapped each other or were pushed off the edge of the screen. I was stuck for weeks (and worked on other projects in the meantime) until cherry blossom season hit Tokyo. Suddenly there was pink everywhere, and it was gorgeous and emotional, and I spent the whole week going to different spots to admire the cherry blossoms. (ﾉ^ ヮ ^)ﾉ*:･ﾟ✿

I lived in Japan for six years as a child, and cherry blossoms are amongst my fondest childhood memories.

Fig.8.8

The flowers with a cutout effect, first experiment with squares and circles (left) and then with leaves (right).

Custom Animations

Animations are an essential part of data visualizations—to convey the passage of time, or to transition from one data state to the next—and it's been extremely helpful knowing how to implement them.

At its core, an animation needs an interpolator, duration, and (optionally) an easing function. To implement a custom animation:

1. Create an **interpolator** with a start and end value.

   Interpolators are functions that calculate all of the in-between values given a start and end value. `D3.js` has a great implementation that not only interpolates numbers, but also colors, strings (including simple SVG path strings and CSS styles), arrays, and objects.

2. Start a **timer**, and pass in `t` (elapsed time divided by animation duration) to the interpolator at each tick.

   The most basic way to implement a timer is to pass a callback function to `requestAnimationFrame` and use `Date.now()` to keep track of the elapsed time. But personally, I like to use `D3.js`'s *timer* module that abstracts that away for me.

3. Optional: use an **easing function** to alter the rate of change in the animation and make it feel more natural, because nothing in nature moves at a perfectly linear rate. For example, when we drop a ball, it bounces a few times before stopping, or when we open a drawer, we might pull it out quickly at first but slow down when it's pulled out enough—we use easing functions when we want to mimic natural motions like these.

4. Terminate the timer once the elapsed time is larger than the animation duration.

I like using `D3.js`'s *transition* module when animating SVG elements and `Greensock`'s `timeline` when I need to chain multiple animations together and manage their timing. When I have to animate in canvas, I'll go full-custom with `d3.timer()` and `d3.interpolate()` functions.

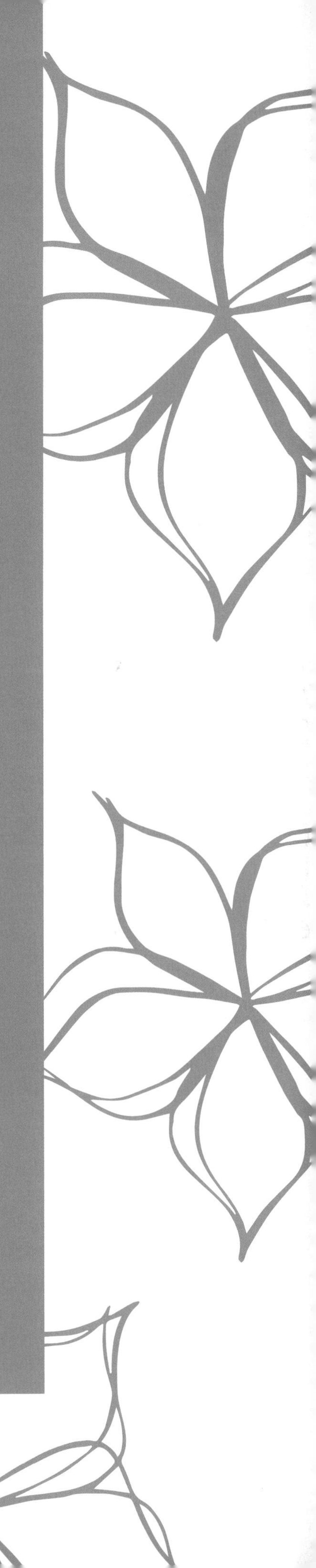

Fig.8.9

A photo I took of the beautiful cherry blossoms in Shinjuku Gyoen.

It was on one of those walks through Shinjuku Gyoen (Figure 8.9), one of my favorite parks in Tokyo, that I realized: I was trying to lay out the flowers in random ways, but flowers grow on tree branches (what a facepalm moment), and tree branches are just fractals. Thinking of the images in terms of fractals made it a very straightforward computer science problem to solve (Figure 8.10).

I again referenced Dan Shiffman's work, this time a chapter on fractals in his book *The Nature of Code*. I adapted his code and drew my flowers on branches, with each tree of flowers representing one day's worth of text messages (Figure 8.11). I loved it immediately.

It took another two months after this point of tweaking to get to the final product. I added a slight curve to the branches to make them look a bit more natural, made the flower petal shape less spiky, fiddled with the timing of the animations, and fixed bugs.

Fig.8.10

Notes on how to program fractal tree branches.

3/27/2018 MoMA

1 "branch" = 1 week

angle

outside = sent at night
inside = daytime

size = # of characters in keyword (more effort).

① go through & calculate leaf or flower, size, angle (perlin noise), & how far down the "branch" (also perlin noise)

② use length of branch to come up w/ bezier curve, place leaves & flowers along it.

3/29/2018 MoMA

Each week as fractal:

① x, y

0, 0

day ← → night

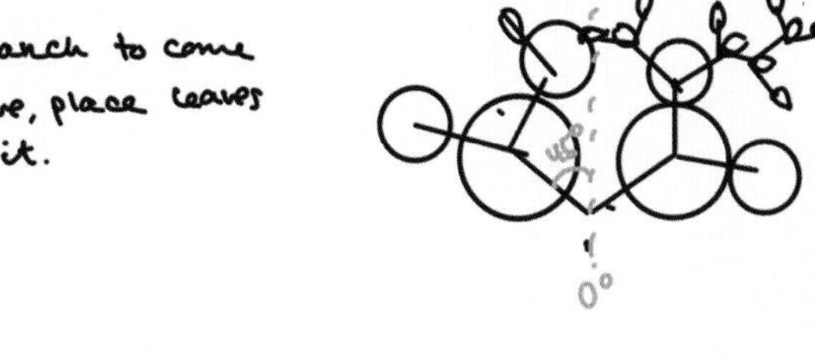

① separate into night vs day

② sort by size

Recursion:

① length (prev + curr. size)

② angle (prev + $\pi/2$ or $-\pi/2$)

③ remember x, y.

④ part into halves ?

Fig.8.11

Flowers laid out with branches.

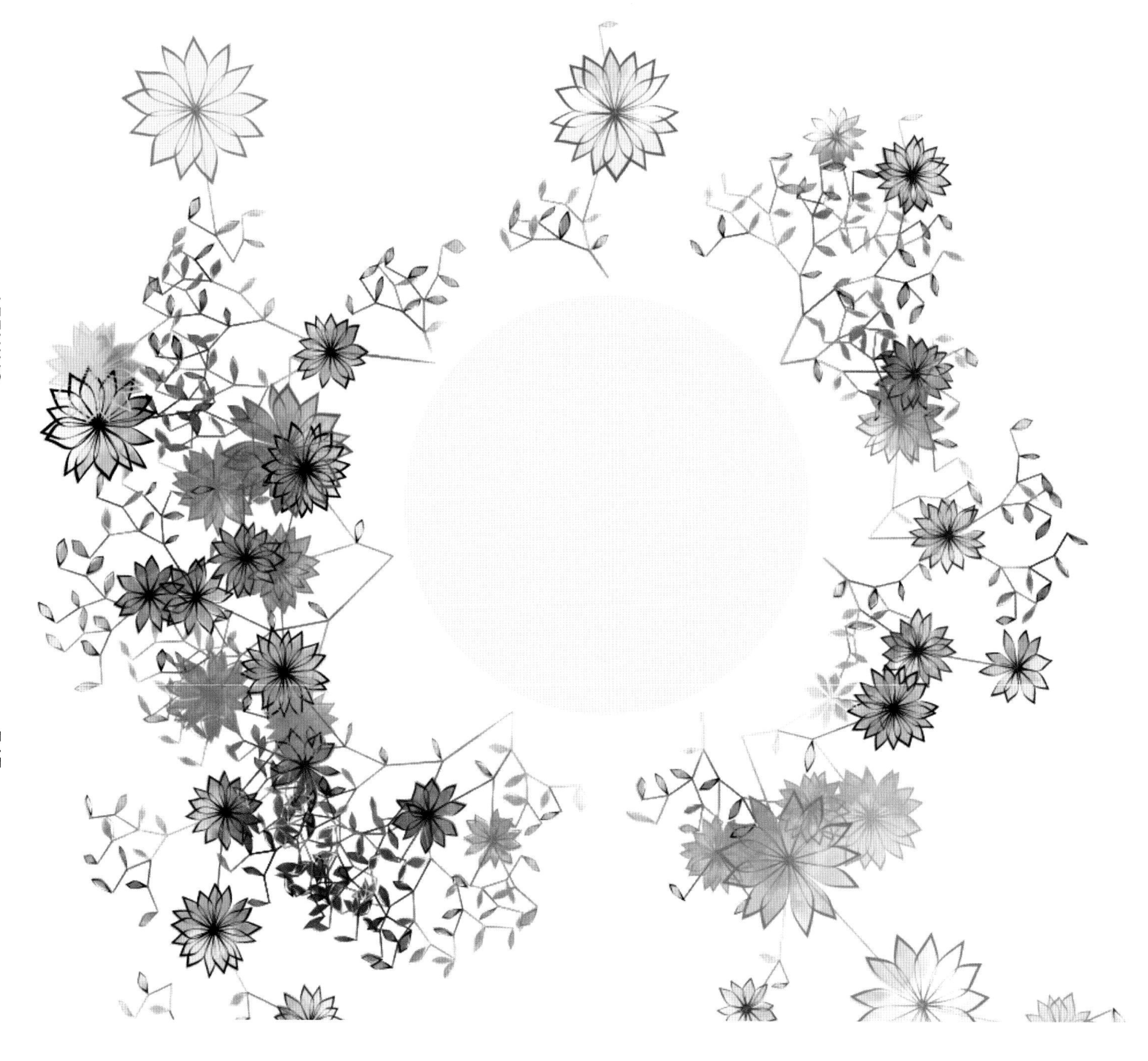

Fig.8.12

Flowers and trees arranged around an artwork in a circular layout. (Artwork covered for copyright reasons)

But most of all, I had a lot of trouble laying out the trees (each representing a day of text messages). I originally wanted to have all the flowers "animate in" around a circle, surrounding the artworks SFMOMA sent back (Figure 8.12). This caused several problems: there was a lot of overlap between days when the trees spread out too far, some of the bigger trees often disappeared off the screen, and most importantly, putting the artworks in the central circle meant cutting off the artwork—something that went against the original artists' wishes.

I decided to try out a horizontal layout next, first with each tree (day) getting the same amount of horizontal space. But with seven days and limited screen space, I couldn't fully display each tree. So I tried a carousel format next, where the selected tree was layered above all of the other trees (Figure 8.13). And because I had so much vertical space, I decided to put in additional information underneath the tree: a log of the keywords texted and their corresponding timestamps. (I liked thinking of this as the tree of flowers sprouting from the soil of that day's text messages. (*´艸`*))

Fig.8.13

Trees arranged horizontally in a carousel format.

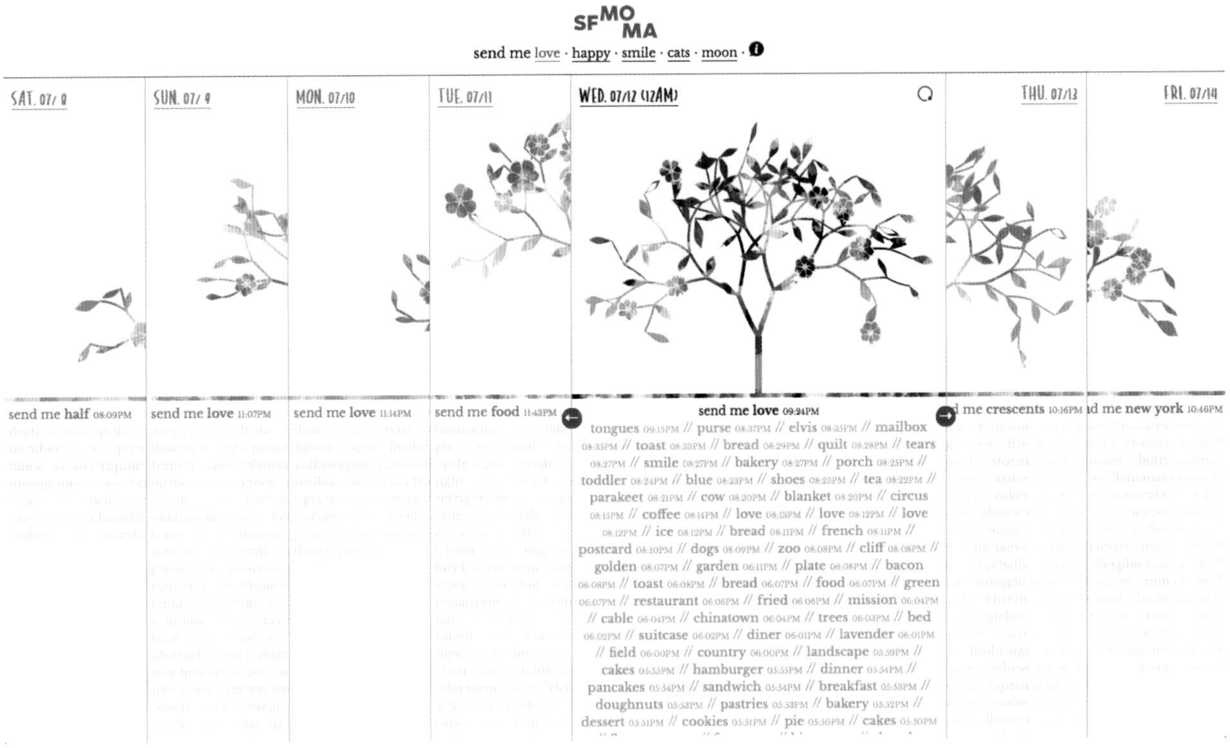

I really liked this layout but missed being able to see all the trees unobscured at the same time, and so I finally adopted a layout where one tree could be expanded and the rest would be minimized. This layout made me the happiest, not only because I could see all the trees at the same time, but also because it was implemented with CSS Grid and was perfectly responsive.

Once I had the layout, I started working on the final touches. I named the five individuals whose text messages I chose to display with the keywords they asked for the most often ("love," "happiness," "smiles," "sunshine," "moons"), and wrote an artist statement. I added a legend and tried to tell a story with it: what mood the individual was in (flower versus leaf), what they got back from SFMOMA (color), and how that might have affected them (Figure 8.14). And to support the last part of how the artwork might have affected them, I devised an interaction where on hover, I drew an arrow from the last text message they sent to the next, and so on. I loved this interaction because I found that I could follow the series of texts the individuals sent, and learned that oftentimes, they *were* influenced by the artworks they got back.

The article "A Complete Guide to Grid" on CSS-Tricks[6] was so comprehensive, it was all I needed to get up and running with Grid.

Reflections

I love this project because I learned so much from it. I successfully used a web worker and learned CSS Grid to lay out all of the trees. It will always remind me of all the experiences I had in Japan, as my stay there deeply influenced the form of this visualization. I'll always have little tweaks I want to do (I'm never 100% satisfied with my projects), but it's one of those projects where I'm really happy with the outcome.

[6] "A Complete Guide to Grid; CSS-Tricks": https://css-tricks.com/snippets/css/complete-guide-grid/

Send Me Love

↳ send-me-love.com

Fig.8.17

Hovering a flower or leaf shows the arrows pointing to the texts sent before and after, as well as the keyword in the text, timestamp, and the artwork that was sent back. (Artwork blurred for copyright reasons.)

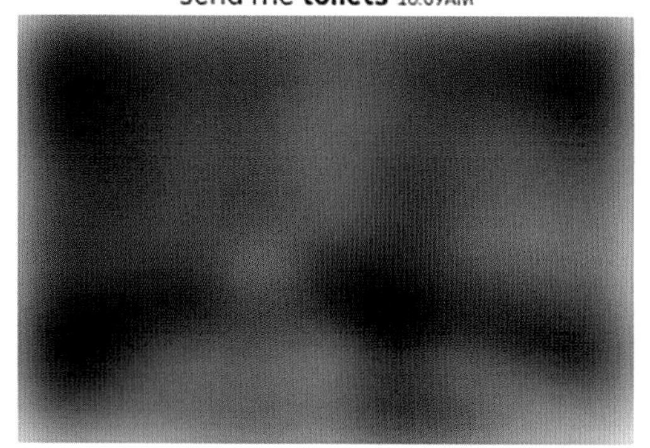

Laurie Simmons
New Bathroom/Woman Standing/Sunlight
1978

Fig.8.18

I loved the flowers so much I made them into postcards.

About

Send Me Love

love · happiness · smiles · sunshine · moons

replay

Day 4: Tuesday (12AM)

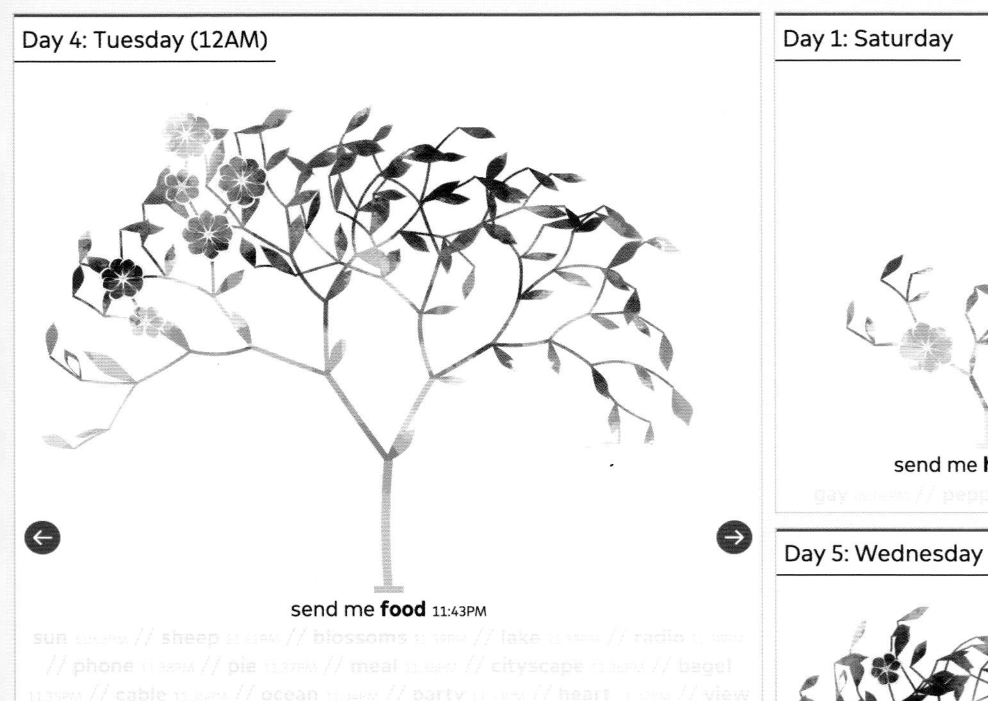

send me **food** 11:43PM

Day 1: Saturday

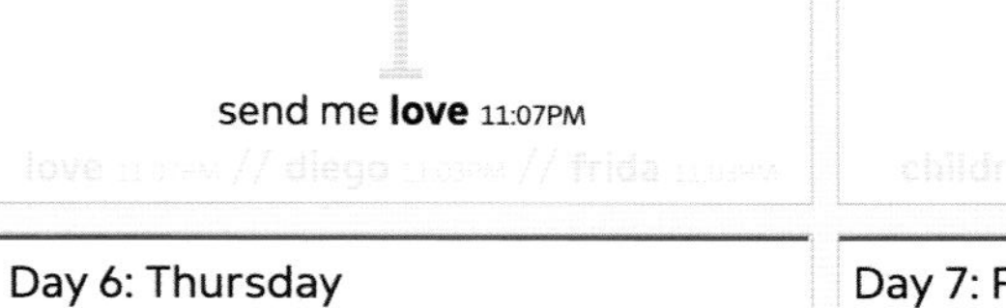

send me **half** 08:09PM

Day 2: Sunday

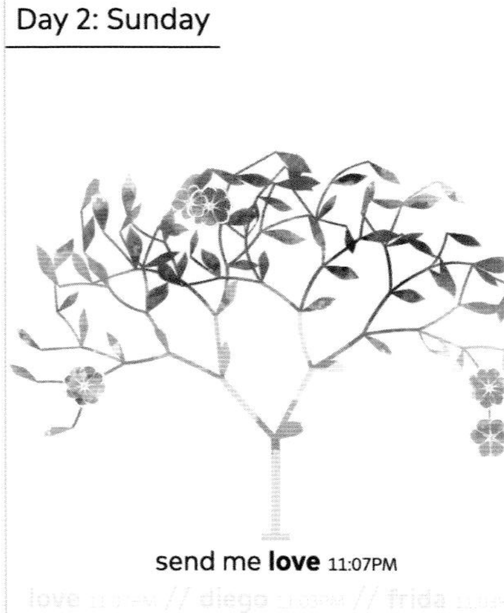

send me **love** 11:07PM

Day 3: Monday

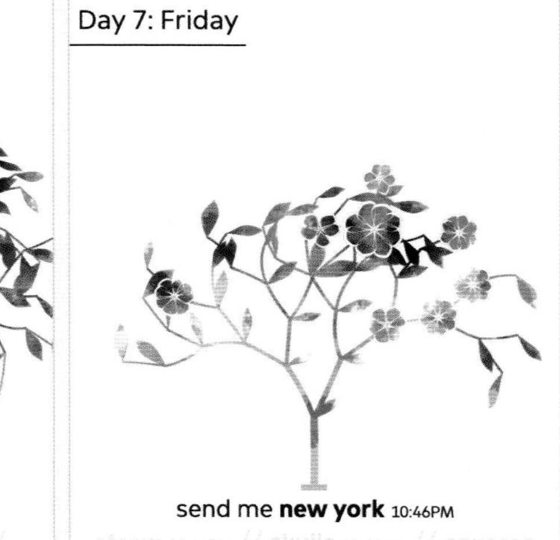

send me **love** 11:14PM

Day 5: Wednesday

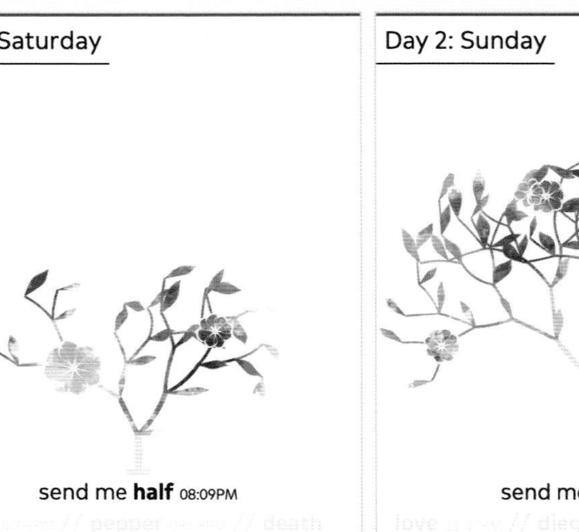

send me **love** 09:24PM

Day 6: Thursday

send me **crescents** 10:16PM

Day 7: Friday

send me **new york** 10:46PM

How to read this visualization

Each visualization is of one individual's interactions with Send Me SFMOMA over the course of a week. Each tree represents a day, and each flower or leaf represents a text they sent and an artwork they got back.

1. What mood were they in?

2. What did they get back?

3. How did it effect them?

CULT

URE

width first name holiday regards reply 60 decis
e chair house proposal well confirmation kitc

colleague colleague colleague

beautiful
thursday
tuesday
wednesday
good
address
place
little
saturday
white

beautiful beautiful beautiful beautiful beautiful beautiful beautiful beautiful beautiful beautiful beautiful beautiful
thursday thursday thursday thursday thursday thursday thursday thursday thursday thursday thursday
tuesday tuesday tuesday tuesday tuesday tuesday tuesday tuesday tuesday tuesday
wednesday wednesday wednesday wednesday wednesday wednesday wednesday
good good good good good good good good good good good good
address address address address address address address address
place place place place place place place place place place
little little little little little little little little
saturday saturday saturday saturday saturday saturday
white white white white white white

beautiful beautiful beautiful beautiful beautiful beautiful beautiful beautiful beautiful beautiful beautiful beautiful
thursday thursday thursday thursday thursday thursday thursday thursday thursday thursday thursday
tuesday tuesday tuesday tuesday tuesday tuesday tuesday tuesday tuesday tuesday
wednesday wednesday wednesday wednesday wednesday wednesday wednesday
good good good good good good good good good good good good
address address address address address address address
place place place place place place place place
little little little little little little little
saturday saturday saturday saturday
white white white white

beautiful beautiful beautiful
business city 20 sunday quantity street description time big dog work
50 friend de
0 available
lleague colleague

Beautiful in English

NADIEH

I still remember Shirley and me getting an email from Alberto Cairo in October 2016 that asked if we'd be interested in doing a *Data Sketches* style project for Google News Lab. Our answer: yes, of course!

We were completely free to supply our own ideas for topics as long as it revolved around Google data. So Shirley and I made a list with four topics, and each of us put our own spin on it as usual. Simon Rogers, our main point of contact at Google, picked the topic of "Culture," something a bit light and fun. Alberto oversaw the general progress of the designs, akin to an art director.

My angle for "Culture" came from something I experience with Google quite often. Being a Dutch native speaker who communicates in English about 95% of my working day, I sometimes need to translate a word into English. I either type "<<Dutch word>> in het Engels" into Google itself or go to Google Translate ("in het Engels" means "into English" in Dutch). I was really interested to know what people from other languages translate into English the most, especially if it's just one word. Do German speakers search for the same kinds of word translations as Spanish speakers? Would it reveal something about their cultures?

Data

To help us get this data, we turned to the Google Translate team. They warned us that the results would probably be quite mundane, but both Simon and I were quite intrigued by that idea.

The Google Translate team wrote a query to get all the *single word translations* that happened on Google Translate between August 2016 and December 2016 for 10 chosen popular languages. (That was about as far back as the translation queries are saved.) Doing the query once took hours and extra resources, and therefore we couldn't get multiple subsets (such as a set per month). Nevertheless, I was super excited about this one dataset, even if no time comparison was possible.

Before getting the actual data, I had already decided only to look at *nouns*. I suspected that common words and phrases like "Hello" or "I" or "Thank you" would probably be the most often translated words. Instead I was interested in the more subtle differences between the words that remained when the most common phrases were removed. So, after receiving the data, I wrote a script in R that could help pick out nouns by tagging each English translation (and if available, tagging the original word as well) with its grammatical form with the `NLP` and `OpenNLP` packages in R.

Nouns are words used to identify any of a class of people, places, or things. Such as "dog" or "chair."

Fig.9.1

The results of the automated word tagging.

	A	B	C	D	E	G
1	language_code	query	translation	frequency	englishTag	rank
2	de	hallo	Hello	309827	Hello/UH	1
3	de	ich	I	259558	I/PRP	2
4	de	schon	beautiful	169808	beautiful/JJ	3
5	de	danke	Thank you	159808	Thank/VB you/PRP	4
6	de	sie	you	156835	you/PRP	5
7	de	auch	also	150519	also/RB	6
8	de	leider	Unfortunately	142815	Unfortunately/RB	7
9	de	über	about	141070	about/IN	8
10	de	bitte	You're welcome	136172	You/PRP 're/VBP welcome/JJ	9
11	de	wie	as	134040	as/IN	10
12	de	noch	still	130294	still/RB	11
13	de	nach	after	127616	after/IN	12
14	de	auf	on	125212	on/IN	13
15	de	vielleicht	maybe	119805	maybe/RB	14
16	de	von	from	118759	from/IN	15
17	de	die	the	118044	the/DT	16
18	de	ihr	her	115582	her/PRP$	17
19	de	aber	but	113404	but/CC	18
20	de	vorname	first name	113347	first/RB name/NN	19
21	de	für	For	112333	For/IN	20
22	de	er	he	110288	he/PRP	21
23	de	trotzdem	Nevertheless	110050	Nevertheless/RB	22
24	de	zu	to	109060	to/TO	23
25	de	wir	we	108078	we/PRP	24

Browsing through the results I noticed three things. First, the adjectives, typically placed before a noun, such as "young" and "intelligent," also held interesting results and seemed quite diverse across languages. Therefore, I decided to keep both nouns and adjectives.

Second, I saw that some languages have masculine and feminine variations of the same word, such as "hermosa" (feminine) and "hermoso" (masculine), both of which mean "beautiful" in Spanish. I also observed the use of synonyms which translated to one word in English, such as "bonito," again, meaning beautiful in Spanish. To get a fair comparison between rankings within languages, I combined the search frequencies for all terms that translated to the same English word manually (per language).

Whenever I say "most translated word(s)" from here on out I mean this to be based on the subset of nouns and adjectives.

And third, as I'd expected, the automatically generated tagging results by R weren't perfect. For example, an English word can be both a verb or a noun, depending on the context. Therefore, to get a top 10 word list per language, I actually ended up looking up each original word that I wasn't sure about on Wiktionary to see what the most probable translation/grammatical form was. It took a long time, but I eventually had a top 10 most translated word list for each of the 10 chosen languages.

To get an overall top 10 word list combining *all* the languages, I looked at the rankings per language to compare relative popularity. I used a point based system where the top word in a language would get the most points, the second most popular word would get one point less, and so on. The overall most translated word would then be the English translation that had the most points from all languages combined.

Sketch

The sketching part of this project actually happened before the data preparation, because Simon and Alberto wanted to see possible designs of visualizations beforehand. As I was dealing with words and languages, I was inspired by the idea of using the words themselves as much as possible. Initially I wanted each element in the final visuals to be comprised of the word that it represented; all the visuals would only consist of words. Figure 9.2 shows the (cleaned-up) designs that I sketched on my iPad Pro and sent to Simon and Alberto.

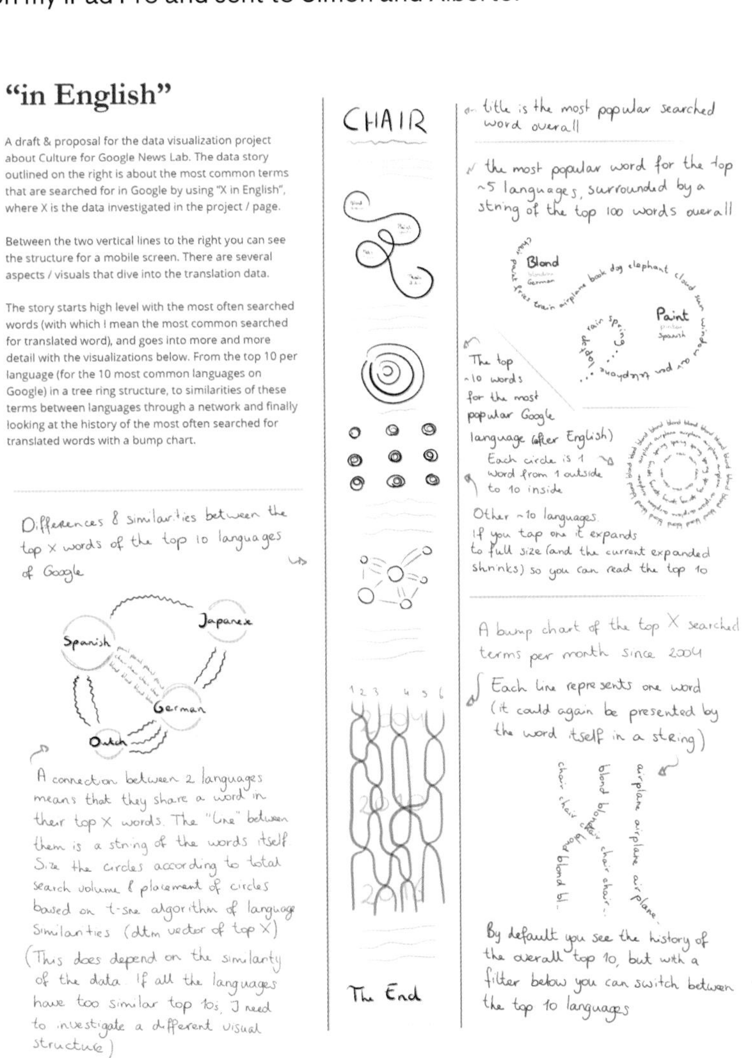

Fig.9.2

The cleaned up design made on my iPad Pro of the different translation-based visuals

I took a "start at the top and provide more details with each new visual" approach. All the way at the top, I'd place the most often translated word across *all* languages. Next, the most translated word per language would appear, which was interlinked by a swirling word string consisting of the top 100 translated words overall. Taking another step deeper would reveal the top 10 per language in a tree ring like structure, comparing all the original words to their English translations. To see similarities between languages I drew a network where each link represented a word that both languages had in common between their top 10 words (such as "beautiful"). Finally, there would be a bump chart comparing the top 10 across time. However, this last bump chart eventually wasn't possible because I only had that one dataset without a time component.

For the first time, I created a secret "mood board" on Pinterest just for a project. Collecting things that inspired me resulted in many black-and-white silhouettes, especially art-like pieces consisting mostly of cut-out hand lettering from children's books.

It was so useful that I've been creating these boards for almost every client project I've done since.

During the project itself I also sketched a lot to help me figure out the math. All of the visuals have some form of text placed along curved paths. And for all of these paths I had to figure out the custom SVG path formula using math. As an added difficulty, the paths had to be constructed so words on crossing paths wouldn't overlap too much and the text had to be placed as upright as possible.

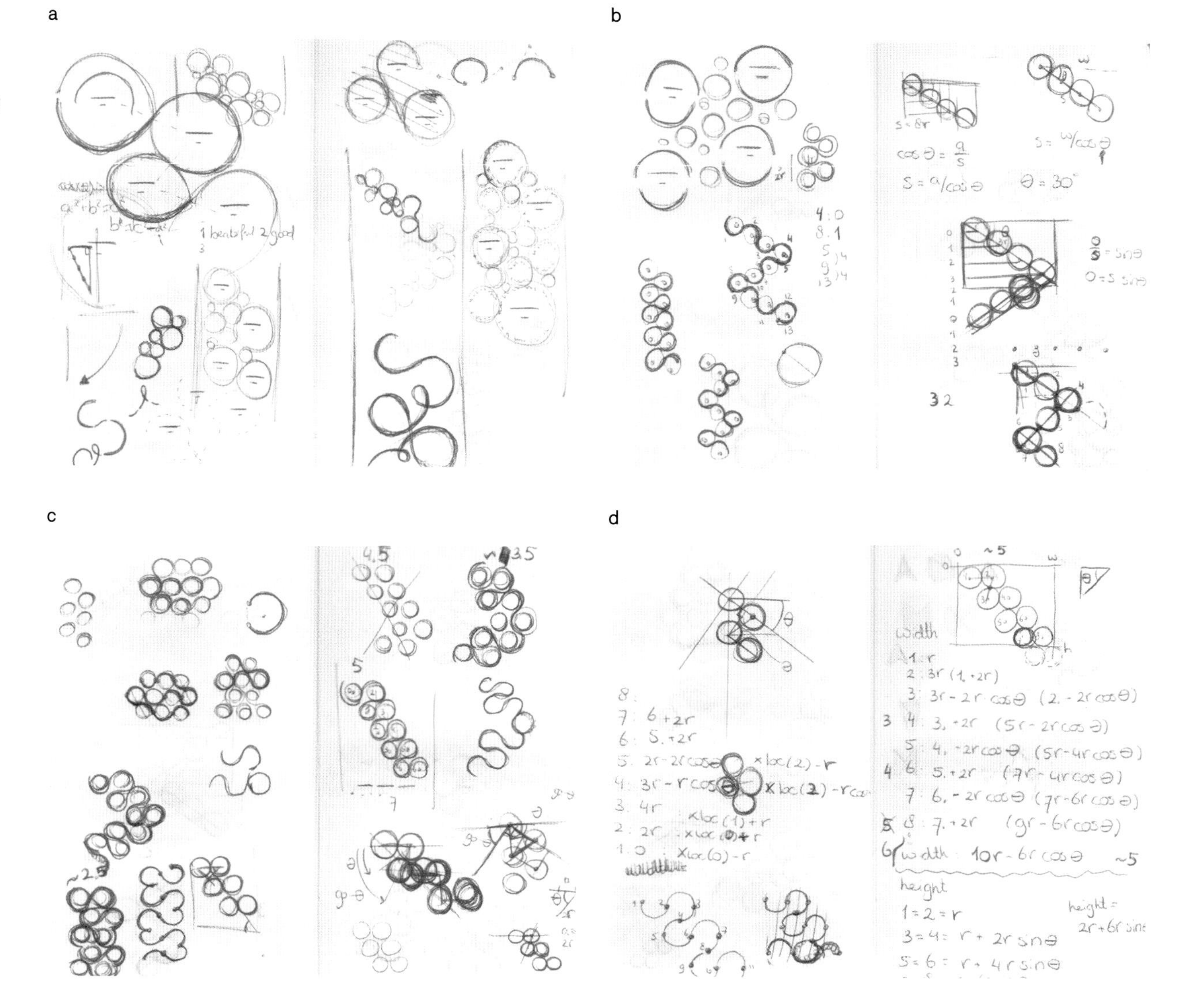

Fig.9.3 (a,b,c,d)

The many different sketches I made to try and figure out which word string concept I could make work.

For the top visual, where a string of words runs through the most translated word of each language, I wanted something that swirled around in a very natural way. However, I also had to make the page responsive and "mobile first". And I just *couldn't* find a programmable/mathematical way to create a natural looking swirl that would work for many different screen sizes.

Therefore, I started to work more with the idea of "beads on a string" which went down along an angle. Figure 9.3 shows that it still took a lot of sketches to figure out which layouts would work and how things should look and function when either two, three, or four circles fit in a row on the screen.

↳ Sketch to Discover and Remove Thinking Errors

By sketching out the swirly path that I had in mind, I realized I couldn't find a way to actually recreate it in code. Thankfully, I was able to quickly iterate and think of other ideas, eventually leading to the "beads on a string" concept. And I can't say this often enough: If you can't make your design work logically on paper, it's definitely not going to work on the computer with the actual data.

For the second visualization of the "tree ring" top 10 words per language I didn't have to sketch that much. The third visual of the network, on the other hand, was more involved. Although the lines inside the network weren't too difficult, the problems arose when an outer circle, representing a language, was clicked on, which made it move towards the center. I'll spare you the details, but it had to do with the text always reappearing in an upright manner combined with the animation of the curved lines the text was drawn on. Figuring out how that was supposed to work took *even more* pages in my little notebook!

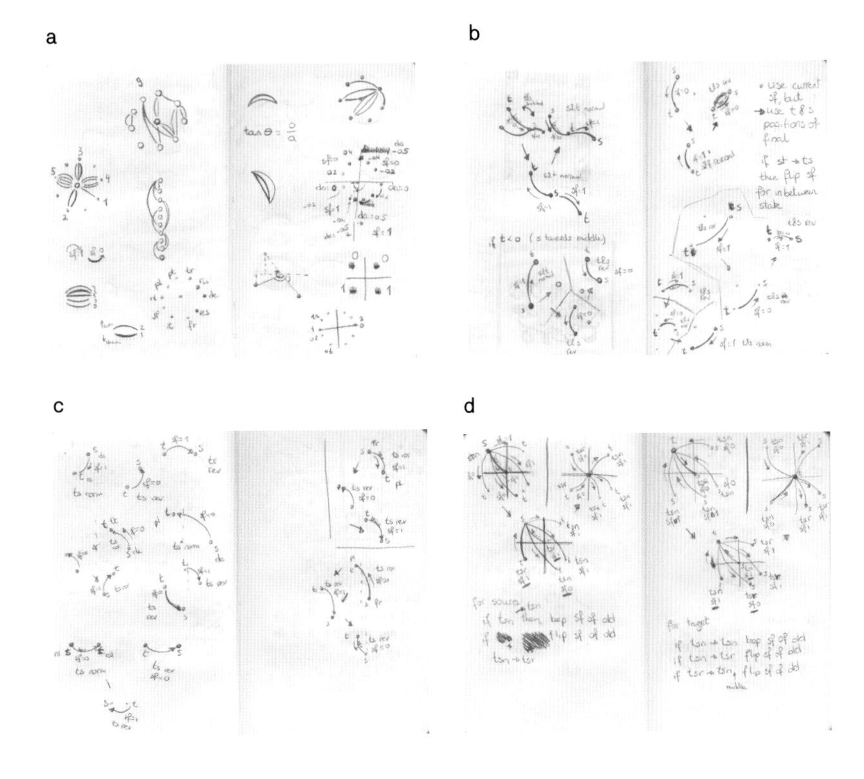

Fig.9.4 (a,b,c,d)

A few of the sketches I made to figure out the language similarity network mathematics.

The final thing I looked into was the overall design of the page. I wanted to recreate the old children's book, black cut-out style imagery. But I couldn't really find a way to incorporate that with the theme of translations (or a responsive page). So I eventually turned to hand lettering and decided to create my own title and headers. Drawing them on my iPad Pro, I started out trying to make the letters look smooth and flow nicely. However, for an inexperienced hand like mine, that was very hard to do. Plus, I wasn't happy with the resulting style. So instead I made them look a bit quirky and imperfect, distinct from anything else I managed to find online.

Fig.9.5

Work-in-progress screenshot of my quirky header titles.

Code

In terms of coding, I started with the "tree ring" visual that displayed the top 10 most translated words of a language. By now I'd worked with SVG arcs often enough in the past to set up the basic forms in an hour or so.

Fig.9.6

The very basics of the tree ring visual.

Next, I focused on how to animate switching between languages. Due to space constraints, especially on a mobile phone, I made one big circle in which the top 10 can be read. The other languages would be squished into tiny, almost flower-like, circles. In my first design, when clicked, a tiny circle would move towards the top and simultaneously open up to reveal its top 10, while the previous big circle did the exact opposite. However, when I had that working, the animation was *extremely* staggered. It was so bad that it became difficult to see that these two circles actually switched places.

If I got a "tribble" every time I had to compromise on my original design due to bad (animation) performance, my apartment would be stuffed by now...

Therefore, I set about trying a different approach: the text in the circle would rotate out of view before rotating back in again in the new chosen language. When I finally overcame all the browser bugs to achieve the effect, I actually liked the result more than the previous idea (Figure 9.7).
In terms of the design, Alberto suggested only making one word stand out to prevent making the visual appear too busy. But as far as I know, it isn't possible to use different styles in one SVG `textPath` element. Therefore I incorporated three different `textPath` elements: a light grey one on the left and a light grey one on the right displaying the original word and a bigger black word with the English translation in the middle (Figure 9.8). Figuring out how to place those three elements so it looked like one string of words that never overlap was another "interesting" puzzle to solve. ಥ_ಥ

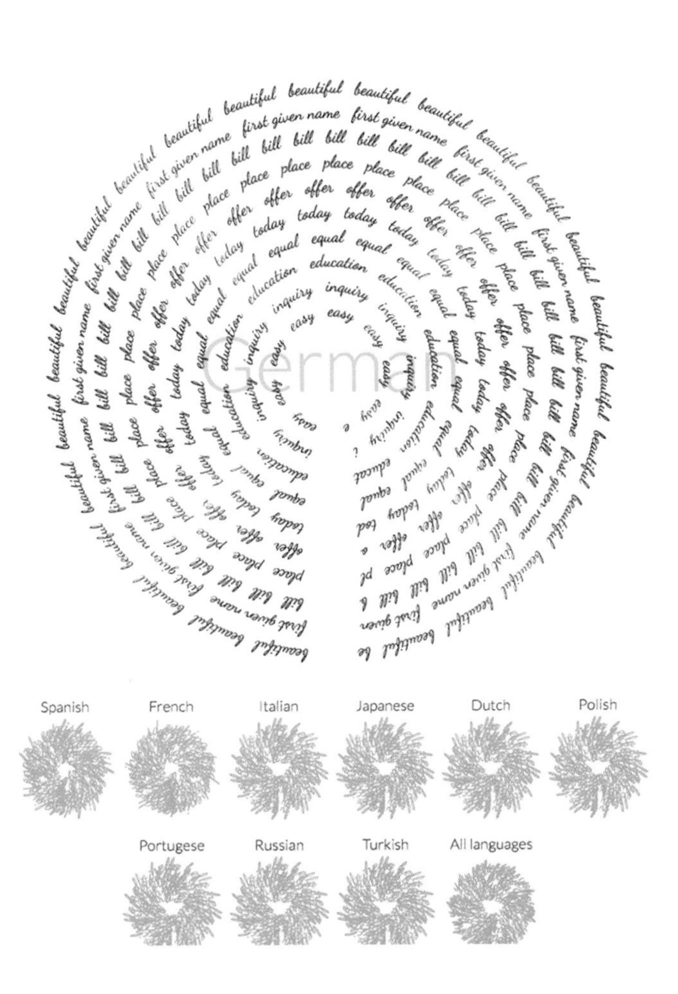

Fig.9.7

An older design of the tree ring visual.

The top 10 words that are translated from Japanese into English*

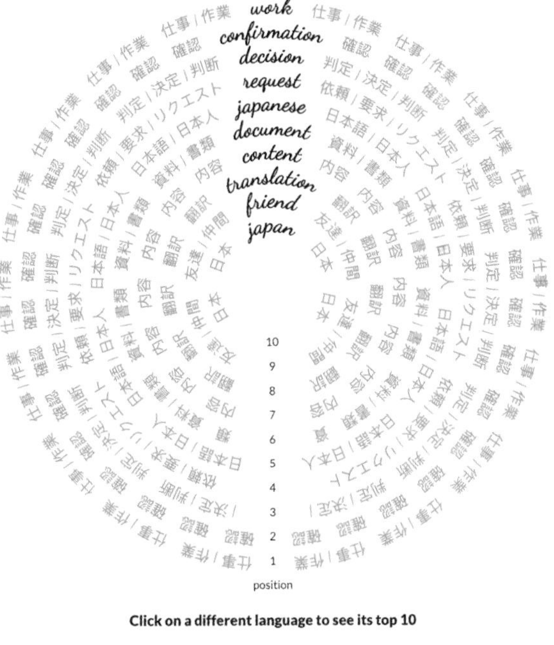

Fig.9.8

The final design, showing the top 10 Japanese words.

Next up was creating the word string, or "word snake" as I started calling it. Because I couldn't mathematically figure out the natural looking swirl while sketching the design, I first tried some other approaches, but I wasn't happy with how those turned out (Figure 9.9). I therefore revisited the swirl idea, this time visualizing the circles as beads on a string—something that I could mathematically solve and create, thankfully. Of course, actually coding up the *correct* SVG path formula took some trial and error...

Due to screen size, I created several options for the "beads" to be positioned. It calculates if two, three, or four circles fit in a row and the rest updates automatically. Personally, I like the "two beads per row" version the most (Figure 9.10).

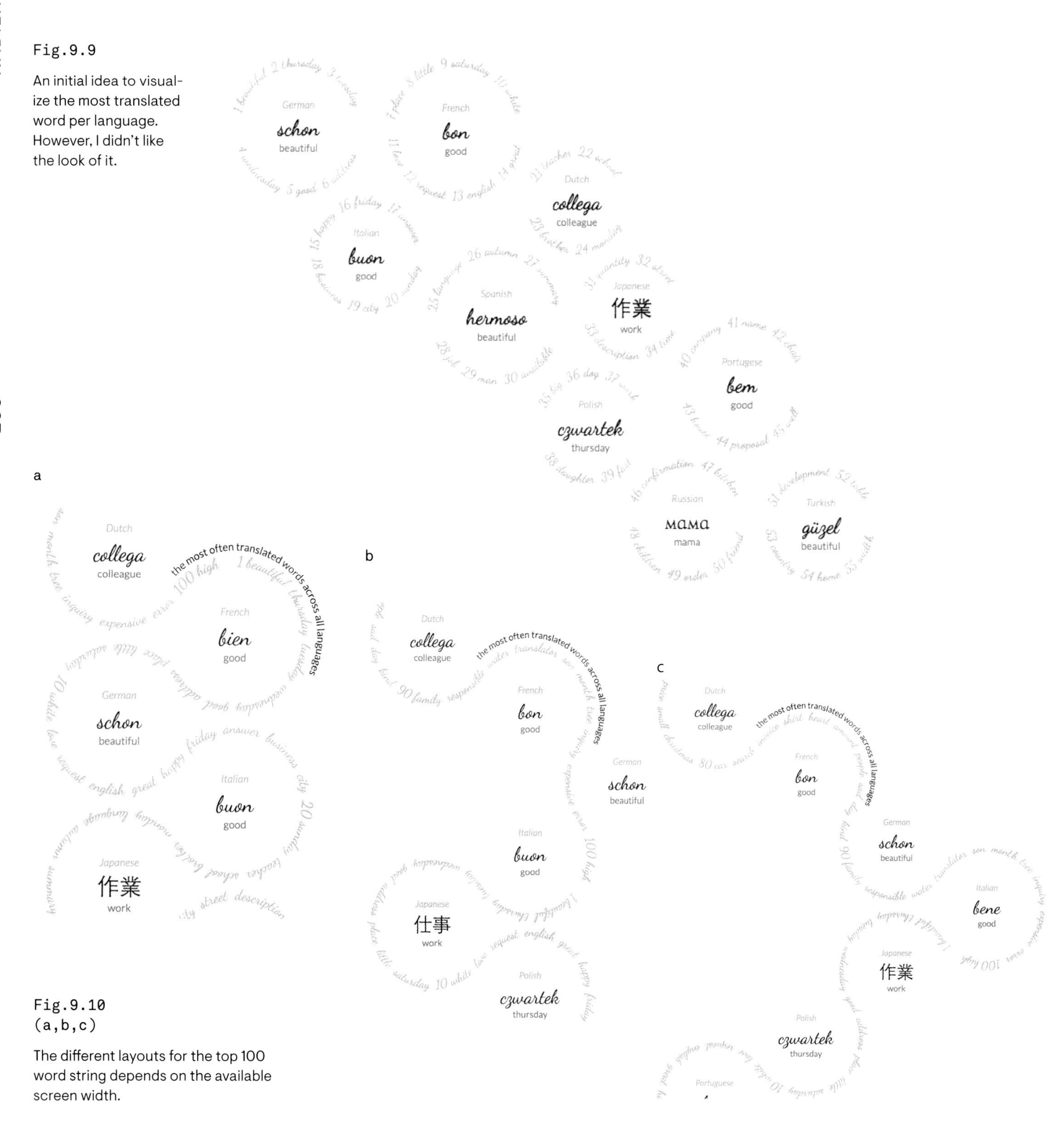

Fig.9.9

An initial idea to visualize the most translated word per language. However, I didn't like the look of it.

Fig.9.10 (a,b,c)

The different layouts for the top 100 word string depends on the available screen width.

↳ Learn to Love Math

You might already love math, but if you don't, I highly recommend at least trying to understand the basics of geometry, especially trigonometry (remember sine and cosine?) and statistics.

When you start moving away from more mainstream visuals, and you want to place the data on your screen according to your own design, chances are that you'll be needing some math to calculate those locations. For example, using a radial layout will involve trigonometry. I'm not talking about difficult math here though—no differential equations—but knowing how to go from the (Euclidean coordinates of) *x* and *y* to the (polar coordinates of) *angle* and *radius*, and back, will become invaluable.

For this project, being able to transform my "beads on a circle" sketches to trigonometric functions that actually drew the swirly line made all the difference!

Alberto and Simon told me that the word string made the inner "language circles" look like they could be clicked on, as if they were buttons. That hadn't been my original intent, but that did make me think about what other interesting information I could reveal on a hover or click to reward the people that actually interacted with the visual.

This brought me to Google Trends, where I looked up the trend of these most translated words and their related queries. Thankfully, all of the words had some interesting peaks and dips throughout the past few years. I thus set about making the most extensive tooltip I've ever made! At the top it showed the worldwide trend of the English word over the last five years. Below the line chart I added a word cloud about the related queries. (The word "mama" was definitely my favorite, with related queries such as "maternal insult" and "yo mama".)

I annotated the line chart in the tooltip using the awesome `d3-annotation.js` library.

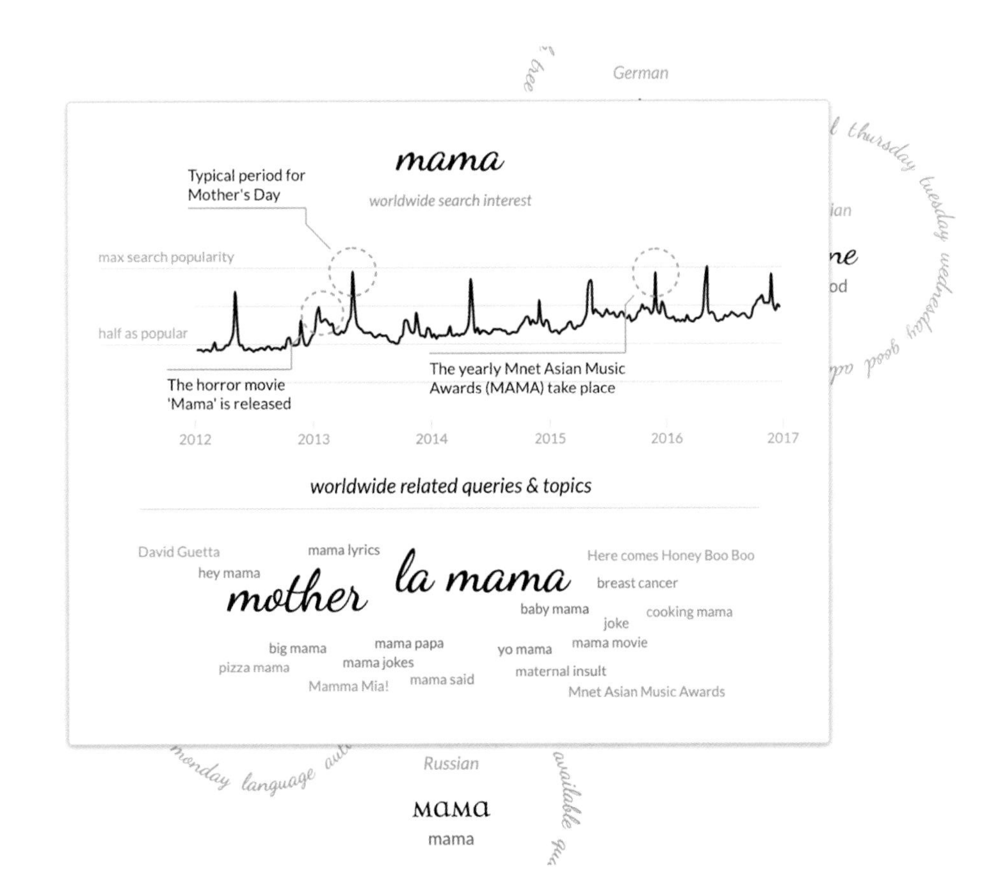

Fig.9.11

The most extensive tooltip I've ever made, which reveals itself when hovering over the most translated word of one of the 10 languages.

The final visual of the full piece was focused on showing the similarities between the top 10 word lists. Each language is a circle and each word that two languages have in common between their top 10 is represented by a line. Interestingly, the highest similarity happened between Spanish, Portuguese, and Italian, and also with Russian and Polish.

I first had to set up a system in which two "nodes" could be connected by multiple "links." Typically people use a thicker line when two nodes are more highly linked. However, because I wanted to eventually replace the lines with their actual words, I had to find something else. Using ever more curved lines seemed like an elegant solution.

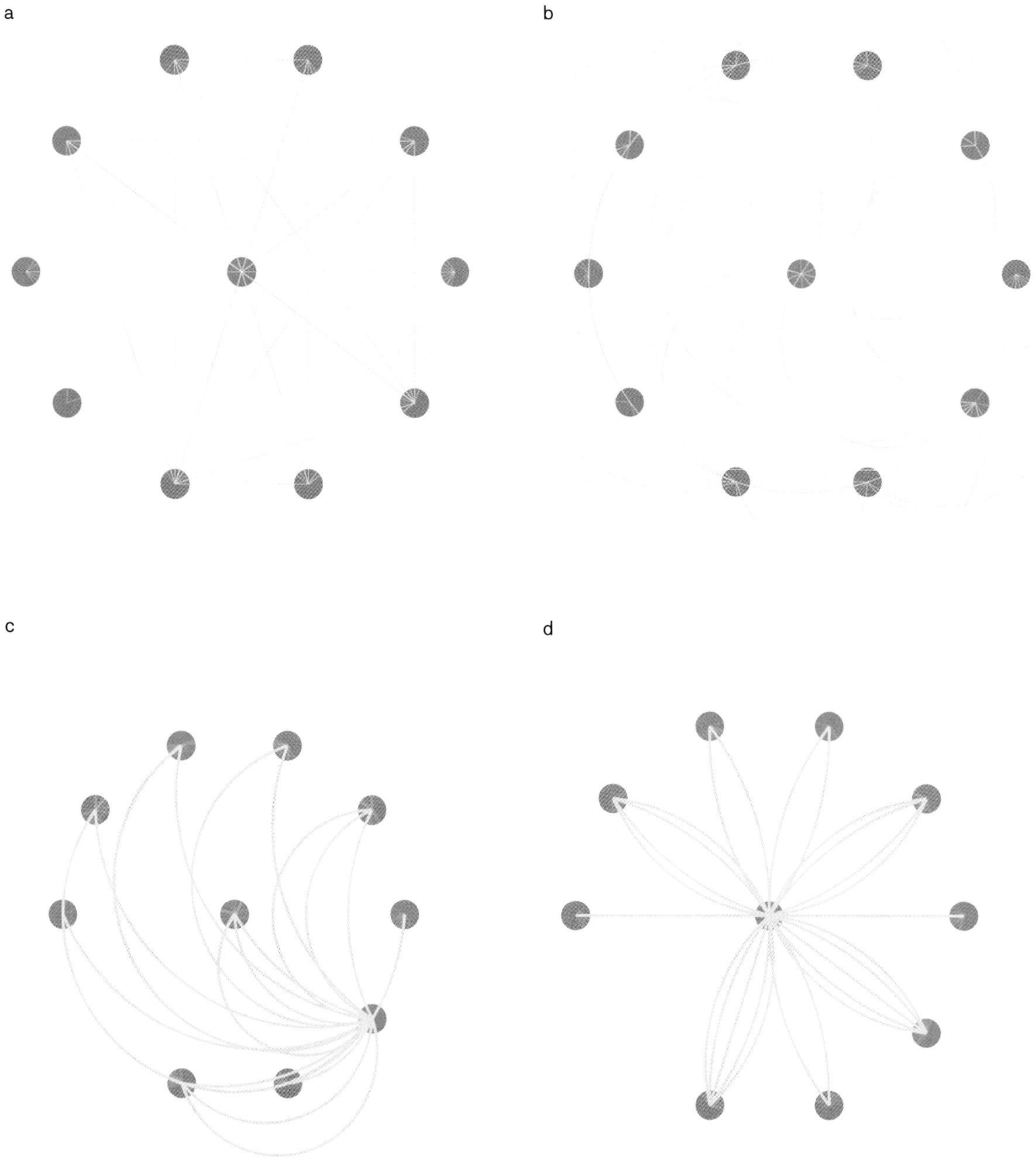

Fig.9.12 (a,b,c,d)

Setting up the network with multiple links between two nodes, getting ever more curved.

a

b

c

d

I then replaced the lines with the words themselves, using a design similar to the tree ring and saw ... chaos (Figure 9.13 a). I totally forgot to think about the fact that many lines would overlap, making it impossible to read the words. Furthermore, the words themselves made the visual extremely full and cluttered. So I put the lines back in and only kept the words on those lines attached to the central circle. Alberto advised to make it even more minimal and only keep the central dark (English) word on the line, to really get the focus.

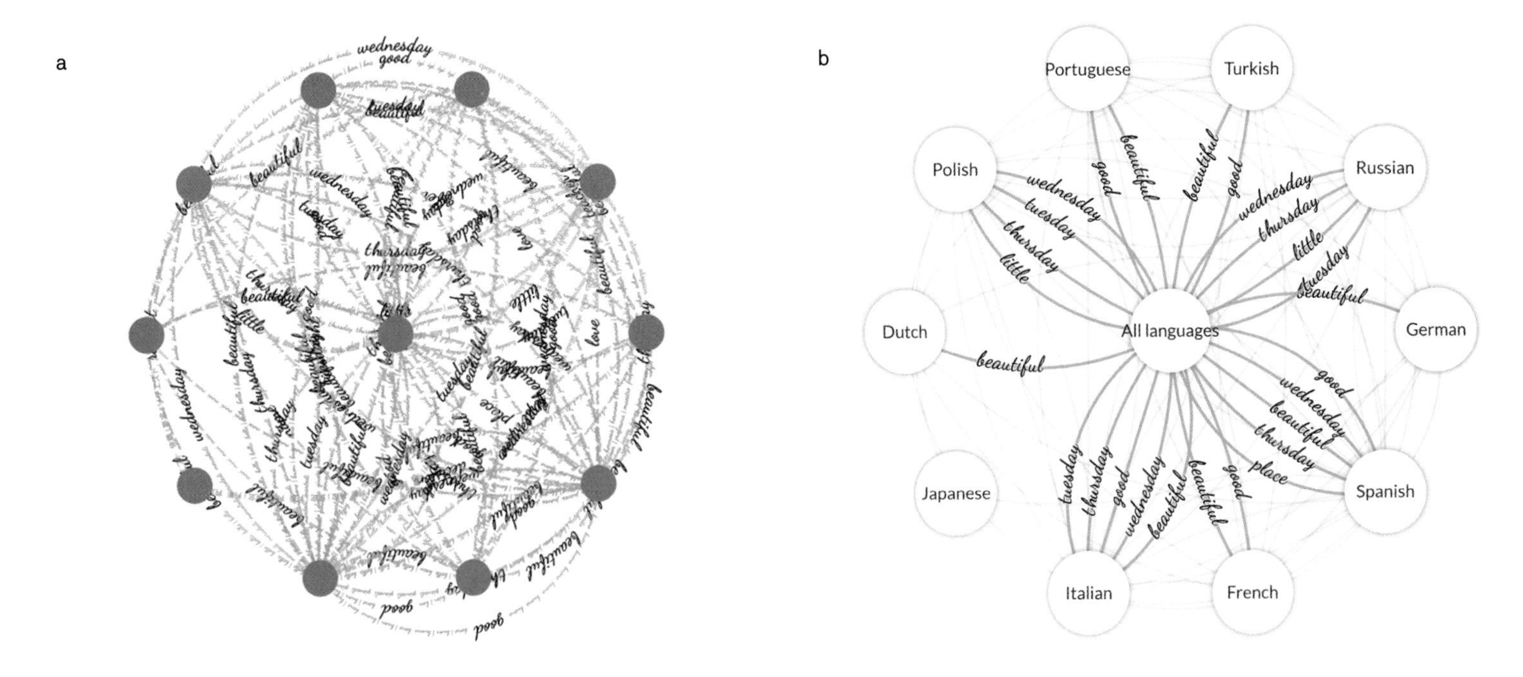

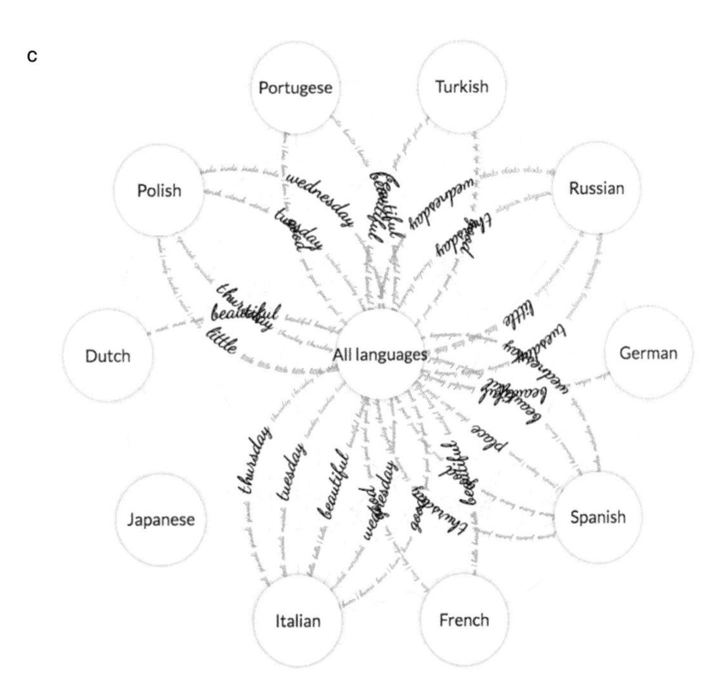

Fig.9.13 (a,b,c)

Going from chaos to an ever more minimal and focused design for words on the network lines.

More time than I'd like to admit later, and the animation to click a language circle and make it move *smoothly* towards the center was working. And with that all of the three visuals were done! I investigated some interesting insights that I'd discovered from each visual and wrote the accompanying text, added in my hand-drawn section headers, to have the whole become my first fully-fledged article as part of *Data Sketches*.

Dataviz for Both Mobile and Desktop

This was the first project where it was important that all the visuals worked well on both desktop and mobile screen sizes. Making a data visualization work on widely different screen sizes is one of the most difficult (and sometimes frustrating) aspects of my work. I totally ignored small screens in all of my previous projects, but I do try and find proper ways to handle the large difference in screen real estate when I have to. What I see a lot on the web is to either scale down the whole visual to fit it on smaller screens, or to stack sections horizontally on large screens and vertically on small screens. Naturally, there are other ways as well. My favorite is more about changing the layout and moving the data itself. Instead of scaling anything down, you move things around to make the most out of the available space. For example, the final visual from this project uses this approach. On large screens the languages are displayed in a circular layout, which is a visually pleasing shape. However, on mobile screens the network becomes a rectangle, so the "language circles" can be placed as far apart as possible .

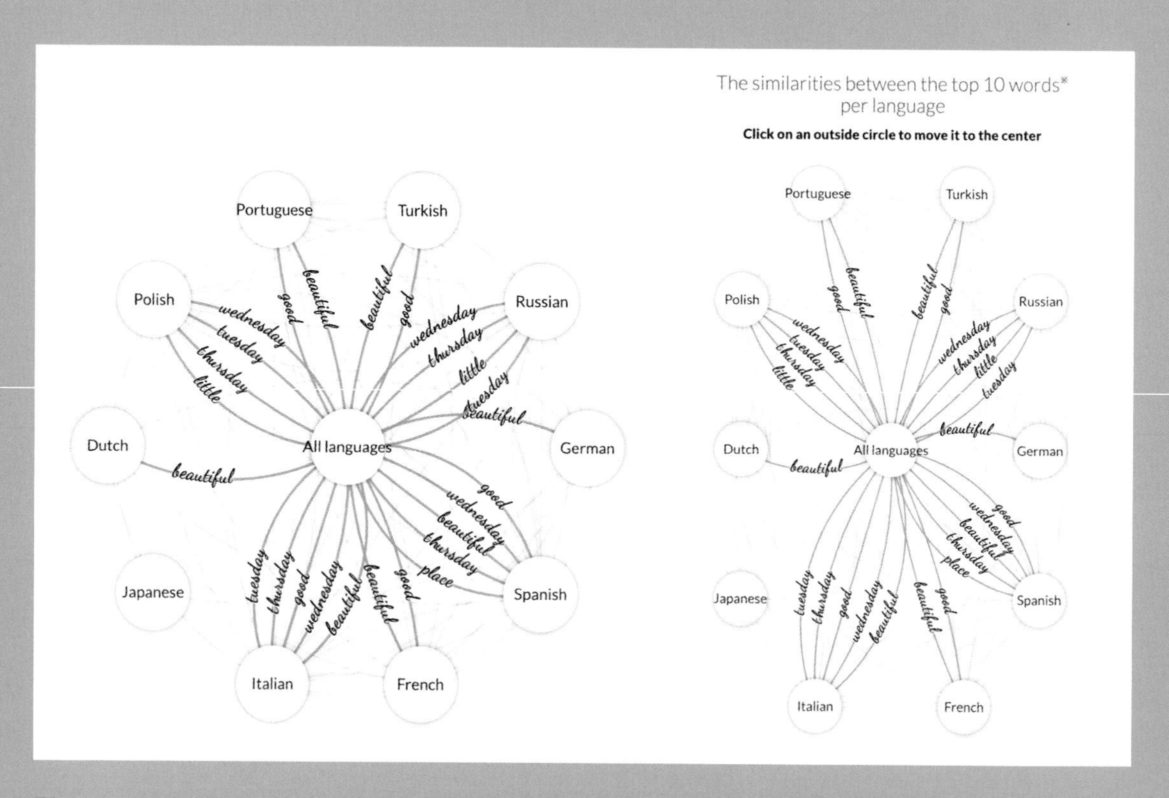

Fig.9.14

The circular version on large screens and the rectangular layout on small screens.

The same idea happens for the word snake in Figure 9.10, where I don't make the language circles any smaller, but instead move them into a smaller and smaller layouts. This required more math to figure out how the swirly line of words should be calculated, with either two, three, or four beads side by side. However, 80% of that was already done by figuring out how to draw the swirly line for one of those options. So keep an open mind when you have to create a data visualization that needs to work well on many different sizes; there's more possible than just scaling and stacking.

Reflections

This was definitely the project that took the most hours to create until now (about 120 hours), but it was also the biggest one, so that makes sense. And, having done this as an *actual*, paid project made it easy to justify working on it. (⌐■_■)

It was really great for Shirley and me to work together with Google and Alberto Cairo. I thoroughly enjoyed having the chance to create such an extensive, full, dataviz-driven story using data that I had specifically requested and that was impossible to find anywhere else!

Beautiful in English

↳ BeautifulInEnglish.com

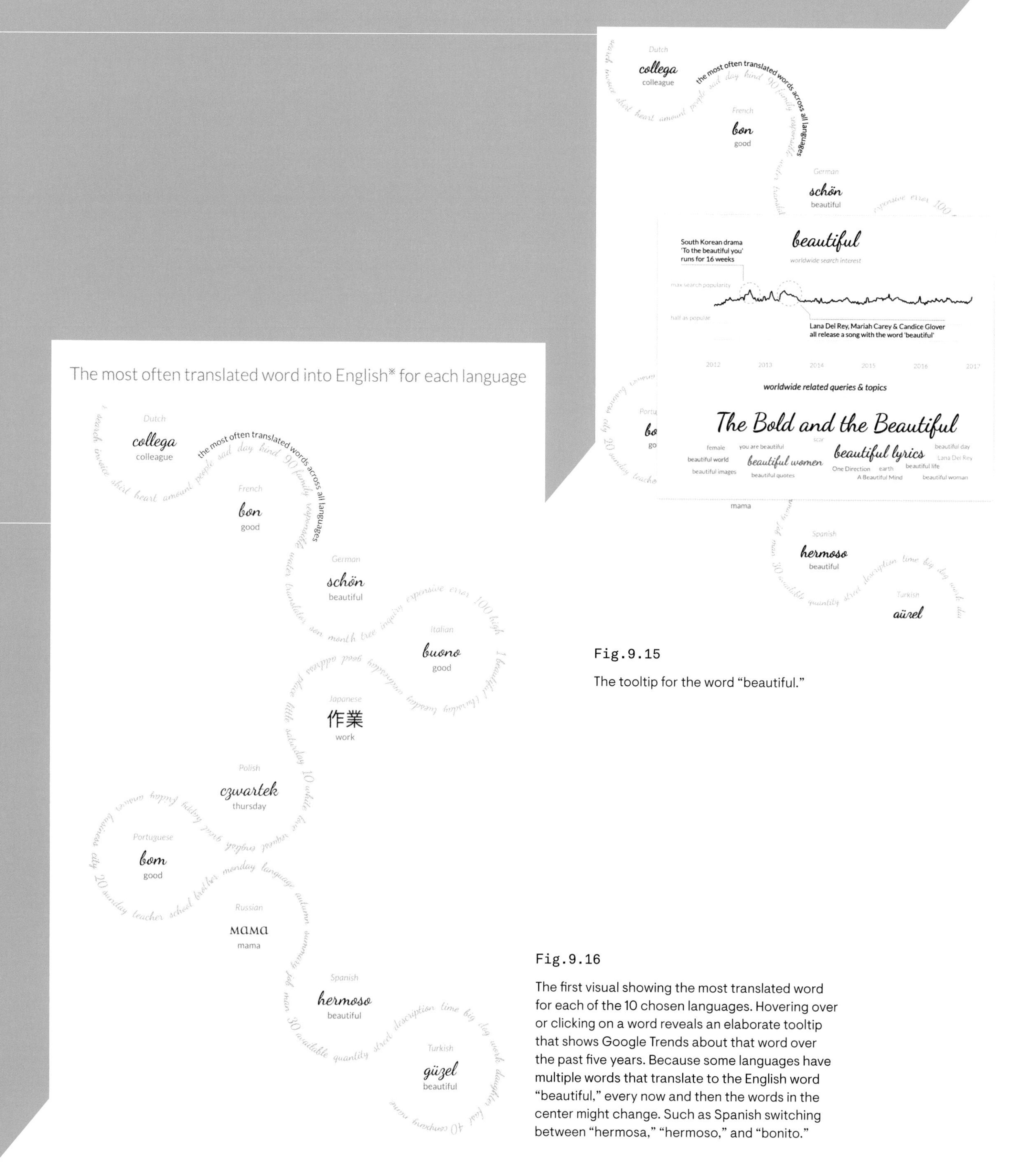

Fig.9.15

The tooltip for the word "beautiful."

Fig.9.16

The first visual showing the most translated word for each of the 10 chosen languages. Hovering over or clicking on a word reveals an elaborate tooltip that shows Google Trends about that word over the past five years. Because some languages have multiple words that translate to the English word "beautiful," every now and then the words in the center might change. Such as Spanish switching between "hermosa," "hermoso," and "bonito."

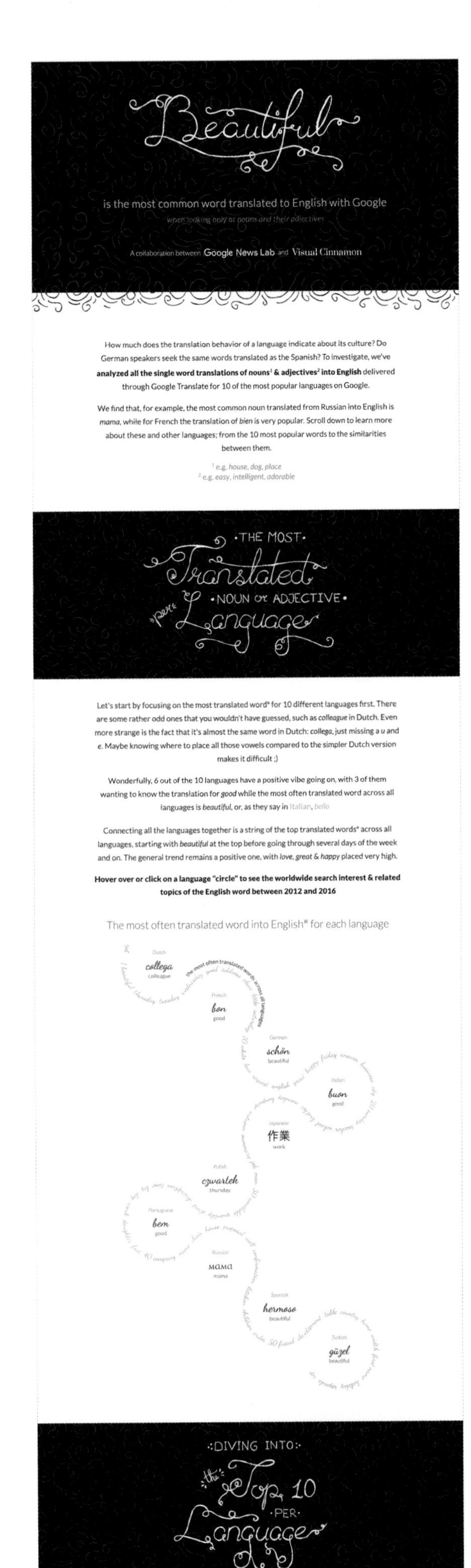

Beautiful
is the most common word translated to English with Google
when looking only at nouns and their adjectives
A collaboration between Google News Lab and Visual Cinnamon
How much does the translation behavior of a language indicate about its culture? Do German speakers seek the same words translated as the Spanish? To investigate, we've analyzed all the single word translations of nouns¹ & adjectives² into English delivered through Google Translate for 10 of the most popular languages on Google.
We find that, for example, the most common noun translated from Russian into English is mama, while for French the translation of bien is very popular. Scroll down to learn more about these and other languages; from the 10 most popular words to the similarities between them.
¹ e.g. house, dog, place
² e.g. easy, intelligent, adorable
·THE MOST·
Translated
·NOUN or ADJECTIVE·
per Language
Let's start by focusing on the most translated word* for 10 different languages first. There are some rather odd ones that you wouldn't have guessed, such as colleague in Dutch. Even more strange is the fact that it's almost the same word in Dutch: collega, just missing a u and e. Maybe knowing where to place all those vowels compared to the simpler Dutch version makes it difficult ;)
Wonderfully, 6 out of the 10 languages have a positive vibe going on, with 3 of them wanting to know the translation for good while the most often translated word across all languages is beautiful, or, as they say in Italian, bello
Connecting all the languages together is a string of the top translated words* across all languages, starting with beautiful at the top before going through several days of the week and on. The general trend remains a positive one, with love, great & happy placed very high.
Hover over or click on a language "circle" to see the worldwide search interest & related topics of the English word between 2012 and 2016
The most often translated word into English* for each language
collega
bon
schön
buon
作業
czwartek
bem
мама
hermoso
güzel
:DIVING INTO:
the Top 10
·PER·
Language

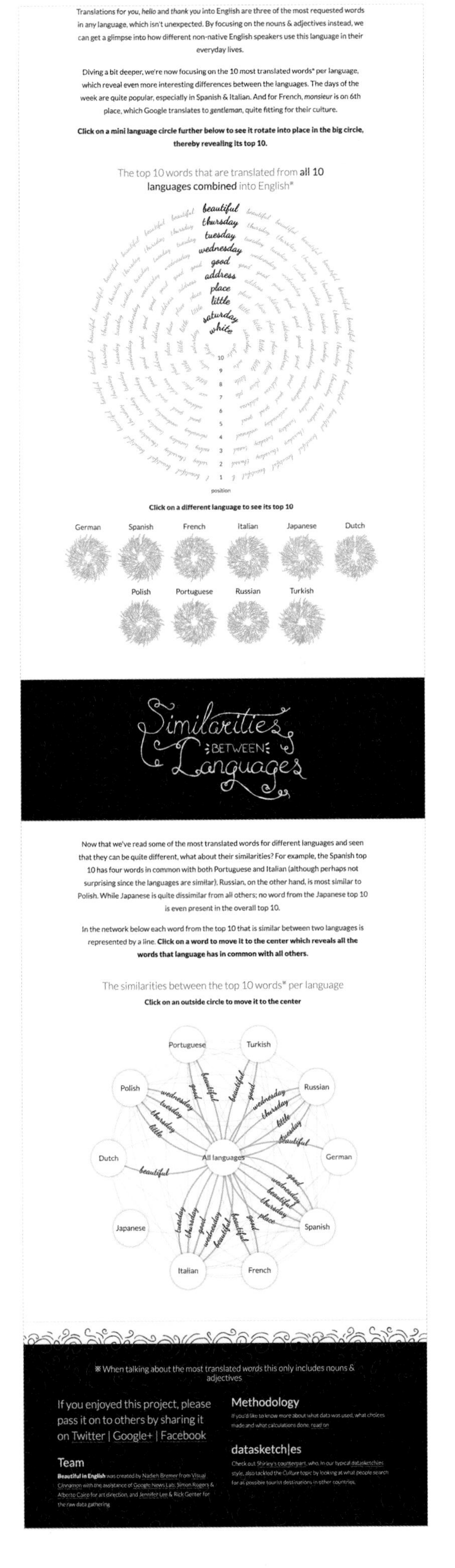

Translations for you, hello and thank you into English are three of the most requested words in any language, which isn't unexpected. By focusing on the nouns & adjectives instead, we can get a glimpse into how different non-native English speakers use this language in their everyday lives.
Diving a bit deeper, we're now focusing on the 10 most translated words* per language, which reveal even more interesting differences between the languages. The days of the week are quite popular, especially in Spanish & Italian. And for French, monsieur is on 6th place, which Google translates to gentleman, quite fitting for their culture.
Click on a mini language circle further below to see it rotate into place in the big circle, thereby revealing its top 10.
The top 10 words that are translated from all 10 languages combined into English*
beautiful
thursday
tuesday
wednesday
good
address
place
little
saturday
white
position
Click on a different language to see its top 10
German
Spanish
French
Italian
Japanese
Dutch
Polish
Portuguese
Russian
Turkish
Similarities
:BETWEEN:
Languages
Now that we've read some of the most translated words for different languages and seen that they can be quite different, what about their similarities? For example, the Spanish top 10 has four words in common with both Portuguese and Italian (although perhaps not surprising since the languages are similar). Russian, on the other hand, is most similar to Polish. While Japanese is quite dissimilar from all others; no word from the Japanese top 10 is even present in the overall top 10.
In the network below each word from the top 10 that is similar between two languages is represented by a line. Click on a word to move it to the center which reveals all the words that language has in common with all others.
The similarities between the top 10 words* per language
Click on an outside circle to move it to the center
Portuguese
Turkish
Polish
Russian
Dutch
All languages
German
Japanese
Spanish
Italian
French
* When talking about the most translated words this only includes nouns & adjectives
If you enjoyed this project, please pass it on to others by sharing it on Twitter | Google+ | Facebook
Methodology
datasketch|es
Team

Fig.9.17

The full overview of the “Beautiful in English” page.

a

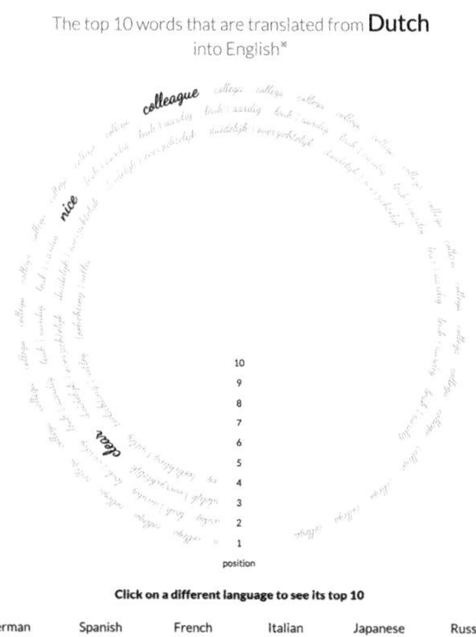

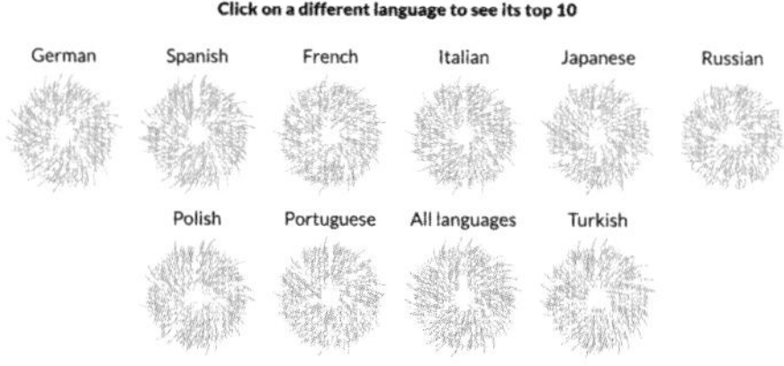

b

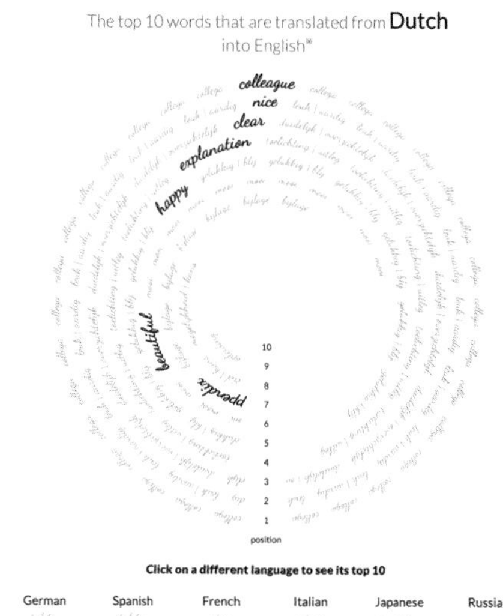

c

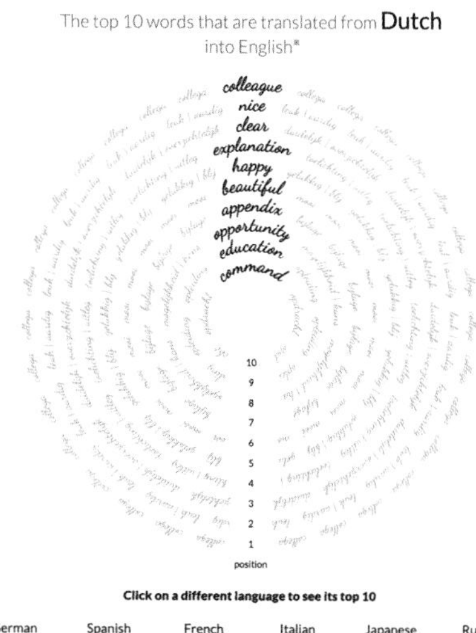

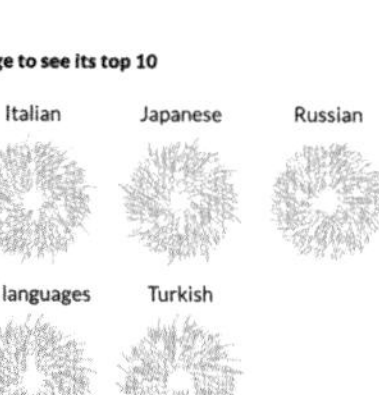

Fig.9.18 (a,b,c)

The tree ring visual while a change between languages happens. First the previous language rings rotate away and afterwards the new language rings rotate back in.

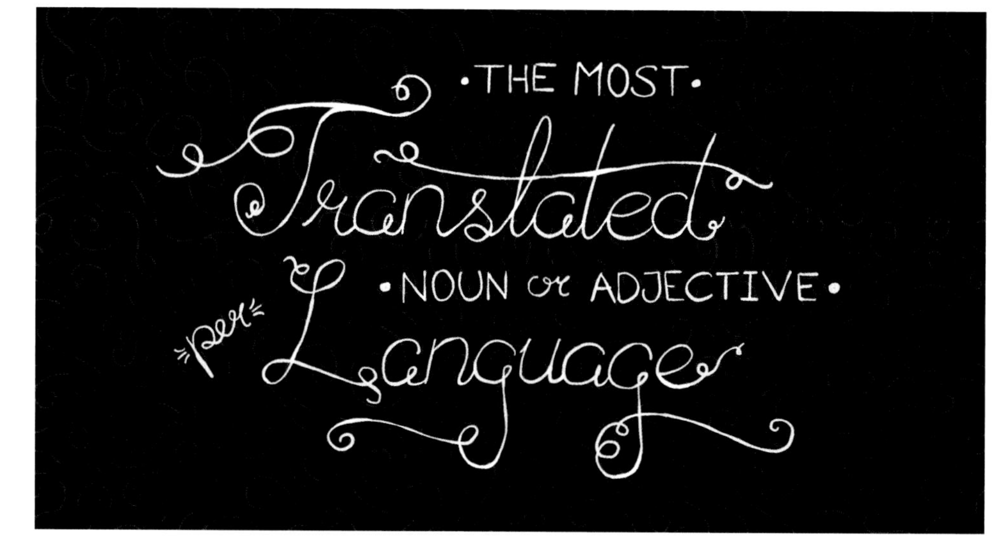

Fig.9.19

The header for the first section that reveals the most translated word for 10 popular languages.

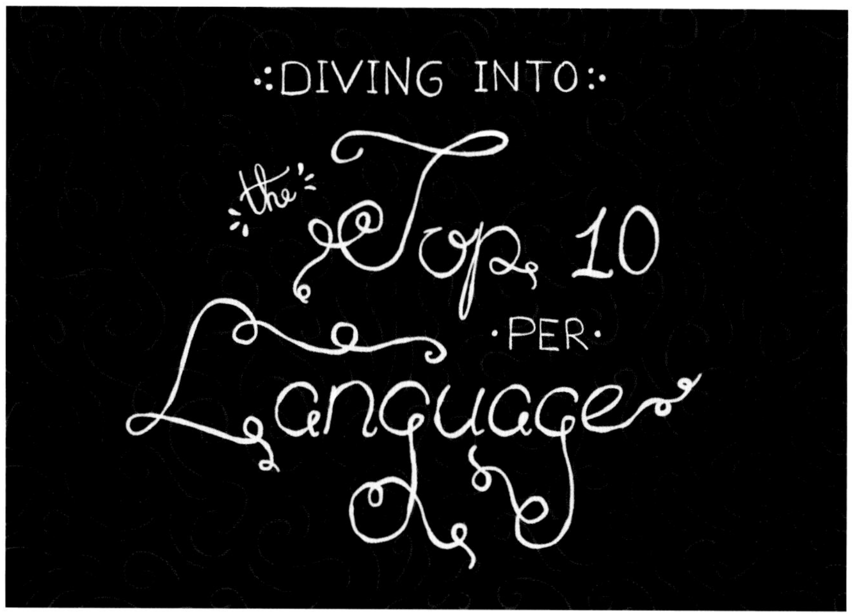

Fig.9.20

The quirky header for the second section of the article that looks at the top 10 per words language.

a

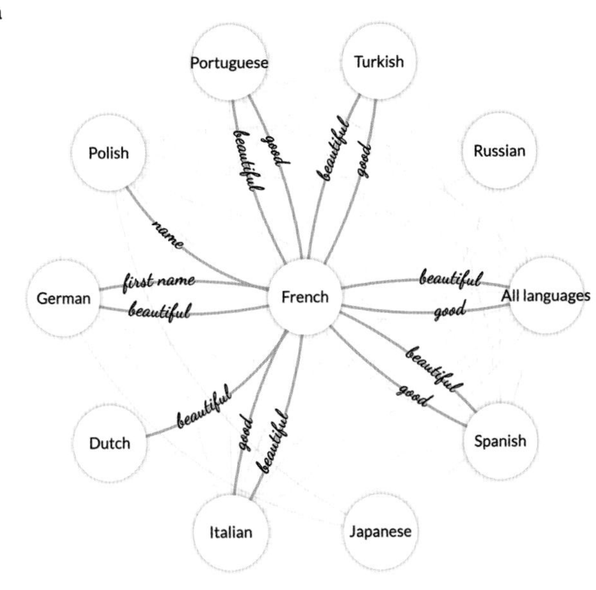

b

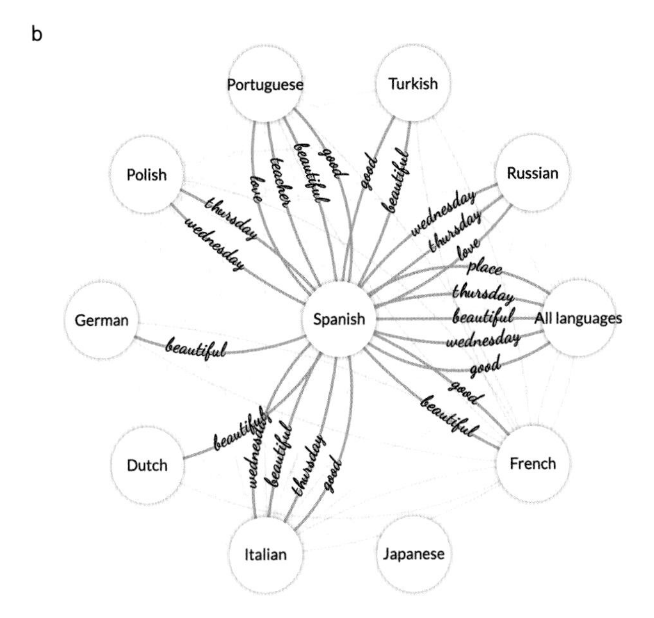

c

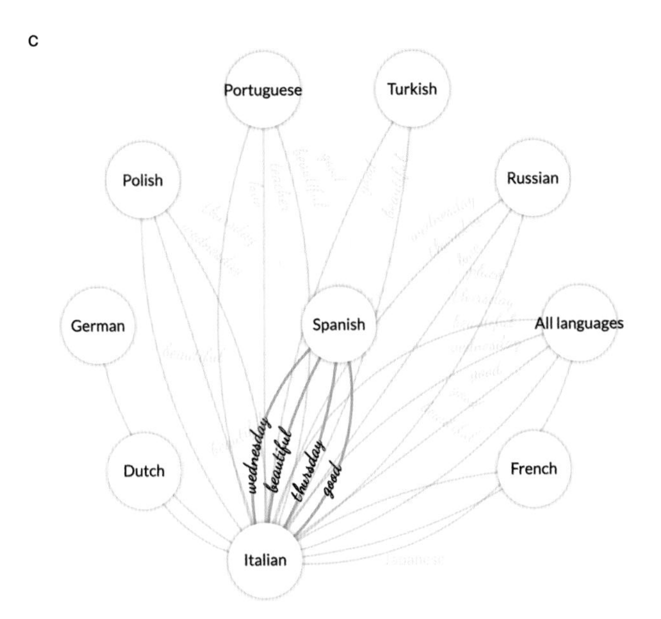

d

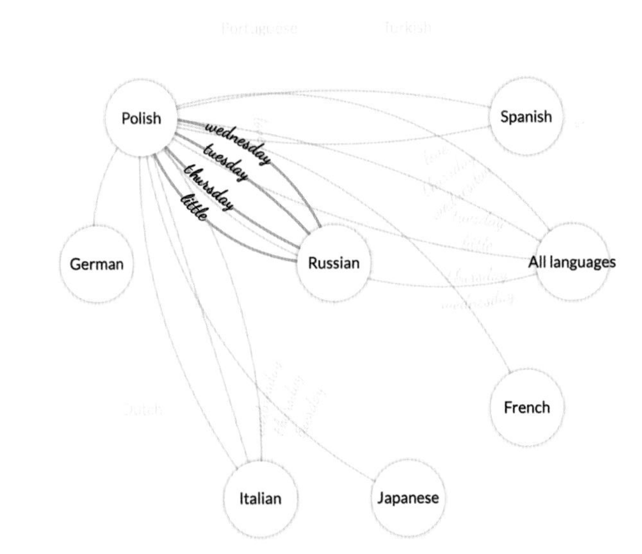

e

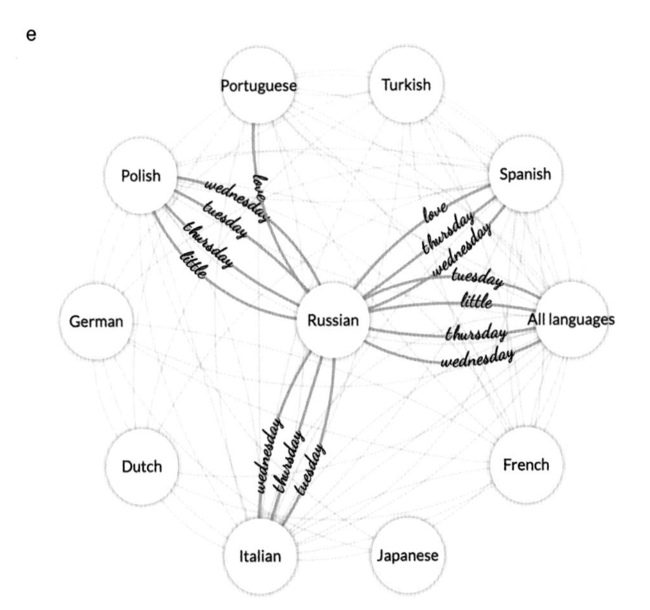

Fig.9.21
(a,b,c,d,e)

Hovering over a language in the similarity network highlights all the lines connected to it to provide a focus for the viewer.

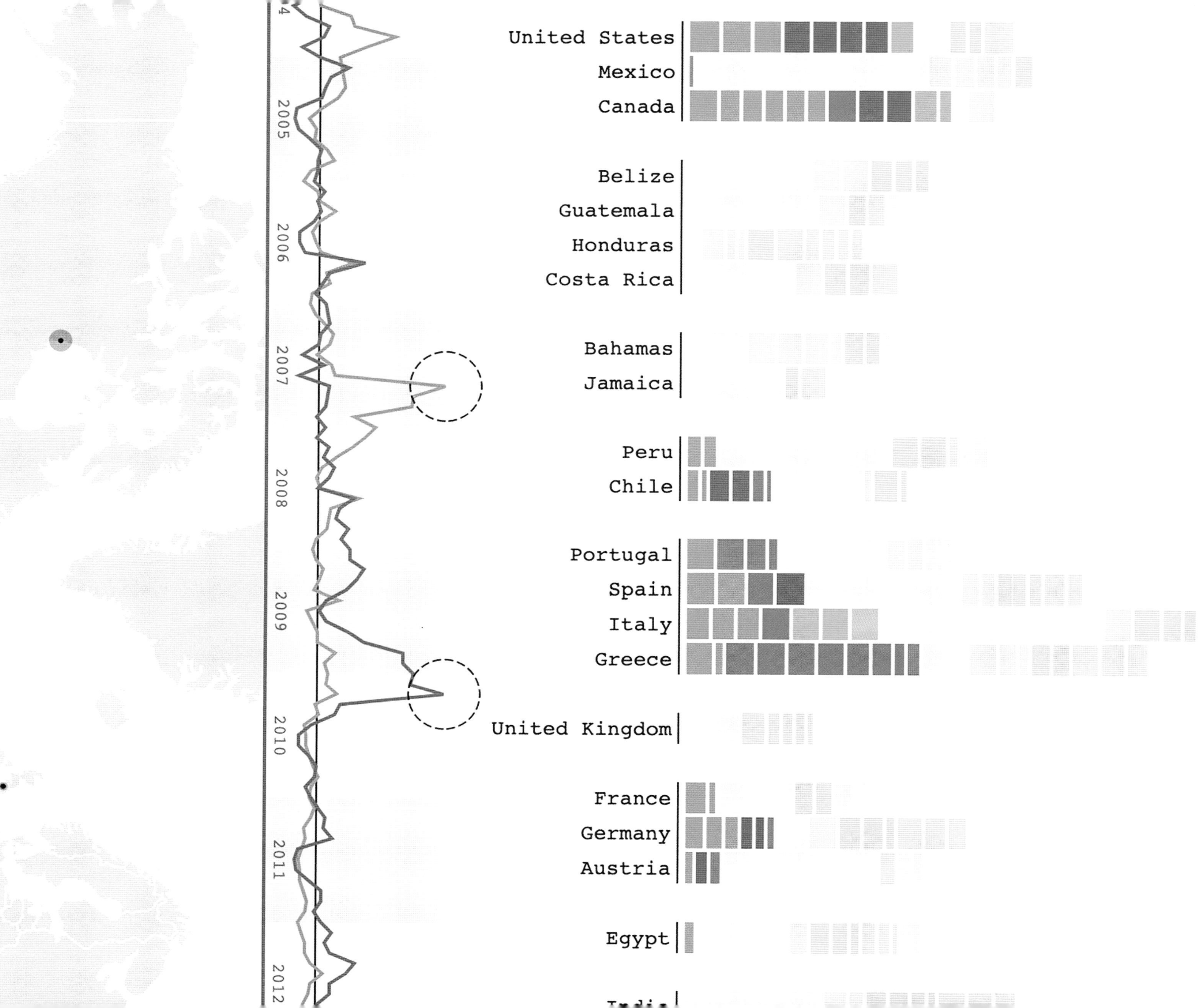

United States
Mexico
Canada
Belize
Guatemala
Honduras
Costa Rica
Bahamas
Jamaica
Peru
Chile
Portugal
Spain
Italy
Greece
United Kingdom
France
Germany
Austria
Egypt
2005
2006
2007
2008
2009
2010
2011
2012

Explore Adventure

SHIRLEY

When Nadieh and I got the email from Alberto Cairo and Simon Rogers to work with Google News Lab, I was ecstatic and beyond intimidated. After all, it was Google, it was Simon and Alberto, and they had search data going back to 2004. They had already published projects in collaboration with Accurat and Truth & Beauty (two data visualization studios whose works I find really inspiring), and I wasn't sure if I could live up to that—but I was determined to try my best.

Nadieh and I explored Google Trends and presented a few ideas, and Simon and Alberto chose our "Culture" proposal. Nadieh decided to look into language and explored the most common words people in other countries searched for to translate into English. I chose to focus on travel and dug into the travel destinations that people in one country searched for in another.

And because Alberto was a fan of our process write-ups, we agreed to include our "Culture" projects and their corresponding write-ups as part of *Data Sketches*.

Data

As Simon preferred having the data live-update, I subcontracted my friend Charles (my very first subcontractor!) to build a database that would periodically pull udpated data from Google Trends and process them into the format my client-side visualizations needed. It was really fun and encouraging to have someone else to work on the project with; not only did he do most of the data collection and clean-up (so I didn't have to (ﾉ≧∀≦)ﾉ), but it was also great to bounce ideas off of someone who was intimately familiar with the data.

To start, we began with a central question: given a country, which other countries searched for that country the most? Were people looking for particular cities in that country? Museums? Specific landmarks? And as I started digging into the data with Google Trends' Explore function, I also started to wonder if people looked for places geographically closer to them. This thought occurred to me when I saw that Australians often searched for places on the American West Coast, which is geographically closer to them than other parts of the United States.

With Google Trends, I can enter up to five search terms and get back their "search interest" (a score out of 100 that describes the term's relative popularity), search interest by region, and related and top topics over a specific time period. I can also filter the results by time range, by a specific "source" country, and by broad, Google-defined categories:

Fig.9.1

The data we were ultimately trying to get. Because of how Google Trends works, we had to approach it from the opposite direction and start with the "target" countries to be able to get their "source" countries, and finally get topics for that pair.

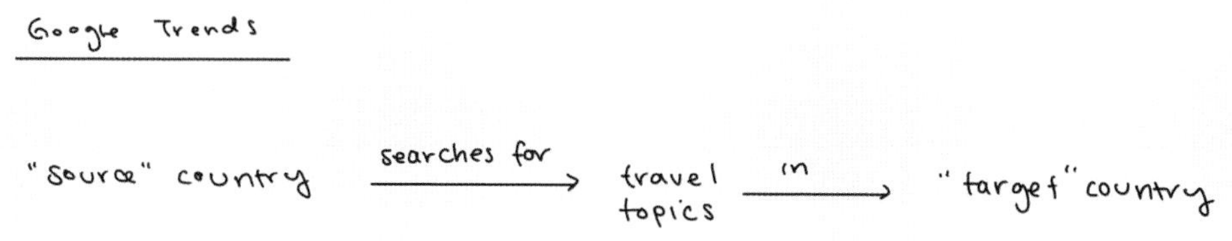

Google Trends
"source" country → searches for → travel topics → in → "target" country

(From now on, I'll call countries that did the searching the "source" countries, and the 20 countries that were searched for the "target" countries.)
To get the data needed, we:

1. Searched for every country using Google's list of country IDs and narrowed the results down by both the category "Travel → Tourist Destinations" and the date "2004 - present."
2. Used the results to aggregate the top 20 most searched for "target" countries.
3. Noted the "source" countries (found in "search interest by region") that searched for each of the 20 "target" countries the most in every quarter since 2004.
4. Noted the top Google-defined topics ("Niagara Falls," "Terracotta Soldiers," "Naples," etc.) for each set of "source" and "target" countries.

The plan sounded reasonable when we first came up with the steps, but the data we got back was overwhelmingly vast. We had thousands of travel-related topics, and in order to make sense of all those topics, we needed a way to meaningfully categorize them.

As the topics were defined by Google, Charles used Google's Knowledge Graph Search to find their details, including images, descriptions, and an array of "types" for each topic. Three of the most common "types"—cities, people, and nature—made sense as categories, but the remaining types were more difficult to define. Some were too broad, like "Thing" or "Place," and others were too narrow, like "LodgingBusiness" or "MovieTheater." After much discussion, we came up with eight categories that could encompass the 252 "types" we got back from the Knowledge Graph: city, region, attraction, nature, person, history, arts, and other:

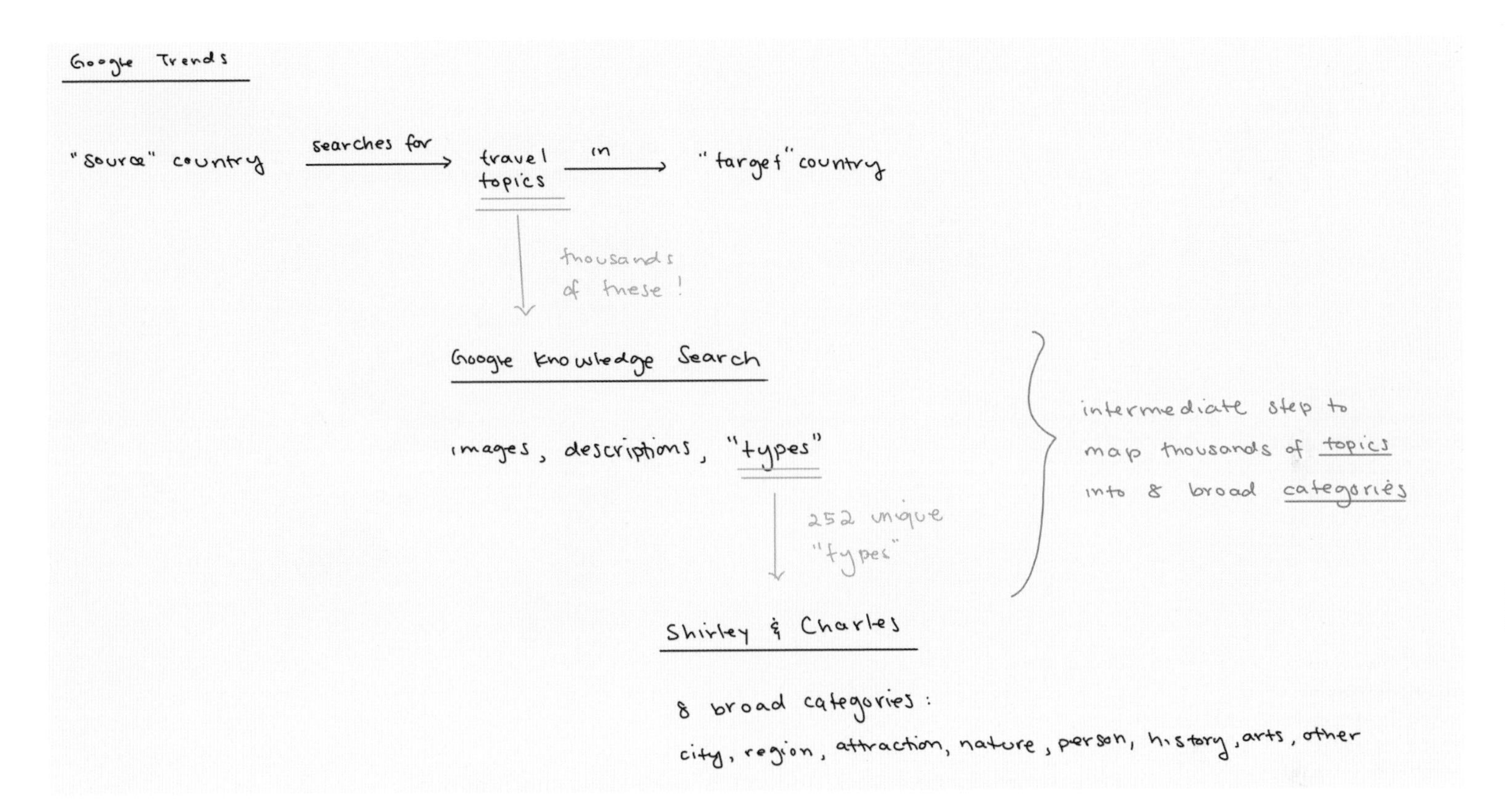

Fig.9.2

Relationship between topics, types, and categories.

From there, Charles produced a mapping of the Knowledge Graph "types" to our eight categories and was able to map most of the thousands of travel topics that way. For the 45 remaining topics that either fell into multiple categories or none at all, we manually assigned their category.

Sketch & Code

Because there was so much data, I switched between sketching my ideas and prototyping those ideas on the screen for this project. I started by looking at Brazil (the top searched country for travel) as well as the countries that searched for it in a given year, defaulting to 2016 (Figure 9.3). Immediately, I could tell by looking at the map that a lot of the countries that searched for Brazil were indeed geographically close to Brazil, proving my theory. But this map only showed 2016 search data, and I wanted a way to see a breakdown of all the countries searching for Brazil each year since 2004.

For the longest time I wondered why Brazil was the top searched country for travel, and I finally have a theory: Brazil was the host country for the Olympics in 2016!

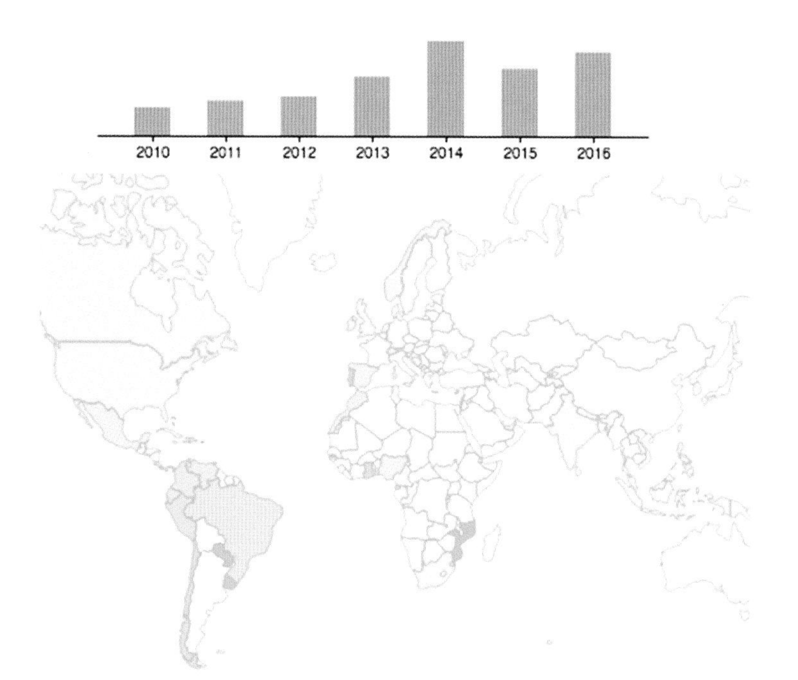

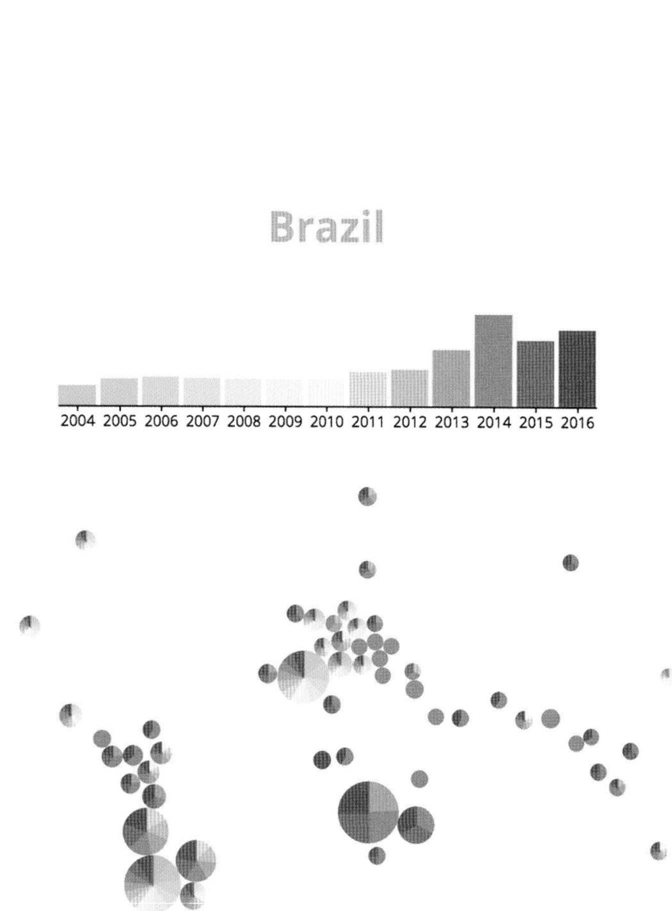

Fig.9.3

Bar chart showing search interest for Brazil across the years, as well as a map for the countries that searched for Brazil in 2016.

Fig.9.4

Pie charts above the centroid of each country showing search interest across the years.

To show a breakdown for each year, I thought of placing pie charts above the centroid (or geometric center) of each country, with the size representing the amount of search interest and the colors representing years (Figure 9.4). The pie charts showed me that some countries have been searching for destinations in Brazil since 2004, but others only started searching recently. This was interesting, but potentially misleading; we couldn't tell from our data if those countries only recently became interested in traveling to Brazil, or if they only started using Google recently. Alberto—who was responsible for our art direction—also advised me against putting pie charts on maps to avoid confusing readers.

I've since learned that I have to be careful with circles as it's hard to judge relative sizes with them, and that pie charts should really only be used for a few values that are part of a whole. Datawrapper, a charting tool for data journalists, has a great article about this on their website.[1]

So, I went back to the drawing board and began my first attempt at categorizing the topics (Charles did the more sophisticated version later on). I wondered if instead of focusing on just one country at a time, I could show all the countries and their topics from the get-go. I represented each topic as an arc and arranged it in a circle around the country it belonged to, colored it by its category, and positioned it clockwise by the year it was searched for (Figure 9.5). This version was certainly pretty, and it was interesting to see which countries were most popular for travel-related topics, but it was also hard to understand. It wasn't easy to compare years that weren't next to each other or see if certain categories went up or down over time. Alberto urged me to try a normal bar chart instead.

[1] Datawrapper, "What to Consider When Creating a Pie Chart": https://academy.datawrapper.de/article/127-what-to-consider-when-creating-a-pie-chart

Fig.9.5

Topics are arranged in a circle around the country they belong to, colored by their categories, and positioned clockwise by the year they were searched for.

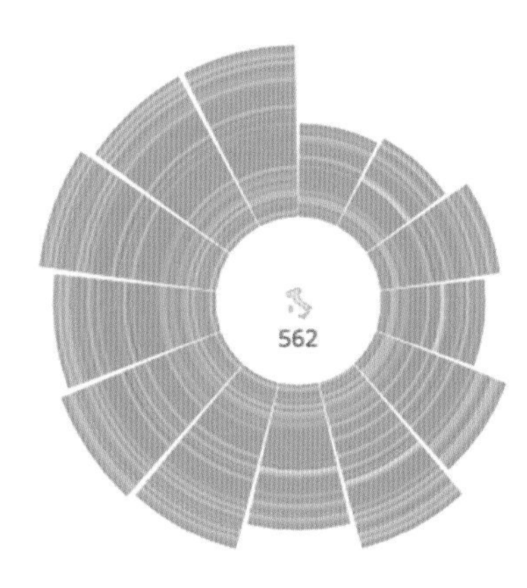

Brazil

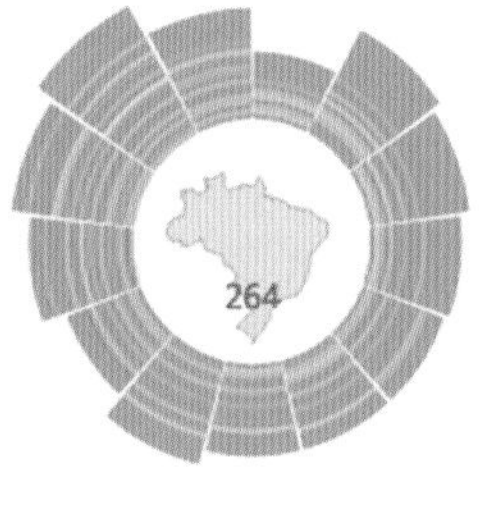

Germany

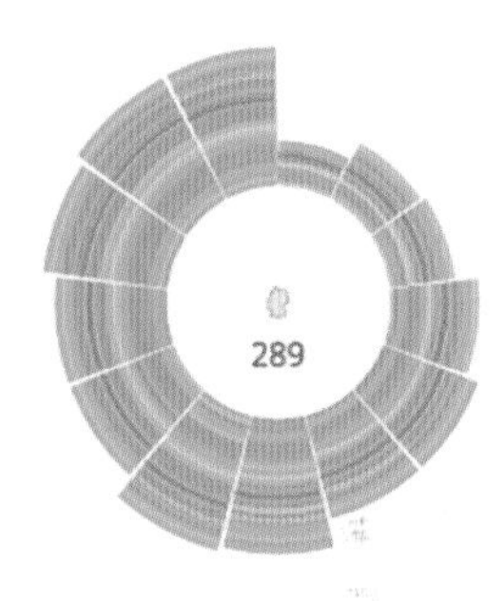

China

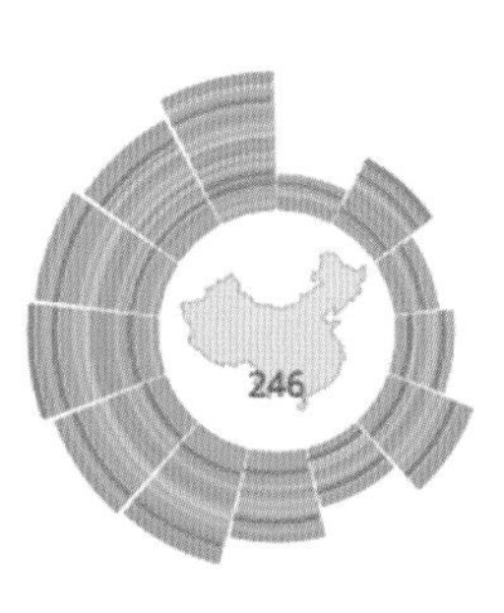

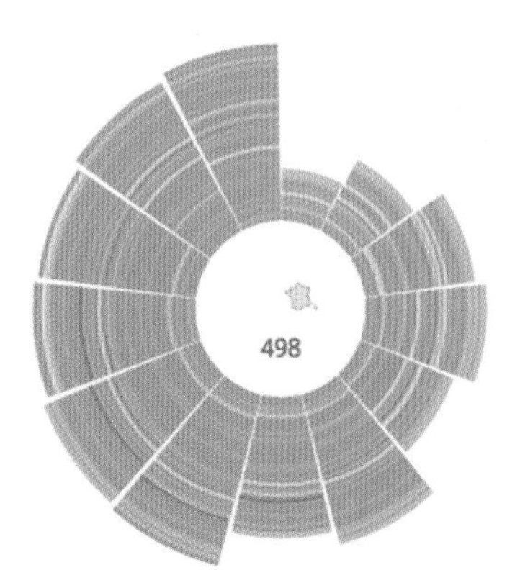

Greece

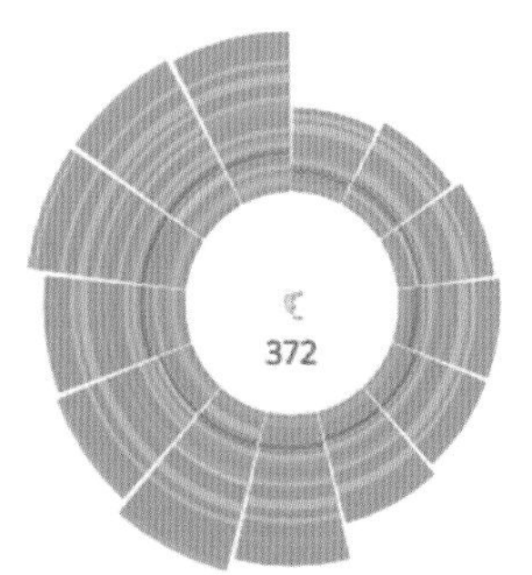

For the bar chart, I decided to represent the topics as blocks, mapped the year to the x-axis, and kept the category as color. I also decided to add another dimension and mapped the search interest (the popularity) of a topic to the width. This was an unfortunate mistake, I call this piece "The Plunger":

Fig.9.6

"The Plunger," 2017. Each block represents a topic, with the *x*-axis being the year, the color being the category, and the width being the search interest.

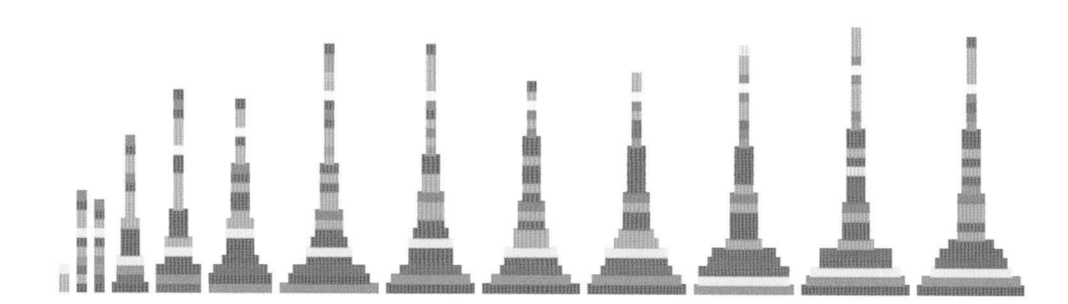

But *I did* like the idea of trying to show the popularity of a topic. So in my next attempt, I swapped out the blocks with circles and added another dimension to it: each circle represented one "source" country searching for that topic, and the radius that country's interest in that topic. This meant that the more overlapping circles there were, the more "source" countries searched for that topic in the "target" country, indicating more international interest (Figure 9.7).

I liked this version the most and decided to expand on it. As I now had a concept of "source" countries per topic, I wanted a way to see which "source" countries were searching for that topic and whether they were geographically close to the "target" country. I added an interaction where clicking on a particular topic would show the "source" countries' proximity to the "target" country along the *y*-axis, and the years searched along the *x*-axis. But I didn't like how some of the countries in Europe overlapped with each other (Figure 9.8, middle), so I tried a heatmap for my next attempt (Figure 9.8 right).

I liked the heatmap enough, so I went on to draft a story around the most searched topic that would also introduce the visualizations. But after digging through the two visualizations for interesting insights and banging my head on the desk for an entire afternoon, I had to face the hard truth: I needed to rethink my visualizations. Even though I found them visually interesting, they weren't actually easy to do any analysis with.

So I went back to brainstorming and asked myself what I wanted to learn about the data. I remembered, in my previous dig through the data, the seasonal nature of some of these topics' search interests, and wondered if there was something there—were certain countries and continents searched for more in summer as opposed to winter or vice versa?

Fig.9.7

Each circle is a "source" country that searched for a particular topic, and the circles overlap for the same topic.

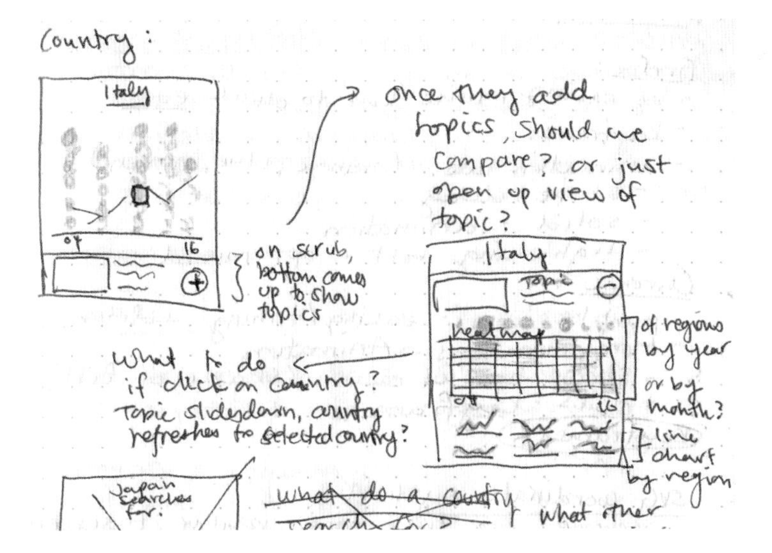

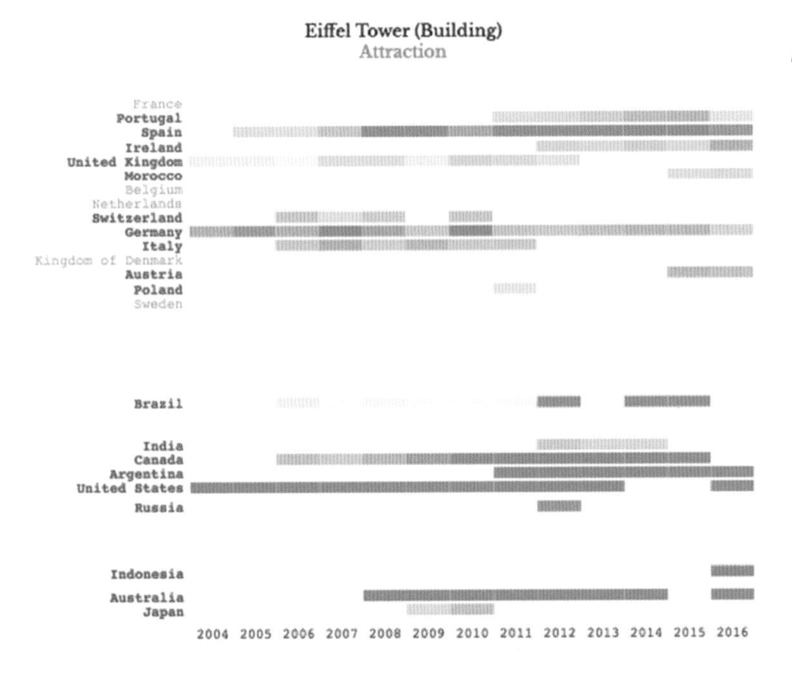

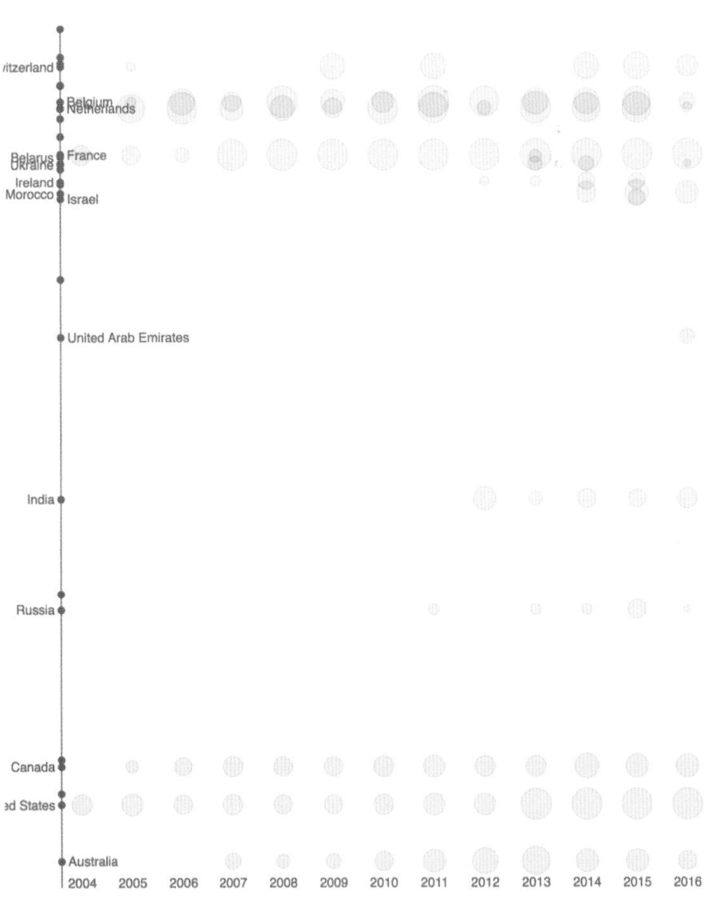

Fig.9.8

Notes on what should be shown when a topic is expanded (top right), first attempt at showing the "source" countries that searched for the selected topic arranged vertically by proximity to "target" country and horizontally by year searched with search interest mapped to radius of circle (bottom right), and second attempt with a heatmap with search interest mapped to color opacity instead (bottom left).

↳ Exploring Data: Ask Questions

I've found that when I mark certain data attributes as interesting, I naturally come up with corresponding questions to explore. I've also found that having a set of questions and hypotheses really helps me keep focused and prevents me from getting distracted by interesting tangents, especially in huge datasets. It's great to note those interesting tangents for later though, so that I can go back to them when all of my hypotheses are proven incorrect or I can't find anything interesting with my questions.

With my "Culture" project, I did all of my data exploration by coding my visualizations from scratch, but I've since learned to use charting libraries instead. (Efficiency!)

For more on starting data exploration, see the lesson "Exploring Data: List Attributes" on page 105 of my "Travel" chapter.

For the next step in data exploration, see the lesson "Explore Data: Use Charting Libraries" on page 336 of my "Community" chapter.

To visualize topics by continent and season, I decided to flip things around. Instead of focusing on travel topics searched for in a "target" country, I visualized search topics by season. Here, the top row represents topics searched for in the spring, and the bottom is summer (Figure 9.9). Each block represents a travel topic people in the United States have searched for, and each topic is grouped by the continent they belong to. The continents are ordered by their geographic proximity to the United States.

And yet again, the visual didn't go the way I was hoping for; it turned out that people search for the exact same topics across all the seasons, so the visualizations for both spring and summer looked almost exactly the same. I was so bummed that I left the cafe I was working from, but on my drive home I realized that even if the topics are exactly the same, their search interests may not be. As soon as I got home, I (very excitedly) set the height of each block by their search interest and finally, *finally*, I had the results that I was searching for: I could see that people in the United States searched for travel the most in the spring (presumably planning their summer getaways), and the least in the fall (Figure 9.10). I rearranged the blocks such that they were grouped first by season, and then by their continent. This made the visualization much more compact, and the seasonal trends stood out even more.

Now that I had a good summary of the topics, I wanted to create a detailed view for each of them. In particular, I wanted to know about the rise and fall of each topic's search interest from 2004 until 2016. My idea was to have the *x*-axis represent the search interest with values out of 100 and each circle represent a given year. An arc above the x-axis meant that searches increased across a year, whereas an arc below indicated a decrease across a year (Figure 9.11). I learned quite a bit from this visual; for example, a lot of topics actually peaked in 2004 and have declined ever since, with a lot of them dipping the most between 2008 and 2011—the years of the global recession. But as much as I liked the visualization and as interesting as those insights were, I had to admit that it took a lot of effort to get those insights from the visualization.

Fig.9.9

Travel topics grouped by continent, colored by category of topic, and organized by season.

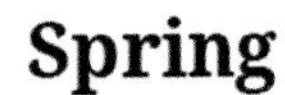

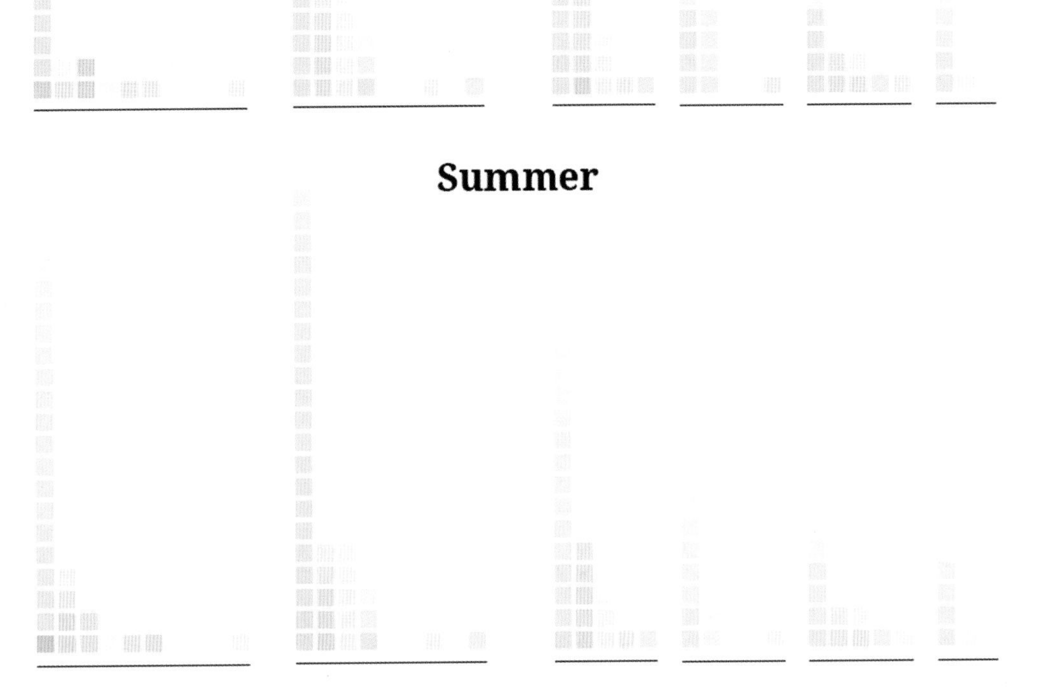

Fig.9.10

Travel topics with their heights set to search interest and grouped by season.

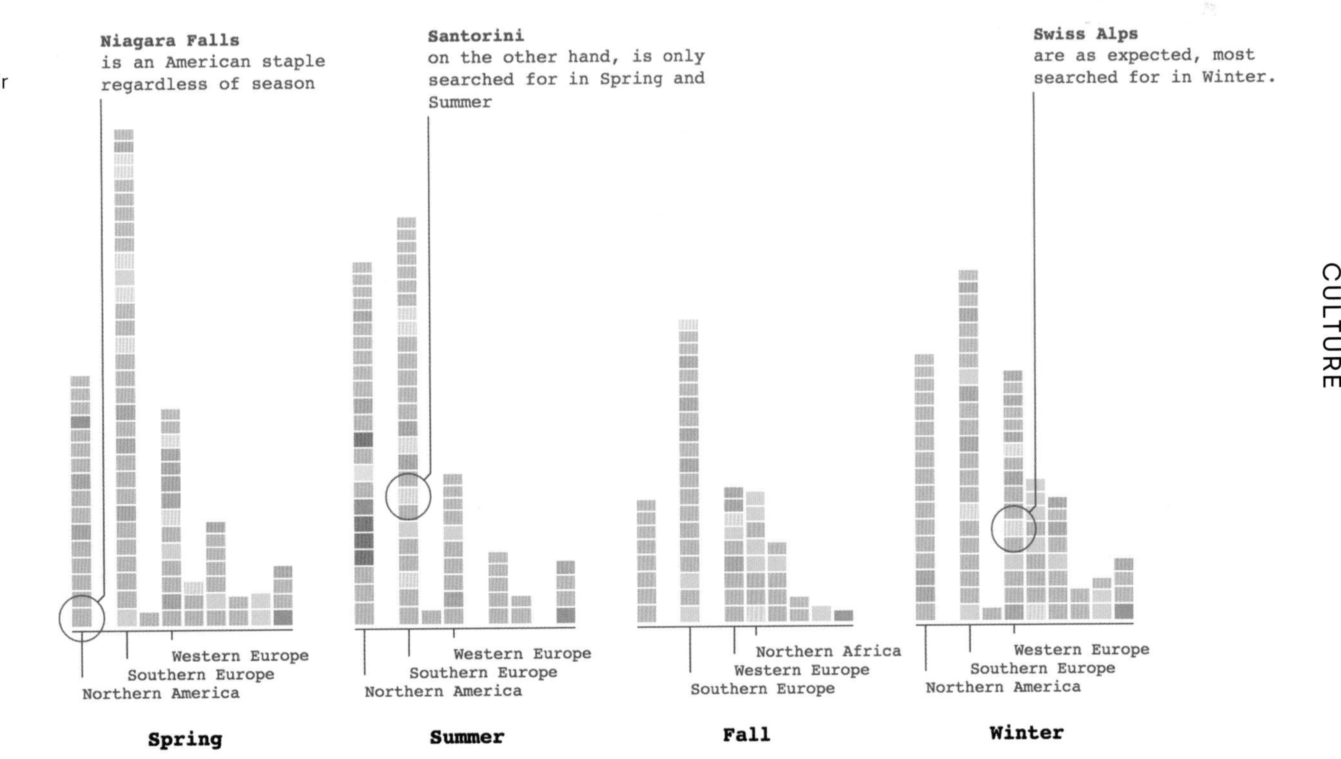

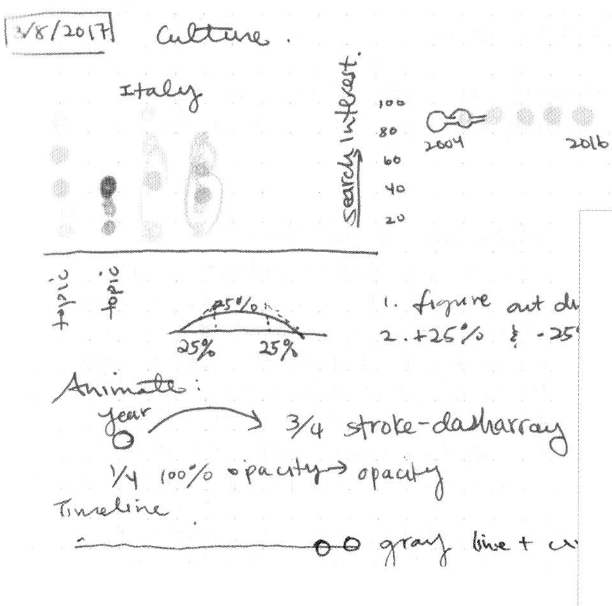

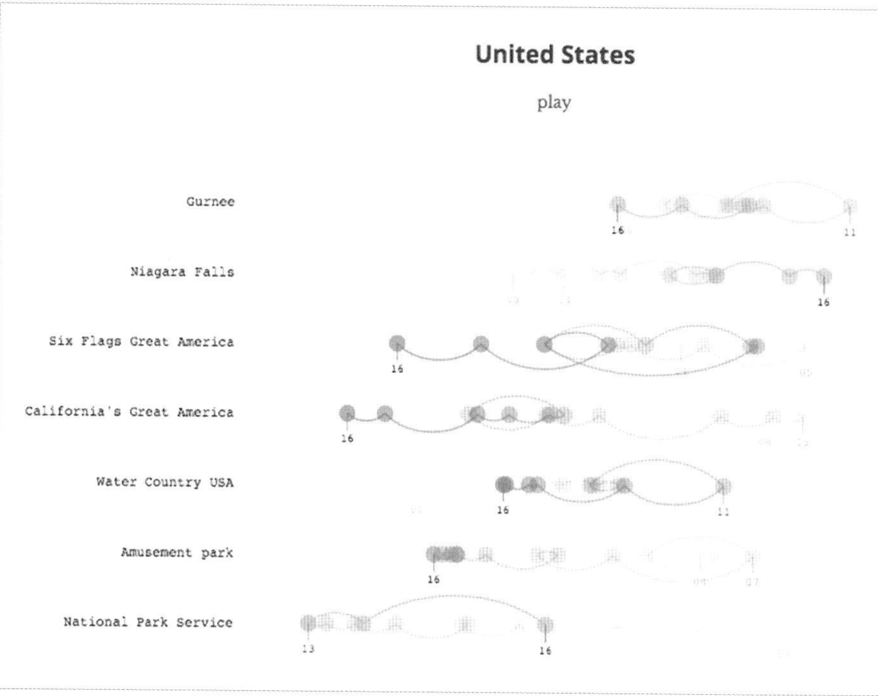

Fig.9.11

Sketches of the detailed topic view, and an image of it implemented. The *x*-axis represents search interest out of 100, and each circle represents a given year. An arc above the axis means the search interest increased across the year, and arc below means that it decreased.

↳ Design With Code

When I first started *Data Sketches*, I tried to sketch out two or three different ideas per project before getting to the code. But I soon found that that was only really viable for smaller, more straightforward datasets (like the dataset for my *film flowers* project). For most of my other projects—especially the ones with larger datasets—I had to get my data onto the screen first, explore it, and use what I learned to inform my design. I've found this to be particularly important when doing a multi-part narrative, where I code section by section. As what I write and visualize in one section can influence the following section, I have to finish the first section before sketching the next.

Nowadays, I mostly sketch to help me work out kinks in the design or remember particular details I want to include in the layout or user interactions (often informed by the code that came before it).

Because I was feeling quite stuck on the detail view, I decided to put it aside and switched gears to work on the story. Around that same time, I took a Web Animation Workshop with Sarah Drasner and Val Head, where I learned the basics of how to animate with the `gsap` (`Greensock Animation Platform`). With that knowledge, I wanted to create "scenes" that explained each visualization in detail, and my first pass used scrollytelling (Figure 9.12). But once I finished implementing the first iteration, I felt dissatisfied with how much vertical space it took to show simple concepts like topics and categories (Figure 9.13).

I've found that if I've been banging my head for a while, it's often more helpful to give it some space and work on something else. When I get back to it a few days or weeks later, I'm always full of fresh inspiration.

Fig.9.12

Plans for the story that would introduce the dataset and explain how to read the visualizations.

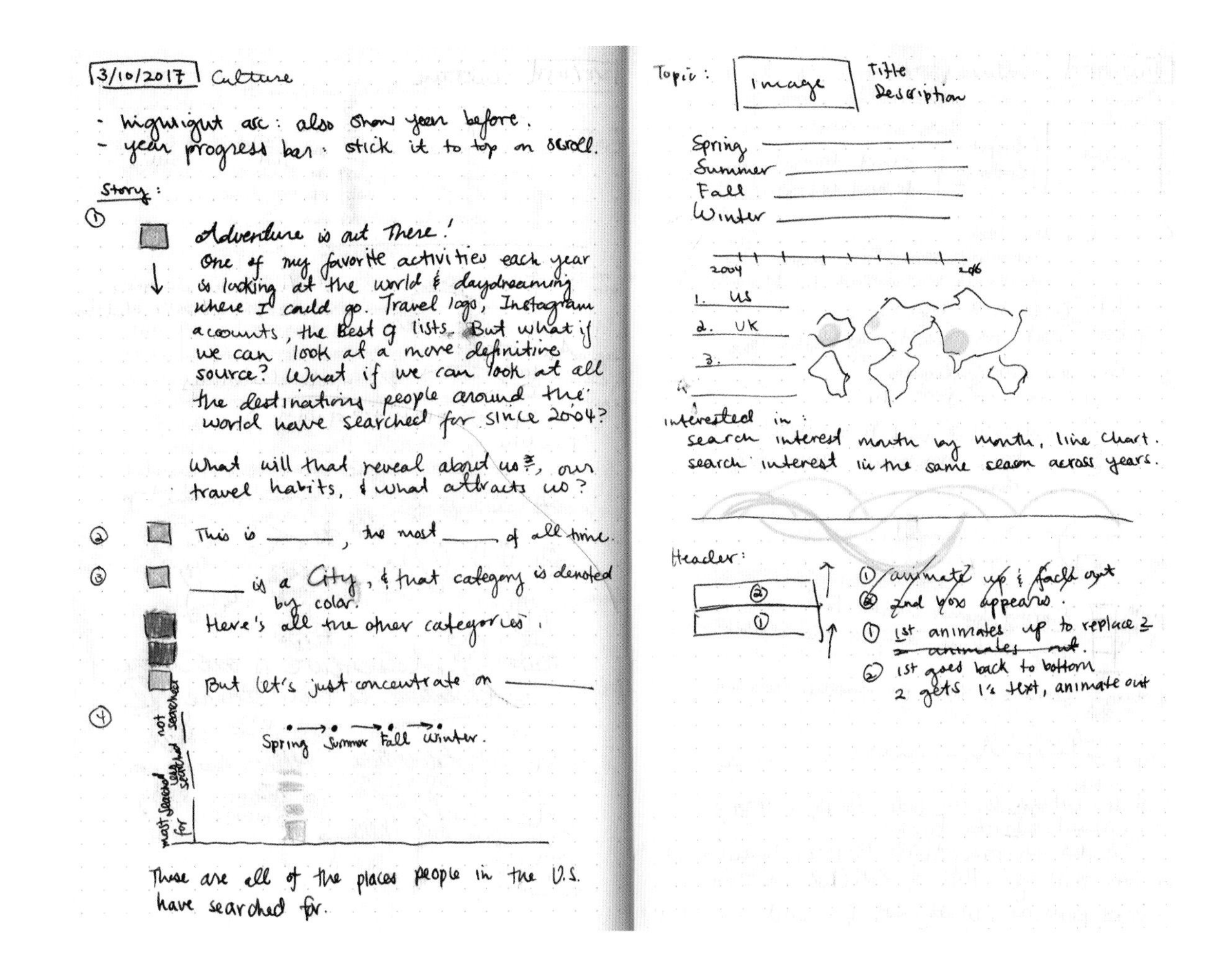

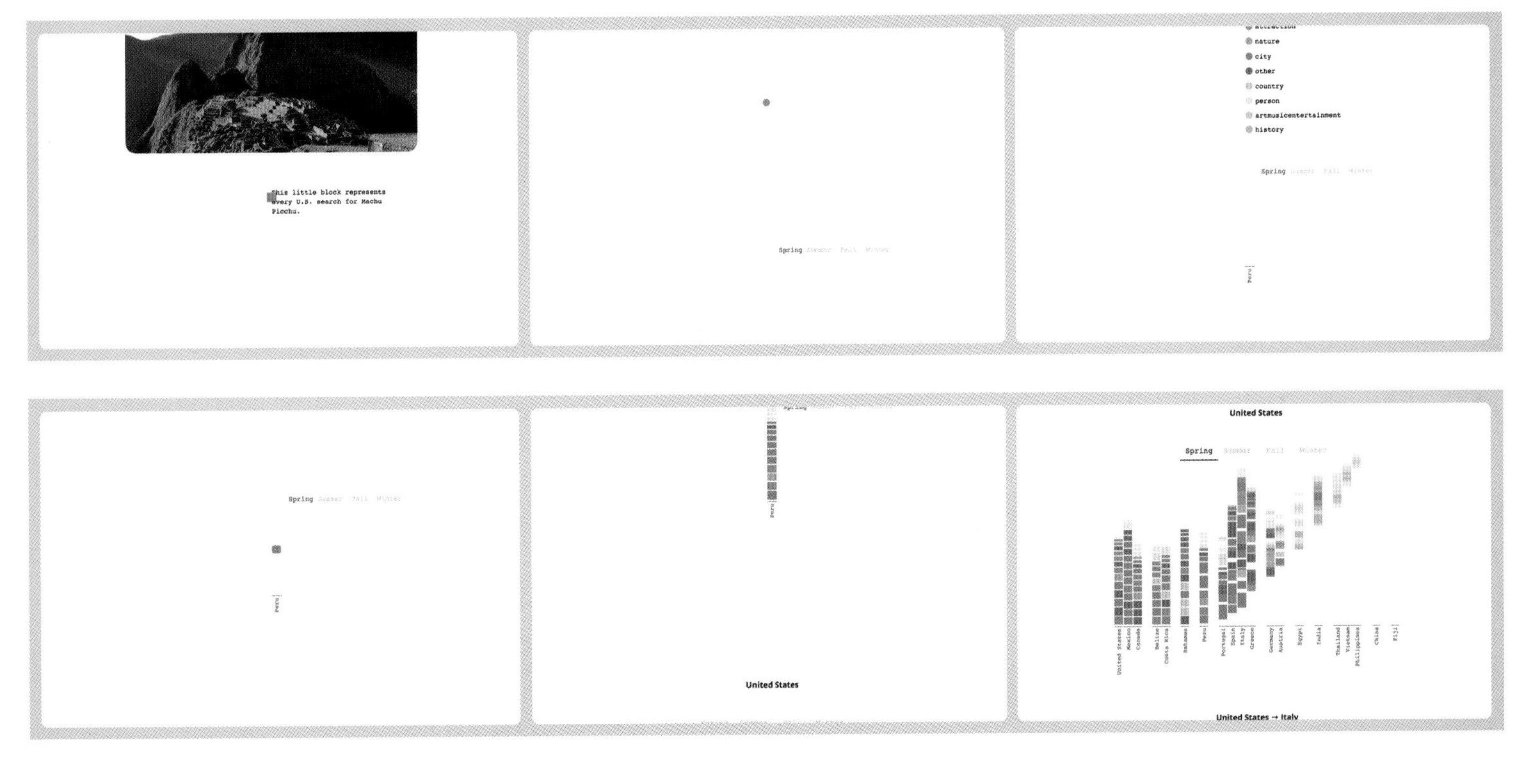

Fig.9.13

First iteration of the scroll-based introduction. Unfortunately it took too much vertical space, and I decided to scrap it.

Despite spending quite a bit of time on the story, I still felt unhappy with it. I switched back to the detail view and asked myself what I wanted to explore and learn from a certain topic. Ultimately, I decided I wanted to see:

1. Search interest across the years
2. Where those searches were originating from

With those goals in mind, I went with Alberto's suggestion of a line chart to show search interest across time and a world map to show the "source" countries (Figure 9.14). I drew circles on top of each "source" country, sized the circles based on search interest, and animated both visualizations across time.

And because line charts and maps are straightforward, familiar charts, they were easy to analyze; I was able to explore and find an interesting story about the similarity in searches for Qin Shi Huang (the first emperor of China) and his Terracotta Army with them (Figure 9.15).

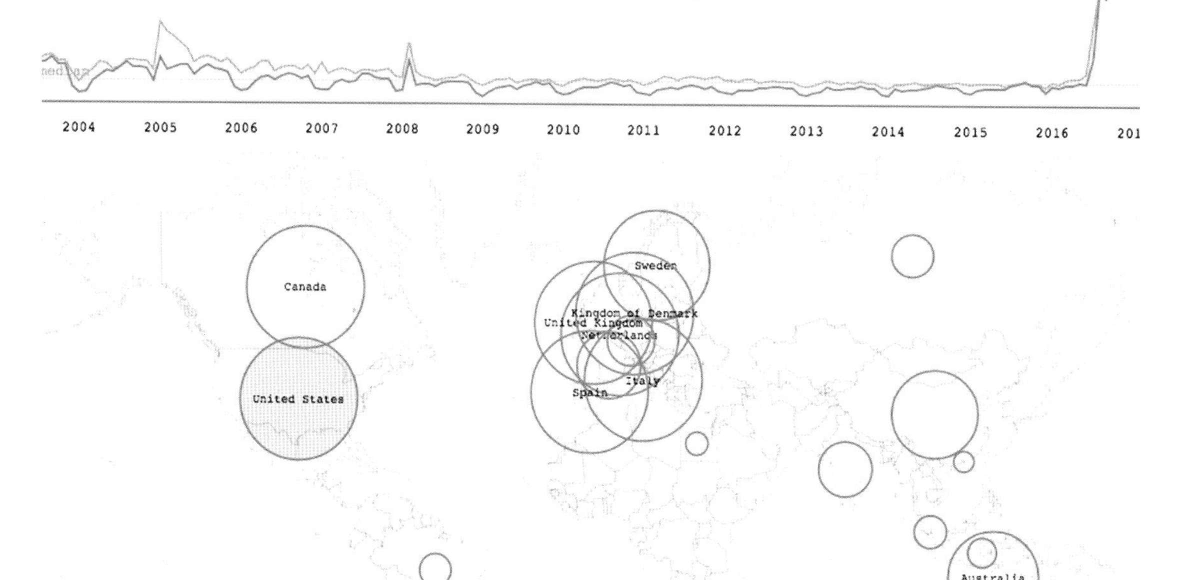

Fig.9.14

Search interest across time for a selected topic and where those searches are coming from. In the final iteration, I animate both visualizations across time.

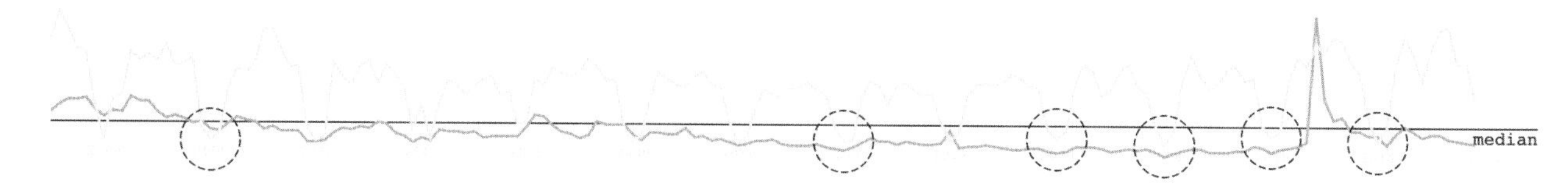

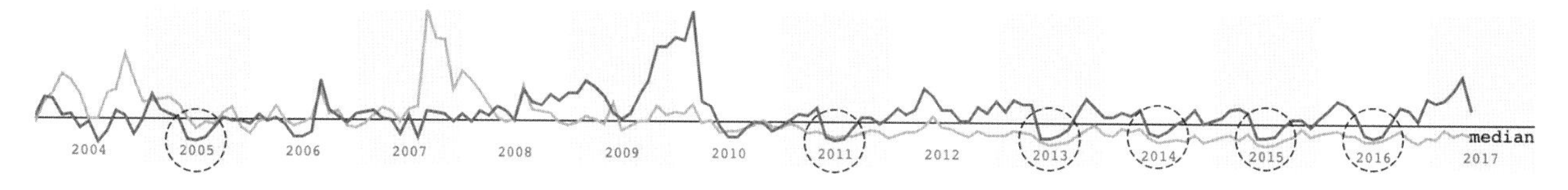

Fig.9.15

Similar seasonal dips in searches for Qin Shi Huang and his Terracotta Army.

With my stories figured out, I was able to concentrate on introducing each of the visualizations and stories in a space-efficient way. I remembered the discussion between scrollytelling and steppers—a technique where instead of scrolling, the user clicks to step through sections of a story—and decided to give it a try. The steppers were great for saving space, as all the visualizations were contained in one place without needing to scroll, but making the user click through each step seemed too much to ask. So I decided to animate each step with `Greensock` to auto-play through all the steps, and included an interaction where clicking a step would trigger the visualization to replay from there (Figure 9.17). That way, the reader would still be in control of the animation pacing between each step.[2]

The scrollytelling versus steppers debate exploded within the data visualization community in the summer of 2016, and our friend Zan Armstrong wrote a great article summarizing the points called "Why choose? Scrollytelling & Steppers."[2]

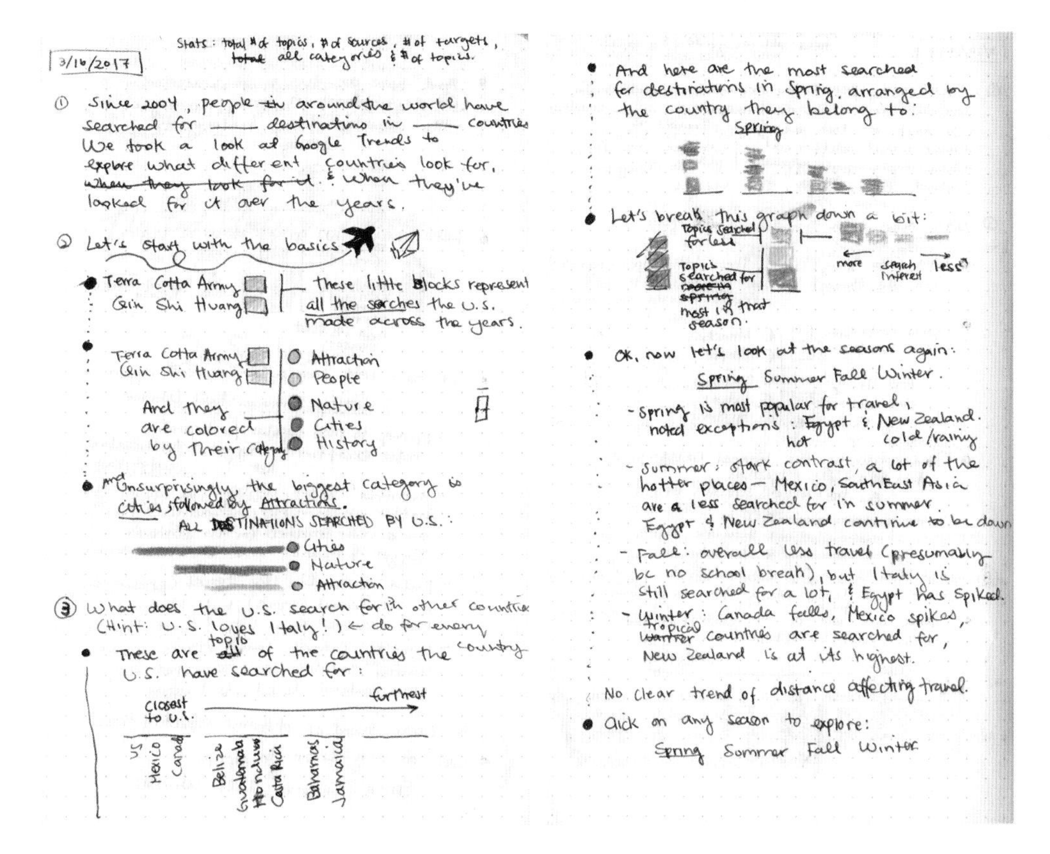

Fig.9.16

Final outline to introduce the concept of search topics and their categories, where those searches came from, and the seasonality of those searches.

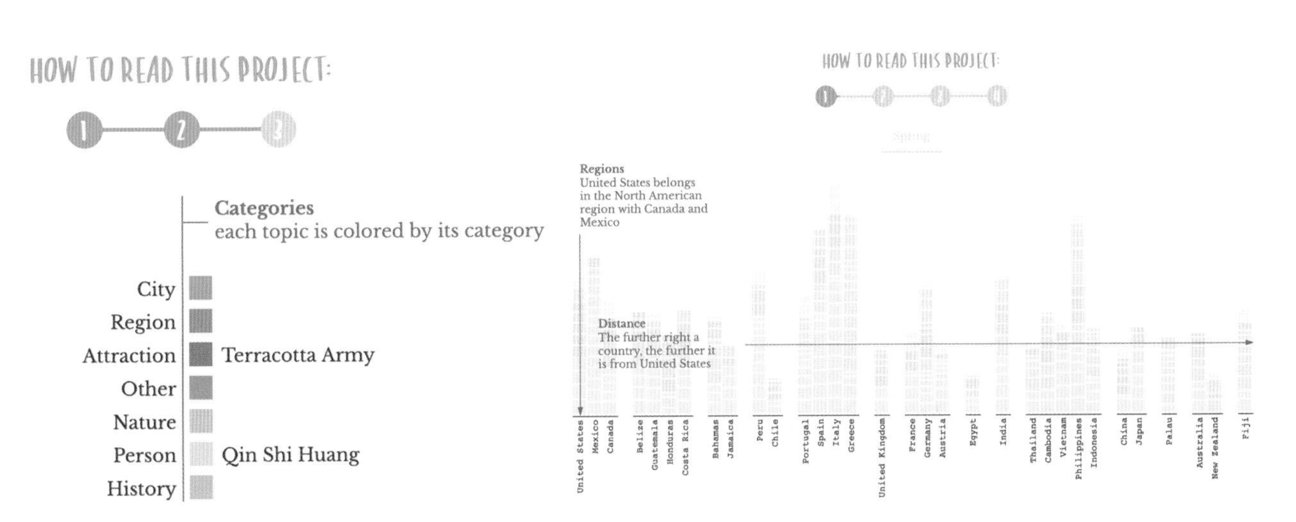

Fig.9.17

Using steppers to introduce search topics and their categories (left) and where those searches came from (right).

[2] Zan Armstrong, "Why choose? Scrollytelling & Steppers": https://medium.com/@zanarmstrong/why-choose-scrollytelling-steppers-155a59dd97fe

As always, the little details (not to mention all the writing) took much more time than I expected, but I'm really proud of two small touches in particular: the animations that automatically start only when the visualization comes into view, and the little paper airplanes that my friend illustrated for me. I like to think that it's all these small, subtle details that show readers how much we care about their experience.

Reflections

This project is, to this day, one of the hardest I've ever completed. The amount of back-and-forth I had with my designs led to a lot of self-doubt. But I also learned a lot from this process, especially about the areas I needed to improve:

1. Streamlining the data analysis process: I lose a lot of time forming hypotheses and coding custom, often time-intensive visualizations from scratch to test them. I've been iterating on my process since.
2. Developing my information design knowledge: I have a lot of ad-hoc knowledge I've collected through the years, and I've since self-studied so I could have a more formalized and systematic way of approaching design problems.

> This was also the first time I began to realize the importance of prioritizing the reader and their understanding of my visualizations, instead of just doing whatever was visually flashy and technically interesting.

I really took this to heart every time Alberto suggested a more straightforward alternative to my flashier and harder-to-read designs (though I do still believe that creativity and freedom of expression is important).

From a technical perspective, I'm really happy that I got to experiment and work with `Greensock` for the first time. Its concept of adding animations to a timeline really makes managing more complex, scene-based animations so much easier and I've used it in almost every project since. I'm also proud of the experience I was able to build, where visitors can interact with it at the level they're most comfortable with, whether it is skimming through the story or deep diving into the exploratory tool to find their own stories.

And finally, I'm really grateful for this project which, along with my *Hamilton* project and *Data Sketches* as a whole, cemented my freelance career. Before this point, most prospective clients weren't willing to pay the rate I asked for. But after I put Google on my resume, I was never questioned again—and I even increased my rate right after!

Explore Adventure

↳ explore-adventure.com

The United States loves traveling in the
SPRING

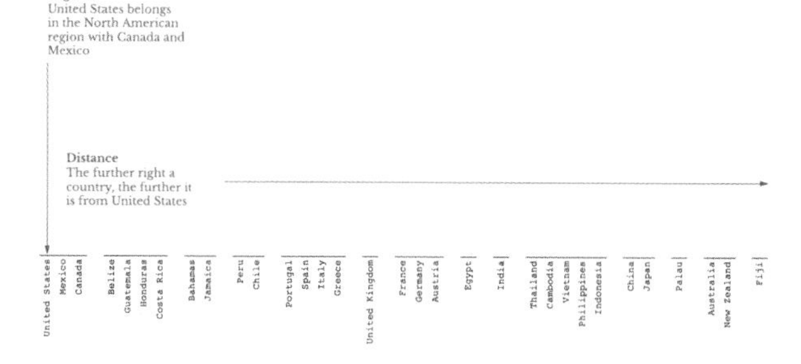

When we first started looking at the countries, we were curious about whether distance affected people's searches; do people in the U.S. search more for places in Canada or Mexico for vacation, or more in faraway "exotic" places?

Turns out, distance doesn't make much of a difference. There doesn't seem to be much of a trend in terms of U.S. searches for Central America versus Europe or Asia (in fact, it seems Europe and Asia have higher search interest overall).

There is, however, a very clear seasonal trend. The U.S. searches for almost every place around the world in the Spring (except for Egypt), presumably for spring break and summer vacation planning. The opposite is true in the Fall, where inbound search interest for Egypt peaks as the weather cools, but drops for everywhere else in the world (most likely because of back-to-school and lack of long breaks in the fall).

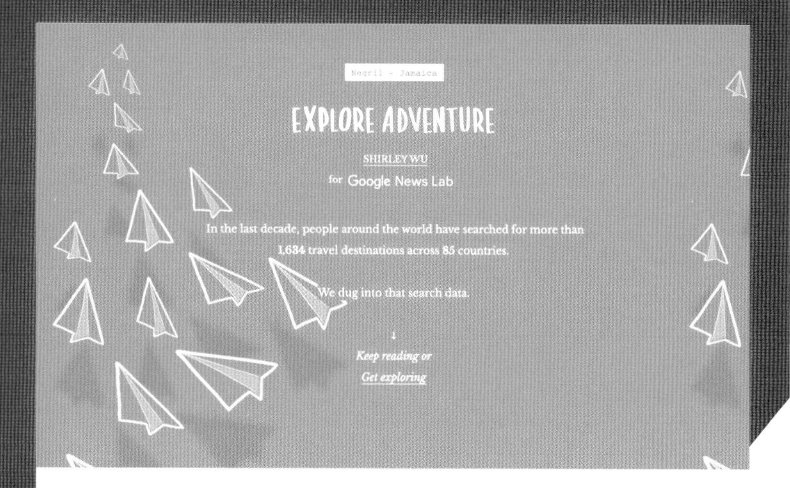

Negril - Jamaica
EXPLORE ADVENTURE
SHIRLEY WU
for Google News Lab
In the last decade, people around the world have searched for more than 1,634 travel destinations across 85 countries.
We dug into that search data.
Keep reading or
Get exploring

Since Google started keeping track in 2004, people have searched for places in Italy, Greece, Austria, Brazil, Thailand...they've searched for cities and islands, for ski resorts and national parks, for Disneylands and modern art museums all over the world.

We looked into travel searches from nearly **40** countries to explore what one country searched for in another; did they search for places closer or farther from them? Did they search more for resorts than national parks, more for Disneyland than for museums?

We wanted to know: what can we learn about these countries from their travels?

HOW TO READ THIS PROJECT:

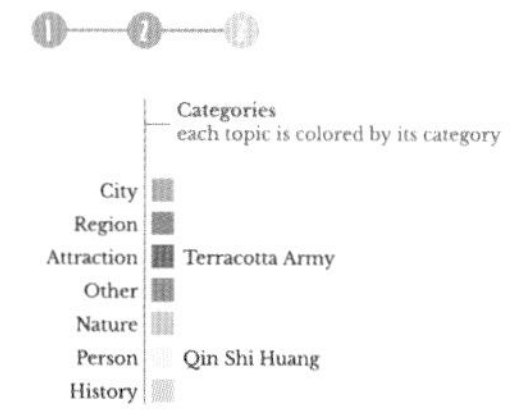

Let's take a look at what the **United States** has searched for.

It turns out, the U.S. searches the most for different cities around the world: Naples (Italy) and Barcelona (Spain) in the Spring, Cebu (Philippines) and Chiang Mai (Thailand) in the Winter, and **206** others over the years.

After cities, the U.S. searches most for attractions like Machu Picchu in Peru, the Colosseum in Italy, castles in Germany, and amusement parks in the U.S. History and the arts were searched for the least.

TOPICS U.S. SEARCHED FOR IN SPRING AND FALL

City Region Attraction Other Nature Person History Arts

Spring

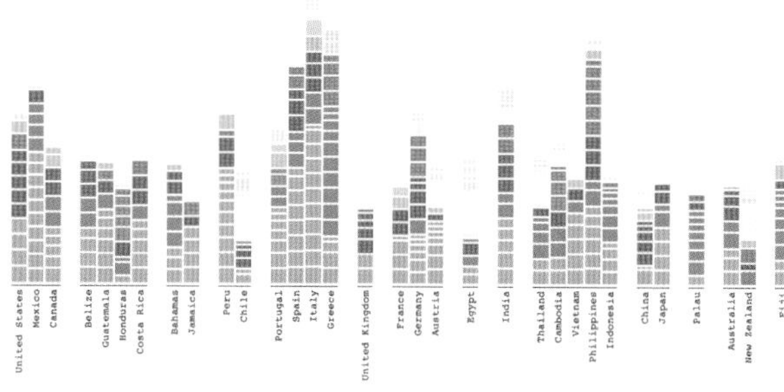

(Try toggling between Spring and Fall !)

Another interesting pattern: the U.S. searches for Canada to the north in the Summer, and Mexico to the south in the Winter. Similarly, hotter regions around the world - Central America, Southeast Asia - are searched much less in the Summer than in the Winter.

TOPICS U.S. SEARCHED FOR IN SUMMER AND WINTER

City Region Attraction Other Nature Person History Arts

Summer

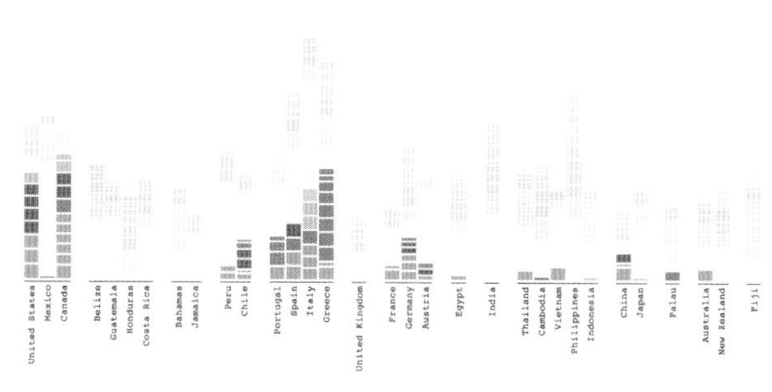

(Try toggling between Summer and Winter !)

And though the U.S. doesn't seem to have a clear trend in terms of distance, there are other countries like Japan, Singapore, and Saudi Arabia that do search primarily for countries closer to them.

On top of the seasonality, we were also interested in seeing if there were any patterns year over year - especially if there were any extreme peaks or dips. We found that most of the trends were economical or political in nature; U.S. searches for more expensive destinations (mostly in Europe) were highest in the early 2000s, and have been declining since 2008, the year of the financial crisis. Searches for Cairo peaked in January 2011 not because of travel interests but because of Arab Spring, and Athens and Beijing peaked in 2004 and 2008 respectively for the Summer Olympics.

But there are also fascinating patterns that have nothing to do with politics. Take for example, the search interest for China's Terra Cotta soldiers:

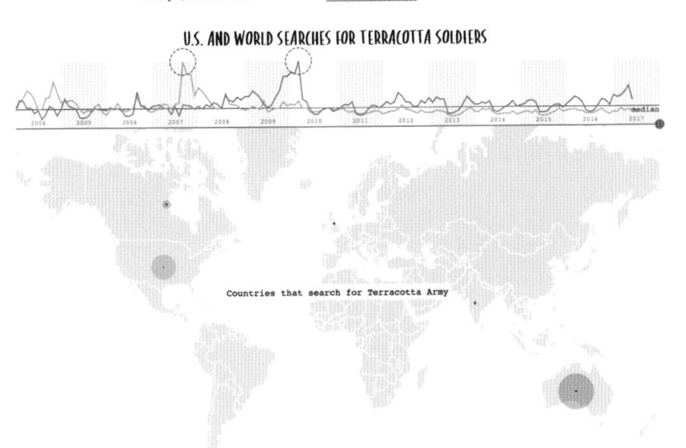

Notice the two spikes in searches, one along the gray World line in late 2007 and another along the red U.S. line in early 2010. The first spike is the United Kingdom searching for the British Museum's special exhibit, "The First Emperor: China's Terracotta Army", that ran from September 2007 to April 2008 (and the U.S. had no interest). The second is National Geographic Museum's exhibit in Washington D.C., "Terra Cotta Warriors: Guardians of China's First Emperor", that ran November 2009 to March 2010.

(Scrub to both 2007 and 2010 with the progress bar and watch the U.K. and U.S. bubbles burst ↑)

Now take a look at the searches for China's first emperor, Qin Shi Huang:

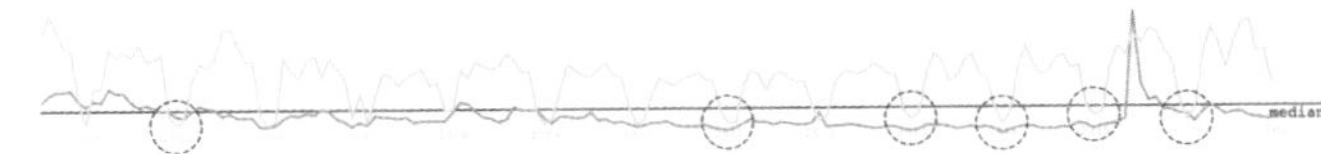

There's a clear seasonality to U.S.'s searches for Qin Shi Huang, with steady interest from Fall to Spring and a dip in the summer, but there isn't a similar trend for the rest of the world. And it's curious: why would there even *be* seasonality searching for a person?

The answer is in the Terra Cotta soldiers' searches:

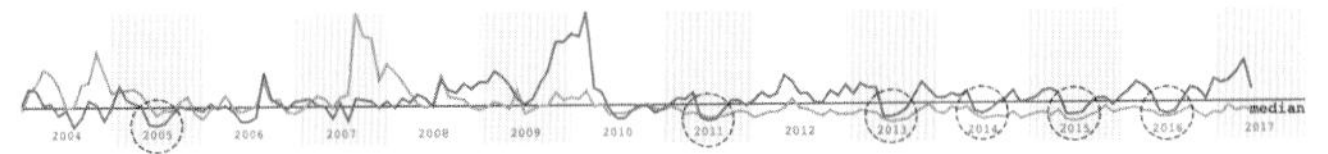

If we look closer and ignore the two spikes, it turns out the U.S.'s searches for Terra Cotta soldiers have a similar seasonality to that of Qin Shi Huang; starting in 2011, there's a dip in search interest every summer. So the United States searches for Qin Shi Huang because they are interested in his Terra Cotta soldiers - which is fascinating, not only because the rest of the world doesn't exhibit this behavior, but also because "Qin Shi Huang" just seems like the harder of the two search terms to remember.

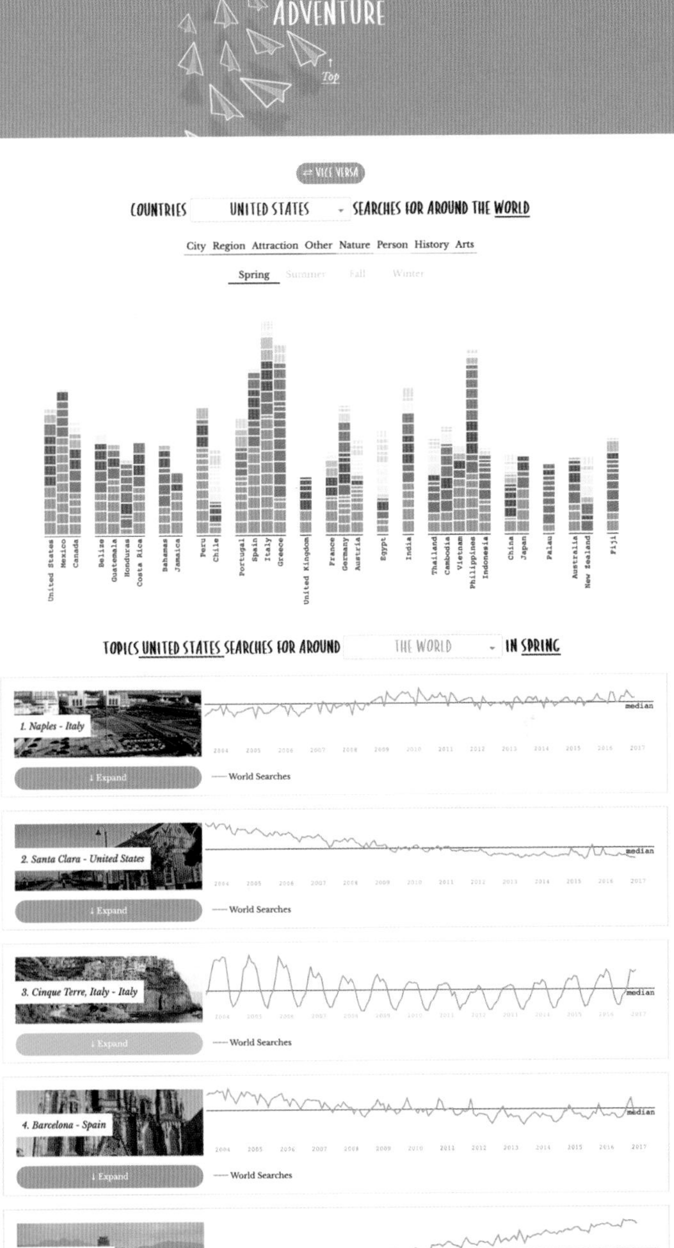

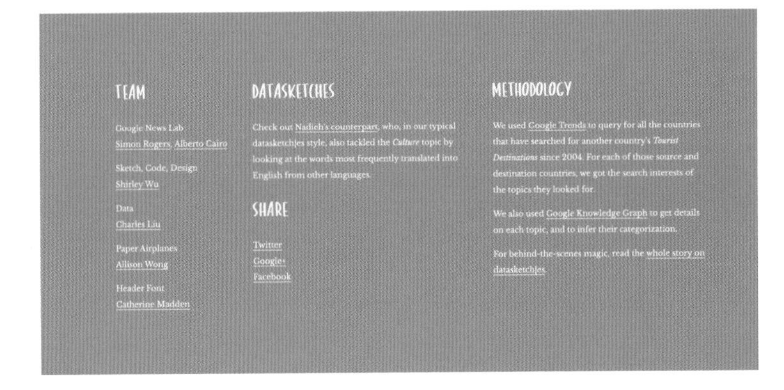

Fig.9.18

The final visual story.

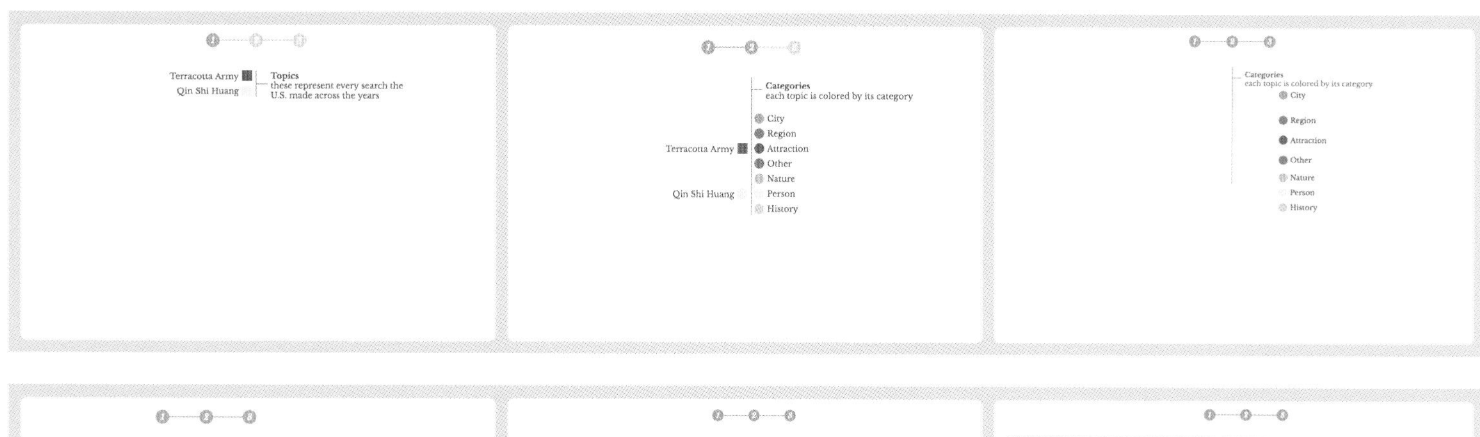

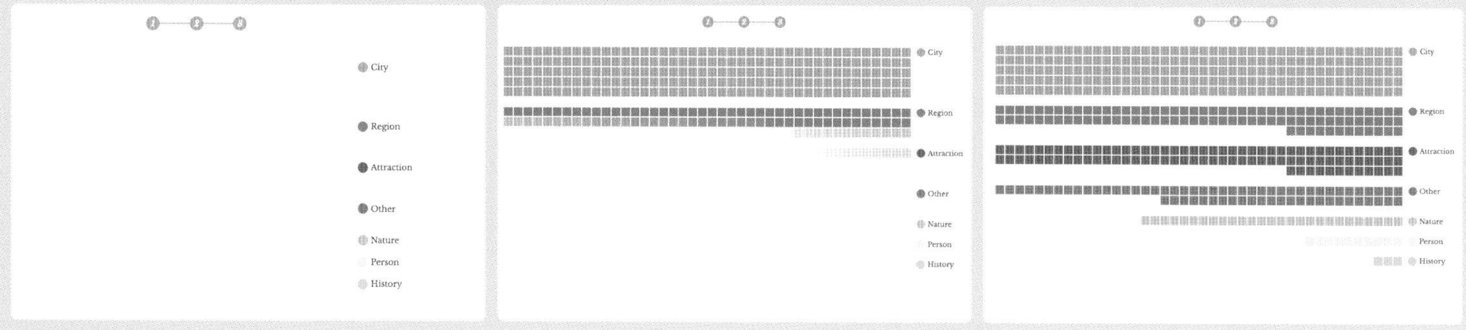

Fig.9.19

Animation explaining search topics and their categories.

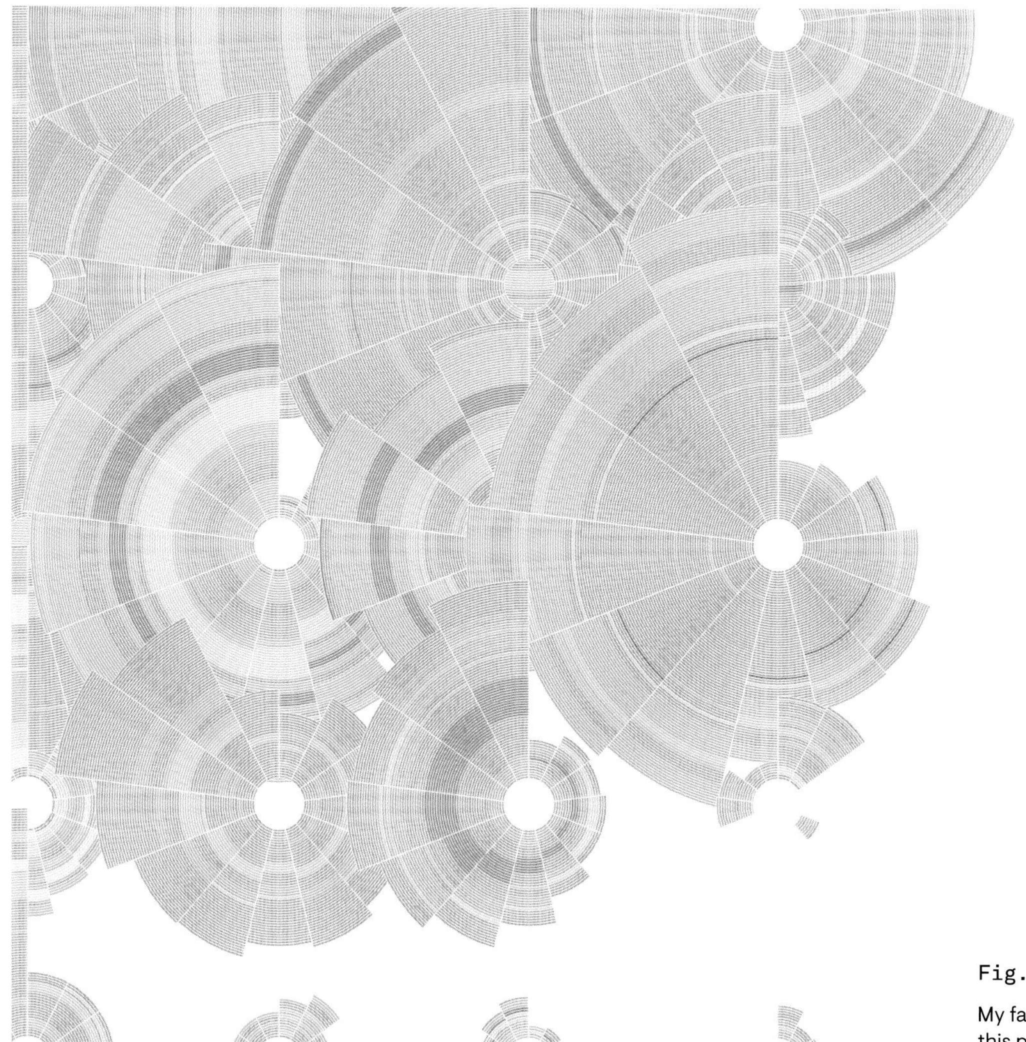

Fig.9.20

My favorite visual bug from this project.

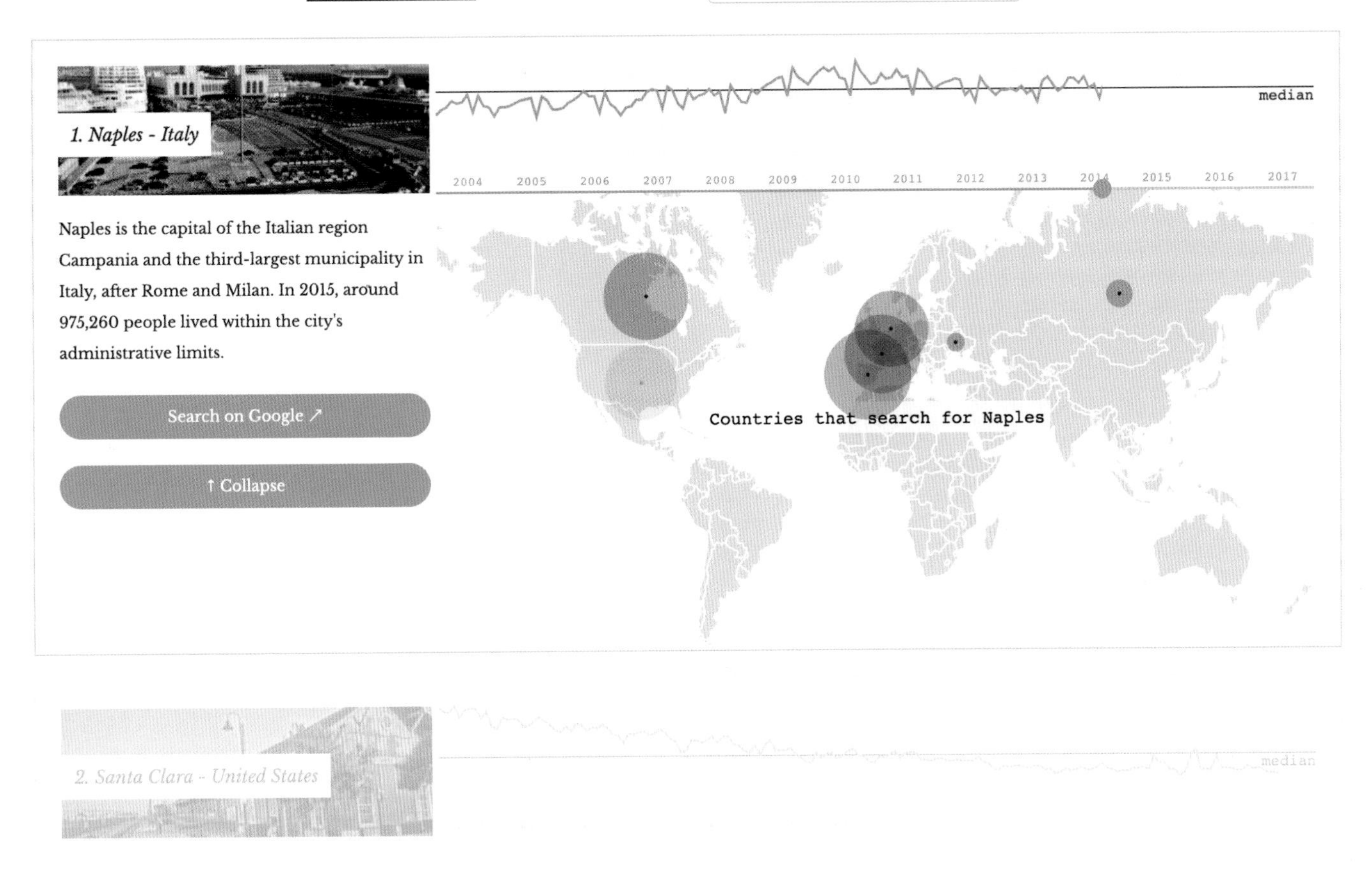

Fig.9.21

An expanded topic.

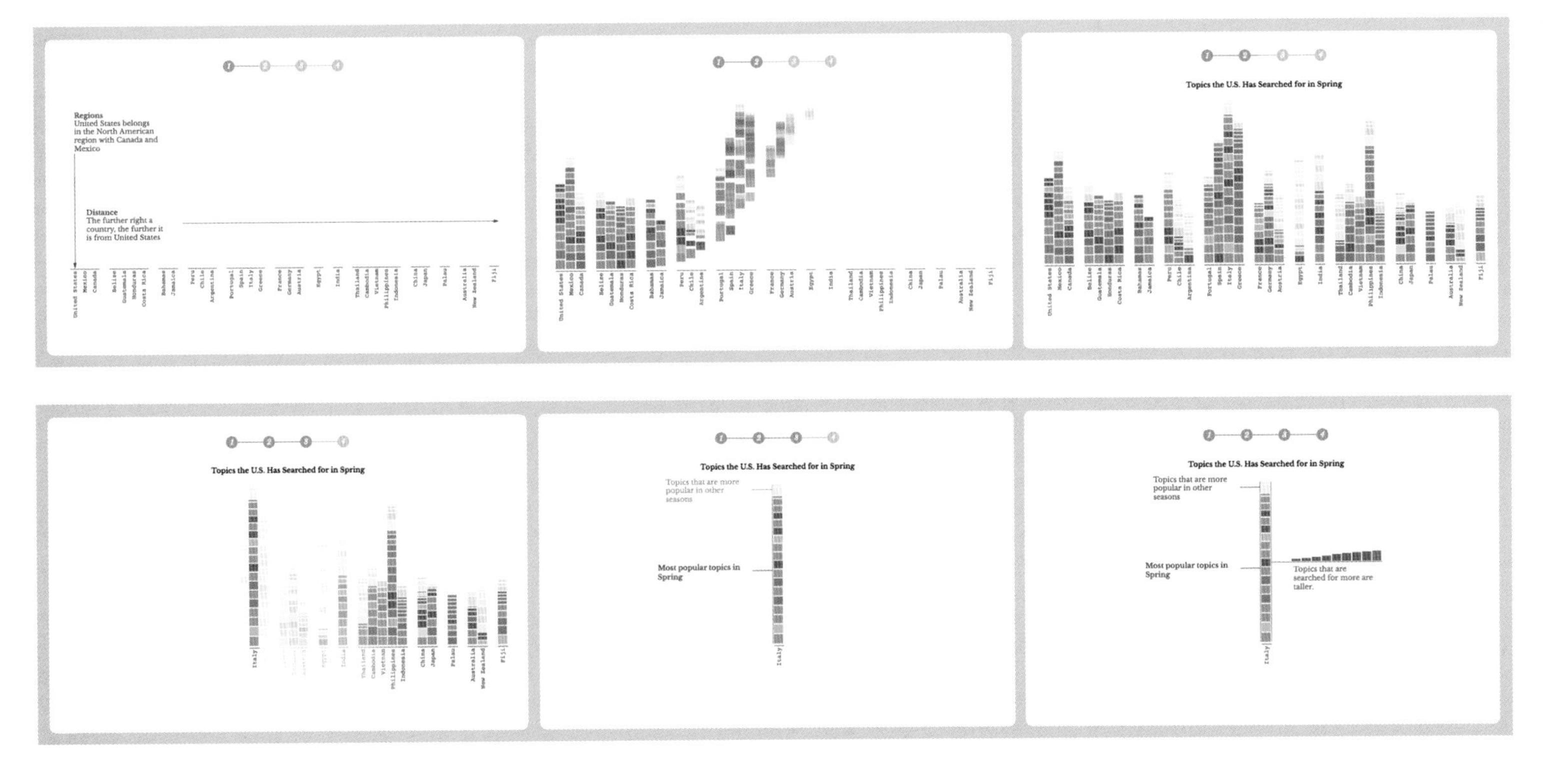

Fig.9.22

Animation explaining how the search topics are arranged geographically, and that each topic's height is mapped to its search interest.

COMM

UNITY

Breathing Earth

NADIEH

One random morning I was reading the World Wildlife Fund's (WWF) magazine, which I receive for being a donor. Suddenly I realized that I would really want to make a visualization that related to something the WWF might do. I pitched the idea to Shirley and we went back and forth a bit on what general topic would work for both of us. And when Shirley found her angle, the data visualization survey that had just come out, we had our topic: "Community."

Data

In early April, I asked Twitter for help with finding datasets that one might associate with the WWF and, thankfully, received a bunch of links and advice. However, due to being in the United States almost the entire month, I didn't get to do anything with the links until I was about to take a flight home to Amsterdam on April 26th.

I received a lot of tracking data links related to either animals or water buoys. But I noticed that the search functionalities of these data repositories were aimed at researchers. I could search datasets based on the ID of a paper or scientist's name. But I couldn't request all tracking data of, say, whales. Another type of dataset that was very prevalent were choropleths, which are filled regions on a map, often representing things such as protected areas or animal habitats.

I started to meander through the links, and I don't know how I got there, but at some point I found myself on the website of NOAA STAR,[1] the Center for Satellite Applications and Research. I was randomly clicking around on their website when I came across an image of the Earth, colored by vegetation health.

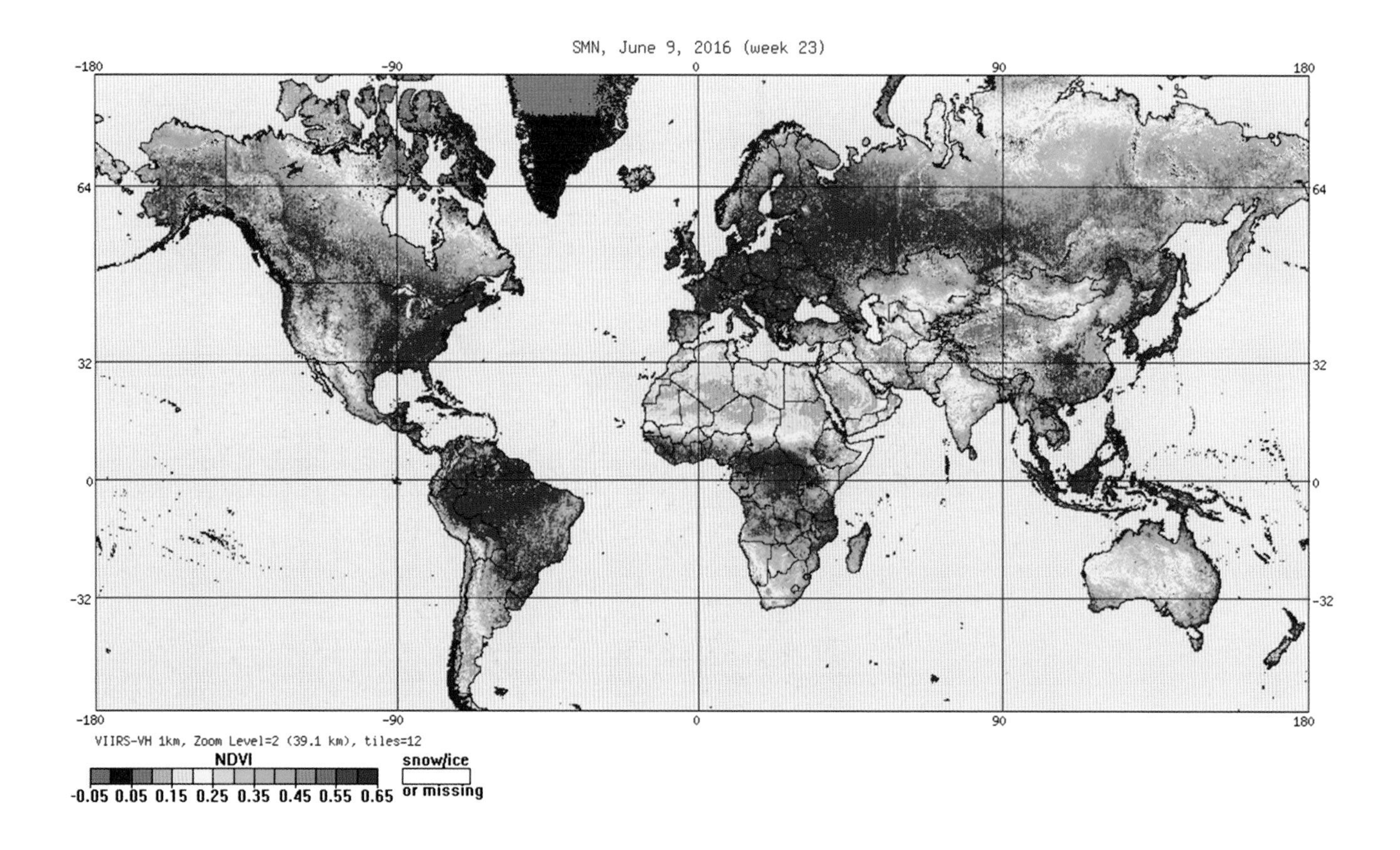

Fig.10.1

NOAA STAR map showing vegetation health for week 23 of 2016. Credit by the NOAA / NESDIS Center for Satellite Applications and Research.

STAR calls it "No noise (smoothed) Normalized Difference Vegetation Index (SMN)" or "Greenness" for short. This is what STAR says about the data: "[Greenness] can be used to estimate the start and senescence of vegetation, start of the growing season, phenological phases."

[1] NOAA STAR website: https://www.star.nesdis.noaa.gov/smcd/emb/vci/VH/index.php

There was a map like the one in Figure 10.1 for every week in the year. Plus, there was also an option that turned a full year's data into a very rough animation, which was basically an automated slideshow through all 52 maps. Even though the animation was crude, and the color palette not optimal, I really liked seeing the changes of vegetation health throughout the year. I knew I wanted to visualize the same thing—a continuously "Breathing Earth"—but do it in my own style. I was *very* happy to see that STAR shared the data behind the images. However, I had never worked with these levels of sophisticated geodata before: the data was formatted as hdf and GeoTiff files. Thankfully, I had just seen a presentation on GDAL (the Geospatial Data Abstraction Library) at OpenVisConf while I was in Boston. According to a Google search, GDAL should be able to open these kinds of files. However, instead of trying to parse the files in the command line, which the original talk was about, I took to Google again to see if there was an R package instead. And of course there was: the appropriately named `rgdal`.

After getting `rgdal` to work, I spent the next few hours trying to understand how to read in a GeoTiff file, what it contained, how I could play with it, and finally, how I could map it.

I also figured out how to switch map projections, although I decided to stick to the projection used by NOAA STAR.

My first goal was to recreate one of the images from the STAR website to ensure that I understood the steps of handling the data. It took about 6-8 hours to complete, but even with the sub-optimal color palette, I think the image in Figure 10.2 is just amazingly detailed.

This was a great start but these images were approximately 22 million pixels/data points *per week*! There was no way I could load that amount of data into the browser 52 times. So I recreated them in lower resolution (sadly). I ran some tests and eventually reduced the resolution to about 50,000 (non-water representing) pixels which looked like a good middle ground. That was small enough for the browser to handle, but high enough to still see interesting details. Finally, I made a few adjustments to the data setup in order to decrease the file size for one week's data to 250kB. (๑•̀ㅂ•́)و✧

Fig.10.2

Recreated map from Figure 10.1 using R in even higher resolution.

Sketch

Sketching was super short this time as my idea was very simple. I wanted to turn the pixel-based data about vegetation health into thousands of circles which animated through the 52 weeks of data, giving the impression the circles were "pulsating." The circles would grow bigger and darker when the vegetation was healthiest and would appear smaller and more yellow-green for low values of "greenness." Apart from the circles, no other types of mapping "markers" (such as country borders) would be used. Our Earth is beautiful in itself.

The more design-based aspects, such as the color gradient, sizes, etc., would be created once I had all of the data on my screen. I actually didn't really sketch at all, but just filled two pages in my small notebook with thoughts outlining the basics of the design and ideas on how to make the final datasets as small as possible.

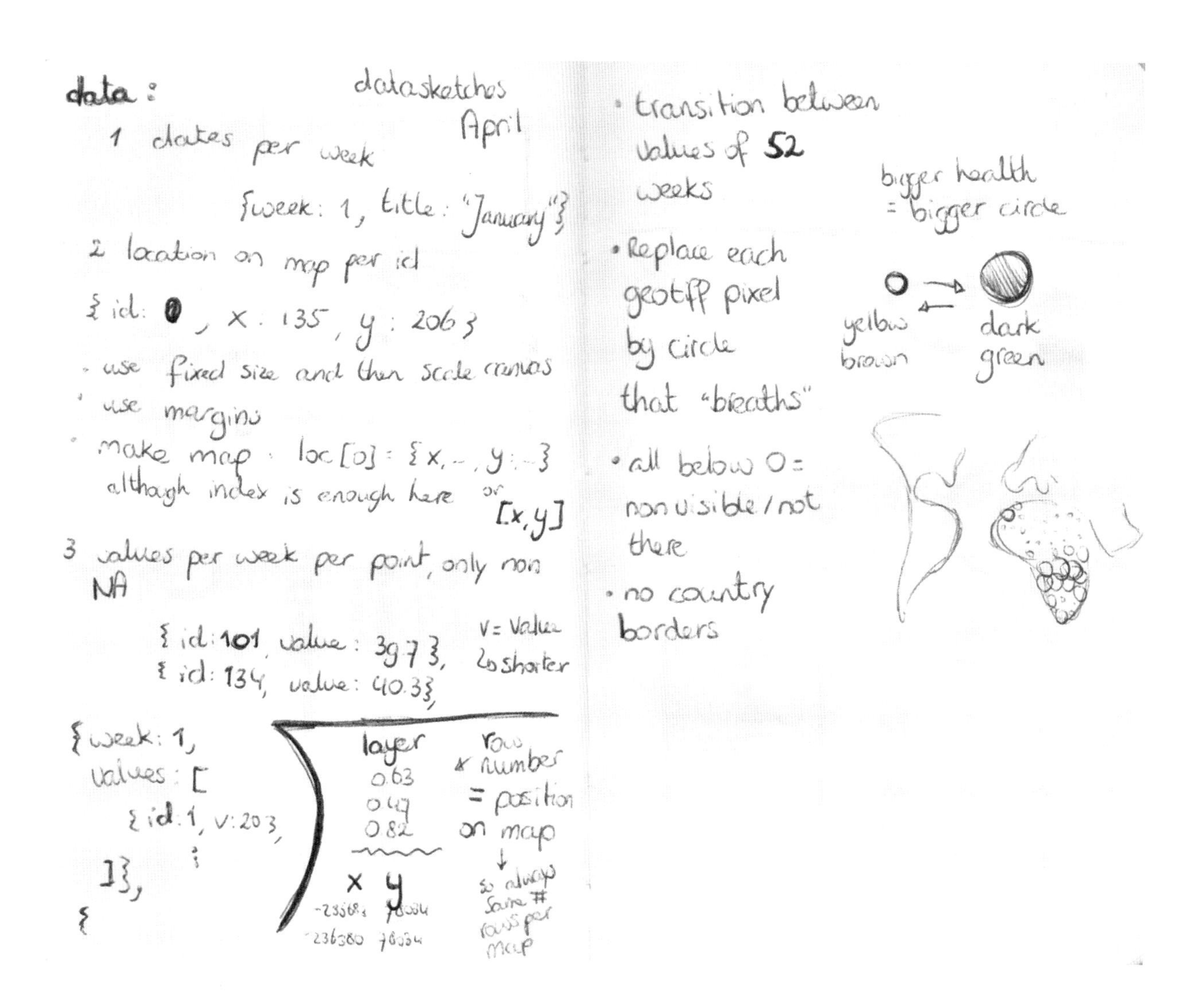

Fig.10.3

Writing down ideas on how to create small files and how to animate the circles.

Code

I started out getting the data on the screen with canvas. I knew that the standard D3.js use of SVGs was going to fail here with so many circles (and especially having them animated). Thankfully, drawing with canvas is actually quite easy and straightforward, especially when only plotting circles at certain locations. Figure Figure 10.4, found on the next page, shows some of the steps in the process: first, placing similarly sized and colored circles in the right location, but with differing opacity based on "greenness;" next, adding a color gradient to the values, adding a *multiply* color blend mode, and finally, making the circle size depend on greenness also.

Color blend modes determine how two layers/images are blended with each other, with multiply resulting in a nice darkening of the overlapping colors

I made a simple interval function that would switch between the 52 maps, quickly testing it by drawing a new week's map as fast as it could. That took about 2-3 seconds per map; not exactly a "frame rate" that I could use for natural looking animations. ఠ_ఠ

Therefore, I dove into Pixi.js, which is a 2D renderer using WebGL. And it doesn't get any faster than WebGL (on the web) as far as I know. I opened up a whole bunch of examples, especially those that I could find on Bl.ock Builder, that combined D3.js with Pixi. Sparing you the coding details, it suffices to say that Pixi was surprisingly easy to pick up. Unexpectedly, however, Pixi was *slow*...

For a short explanation of WebGL, please see "Technology & Tools" at the beginning of the book.

Unable to find a solution to make Pixi perform faster for my specific case, I asked Twitter and received replies with ideas and even some sandbox examples! Some solutions suggested using regl or Three.js, but I also got some interesting ideas for Pixi itself. For example, I learned faster performance with Pixi was possible with something called "sprites." You can think of this as small images. A popular example of sprites shows how to make *hundreds of thousands* of the same bunny image bounce around. For my case, I used a small white circle for my image (or sprite) and then applied a specific color and opacity to it for each of the 50,000 locations. But when I looked at them more closely, I noticed the circles weren't perfectly circular, especially the smaller ones, and they looked rather pixelated (see Figure 10.5). Bummer! (≧Д≦)

Fig.10.5

Somewhat pixelated circles with Pixi sprites.

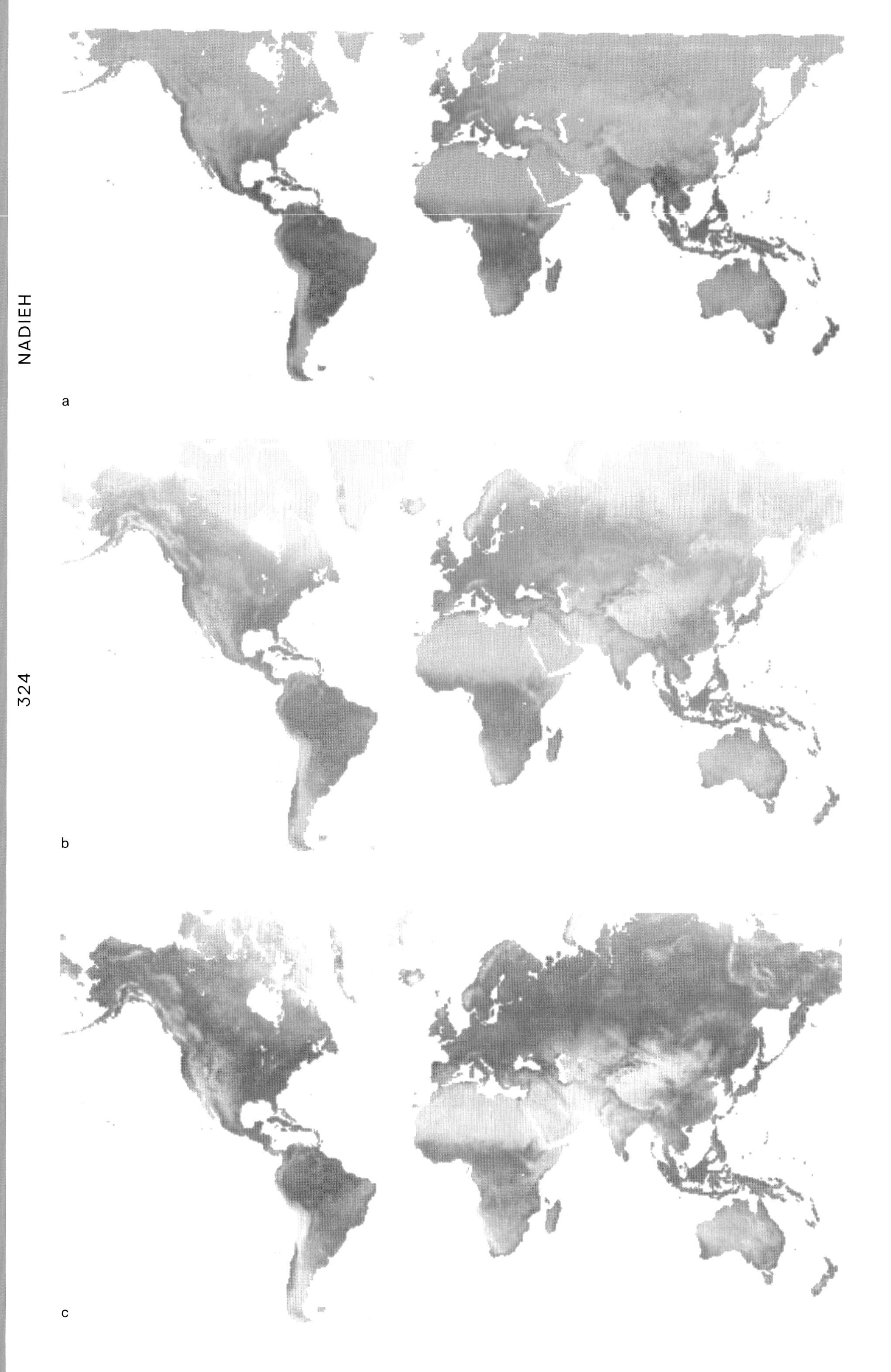

Fig.10.4 (a,b,c)

Building up the map of greenness in canvas.

The Extensive Applications of D3.js's Functions

The magic of `D3.js` isn't only in connecting data to (SVG) elements that will appear on the page; it's also in all the data preparation functions that it offers. For example, even when the final visual is made with canvas, I still always use `D3.js` to create my scales. Going from whatever values are in the data to pixel values on my screen (e.g., locations or size), and I use the wide variety of color interpolations to create color scales. `D3.js` also has several functions that perform mathematical operations, such as finding the minimum, maximum, range, standard deviation, mean, and more that can save you from having to load an extra mathematical library.

I also use `chroma.js` when I need even more specialized control over my colors. To illustrate a more advanced example, I would use `D3.js` just for the power of its `d3.delaunay()` function which gives me a simpler way to handle interactions on canvas, and for its `d3.stratify()` functionality to turn my data into a hierarchical nested variable, which I can use to create hierarchical, tree-like, visuals.

In summary, if you work with `D3.js`, I advise you to explore the wide variety of functions that are available through `D3.js` to speed up the data visualization process and do more advanced things.

That's when I decided to give `regl` a try, which helps to simplify programming with WebGL. Also at OpenVisConf, I saw an inspiring presentation about `regl` featuring bouncing rainbow bunnies. And when I found a blog post[2] that explained how to animate 100,000 points with `regl` I knew it was enough to start with.

I decided not to go into `Three.js`. It was just too much to handle in one week—so many new programming libraries!

At first I hoped to get the hang of `regl` by going through examples. But after an hour or so I acknowledged the fact that I *really* didn't understand anything yet and that I had to read some introductions to WebGL, but also to *GLSL* and *shaders*, two rather complex parts of drawing visuals with WebGL. It took a while, but my brain *slowly* started to grasp the main concepts.

Feeling somewhat more enlightened on WebGL, I started with the code from a simple example[3] to create a single gradient colored triangle in `regl`. Next, I slowly adjusted it to show circles on a map instead.

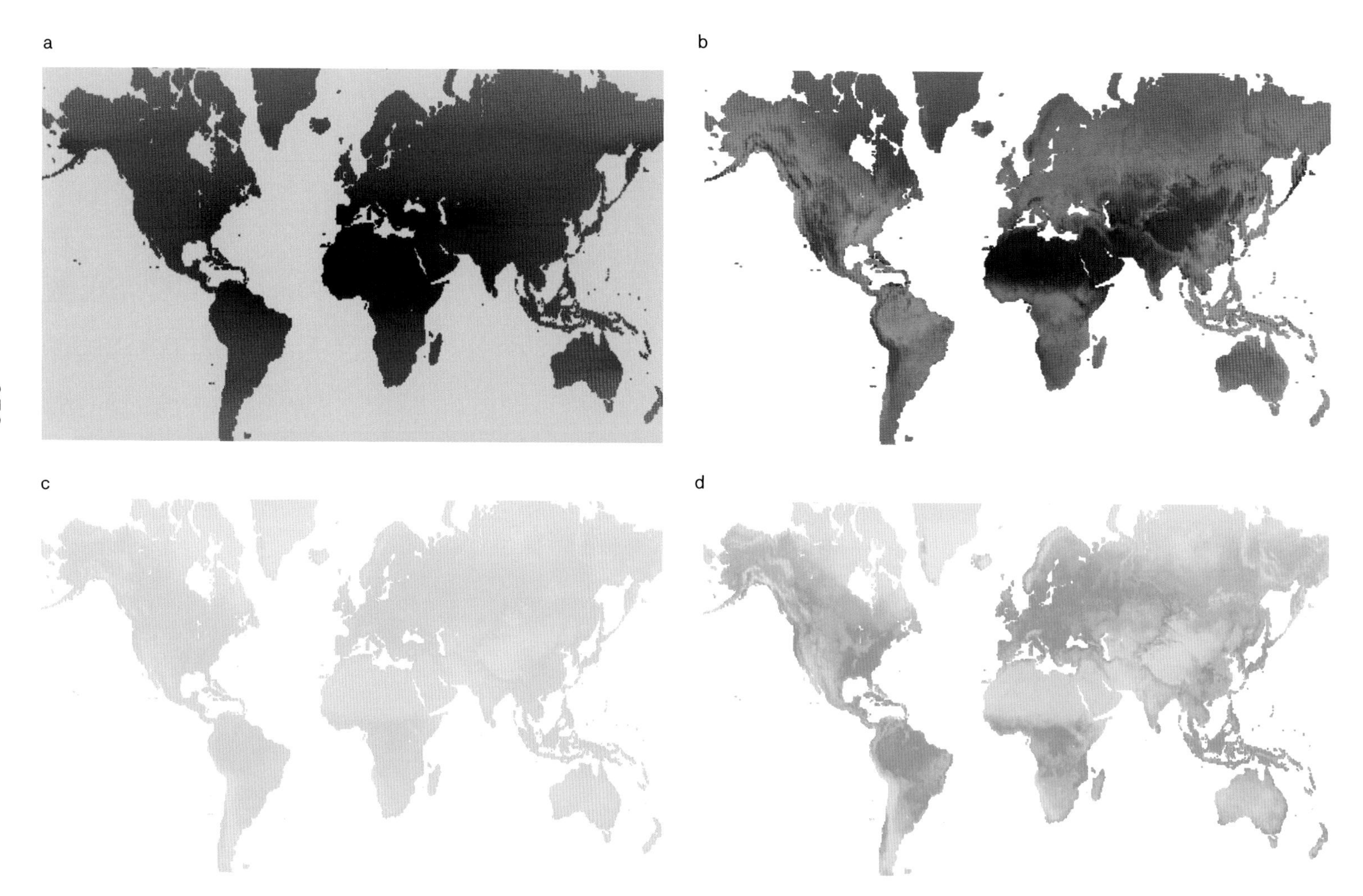

Fig.10.6 (a,b,c,d)

Several steps showing how I transformed the regl based map from the initial example's colors to the green colors I wanted to use.

Well, it took *a lot* of browsing through example code, but eventually I had a map in `regl` with correct circles and opacities. The one thing that I just couldn't manage was adding that final touch of a multiply blend mode, while combined with a circle shape *and* semi-transparent circles. However, even without the multiply effect, if I zoomed in, I saw the same pixelated effect as with `Pixi`! (⋟Ω⋞)° I did notice that it rendered faster than `Pixi` though.

[2] "Beautifully Animate Points with WebGL and regl" by Peter Beshai: https://peterbeshai.com/blog/2017-05-26-beautifully-animate-points-with-webgl-and-regl/
[3] Regl-demo by Adam Pearce: https://bl.ocks.org/1wheel/e025cbd91ac499d360a8b3346cb6f9e7

Fig.10.7
(a,b,c)

Failed attempts in trying to get circle shapes, with differing opacities, plus having a multiply color blend working with `regl`.

While trying to find information on creating anti-aliased circles (giving them smooth edges) in `regl`, I came across a code snippet that showed that `Pixi` actually has an “anti alias” setting! And, on one of my continuing Twitter chats, I received an *animated* `Pixi` example that I tested with 50,000 circles which still seemed to work smoothly. These two interesting avenues to explore brought me back to my `Pixi` based map. Another hour or two of work adjusting the animated `Pixi` example to my data, playing with some anti-aliasing things, and I was finally looking at a smoothly changing map; yay! (^▽^)

After I got `Pixi` working, I sent out another Twitter request to help me with the anti-aliasing and multiply blend mode in `regl`, as `regl` was still faster than `Pixi`. It wasn't long before several wonderful people sent me sandbox examples[5] to try and tackle my issues. These examples, plus some more I had found while traversing the web, increased my understanding of how to tackle the anti-aliasing in `regl`. And I was finally able to get the circles to look like *actual* circles in `regl`, too! (Figure 10.8)

Fig.10.8

Finally! Smooth anti-aliased circles in `regl`. No multiply color blend mode though.

[4] Pixi circle animation example by Alastair Dant: https://bl.ocks.org/rflow/55bc49a1b8f36df1e369124c53509bb9
[5] Animate 100,000 points with regl by Yannick Assogba: https://bl.ocks.org/tafsiri/dba04b04ae949760f96f97a2fba23ba6
ReGL circle animation example by Alastair Dant: http://bl.ocks.org/rflow/39692bd181fb1eb0b077a4caf886b077
Shapes and WebGL tweening by Robert Monfera: http://bl.ocks.org/monfera/85aa9627de1ae521d3ac5b26c9cd1c49

a

b

c

d

Fig.10.9
(a,b,c,d)

Even more failed attempts in trying to get circle shapes, with differing opacities, plus having a multiply color blend working with `regl`.

That *only* left the multiply blending that was missing from the `regl` version. However, that almost turned out to be one step too far. I couldn't find a single example of a multiply blend in WebGL, specifically where it was based on many elements overlapping (not two predefined images). I did get a lot of interesting other color combinations though (see Figure 10.9).

But thanks to the help and perseverance of several great people (and experts in WebGL) who provided me with demos[6] through my ongoing Twitter chats, I was eventually looking at a `regl` based map that had it all working—opacities, anti-aliased circles, and multiply!

↳ Know When to Ask for Help

Although I've asked for help on Twitter before, no project relied on the level of assistance I received from many generous people like this one. I received help with demos and was given multiple resources and possible solutions. I literally don't think I could've managed this project without having asked for help.

Don't be afraid to ask on social media or in dedicated places such as Slack channels or Stack Overflow (or even in real life) in case you get stuck or are looking for advice. The Internet can still be a great place from time to time with people willing to help you!

Finally, all three map versions (canvas, `Pixi`, and `regl`) looked the same. And the `regl` based map was definitely the fastest. I even had to slow it down to avoid it animating through a year too quickly! I added links to each of the three options on the main project page to give technically curious viewers the chance to compare performance.

In terms of the page design itself, I kept it very minimal. I wanted the focus to be on the map, and I felt that it needed virtually no explanation. All I created were a simple title, legend, and a few paragraphs of text explaining the data.

Reflections

From ideation, data, sketching, and coding (3x even; canvas, `Pixi`, and `regl`) this project took almost 60 hours to complete. In short, this was a *very* technical project for me. Sure, the visual in itself isn't as out-of-the-box as some of the other projects. But I had never learned so many new programming languages and libraries within such a short amount of time before.

I'm guessing the `regl` multiply color blend issue took at least 10 hours.

And I couldn't have done it without the help of a lot of great people that came to my aid on Twitter and through other channels. Thank you to everyone who offered a suggestion and sandbox examples!

[6] Regl multiply blend weights by Ricky Reusser: https://codepen.io/rsreusser/pen/YVRXzy?editors=0010

Breathing Earth

↳ BreathingEarth.VisualCinnamon.com

A Breathing Earth

50.000 circles moving through a year of data on planetary vegetation health

Week 17, April & May, 2016

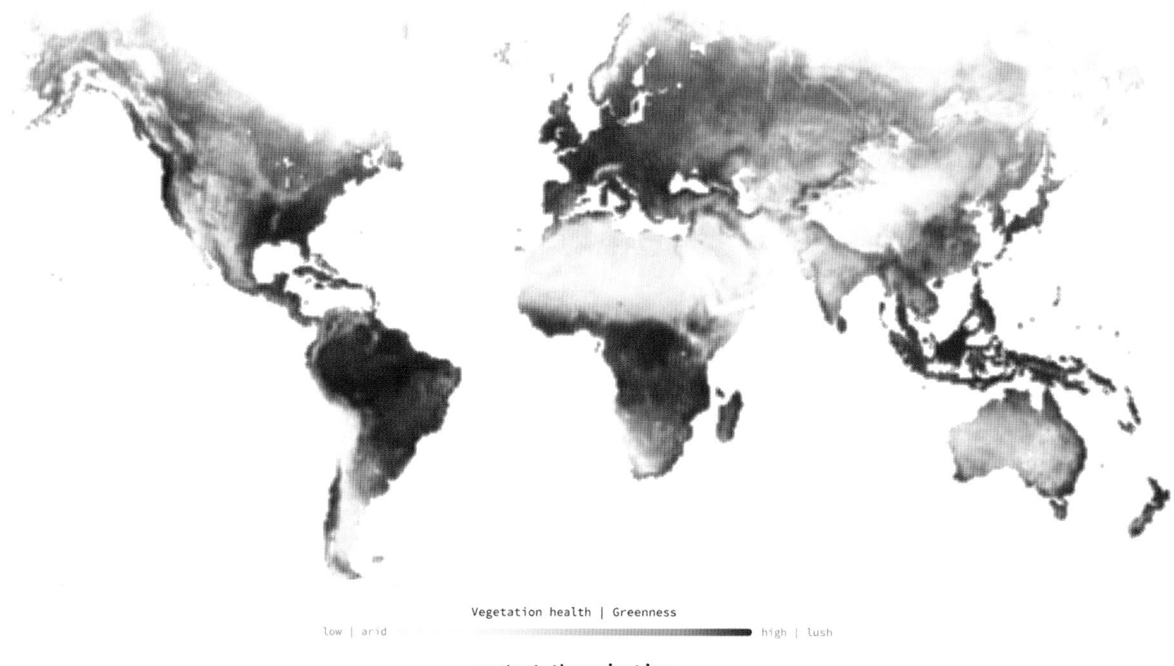

Vegetation health | Greenness

low | arid high | lush

restart the animation

We all know that our Earth goes through a seasonal cycle, especially for the latitudes farther away from the equator. In the fall the "deciduous" plants lose their leaves only to grow back again in the next spring.

The animation happening in the map above, through all 52 weeks of 2016, visualizes these seasonal cycles. The rise and fall of the growing season in the Northern Hemisphere is particularly visible. However, when focusing on different parts of the planet other cycles & different seasons become noticeable as well; the Southern regions of Africa, Brazil, and New Zealand, having the reverse cycle as the North, or India getting drier and drier up until the July when the monsoons start. The more often you watch the year go by, the more small details will start to stand out.

Watch and see our Earth "breathing" throughout the year. All living organisms depend on these cycles in the growth of plants; for food, for oxygen, and more. Although we humans have started to affect these cycles, hopefully this gorgeous spectacle will be part of our lives for millennia to come.

The data comes from NOAA STAR, the center for satellite applications and research, who use the VIIRS (Visible Infrared Imager Radiometer Suite) sensor onboard Suomi National Polar-orbiting Partnership (SNPP) satellite to get detailed vegetation information of our Earth every week.

The specific variable visualized in the map is called 'Greenness', or in more scientific terms; the no noise (smoothed) Normalized Difference Vegetation Index (SMN). Greenness can be used to estimate the start and senescence of vegetation, start of the growing season and phenological phases. For areas without vegetation (desert, high mountains, etc.), the displayed values characterize surface conditions.

Animating 50.000 circles in the browser, all separately from each other through different sizes, colors & opacities became a real technical challenge. In the end 3 different methods were tested to see which one would appear most natural. For those interested to see the differences, you can find the (slow) pure canvas version here, the pixi.js WebGL version here and the regl WebGL version here.

Fig.10.10

The full page of "Breathing Earth."

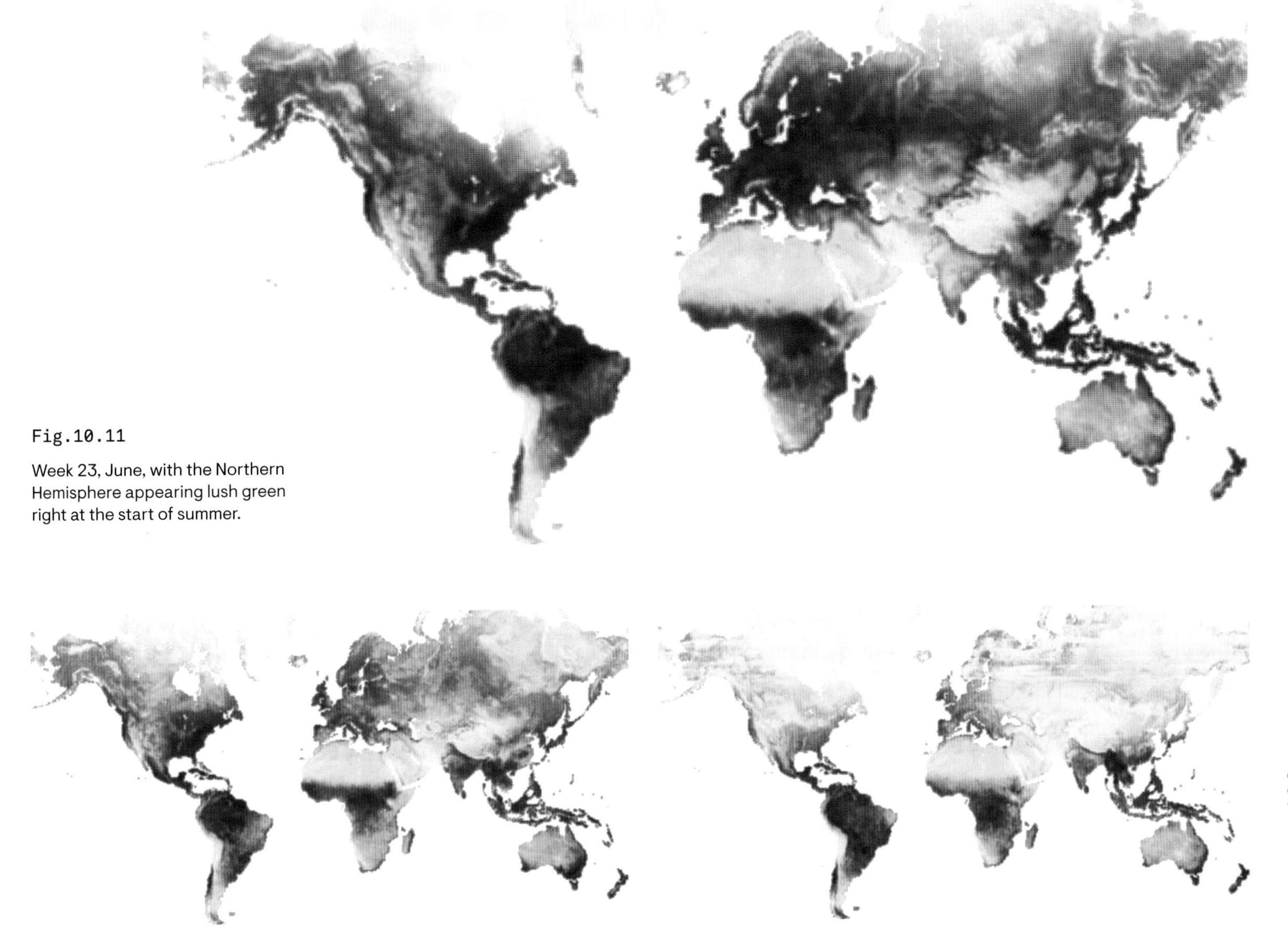

Fig.10.11

Week 23, June, with the Northern Hemisphere appearing lush green right at the start of summer.

Fig.10.12

Week 40, October.

Fig.10.13

The start of winter in the Northern Hemisphere, week 51, December.

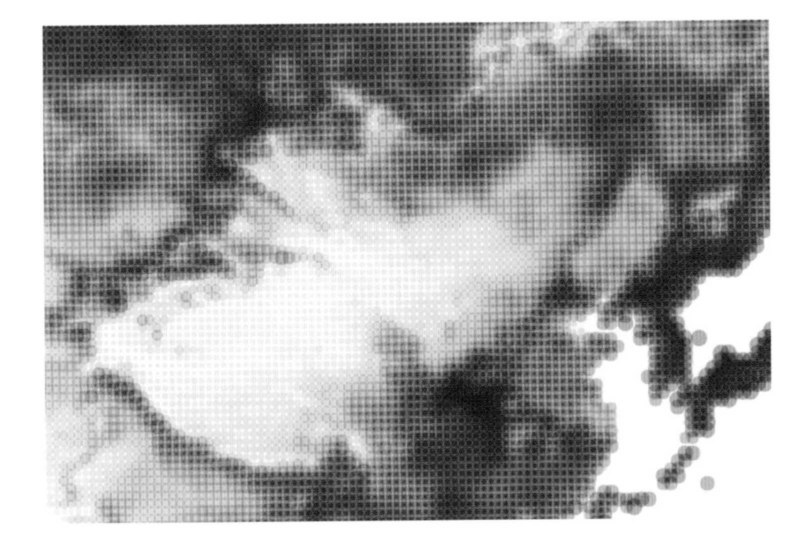

Fig.10.14

A zoom in on Asia, week 23.

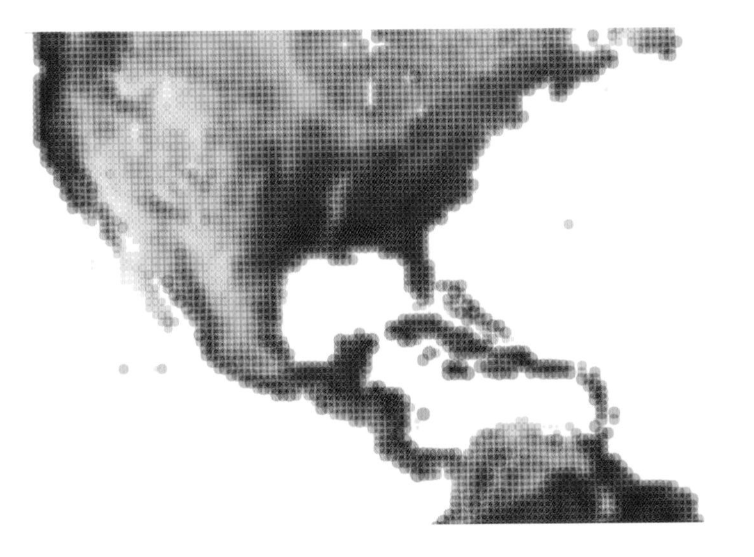

Fig.10.15

A zoom in on North and Central America, week 16.

Primary

249 / 249 (100%)

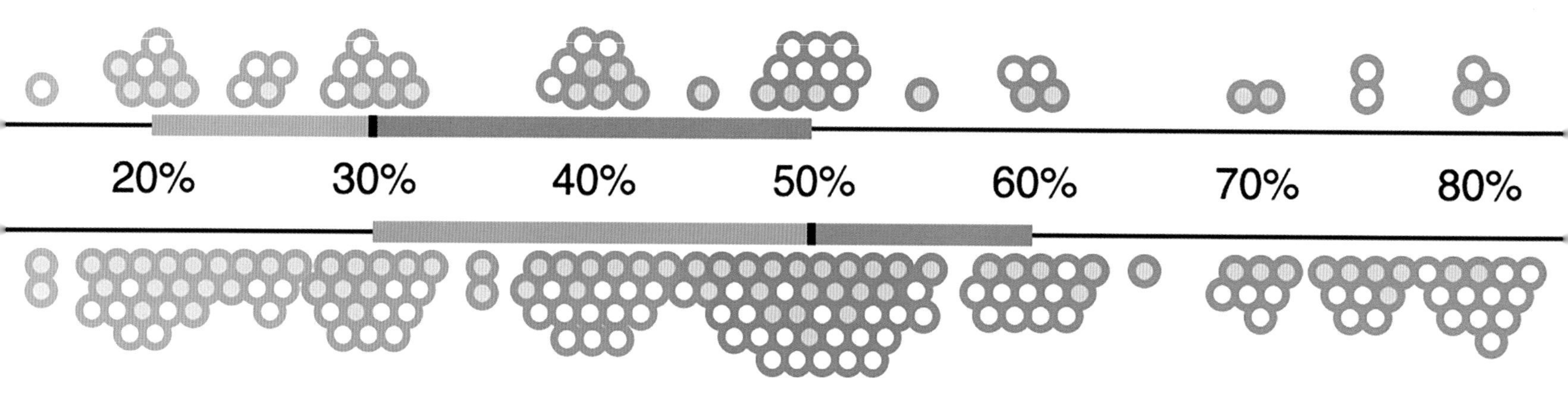

Secondary

383 / 384 (100%)

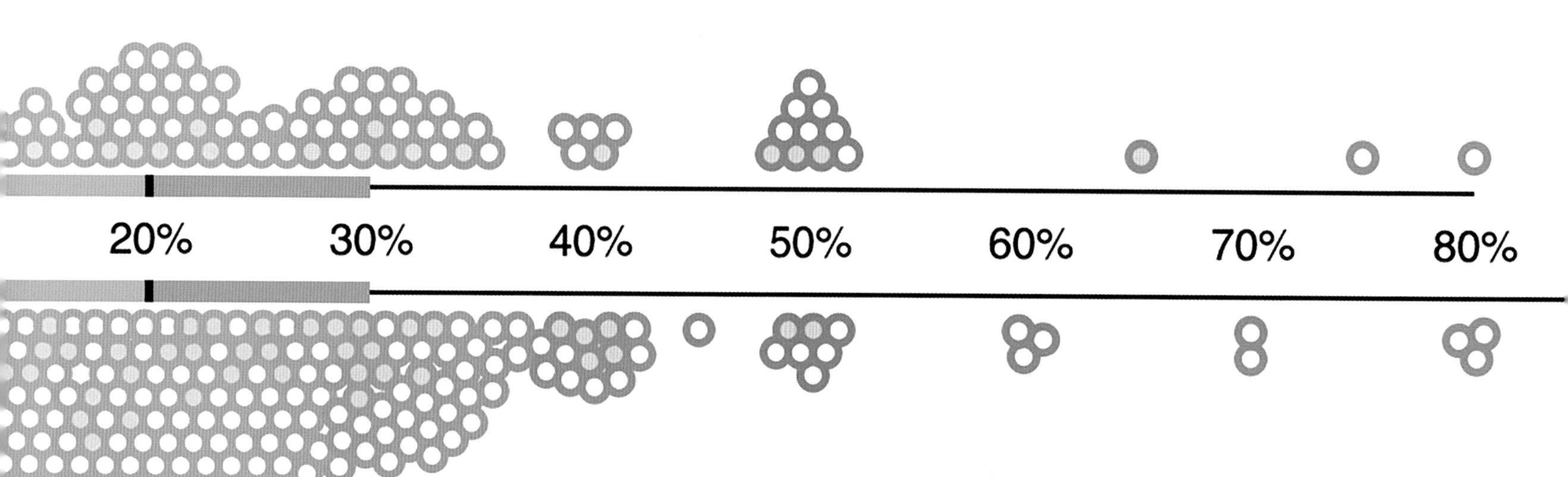

One of several

347 / 347 (100%)

655 Frustrations Doing Data Visualization

SHIRLEY

In February 2017, our friend Elijah Meeks made a bold claim: most people in data visualization end up leaving because there's something wrong with the current state of the field. That statement stirred up quite a bit of conversation and resulted in a community survey with 45 questions and 981 responses. By mid-March, Elijah had cleaned, anonymized, and uploaded all the data onto GitHub.[1] And I knew I had to do something with that data.

[1] Data Visualization Survey, 2017 Responses: https://github.com/data-visualization-society/data_visualization_survey/blob/master/data/cleaned_survey_results_2017.csv

Data

This was probably one of my favorite projects in terms of data because I didn't have to do any manual data collection or cleaning. Elijah had already cleaned up all the survey responses and put them into a nicely formatted, (let me repeat, cleaned up) CSV file. Honestly, I don't think I've ever had it easier. It's probably why I wanted to do the project in the first place. ヽ(0 ∀ 0 *★)° *・.。

So with the data collection and cleanup already done (hehe), I moved on to analyze and explore the data. The very first thing I did was to create a list of the survey questions I was interested in, and then grouped them by theme. (Figure 10.1, left) .

As the premise of the survey was Elijah's claim that practitioners were leaving because of the state of the field, my primary question was: Why might people want to leave? But there was no such question in the survey, and those that had already left definitely wouldn't have participated in the survey. So I decided to go with a proxy instead: "Do you want to spend more time or less time visualizing data in the future?" I then organized the relevant questions into categories (Figure 10.1, right):

The basics of their data visualization jobs.

- Were you hired to do data visualization only?
- What focus is data visualization in your work?
- Is your total compensation in line with Software Engineers and UX/UI/ Designer roles at your level?

The aspects of their role that might affect their job satisfaction.

- Is there a separate group that does data visualizations or are you embedded in another group?
- Are data visualization specialists represented in the leadership of your organization?
- What knowledge level do your consumers have of the data you are visualizing?
- How often do they consume your data visualizations?
- How would you describe your relationship with your consumer?

And their biggest frustration doing data visualization in their jobs.

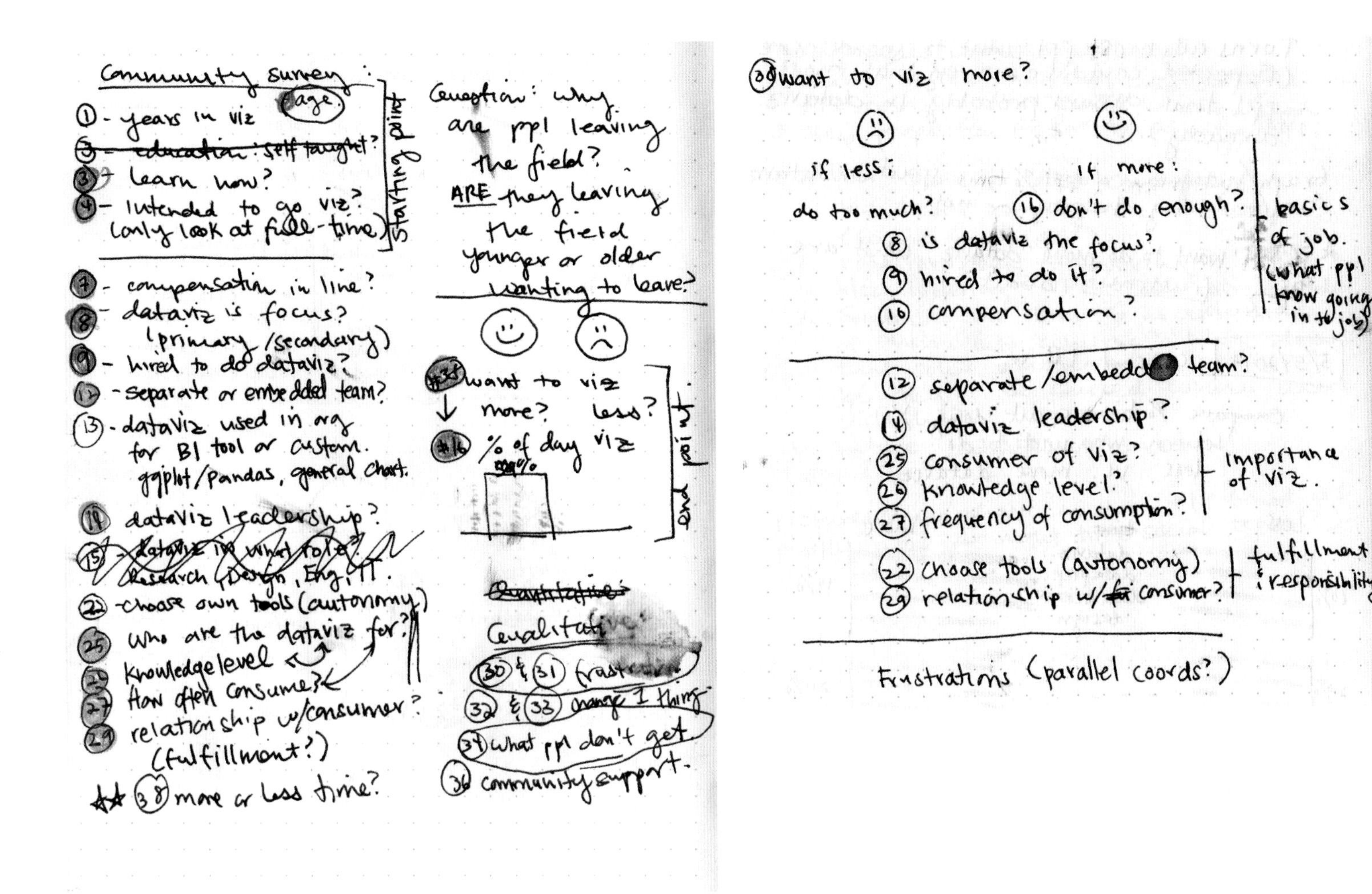

Fig.10.1

A list of survey questions grouped into overarching themes.

One of the talks at Openvis Conf 2017—which I attended right before starting on this project—was about Vega-Lite, a JavaScript charting library for quickly composing interactive graphics, and I decided to give it a try for my data exploration. I used histograms to look into what part of the creation process (data preparation, engineering, analysis, design, visualization) the respondent spent the most time doing at their jobs and learned that most people don't focus on just one part and instead juggle most or all of the creation process. I also plugged some of the qualitative survey questions into bar charts and explored what technologies the respondents used to visualize their data, who they made visualizations for, and so on:

For explanation of Vega-Lite, see "Technologies & Tools" at the beginning of this book.

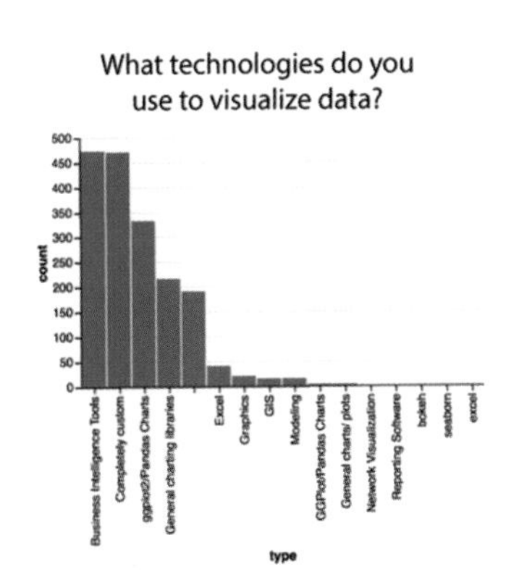

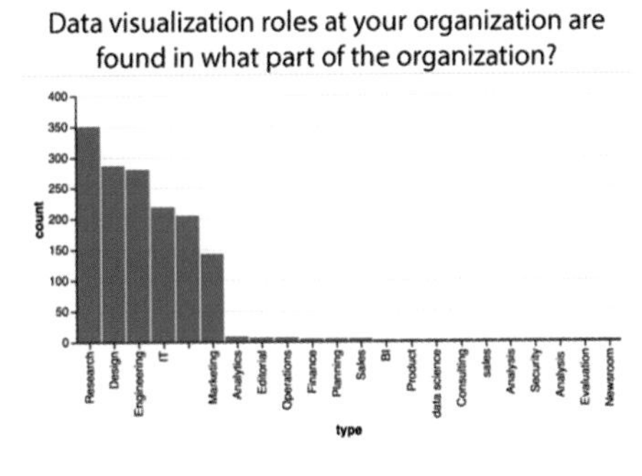

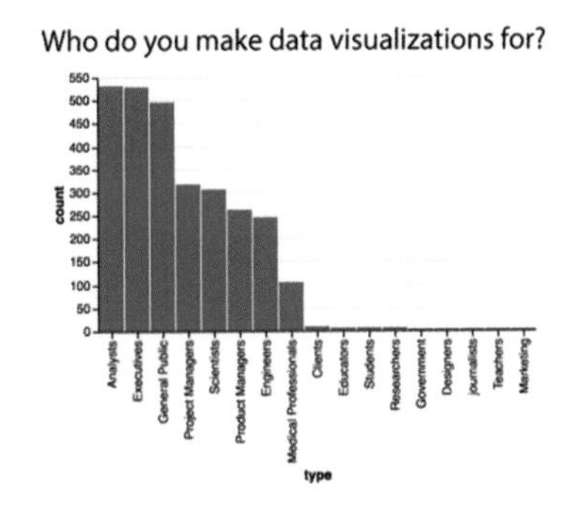

Fig.10.2

Using bar charts to explore some of the qualitative survey questions.

I never got beyond bar charts and histograms with Vega-Lite for this project, but the exploration did allow me to quickly understand the structure of the survey questions and responses. It taught me to work with the quantitative or multiple choice questions, instead of the open-ended ones whose answers were too many and too varied for me to analyze efficiently for this project. It also helped me see the value in using Vega-Lite as a quick exploration tool—a big step-up from the last project, where I built visualizations from scratch for every theory I wanted to test.

If I were to come across similarly open-ended text nowadays, I'd consider looking into a Natural Language Processing (NLP) algorithm to analyze the data.

↳ Explore Data: Use Charting Libraries

Once I have my attributes listed and my questions and hypotheses formulated, I like to quickly explore the data with Observable and Vega-Lite. Observable is an online notebook tailored for visualizations, and it lets me experiment with ideas without the pressure to write "beautiful" production code. I import my data into a notebook and use Vega-Lite to visually test my hypothesis.

Some common charts I use in my exploration include bar charts for comparisons, box plots and histograms for distributions, scatterplots for correlations, node-link diagrams for relationships, and line charts for temporal trends. I then note anything interesting in my notebook, and use those notes to inform my designs.

For more explanation of Observable, see "Technologies & Tools" at the beginning of the book.

This is why I like to list data types (quantitative, nominal, ordinal, temporal, spatial) next to the attributes in my first step , because they inform the charts I should use for exploration.

Sketch

As my goal was to figure out why people might want to leave the field, I wanted to know if there was any correlation between how much time respondents spent on creating data visualizations, and whether they wanted to do more or less of it in the future.

For my first pass, I decided to use a stacked bar chart (Figure 10.3). The y-position represented the percentage of one's day spent on data visualization, arranged in order from least amount of time (10%) to most amount of time (100%). I used color to represent whether they wanted to do more or less dataviz, with bars to the left of the gap being "much less" or "less," and to the right being "same," "more," or "much more."

It turned out that the majority of respondents wanted to do the same or more data visualization going forward. I wondered if that sentiment changed if they were more focused on other parts of the data visualization process, such as data preparation, engineering, science, or design. I also wondered if I could pinpoint someone's frustration with the dataviz process by examining how much of their day was dedicated to a specific task, and whether they wanted to do more or less of it. Unfortunately, because I only thought about *how* to mash those three questions together and gave no thought to readability, the prototype ended up being really hard to understand. What I did realize from prototyping the stacked bar chart was that I was much more interested in showing the individual responses, rather than trying to aggregate them all together into a summary. It also taught me that too few people responded that they wanted to do "less" dataviz, which meant that it wouldn't adequately answer the question: "Why might people leave?"

In retrospect, this makes a lot of sense; most people answering a survey about data visualization would probably want to do more of it.

Fig.10.3

Sketch of my idea to explore how much time respondents currently spend on data visualization versus how much they want to do it in the future.

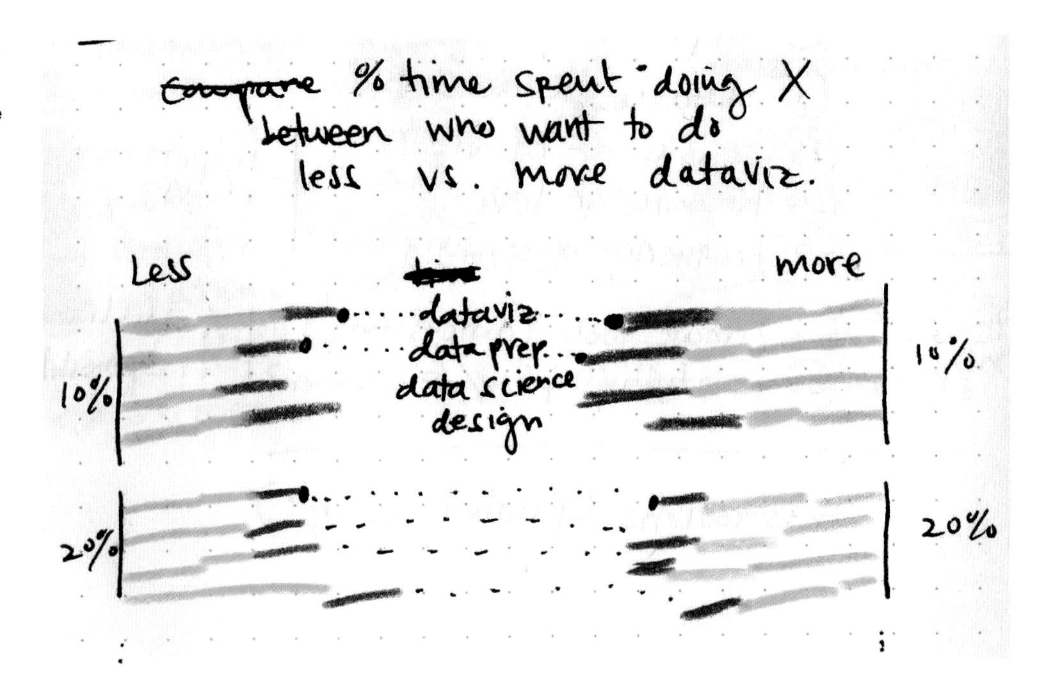

Code

For my second iteration, I decided to try a *beeswarm* plot because I could feature the individual responses as dots centered around certain attributes. And I liked that because they use dots, beeswarm plots tend to be compact and easy to glance through for the big picture.

I also decided to use the open-ended question, "What is your biggest frustration with doing data visualization in your job?" as the proxy for whether a person might leave the field, and noted whether they answered with any frustration or left the question empty. I placed those who answered with frustrations in the left column and those who didn't answer in the right. I also visualized whether dataviz was a focus of their work and placed the dots vertically by the answers they gave: top row for "primary," middle for "secondary," and bottom for "one of several." I colored and positioned them horizontally by their years of experience and filled the dots if they meant to go into dataviz in the first place:

Fig.10.4

Beeswarm plot showing every respondent. Those who responded with frustrations are placed to the left and those without are to the right. They are grouped vertically by what focus dataviz is within their work, with top being "primary," middle being "secondary," and bottom being "one of several."

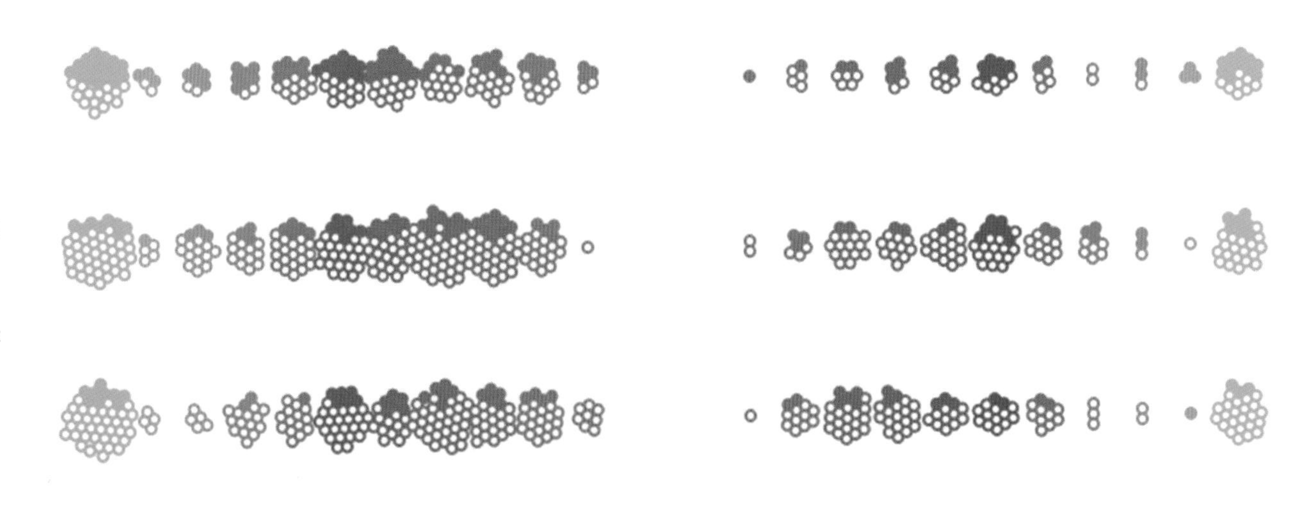

I liked the beeswarm, but didn't like how hard it was to compare those who responded with and without frustrations, so I stacked the two on top of each other. And I realized that my *x*-axis (how many years they had been working in the industry) probably had little relation with the questions I was interested in (whether dataviz was their primary focus, or whether dataviz was represented in their leadership) and whether they responded with frustrations. So I updated the colors and *x*-axis to represent percent of day focused on dataviz instead. Finally, I wanted to make it easy to compare across answers, so I added a box-and-whisker plot on top of the beeswarm to mark the median and first and third quartites (Figure 10.5).

I really like showing individual data points and layering summary metrics on top of them, and find that I do it often in my projects.

When I showed this iteration to my friend RJ Andrews, he immediately suggested putting the box-and-whisker plot in the middle, the dots representing those with frustrations below the box-and-whisker plot, and those without frustrations above it. He explained the importance of visual metaphors; those with frustrations should "drip down" like they're being weighed down, while those without frustrations should "rise up" because they were unburdened (Figure 10.6).

I know I can't assume that everyone who didn't answer with frustrations was happy with their jobs (there's definitely a percentage that just didn't want to answer), but it's the best proxy I had.

¯_(ツ)_/¯

↳ Visual Metaphors

I first learned about visual metaphors from my friend RJ Andrews, and it's really changed the way I think about designing visualizations. Visual metaphors take advantage of what people might already be familiar with—like having a negative feeling (frustrations) move downward—which can reinforce what the underlying data is, what the visual is trying to communicate, and potentially reduce the learning curve for an unfamiliar visualization.

It's an additional step I like to take after considering what chart type might work best for what I'm trying to communicate, and it definitely makes the visualization more approachable and visually interesting.

I implemented the `beeswarm` with `D3.js`'s *force* layout, using the positioning (`d3.forceX()` and `d3.forceY()`) and collision (`d3.forceCollide()`) forces to calculate the position of each dot. To create the middle split, I "nudged" any dots back if they went past a certain vertical position (called the "bounded force layout"). The box-and-whisker was more straightforward to implement, but still took many iterations until I was happy with the design.

After I had the visualization down, I went back to the eight questions I had first outlined. I wanted to be able to easily compare data between questions and see if there were any correlations. To do this, I decided to display two questions at a time and place them side by side. I added a dropdown menu to let users switch to different questions, and a brush—an interaction where the user can draw a bounding box within the visualization—on the beeswarms to filter survey responses. I implemented the brush interaction with `D3.js`, and used `React.js` to link the dropdowns with the beeswarms, and the beeswarms with each other. The goal for linking the two questions (beeswarms) via the brush filter was so that I could brush and filter a particular answer for one question ("those whose primary focus is data visualization, and who spend more than 50% of their day on it") and see how the same people answered the other question ("for those same people, the majority of them perceive their compensation to be either in-line with or higher than software engineers and designers at the same level") (Figure 10.7).

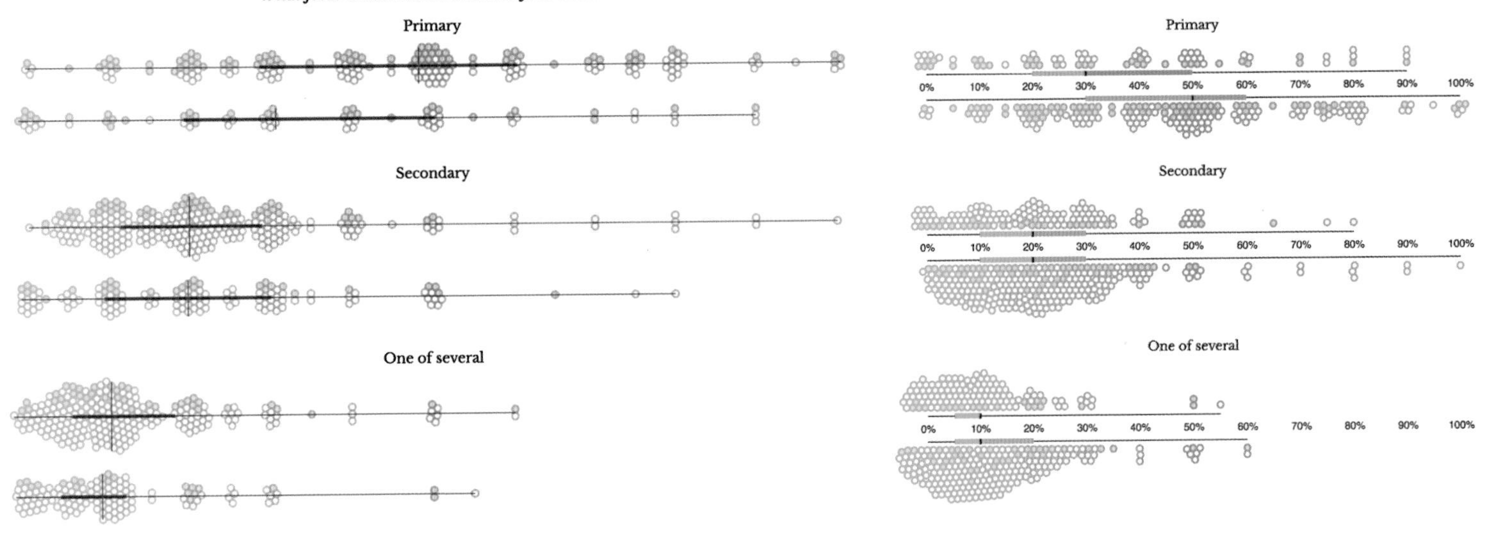

Fig.10.5

Adjusted beeswarm plot. Those with and without frustrations are placed on top of each other instead of side-by-side, and box-and-whisker plots are overlaid to give additional context.

Fig.10.6

Final beeswarm plot with respondents rising or dripping based on whether they answered with frustrations.

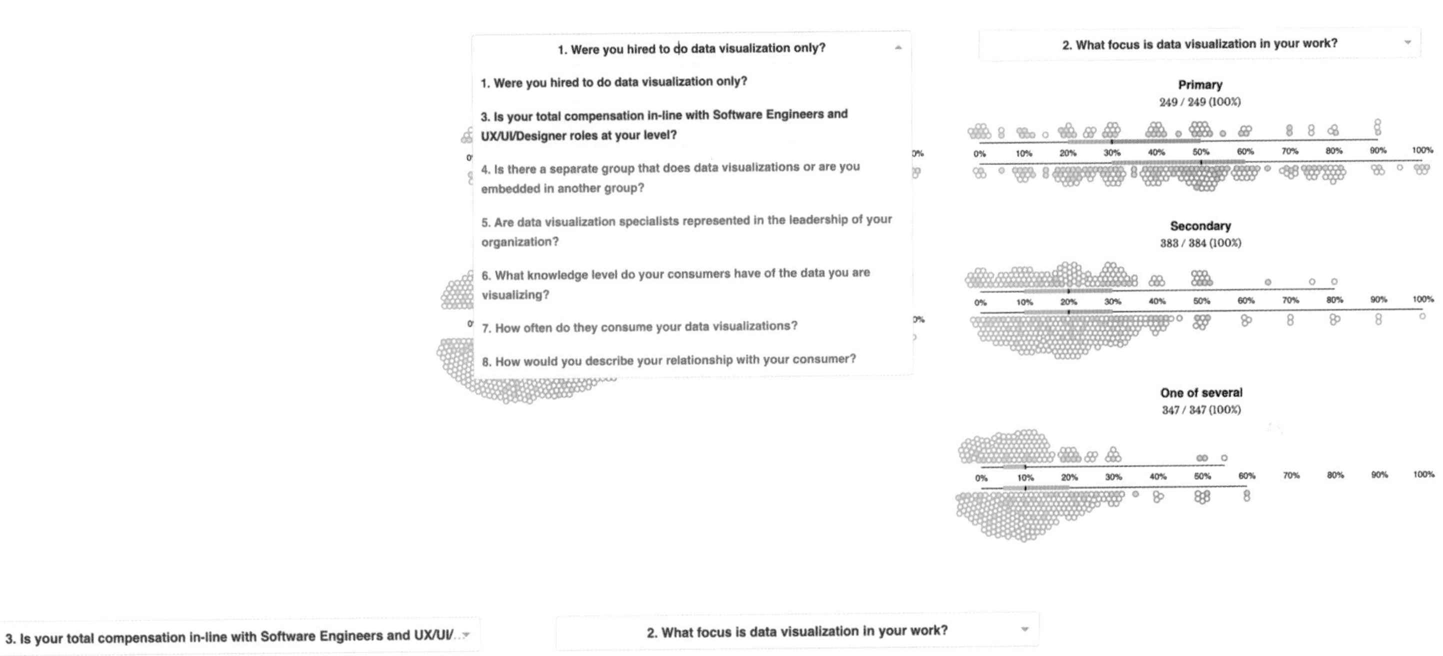

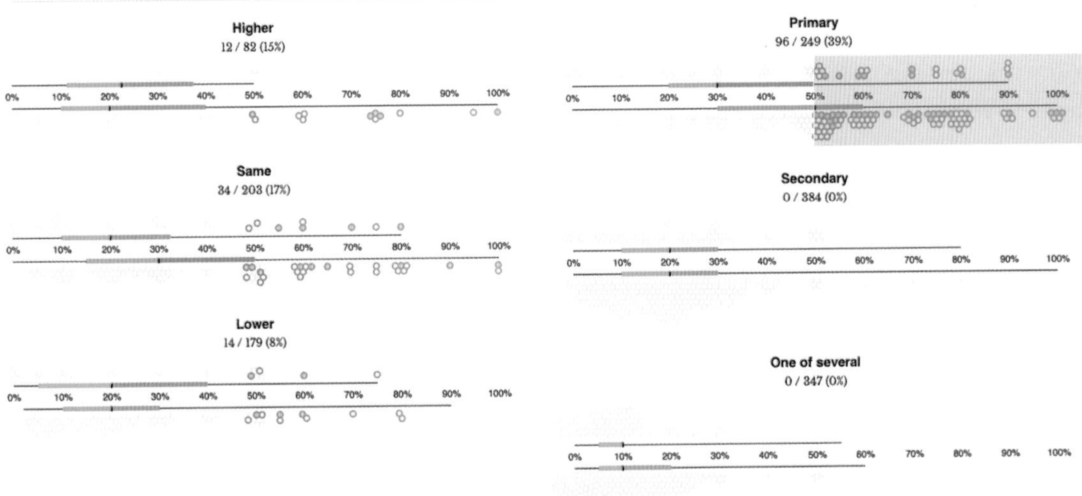

Fig.10.7

Selecting a different question from the dropdown updates the corresponding beeswarm (top), and brushing the beeswarm fades out all other respondents whose answers didn't fall within the brush's bounding box (bottom).

I then used my newly completed exploratory tool to try and answer my original question: "Why are people leaving the field?" I went through each of the eight questions, filtered their answers, and jotted down any interesting things I noticed.

After doing the analysis, I realized I could center my observations around what a "successful" data visualization role might look like: a role with a higher perceived salary that allowed the person to spend a large percent of their time on creating data visualization. I used those two metrics and looked through the answers for what might correlate with "unsuccessful" dataviz roles that might cause someone to leave. I found that those roles were typically found:

- On an embedded team,
- Not hired to work on data visualization,
- Does data visualization as only one of several tasks, and
- Has a subordinate relationship with the stakeholder of their visualization.

I filtered for the respondents that fell within those four situations, collected their frustrations, and tried my best to categorize their frustrations (Figure 10.8).

I wrote a blog post[2] centered around what I learned, where I outlined what factors contributed to more or less time spent working on data visualizations (being on a dedicated dataviz team, as opposed to on an embedded team), and what led to higher or lower perceived salaries (primary focus on dataviz, with a collaborative relationship with stakeholders). I then outlined the most common frustrations that I came across and put them into two categories: those stemming from working relationships in the organization ("coworkers do not understand what is possible," etc.), and those related to working with the available technology. Finally, I presented what we could potentially do to alleviate those frustrations: to educate organizations on how effective visualizations can benefit them, and to provide resources on continuous education for dataviz practitioners.

Reflections

Even though I *really*, really abhorred all the writing, I'm happy with all the research and analysis I put into my blog post. And though there are always things I want to improve, I'm also satisfied with my final visualization. I was able to take away two important lessons:

1. Use `Vega-Lite` or other similar charting libraries to quickly explore the data.
2. Use visual metaphors to better communicate the nature of the dataset (and more often than not, make the visualization more interesting).

Most importantly, I'm glad I learned so much while analyzing the survey data and writing about the community's frustrations. It has informed a lot of what I do inside and outside of my work—from prototyping ideas and teaching my clients how to think about data to getting better at designing effective visualizations. It has motivated me to create workshops and talks for front-end developers with the end goal of making `D3.js` and data visualization more approachable.

[2] "655 Frustrations of Doing Data Visualizations": https://medium.com/visualizing-the-field/655-frustrations-doing-data-visualization-e1087c8176fc

- ~~external~~ lack training/mentorship
- ~~external~~ lack understanding + education
- data difficult
- difficult tools
- ~~lack respect for~~ external lack respect for viz esp. lack time
- personal training/education

I don't have someone else to offer ~~advice~~ on best practices or suggestions for data visualizations.\

Lack of understanding from supervisors as to the extent of the data prep process and how long it can take.

Lack of training

The data itself is often quite disparate- incomplete- and skewed towards categorical.

reluctance to accept new visualizations that could be useful

Messy and poorly documented DB

it's the last step and therefore gets the least amount of time. most time is in data exploration

Not enough time

Too much ETL (extract, transform, load)

Too many options- too short a deadline

Not doing it enough time

Multiplicity of tools and approaches- but in an academic university this is difficult to dictate at an organizational or institutional level.

no general easy access flexible and interactive tools

Tools used are slow

Lack of opportunity for independent exploration or creativity

Lack of options for delivery

Too many libraries to work with

Not enough time for data viz. Data Science/Engineer colleagues do not see the value of it.

Inconsistencies in data quality/integrity produced by colleagues necessitate extensive cleanup and thwarts implementation of automated visualization processes.

it is never a priority

Ensuring accurate data

Time to easily create d3 plots

Coworkers do not understand what is possible

The data is insanely messy

the structure and organization of the underlying data

Deciding which visual to use

~~***Lack of components in some of the BI tools***~~

Own limitations in implementation (e.g. knowledge of d3.js)

not enough design skills

- lack external mentorship
- lack external understanding/education
- external lack of respect for viz (no time given)
- difficult tools
- difficult data
- personal training/learning insufficient

Poor training of others in recognizing quality issues with visualizations

Lack of infrastructure and policy guidance.

Building info-graphics- we often find ourselves spending time building out info-graphics by screenshotting images in a dashboards. Having info-graphic templates would be helpful.

Incredible amount of data cleanup work reduces time spent on statistical modeling and visualization design.

One of our senior executives doesn't appreciate data viz and considers herself a "spreadsheet person." So I am discouraged from having fun with viz and extending beyond the basics.

clients not willing to pay for visualizations that work

Lack of large color printer

Not having enough time to speak with the people I'm building dashboards for. does not feel collaborative.

people don't understand it takes a lot of time

Have not implemented more dynamic for query and display the data

tools limitation (excel)

Time crunch because they get the data to you right before the final product is due leaving no time for my work.

Not enough time

Time consuming

The number of managers that ignore the data because their gut has gotten them this far.

Not included in the initial design of the whole project- sometime data are not enough- no connection between data to be included(missing links/common values) and also goals are vague.

Upper Management not being data savy

not good enough at d3 to fully exploit it yet

It is often overlooked

I am not that experience on javascript so sometimes when I use D3 I can be stuck for hours.

Too little time and too little information about some projects to create effective visualization.

It's non-essential and done in spare time.

Lack of trust that I am selecting the best and most appropriate method to visualize the data.

Fig.10.8

I printed out all the frustrations that fell under the four situations (embedded team, dataviz only one of several responsibilities, subordinate relationship with stakeholders, perceived lower salaried) and tried to categorize their frustrations.

655 Frustrations

↳ shirleywu.studio/projects/community

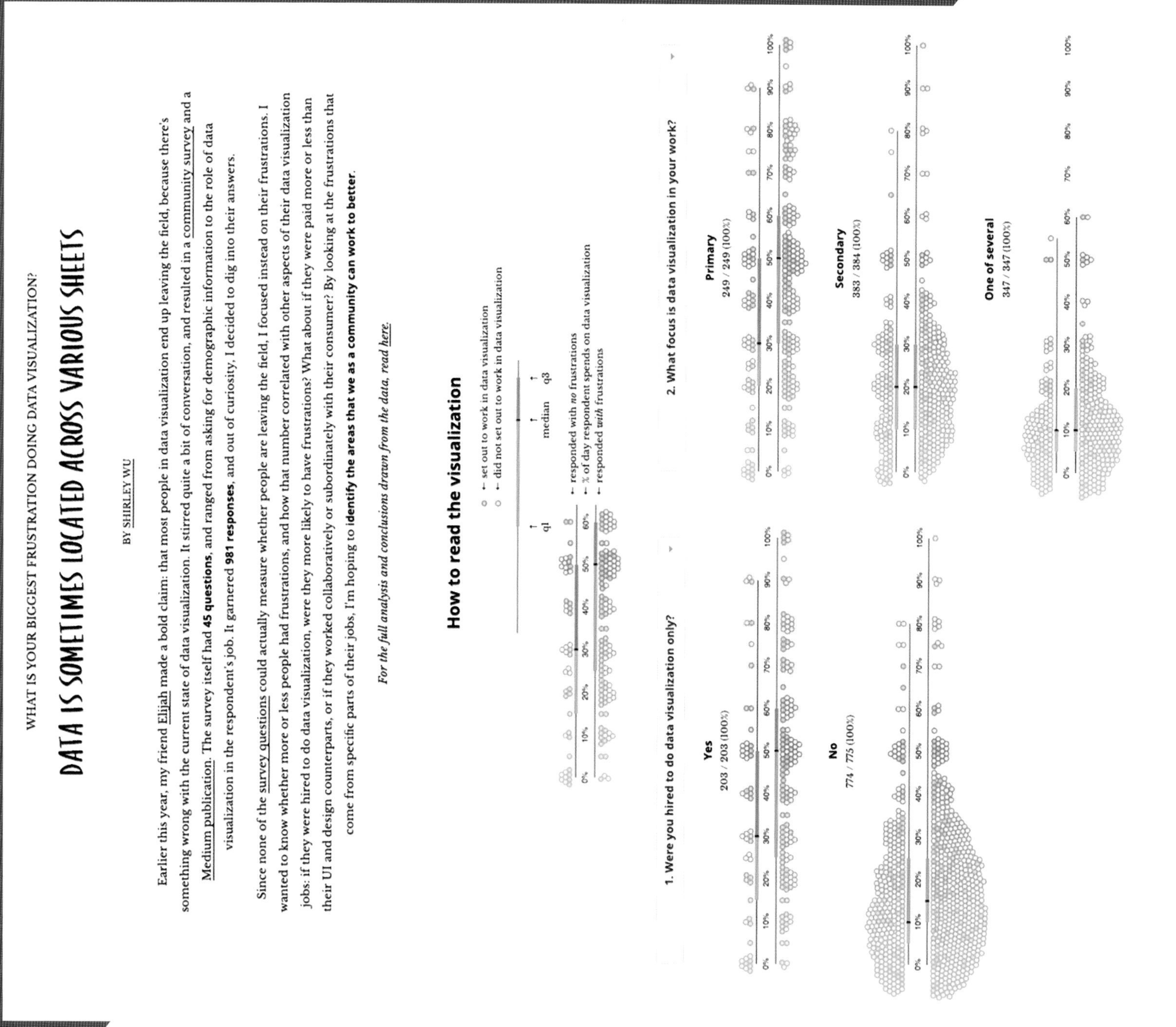

WHAT IS YOUR BIGGEST FRUSTRATION DOING DATA VISUALIZATION?

DATA IS SOMETIMES LOCATED ACROSS VARIOUS SHEETS

BY SHIRLEY WU

Earlier this year, my friend Elijah made a bold claim: that most people in data visualization end up leaving the field, because there's something wrong with the current state of data visualization. It stirred quite a bit of conversation, and resulted in a community survey and a Medium publication. The survey itself had **45 questions**, and ranged from asking for demographic information to the role of data visualization in the respondent's job. It garnered **981 responses**, and out of curiosity, I decided to dig into their answers.

Since none of the survey questions could actually measure whether people are leaving the field, I focused instead on their frustrations. I wanted to know whether more or less people had frustrations, and how that number correlated with other aspects of their data visualization jobs: if they were hired to do data visualization, were they more likely to have frustrations? What about if they were paid more or less than their UI and design counterparts, or if they worked collaboratively or subordinately with their consumer? By looking at the frustrations that come from specific parts of their jobs, I'm hoping to **identify the areas that we as a community can work to better**.

For the full analysis and conclusions drawn from the data, read here.

How to read the visualization

Brush one or both questions to filter

← 1 - 12 →

out of 980

1.

Were you hired to do data visualization only?
Yes
What focus is data visualization in your work?
Primary
Percent of your day spent on data visualization?
90%

Biggest frustration(s)
waste too much time filling in sports scores and other trivial stuff rather than investing time in exploiting the investigative potential of data journalism.

2.

Were you hired to do data visualization only?
Yes
What focus is data visualization in your work?
Primary
Percent of your day spent on data visualization?
70%

Biggest frustration(s)
Gap in high-level and low-level programming and customizaton

3.

Were you hired to do data visualization only?
Yes
What focus is data visualization in your work?
Primary
Percent of your day spent on data visualization?
50%

Biggest frustration(s)
Not knowing enough about basic graphic design.

4.

Were you hired to do data visualization only?
Yes
What focus is data visualization in your work?
Primary
Percent of your day spent on data visualization?
50%

Biggest frustration(s)
Editing from non data-viz experts on the data-viz end

5.

Were you hired to do data visualization only?
No
What focus is data visualization in your work?
Secondary
Percent of your day spent on data visualization?
40%

Biggest frustration(s)
Executive buy in is spotty

6.

Were you hired to do data visualization only?
No
What focus is data visualization in your work?
Secondary
Percent of your day spent on data visualization?
35%

Biggest frustration(s)
N/A

7.

Were you hired to do data visualization only?
No
What focus is data visualization in your work?
Secondary
Percent of your day spent on data visualization?
25%

Biggest frustration(s)
Antiquated data architecture

8.

Were you hired to do data visualization only?
No
What focus is data visualization in your work?
Secondary
Percent of your day spent on data visualization?
20%

Biggest frustration(s)
Tooling

9.

Were you hired to do data visualization only?
Yes
What focus is data visualization in your work?
Secondary
Percent of your day spent on data visualization?
20%

Biggest frustration(s)
Still working out the most efficient way to publish the vis

10.

Were you hired to do data visualization only?
No
What focus is data visualization in your work?
Secondary
Percent of your day spent on data visualization?
10%

Biggest frustration(s)
Messy and poorly documented DB

11.

Were you hired to do data visualization only?
No
What focus is data visualization in your work?
One of several
Percent of your day spent on data visualization?
5%

Biggest frustration(s)
Our data viz efforts are not of uniform quality-don't really have folks focused on maintaining rigor across the board.

12.

Were you hired to do data visualization only?
No
What focus is data visualization in your work?
One of several
Percent of your day spent on data visualization?
5%

Biggest frustration(s)
people asking for reports and not regularly utilizing them

Fig.10.9

The final visual tool for exploring the 2017 Data Visualization Community Survey.

MYT
LEG

IS &

ENDS

Figures in the Sky

NADIEH

This project took a *long* time to figure out topic-wise and to create; I even finished the next scheduled project (about "Fearless") before this one. There were several avenues that Shirley and I investigated (Cinderella, Disney), but they didn't pan out. And so, many months after my previous project, while at OpenVisConf in Paris, I decided to look for completely different ideas. The talks definitely inspired me, especially one about Google's Quickdraw dataset. I thought, maybe I'd make something about the "mythical" creatures from the Quickdraw word list and how they're drawn, like dragons and mermaids? Something about dragons in general? Or about myths from many different cultures and their timelines and similarities? Unfortunately, that would probably mean a lot of manual data gathering. But myths across cultures ... that suddenly reminded me of constellations! Many constellations have been named after characters from certain myths and legends. My favorite constellations are Orion and the Swan (officially known as Cygnus). But what did other cultures make of those same stars? What shapes and figures did *they* see in the same sky?

> **That idea sparked a feeling of enthusiasm and wonder in me in such a way that I knew it felt right.**

As an astronomer, it also felt kind of appropriate to have my final *Data Sketches* project to be connected to actual stars.

Of course, that idea still hinged on data availability. I thought that the subject I had chosen would be specific enough for Google. But alas, trying to search for constellation data was heavily intermixed with astrology. (—_—*;)

I found some promising information about the "modern" 88 constellations, but nothing about constellations from multiple cultures. That is, until I came across Stellarium, an amazing open source 3D planetarium software and all its data can be accessed on GitHub. The giant cherry on the cake is a folder[1] called "skycultures" which contains information on constellations from ±25 different cultures from across the world, including Aztec, Hawaiian, Japanese, Navajo, and many more. This data was *exactly* what I needed, but it wasn't available in a simple CSV format, nor in the shape that I wanted for my visualization. Luckily, Stellarium has a very extensive user guide[2] that explains exactly how to interpret the data.

For example, Figure 11.1 shows the data to create "stick figures," or the lines between stars. Each row is one constellation, with the constellation's ID listed in the beginning, followed by the number of connections (lines) in the constellation. After that come the so-called Hipparcos (HIP) star IDs. Each pair of HIP IDs defines a line between those two stars.

I converted these files into something very similar to the typical *links* file of a network, with a `source_id` and `target_id` per row, having one row for each line to draw in the stick figure/constellation.

I pulled the full names of the constellations from a different Stellarium dataset and created another file that contains all the constellation IDs that a specific star is connected to. However, there was still one important "subject" that I was missing in terms of data: the stars.

Thankfully, that's a dataset I'm already familiar with and have used in a few other astronomer themed visualizations. The HYG database[3] contains lots of information about many, many stars. From that database I took the *right ascension* and *declination* so I could place the stars on a map (you can think of these as the *latitude* and *longitude* of the sky). But I needed more information, such as the HIP ID, to connect them to the constellation data from Stellarium. I also found the *apparent magnitude*, which represents how bright the star looks, to use as a star's size, and finally the star's *color index* to get an effective temperature, which could then be mapped to a color for the stars. It would be a shame not to color the stars the way they actually appear to us.

The HIP ID was the unique key that made it possible to link the constellation data from Stellarium to the star data from the HYG database.

```
2   And 5  677 3092 3092 5447 9640 5447 5447 4436 4436 3881
3   Scl 3  116231 4577 4577 115102 115102 116231
4   Ara 7  88714 85792 85792 83081 83081 82363 82363 85727 85727 85267 85267 85258 85258 88714
5   Lib 5  77853 76333 76333 74785 74785 72622 72622 73714 73714 76333
6   Cet 20  10324 11484 8102 3419 3419 1562 3419 5364 5364 6537 6537 8645 8645 11345 11345 12390 12390 12770 12770 11783 1178
7   Ari 3  13209 9884 9884 8903 8903 8832
8   Sct 5  92175 92202 92202 92814 92814 90595 90595 91117 91117 92175
9   Pyx 2  42515 42828 42828 43409
10  Boo 9  71795 69673 69673 72105 72105 74666 74666 73555 73555 71075 71075 71053 71053 69673 69673 67927 67927 67459
11  Cae 2  21060 21770 21770 21861
12  Cha 2  40702 51839 51839 60000
13  Cnc 5  43103 42806 42806 40843 42806 42911 42911 40526 42911 44066
14  Cap 9  100064 100345 100345 104139 104139 105515 105515 106985 106985 107556 105515 105881 105881 104139 100345 102485 10
15  Car 14  45238 50099 50099 52419 52419 52468 52468 54463 54463 53253 53253 51232 51232 50371 50371 45556 42568 41037 41037
16  Cas 4  8886 6686 6686 4427 4427 3179 3179 746
17  Cen 16  71683 68702 68702 66657 66657 68002 68002 68282 68282 67472 67472 67464 67464 65936 65936 65109 67464 68933 67472
18  Cep 6  109492 112724 112724 106032 106032 105199 105199 109492 112724 116727 116727 106032
19  Com 2  64241 64394 64394 60742
20  Cvn 1  61317 63125
```

Fig.11.1

Stellarium's data to create "stick figures" between the stars.

[1] Stellarium skycultures folder: https://github.com/Stellarium/stellarium/tree/master/skycultures
[2] Stellarium user guide: https://github.com/Stellarium/stellarium/releases/download/v0.19.1/stellarium_user_guide-0.19.1-1.pdf
[3] HYG database: https://github.com/astronexus/HYG-Database

To stay true to how we see the night sky, I filtered the stars to only include those that are bright enough to be seen by the naked eye, which is an apparent magnitude smaller than 6.5.

In astronomy the smaller the apparent magnitude, the brighter the star appears to us.

In addition to sky maps with constellations, I also wanted to display something more "statistical" using a bigger set of data. What sparked my interest was seeing how the data looked when I plotted a star's brightness versus the number of constellations that each star is used in. Was there a trend? If so, which stars deviated and why? I made a quick plot in R using `ggplot2` (Figure 11.2) that revealed some interesting insights, specifically, insights around which stars deviated from the general trend of "the brighter a star, the more constellations that use it."

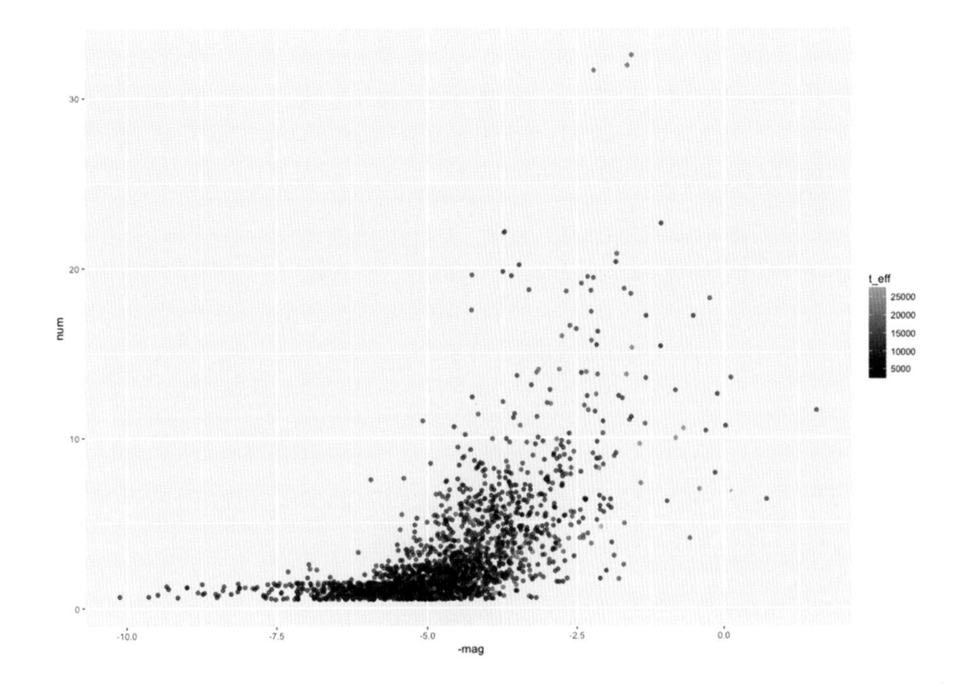

Fig.11.2

A scatter plot made in R showing apparent magnitude versus the number of constellations a star is part of, for approximately 2,200 stars.

However, while investigating this scatter plot more closely, I noticed that my star data was missing many proper star names. Almost all nine stars of the Pleiades were not named! I searched for a bigger list of named stars and found a sort-of official list of ±350 stars on Wikipedia.[4]

By proper names of stars I mean their popular/common names instead of their catalogue IDs, such as the star names Betelgeuse and Sirius.

However, these only contained the names themselves, not the HIP IDs needed to connect them to my data. Thankfully, there is a website called the Universe Guide[5] where the URLs are based on the star's name, while the page itself contains the HIP ID in the HTML's `h1` header (title) of the page. I therefore used the `rvest` package in R to download the Universe Guide page of all of the stars on the wiki list, grabbed the `h1` from the HTML, and only kept the HIP `id`. I only had to do a few manual lookups for names that didn't return results from the Universe Guide through my script. Finally, I merged this "proper star names" dataset into the original HYG dataset for a much more complete set of star names.

I copied the Wikipedia list of 350 star names into Excel using its "data from web" import option.

A final note about the data: there are no officially declared constellation figures. There are indeed 88 official constellations, but the only thing that is recorded is what area of the sky that constellation takes up (kind of like how the US states divide up the land). There is no official consensus on how the stick figure part of the constellation should be drawn. I've therefore decided to use the data from Stellarium as my "single source of truth."

[4] List of proper names of stars: https://en.wikipedia.org/wiki/List_of_proper_names_of_stars
[5] Universe Guide website: https://www.universeguide.com/star/atlas

Sketch

This project was pretty light on actual "design" sketching. That's mostly because the basic idea was quite simple: I would focus on one star and visualize all the constellations that use that star. I created a sketch that surrounded a star with a donut-like mini chart that would show the constellations it appeared in. You can see the tiny sketch from my brainstorm in Figure 11.3 (bottom of the left page).

That's three projects in a row with very little sketching—definitely unusual for me!

I wanted the star map to look like a combination between current and ancient sky maps. An example of the latter is the exquisite illustration made by the author Alexander Jamieson in 1822 (see Figure 11.4).

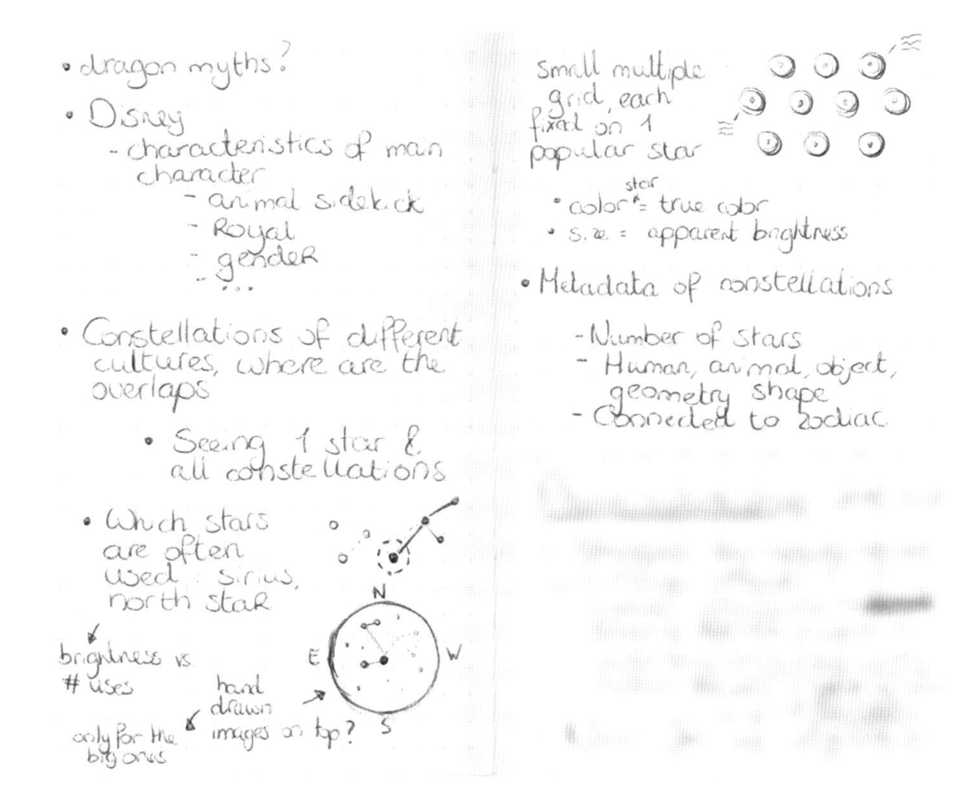

Fig.11.3

Brainstorming concept ideas and data to use for this project.

Another part of the project that took up more pages in my little notebook was math (as usual). If there were multiple constellations with a line between the same two stars, I wanted to draw those lines *side by side.* I started out thinking it would require some trigonometry, with four different cases. In Figure 11.5 you can see how I tried to conclude if there were different solutions for the four orientations. However, it eventually all came down to a little vector math; I had to find the *normal vector*, which is simple to calculate, and I needed each new line from a constellation to move up a little farther along the normal vector. The red sort-of-circle in the middle-left of Figure 11.5 b shows when I finally realized the normal vector was all I needed.

The four were cases being defined by how the target star was positioned relative to the source star, in the top-right quadrant, or any of the other three quadrants.

The normal vector lies perpendicular to the line you're focusing on, making a straight angle.

↳ Learn to Love Math

Math is truly your best friend in creating more unique data visuals. And although it's usually trigonometry, this time knowing just a little bit of linear algebra/vector math helped turn a rather difficult problem into a straightforward approach.

Fig.11.4

Plate 1 from A Celestial Atlas by Alexander Jamieson from 1822.

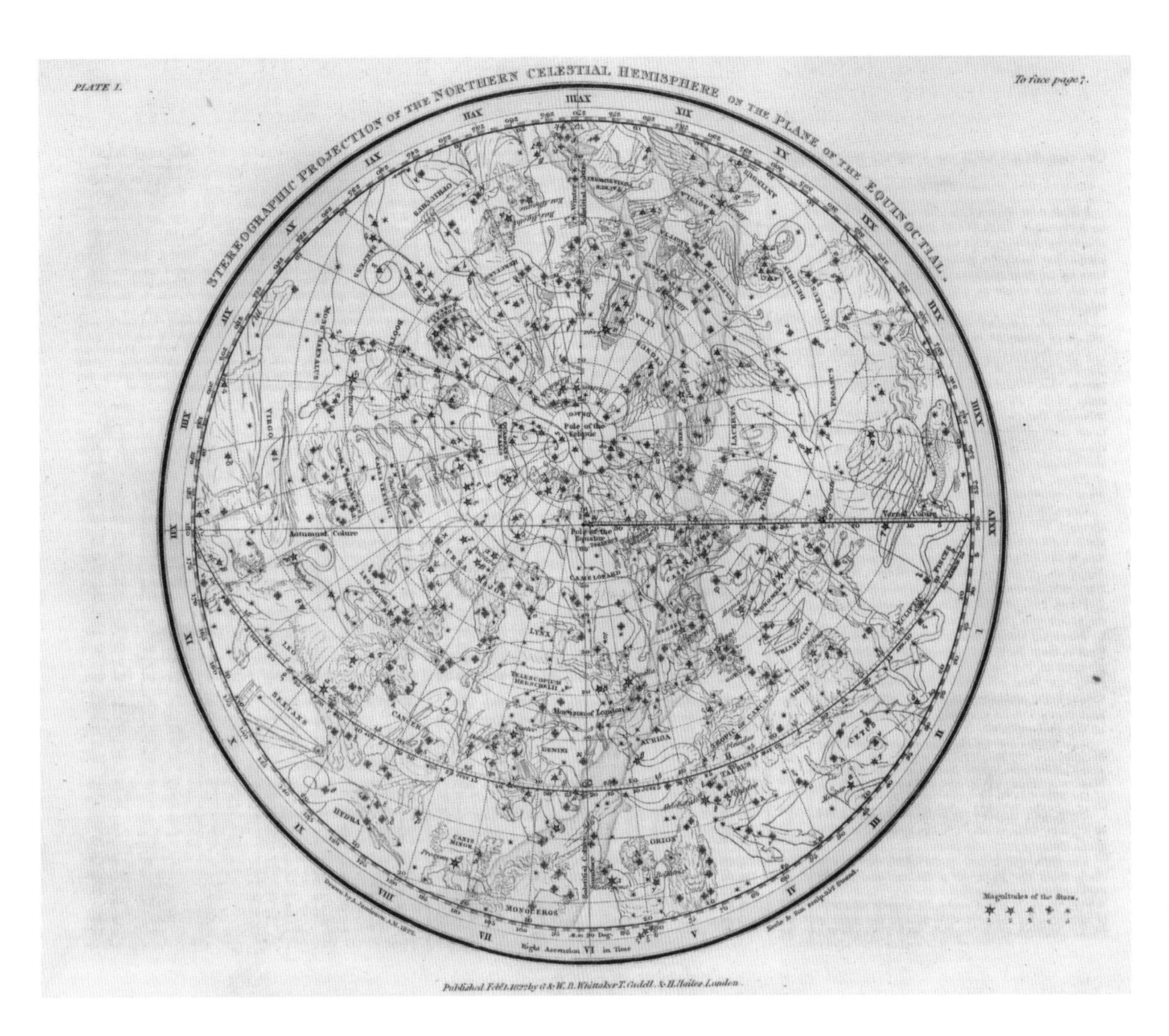

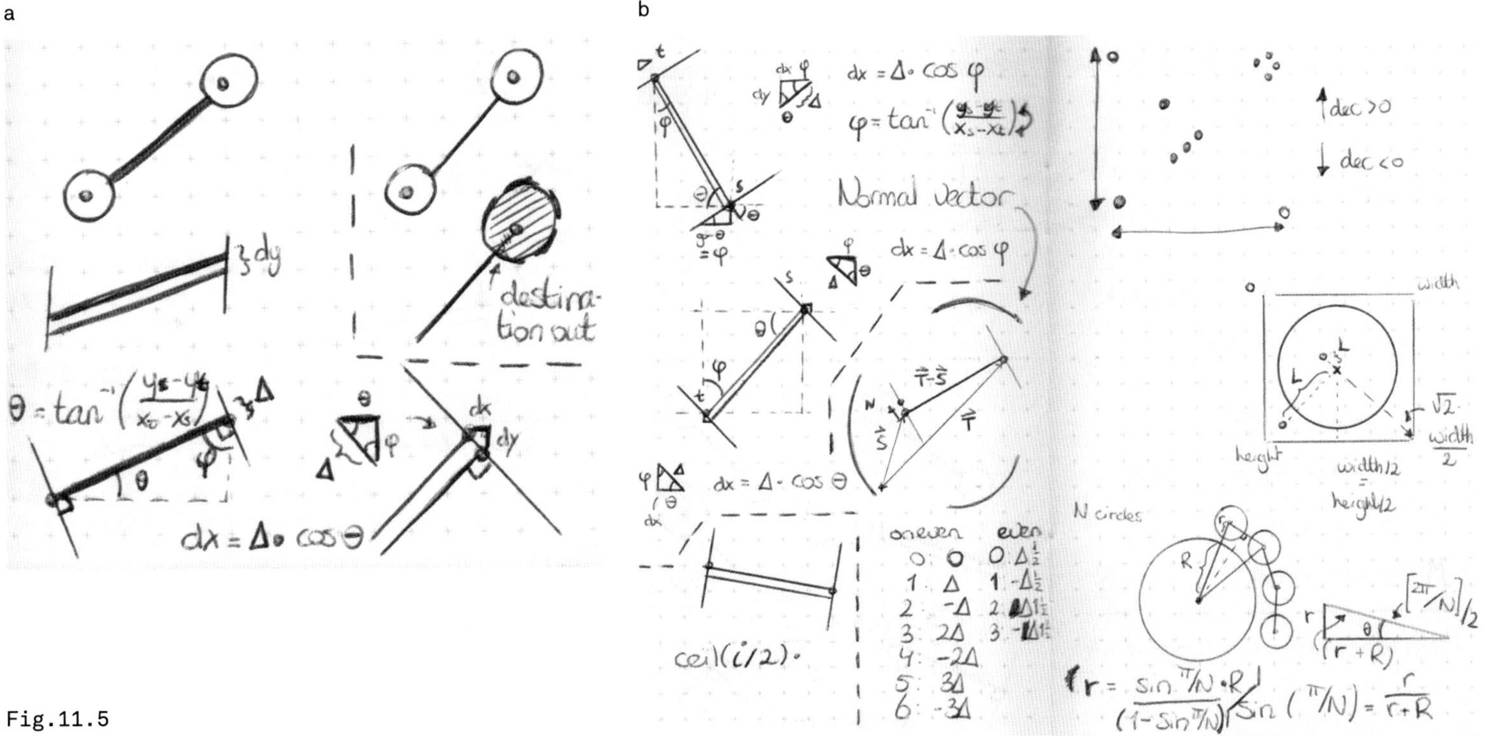

Fig.11.5 (a & b)

Sketches about figuring out how to get lines next to each other.

Finally, I made some sketches of the general page layout. The title would stand out nicely against a background of the night sky. With the sky maps themselves already providing more than enough aesthetically, I wanted to simplify the other elements of the text and layout.

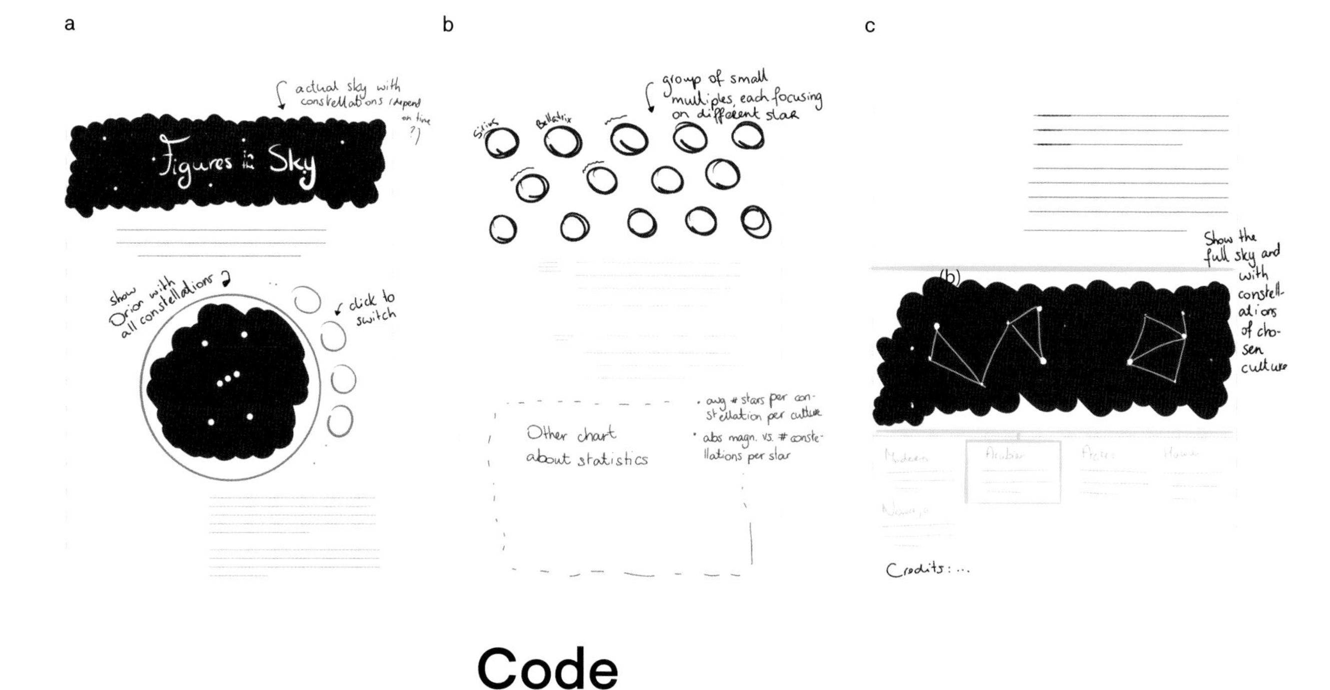

Fig.11.6 (a,b,c)

Sketching out the layout of the page in later stages of the project.

Code

My first goal, before focusing on the *actual data* visualization side of things, was to create a "base map" of the sky. I've never created one, so I did a little research on what kind of map projection is typically used for sky maps. (I decided to go with a *stereographic* one.)

I never truly worked with projections before outside of the default Mercator projection.

Given this map would include 9,000 stars for the full sky, mini donut charts, and many constellation lines, I set out to create this project with canvas due to its better performance. I loaded my star data and set up my code following several `D3.js` based examples of sky maps.[6] But all my code produced was a thin stripe of stars. ಥ﹏ಥ (see Figure 11.7 a).

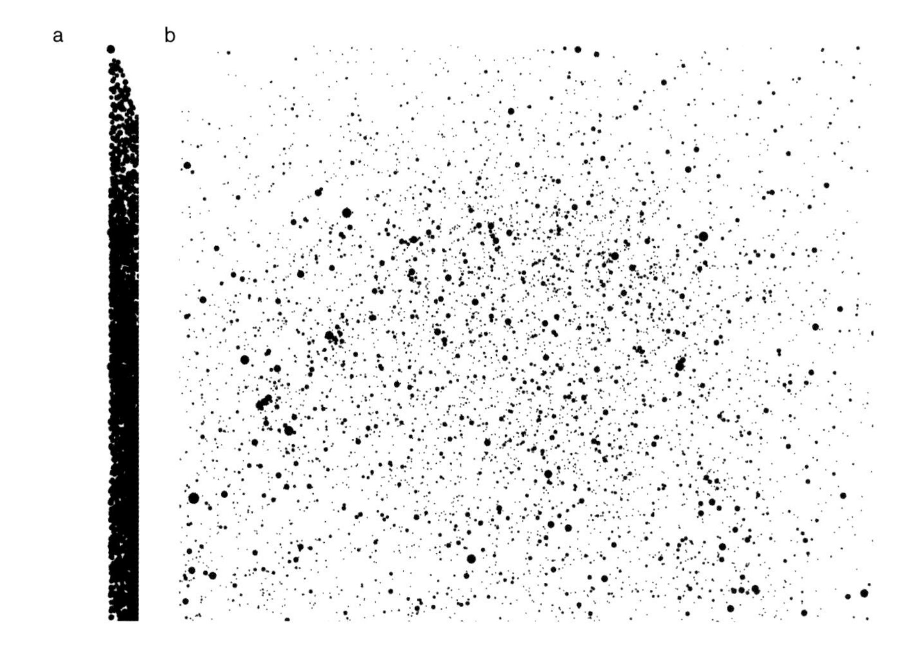

Fig.11.7 (a & b)

The very first results on screen, failing first, but getting it right the second time.

After a more careful comparison of the other sky map examples and mine, I realized I forgot an easy-to-miss transformation calculation of the RA (right ascension) and declination. A few code adjustments later and I had the map from Figure 11.7 b. However, it was still too abstract for me to see if it was correct or just a random collection of points. I felt that the one thing that would probably help me realize if the stars were correctly plotted was to add the *graticule lines*, the background grid. Thankfully the `D3.js` based sky map examples helped out again. When I rotated the projection to face North, added the stick figure lines for the modern constellations, and even recognized a few, I *finally* knew for sure that the stars were in the correct location (Figure 11.8).

For the rest of the sky map, I focused mostly on the central star of Orion, my favorite constellation for a variety of reasons: it contains many bright stars, is easy to pick out, and I could look at it from the living room window of my childhood home for many winters.

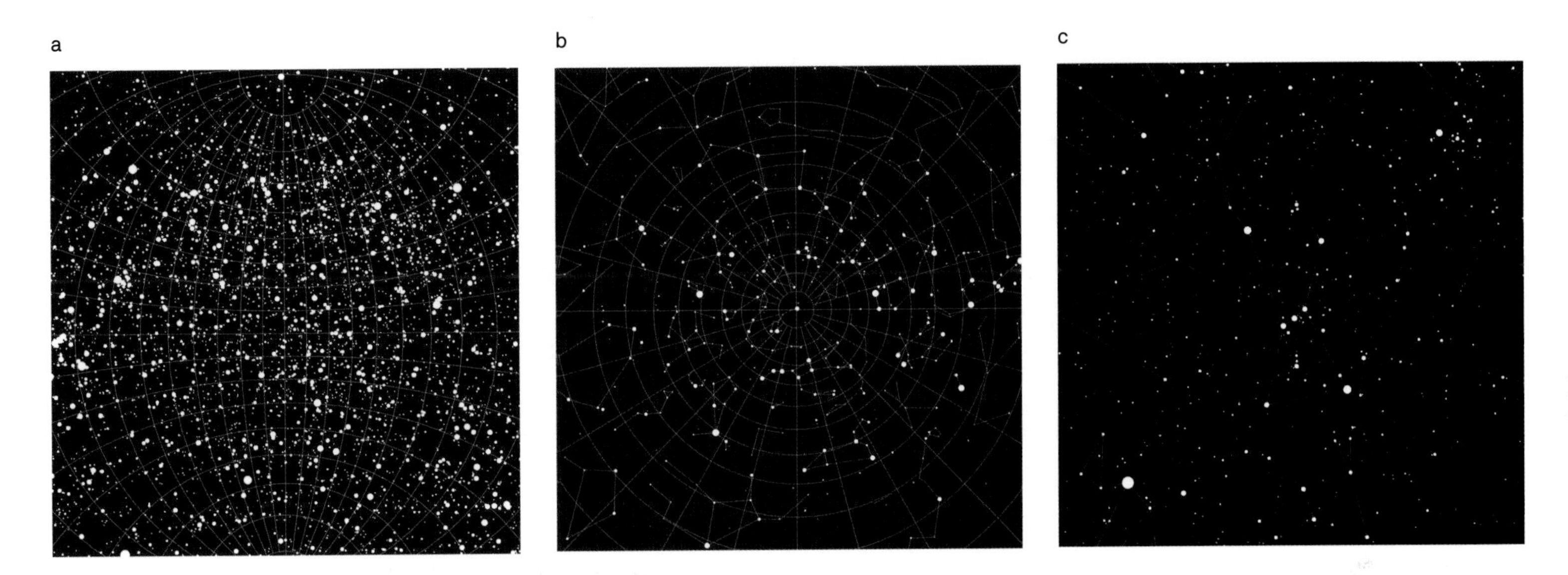

Fig.11.8
(a,b,c)

Steps in setting up the sky map, such as adding a background graticule, rotating it to face North, and zooming in on Orion.

↳ Remix What's Out There

Mentally, it's sometimes easier to get started on a project when you don't have to do everything from scratch. Even though the code for the `D3.js` based sky map examples wasn't very long or intricate, I felt it was just the right thing for me to use as a base to start working from. Beginning from some basic code examples can help unblock you and allow you to build on something. And I don't mean just plug your own data into the base code and calling it finished! I mean to truly remix it and turn it into a new visual that takes the quirks of your data into account and has its own style. (Think "inspired by" instead of "cheap knock-off".)

[6] Star Map by Mike Bostock: https://observablehq.com/@mbostock/star-map and Sky by Matteo Abrate: http://bl.ocks.org/nitaku/9607405

As you can see in the images of Figure 11.7 and 11.8, I was already using the magnitude to scale the radius of each star; the brighter they appear to us, the bigger the circle. Now it was time to look at the *colors* of these circles. I started with a temperature-to-color-scale I'd already investigated and developed to be very similar to the star's actual perceivable colors for a previous astronomy-related project. For a while I played around with making the colors more vibrant, but that made the sky *way* too colorful.

Fig.11.9

A much too saturated colorful sky.

But even with the more real and nuanced colors and adding a glow to each star (with canvas' useful `shadowBlur` property), the bigger stars looked a bit flat. In reality, our Sun looks a bit brighter in the center and dimmer around the edge. Luckily, canvas has the option to create radial gradients, which I could use to set up a unique gradient for each star: I'd have a base color for each star using a slightly lighter color in the center and a darker color at the edge (I used `chroma.js` to create the lighter and darker color from the base color). It took some experimentation to figure out the best settings. Figure 11.10 a, where I made the stars bigger to better assess the gradients, was *definitely* not correct.

Fig.11.10
(a & b)

Trying to create a unique (sometimes tiny) gradient for each star

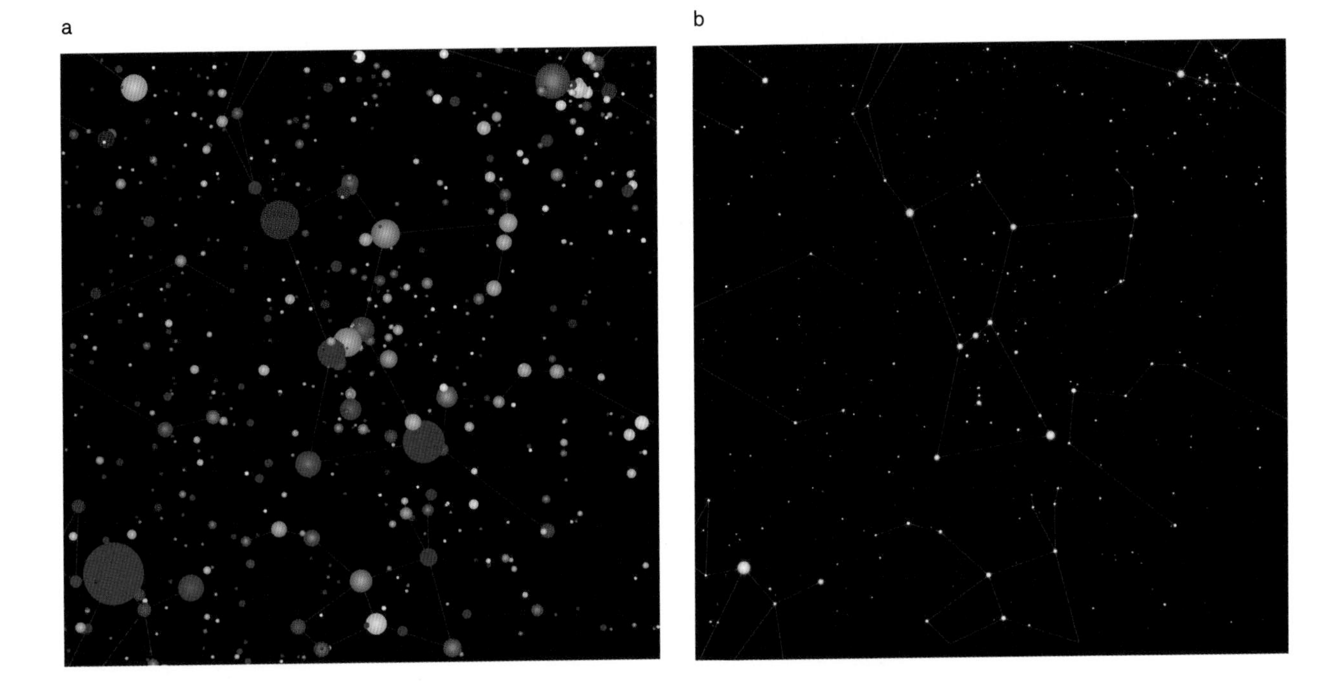

I clipped the sky map to a circle with a dashed line around it and added a separate dashed line inside the map to show the so-called *ecliptic*, for embellishment. This is the path that the Sun makes across the sky relative to us. Because our Earth's rotational axis is tilted by 23.44° with respect to the plane in which the planets go around the Sun, the ecliptic is not a straight line across the image (following the 0° declination line).

For the points where the background lines touch the sky map's edge I added the RA and declination degrees. I really liked the use of the zodiac signs in Alexander Jamieson's map from Figure 11.4 and replaced the degree number by the corresponding zodiac signs for 12 "major-RA" lines.

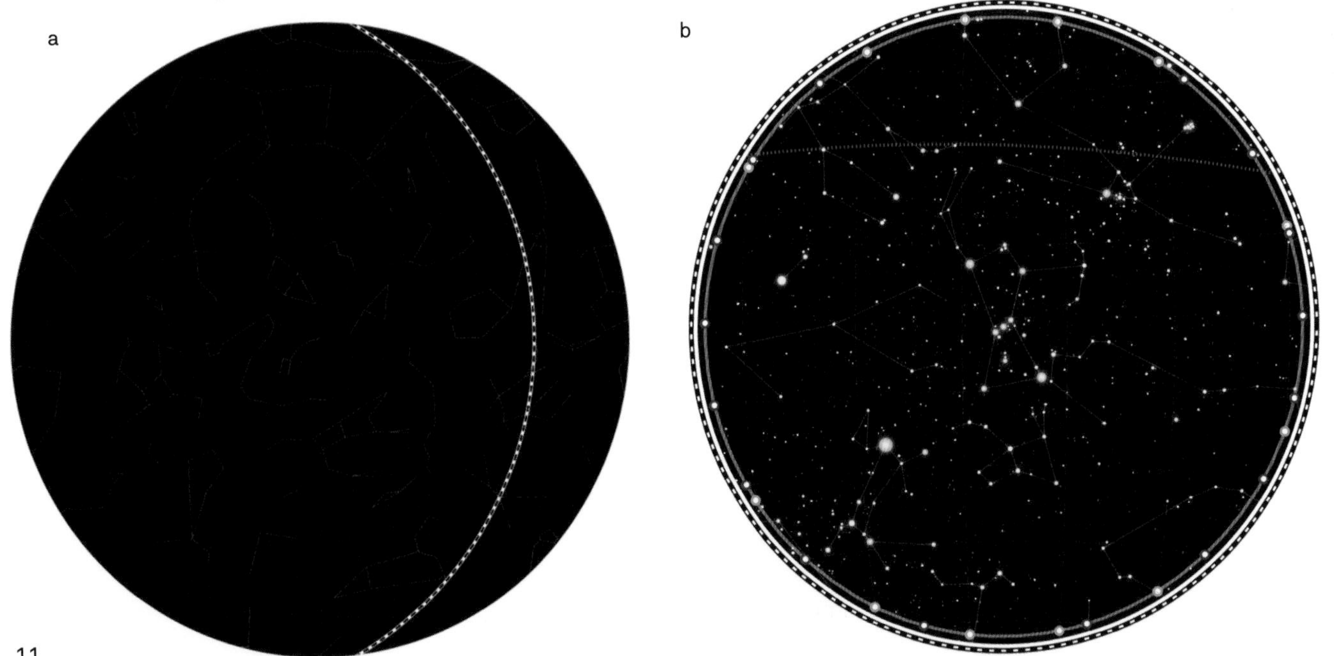

Fig.11.11
(a & b)

Clipping to a circle and adding the ecliptic path on the left, while calculating where the background grid touches the circle boundary for degree notations and zodiac signs on the right.

Having finished with the lines, it was time for the final big aspect that I had in mind for the sky map base: space itself. The background color ("space") up until this point was a dark blue. I really wanted to add some depth to each image by mimicking lighter colored streaks across the background, inspired by the gorgeous streaks of the Milky Way. Not long before, I'd seen some great experiments and bloopers from a friend that was using the contouring options of `D3.js` and remembered that it gave *exactly* the kind of feeling I was looking for with my "outer space." I started to experiment with the contour functionality and, after several iterations, built something I was happy with (see Figure 11.12)

A North and South pointer as the final ornamental element and my night sky base map was done! (See Figure 11.13) I was finally ready for the *actual* data visualization part of this project. (๑•̀ㅂ•́)و✧

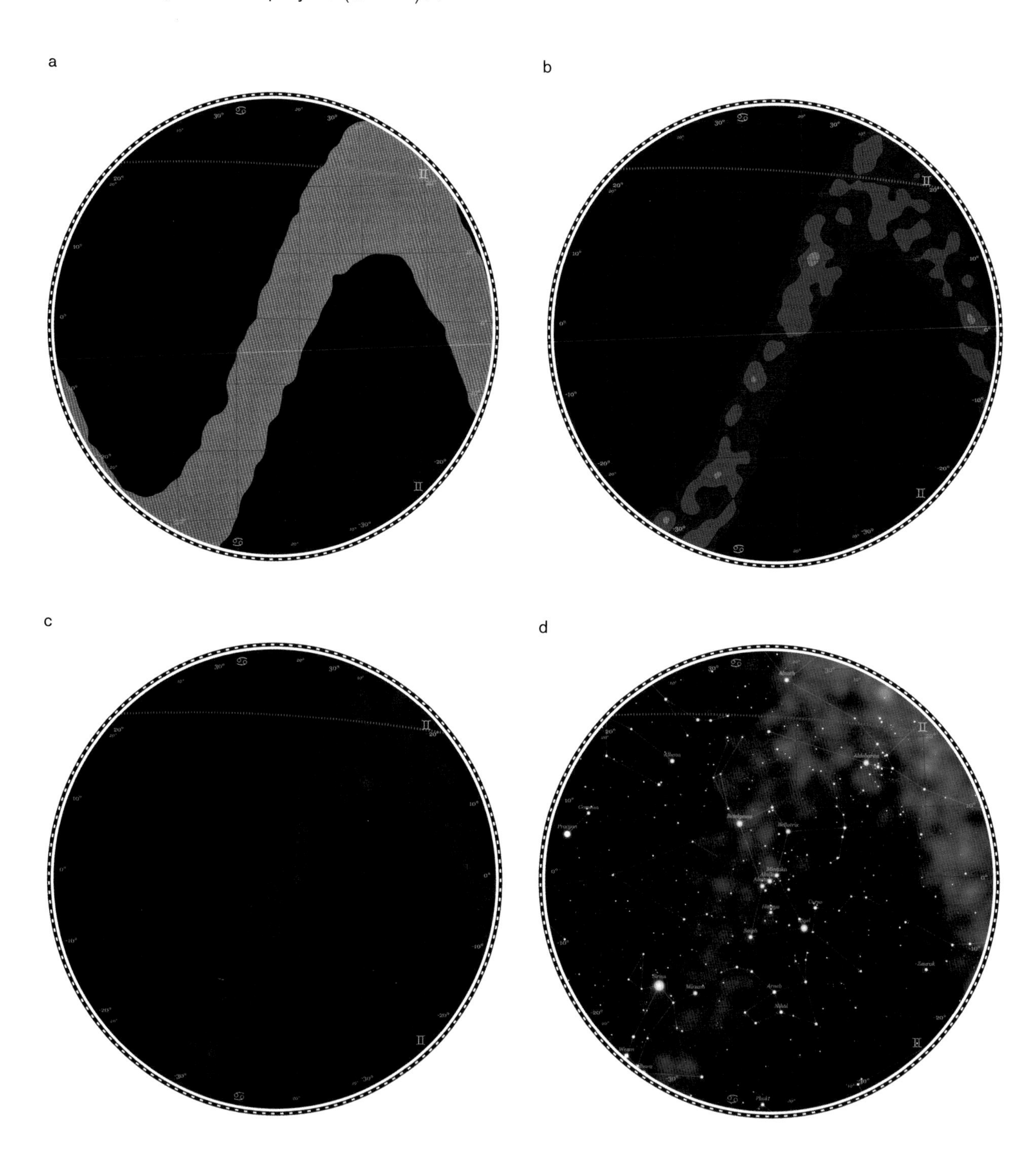

`Fig.11.12 (a,b,c,d)`

Slowly building up a "swirly background" using the contour function of `D3.js`.

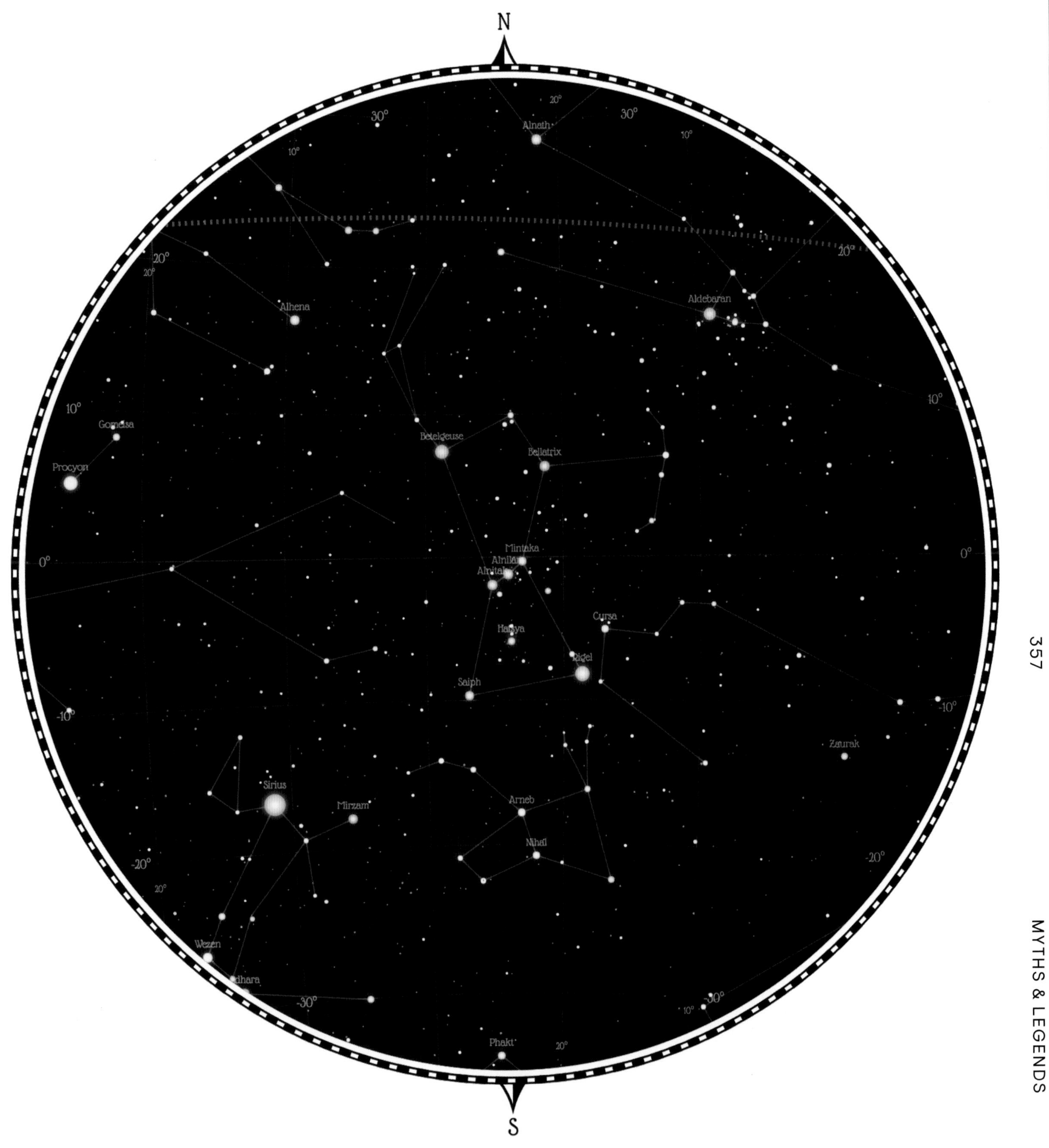

Fig.11.13

The final result of the base sky map. Now the data visualization would still need to be overlayed.

The end goal was to visualize how many constellations used each star. I started with creating a small donut chart around each star that was part of a constellation. Even though one star would be chosen to focus the visualization on, many neighboring stars would also be included in the different constellations. I first created simple donut charts featuring only white slices. When that was working and looked good I turned it into a colored version with rounded edges and a bit of padding.

I chose to switch and focus on the brightest star in the night sky, Sirius.

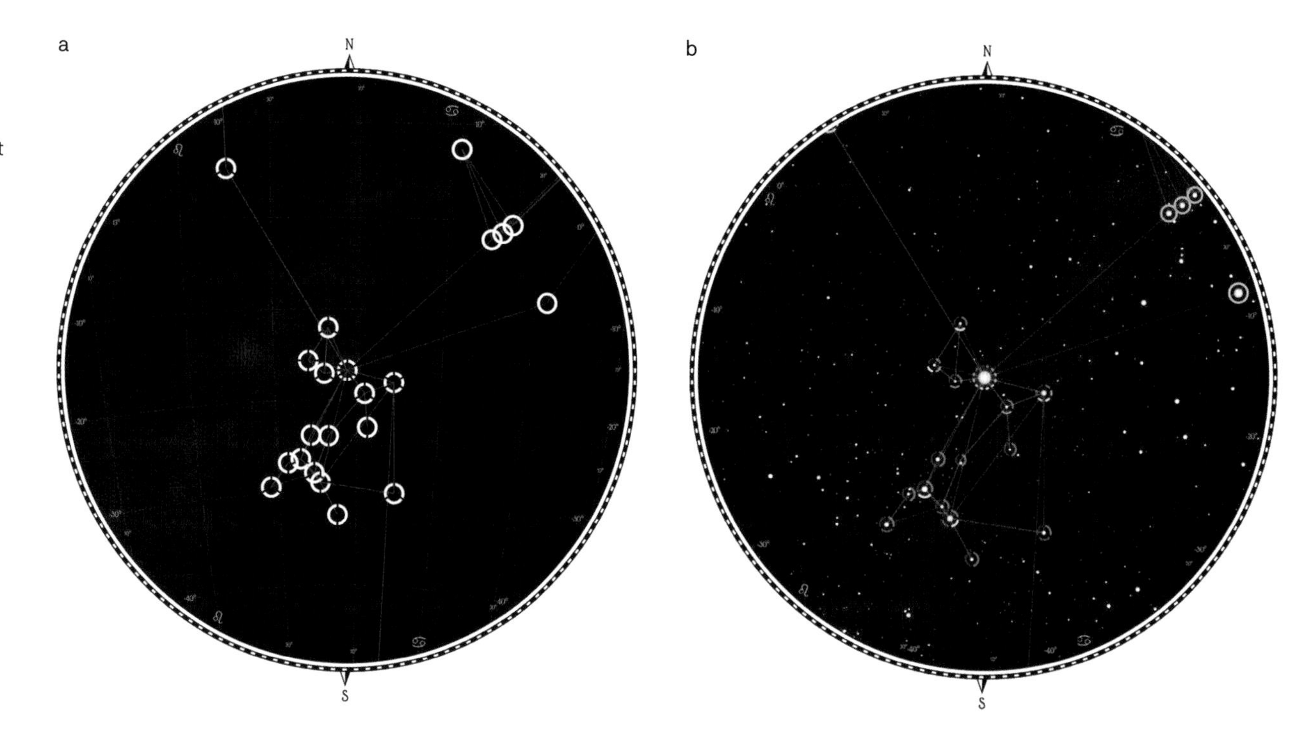

Fig.11.14 (a & b)

Creating a mini donut chart around each constellation star.

The lines in between the stars were a bit of a different story. I wanted these lines to be placed alongside each other, but I only had the exact center location of each star, so calculating the offset in the x- and y-direction that each extra line would need wasn't trivial, I thought. Until I finally remembered not to think in trigonometry, which created the wrong image of Figure 11.15 a, but in vectors and the *normal* vector (resulting in Figure 11.15 b).

You could already see my (mostly useless) math in the "Sketch" section earlier.

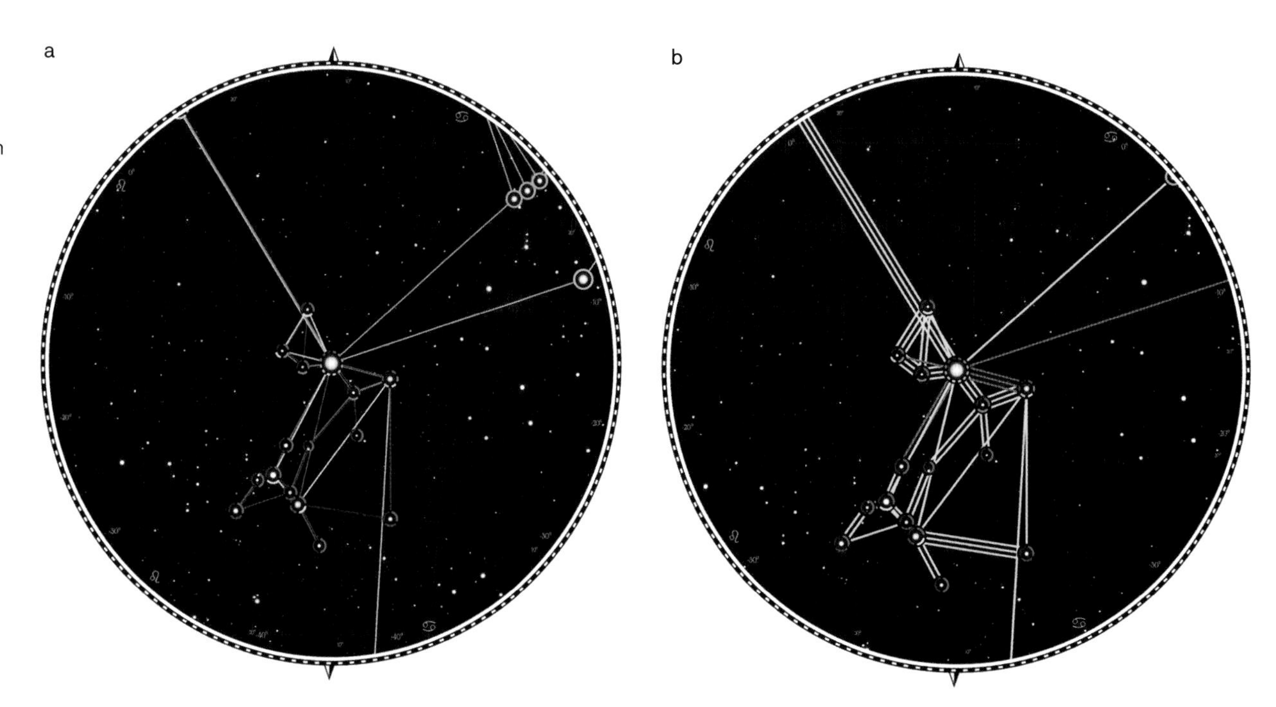

Fig.11.15 (a & b)

Positioning the lines to run parallel between the stars—doing it wrong at first and corrected eventually.

Fig.11.16
(a & b)

Focusing on (the white circle) Betelgeuse (in the modern constellation of Orion) and on Sirius in a star-line from Hawaiian culture.

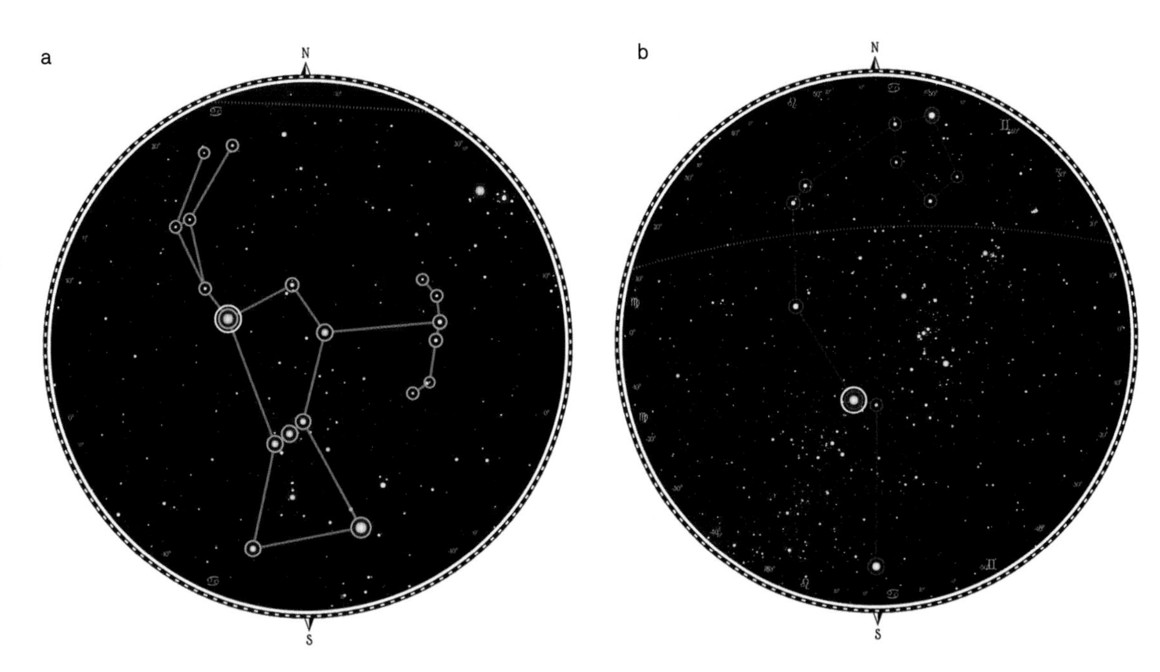

Working with the different constellations showed me that using one particular zoom level and center would definitely not work to properly reveal each of the separate constellations. It took me a while to finally figure out the logic and to automatically calculate the optimum zoom level, rotation, and center that would nicely fit any constellation that I would give the program (see Figure 11.16).

To make it easier to select and see the full shape of each separate constellation (using the same star), I added all of the separate constellations in a ring around the version that showed all of the constellations at once.

Fig.11.17

The separate constellations in a ring around the main sky map.

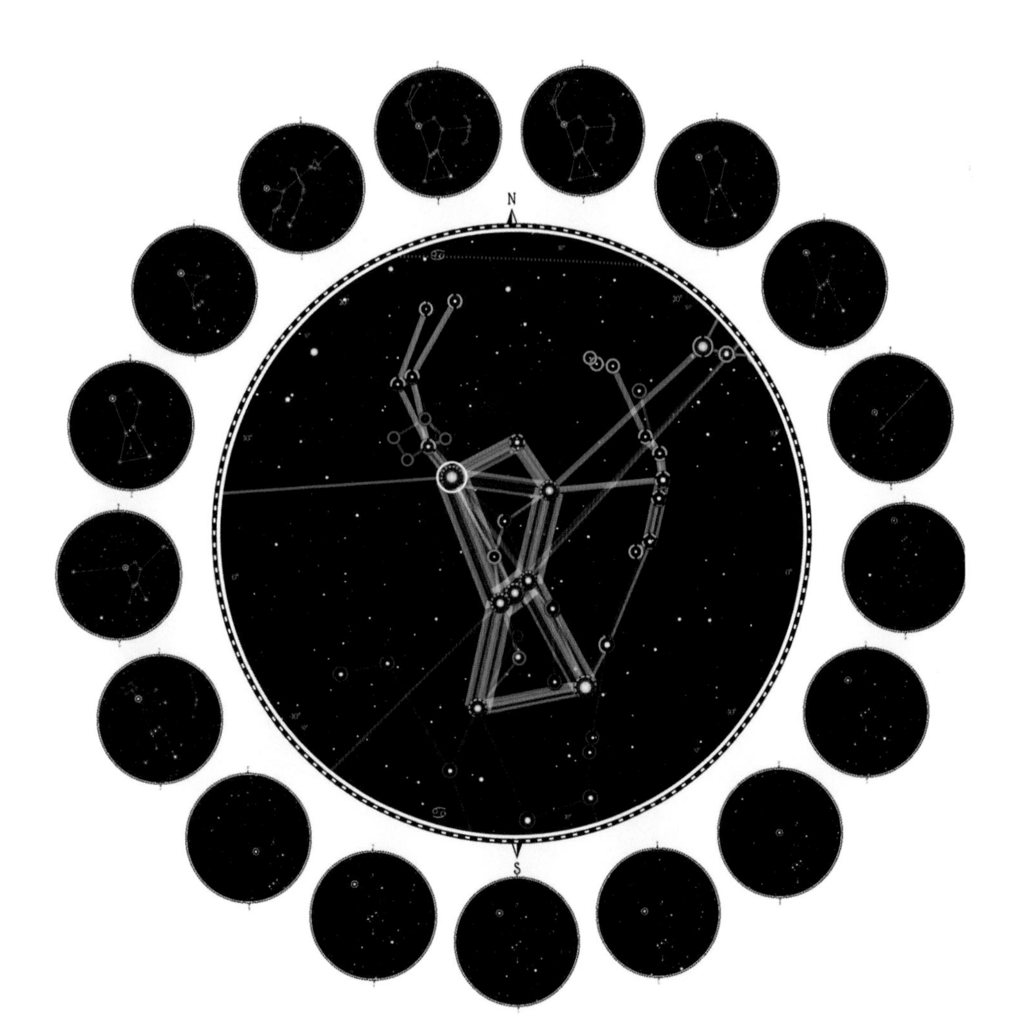

That immediately showed me two things. For one, this was *excruciatingly* slow! But also, the complete sky map on the mini circles wasn't needed at all. They were too small to really have any visual effect, and they were too distracting from the central map. Luckily, removing elements from the mini maps would also make them faster to load. The main thing to see were the constellation shapes anyway, which was the most performant part of the sky map's three layers: the glowing stars, the constellation lines and donut charts, and the entire background. After some fiddling around, I got it all working and ended up with the version from Figure 11.18. Finally, I added an interaction that allowed visitors to click any of the outside mini circles to see it drawn properly, and bigger, in the center (see Figure 11.30 for an example of selecting an outside mini circle).

I could've stopped there. The sky map was a complete visualization in its own right. But just showing the star Betelgeuse felt so incomplete! I had so much more data that I could use to tell a fuller and more interesting story. So even though I had already racked up way too many hours to get to this point, I decided that this project would become a complete article; with beginning, middle, and end. (In other words, *even more visuals*. (◍•﹏•))

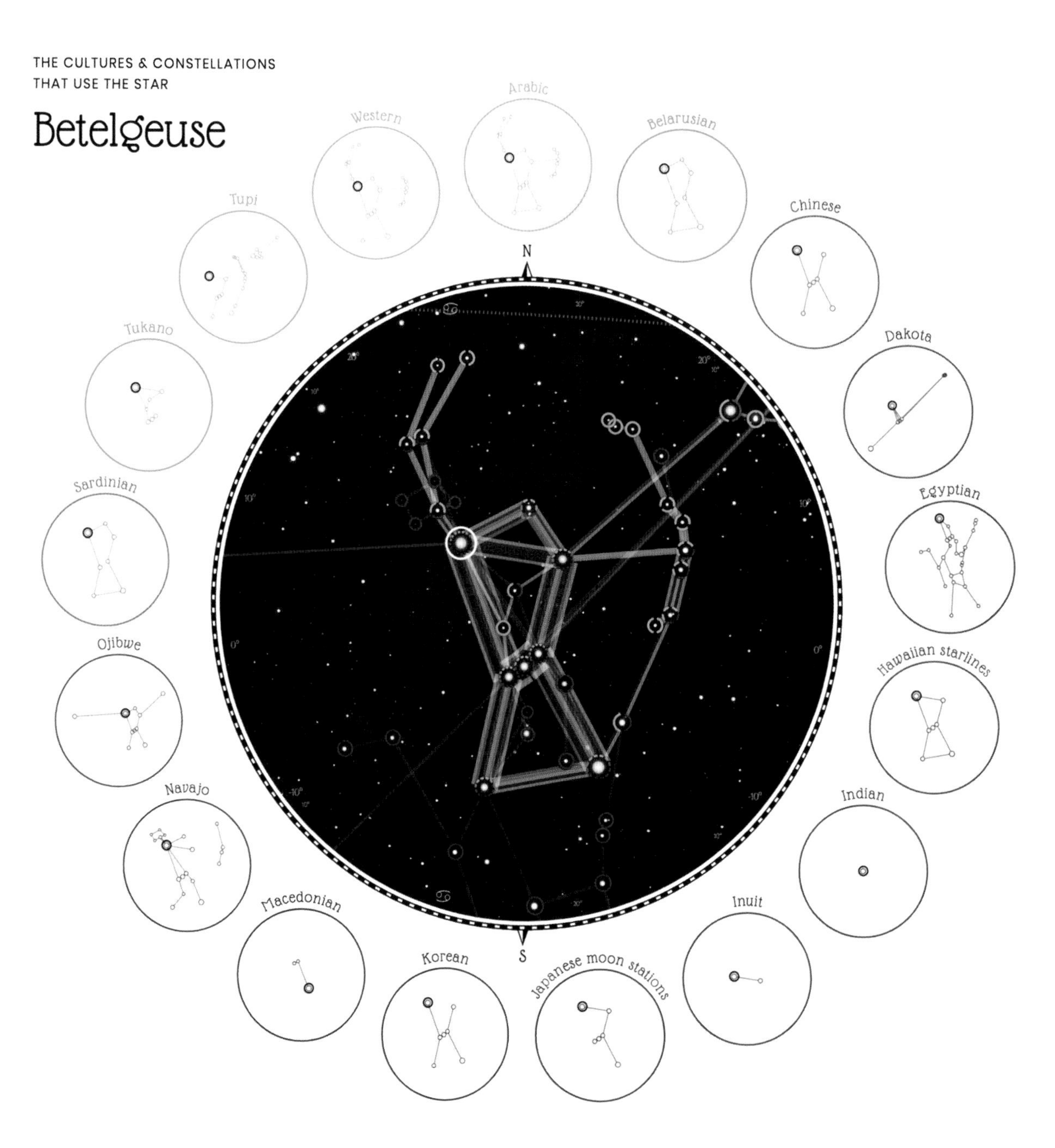

Fig.11.18

The final look of the circular sky map with the mini circles around it.

An (almost) full sky map that would show *all* of the constellations of one chosen culture was the first on my list. It would allow users to select a culture and view its constellations separately. Given how complete my circular sky map function was, setting up the base for this was quite easy. In essence, the only change I had to make was to adjust the projection from *stereographic* to an *equirectangular* one (while also using a different width/height and not clipping the visual to a circle). For this full sky map I made sure to have the background fuzzy patches follow the *actual* rough location of the Milky Way.

I was quite happy with the result, so I decided to use it for the header of the full article as well.

Fig.11.19

The full sky, now with the background fuzzy patches sort-of following the shape of the Milky Way.

Betelgeuse might be a fascinating star, but I wanted to reveal many more interesting stars and constellations! The function to create the full sky map with all of its constellations could be used for any star, thankfully. What *did* end up taking several hours was the exact design of these extra sky maps on the page and manually going through about a hundred stars and selecting the ±15 I thought looked the most interesting and diverse.

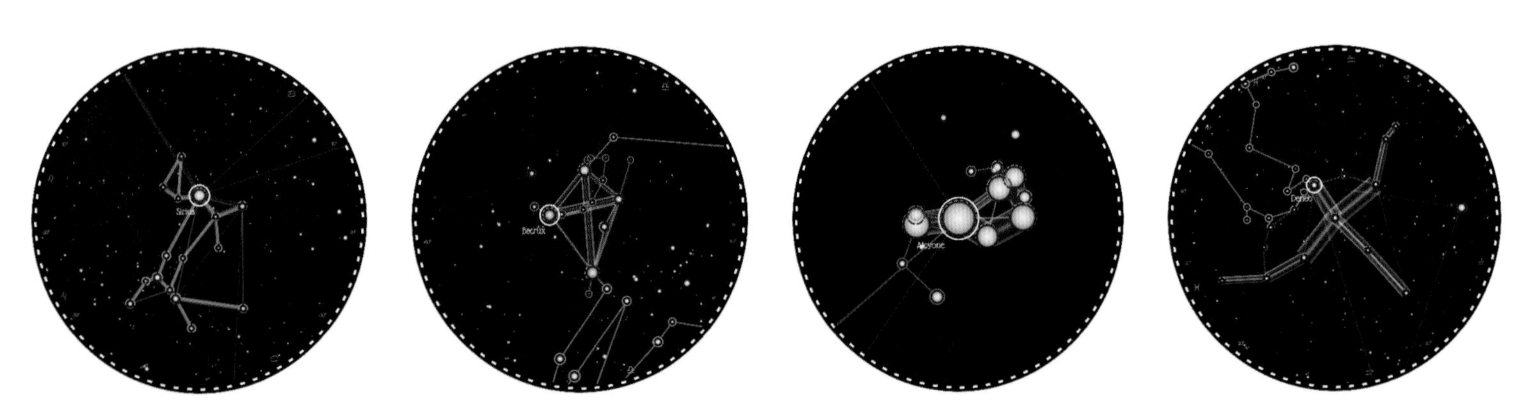

Fig.11.20

An early look of several smaller sky maps, each focusing on a different star.

The final visual pieces to add to the page were the statistical charts, starting with the scatter plot that displayed a star's brightness versus the number of constellations it was a part of. As ±2,200 stars were included in at least *one* constellation, I went for canvas as the base. However, I used a separate SVG on top for all the axes, text, interactivity, and annotations. Using canvas made it easy to reuse the same coloring of the stars as I had in the sky maps. However, with the white background those colors looked *much* too soft, and the gradient effect was too distracting (Figure 11.21 a). Removing the gradient and adding a multiply effect to darken any overlapping stars helped to make it visually more appealing (see Figure 11.21 b). However, I felt that the colors were still too soft. So I made them more vibrant, added a bit of "glow" around the edges, cleaned up the axes a bit, and the visual style of the scatterplot was done (see Figure 11.21 c). Eventually, I also added a mouse hover and textual annotations.

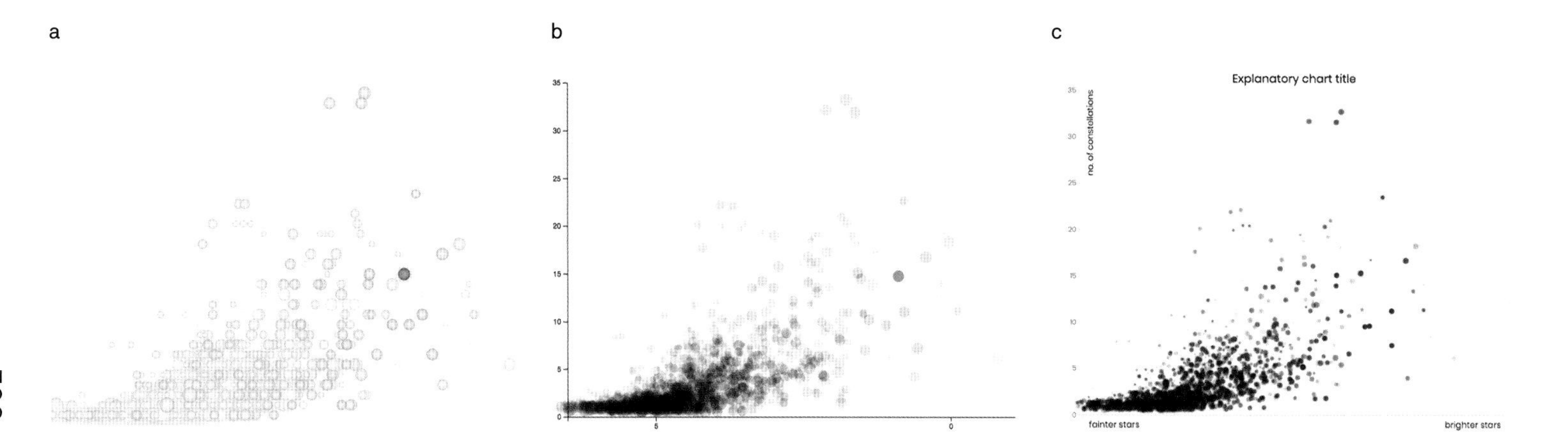

Fig.11.21 (a,b,c)

First using the same star colors and gradients as I did for the sky maps, but later going for darker colors and no gradient, and finally even more vibrant and glowing stars.

↳ Annotations Are of Vital Importance

Often overlooked, annotations are one of *the* best ways to make a chart understandable to an audience. Underutilized in many data visualizations, annotations are the ideal way to highlight *exactly* those things that you, as the creator, want the audience to pay attention to. My current go-to is the wonderful `d3-annotations` library. And I think the scatter plot from this project has the most intricate placement of annotations that I've ever applied (see Figure 11.26 later in this chapter for the fully annotated version).

At the bottom of the article I added a section that tells more about each culture. Selecting a culture results in the full sky map updating to show all of the constellations from that culture. And a bar chart that I had in mind with the average number of stars per culture eventually ended up as a small mini bar in each of the culture "boxes" (Figure 11.22).

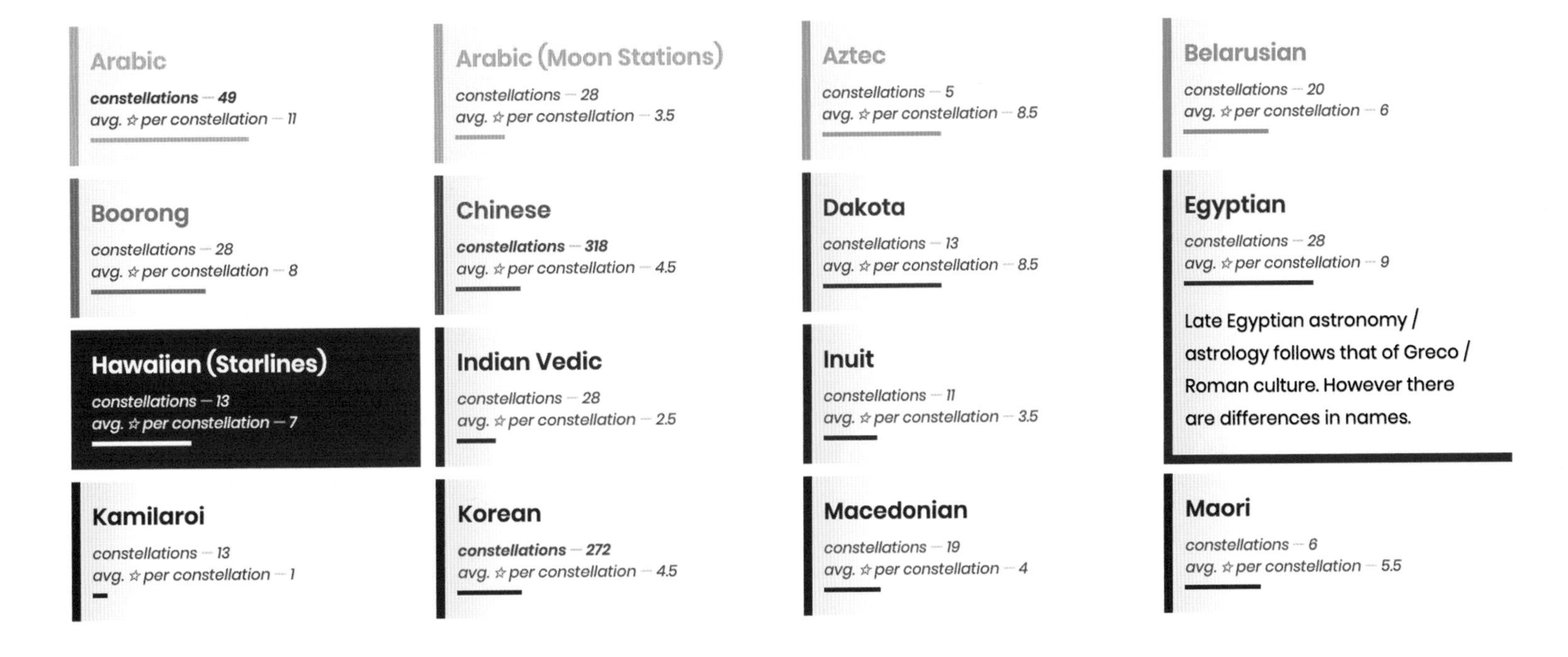

Fig.11.22

All the sky cultures in their own overview.

And then! Then I replaced as many of the visuals as I could with images. (•__•) Images are much easier to load than doing the heavy sky map calculation, and the sky maps wouldn't change anyway.

Adding in text between all the different visuals, and my second (and thankfully last) full article style data visualization was finally done!

Reflections

This was my longest project in terms of hourly investment. I clocked about 110 hours, but estimate I spent more than that, due to not always timing myself whenever I thought something would take five minutes to do, and suddenly, I was an hour in. Some parts took an unexpected amount of time to work on, such as setting up the functionality to create a base sky map that could handle any star and constellation combination. I am generally less enthusiastic about working on overall page layouts, but I spent extra time trying to perfect this layout since it was a vital part of the story.

"Beautiful in English" isn't far behind in terms of hours spent.

Even though it took so long, I'm super happy to have created a project that combines my love of astronomy with my passion for dataviz! Especially since this was, for me, my farewell to the creation of new visualizations as a part of *Data Sketches*. I'm amazed at all the things that I've learned about making data visual across the 12 topics. And it's fascinating to look back at my skills for the very first project and comparing that to the full-length article that my final project became. I'm exceptionally happy to have been a part of *Data Sketches*. It has opened doors to opportunities that I didn't even know I was looking for!

I ain't ever doing it again though. Damn, such work!! [¬°-°]¬

Figures in the Sky

↳ FiguresInTheSky.VisualCinnamon.com

Figures in the Sky

How cultures across the World have seen their myths and legends in the stars

No matter where you are on Earth, we all look up to the same sky during the dark nights. You might see a different section of it depending on your exact location, time & season, nevertheless the stars have fascinated humans across time and continents.

Our own creativity combined with stories about local legends and myths have created a diverse set of different constellations. And even though the stars don't change, people have found many different shapes in the same sky. From humans, to animals, to objects, and even abstract concepts.

The same sky, different figures

Let's compare 28 different "sky cultures" to see differences and similarities in the shapes they've seen in the night sky. Ranging from the so-called "Modern" or Western constellations, to Chinese, Maori and even a few shapes from historical cultures such as the Aztecs.

Take the star Betelgeuse. This red supergiant is one of the brightest stars in the night sky. In proper darkness, you can even *see* that it shines in a distinctly red color. It's part of one of the easiest to distinguish modern constellations known as Orion, named after a gigantic, supernaturally strong hunter from Greek mythology.

Fig.11.23

The start of the "Figures in the Sky" article.

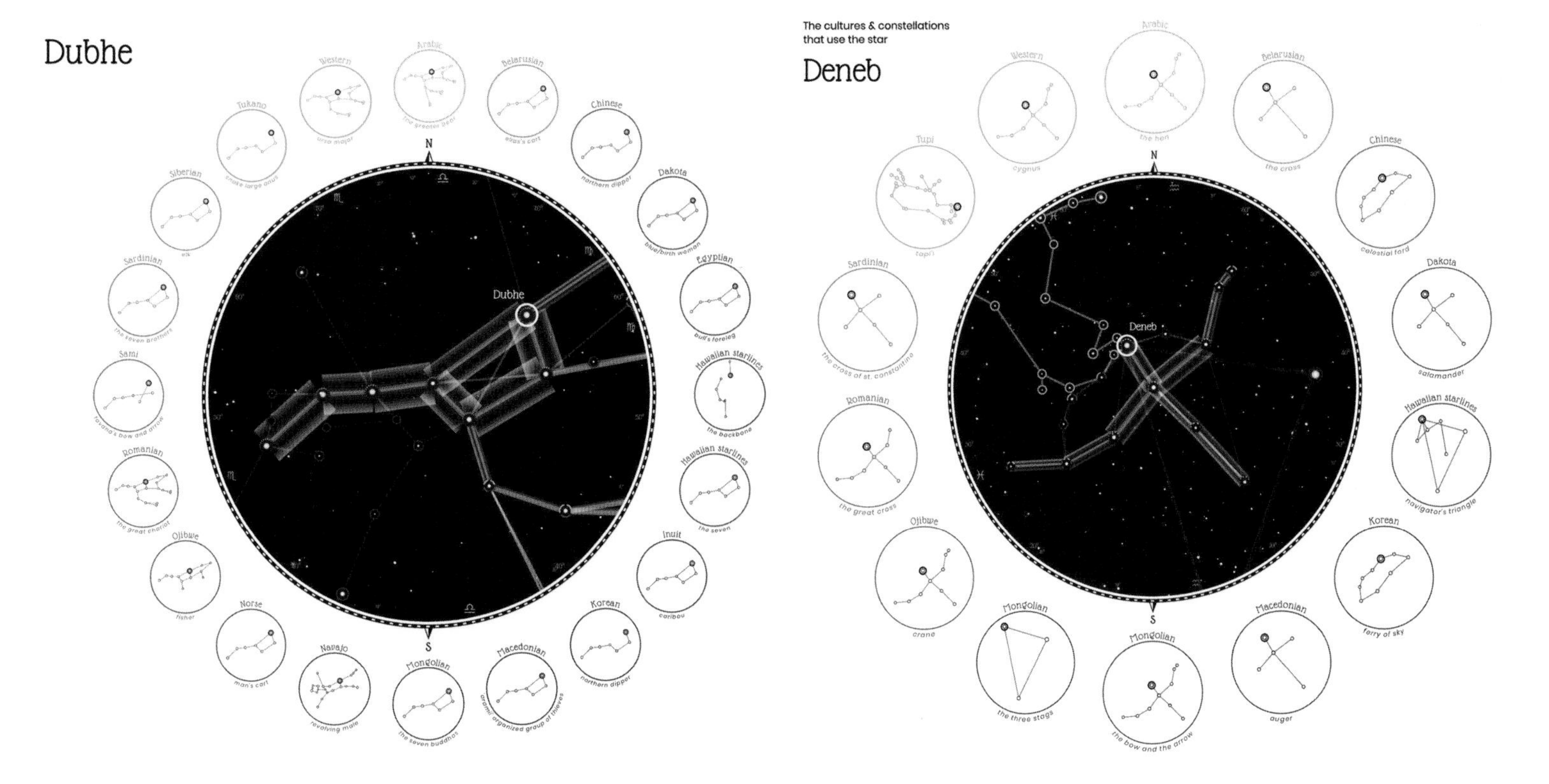

Fig.11.24

The many constellations that use the star Dubhe, part of the well-known Big Dipper.

Fig.11.25

All constellations that are connected to the star Deneb. In Western cultures its constellation is known as Cygnus (the Swan), but it is also part of the "Summer Triangle," a very easy-to-make-out group of three stars in the high of a Summer night on the Northern hemisphere.

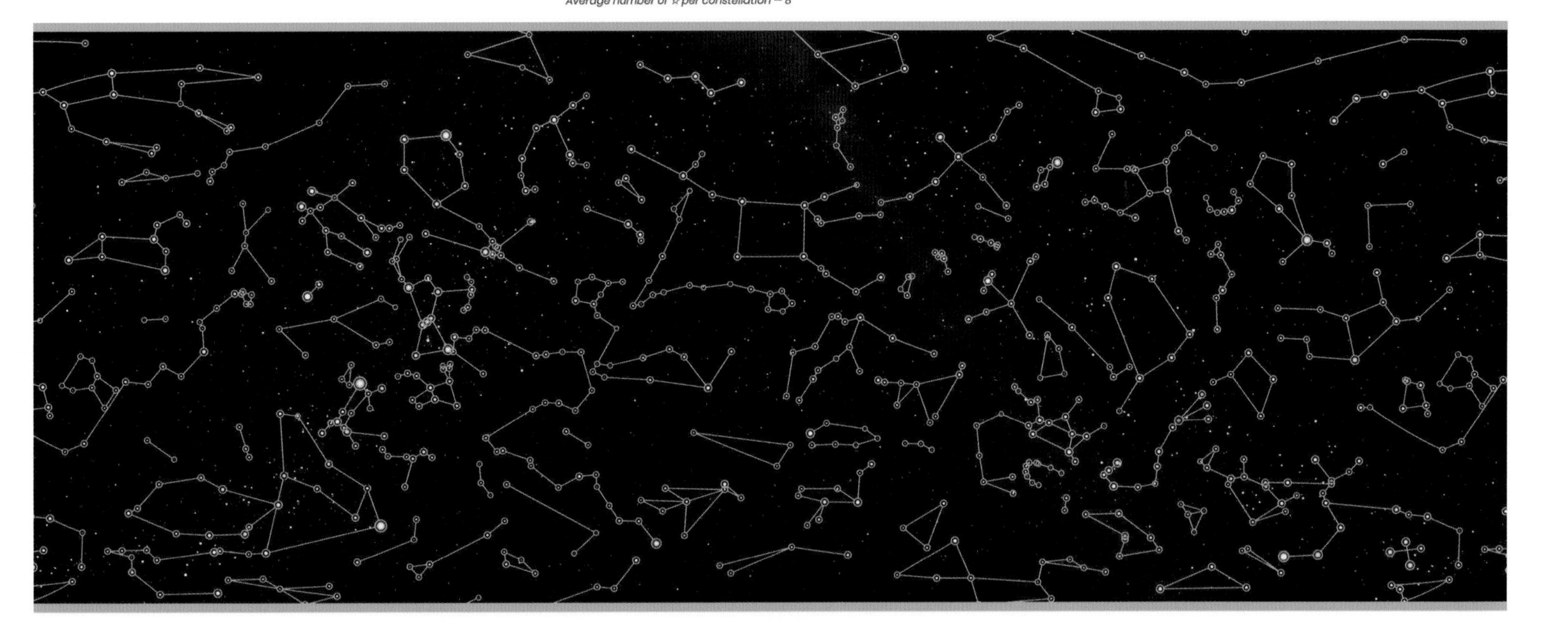

Constellations in the night sky as seen by Western culture
Number of Constellations — 88
Average number of ☆ per constellation — 8

Fig.11.26

The full sky map showing all 88 "modern" (Western) constellations.

Fig.11.27

This beauty of a constellation comes from severa tribes in South America and is called Veado (whic Google tells me is similar to "deer"). I would say that it seems a bit too specific for a constellation that can "easily" be found in the sky, but that's perhaps my own bias of having lived in very light-polluted areas all my life.

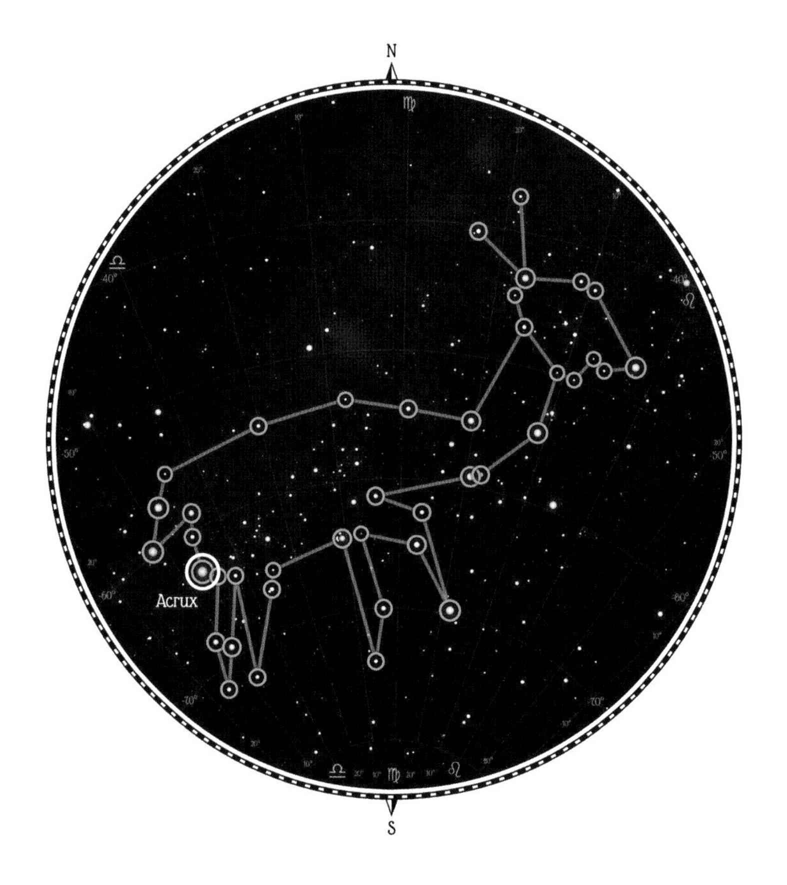

Acrux
Located in the Southern Cross constellation, quite near the South Pole
Aldebaran
Part of Taurus, this star is used most often across cultures after Orion's 'belt'
Alphekka
A great number of things are seen in this half circular 'Corona Borealis'
Altair
This star is easiest to find as the bottom of the 'Summer Triangle'
Aludra
This far-away star shines more than 176,000 times brighter than the Sun
Atlas
Part of the Pleiades, a tightly packed 'star cluster' of 9 relatively bright stars
Antares
A distinctly red star that is known by many cultures as The Heart
Canopus
The second brightest star, but no clear shape appears across cultures
Capella
Interestingly known as 'the Goat star' across several cultures
Deneb
Meaning 'tail', it belongs to both the Swan & the Summer Triangle
Mirphak
Ascribed to fascinating animal shapes, such as a puma, deer, elk and bird
Polaris
The famous North (Pole) star and part of the Little Dipper (and Ursa Minor)
Pollux
The 'heavenly twins' (together with Castor) and the zodiac sign of Gemini
Regulus
The brightest star (actually 4 stars together) of the zodiac Leo, the Lion
Spica
Derived from 'the virgin's ear' in Latin, it's part of the zodiac constellation Virgo

Fig.11.28

After highlighting the constellations of Betelgeuse, Sirius and Deneb, the article lets the viewer inspect 15 more stars by clicking on any of the mini images.

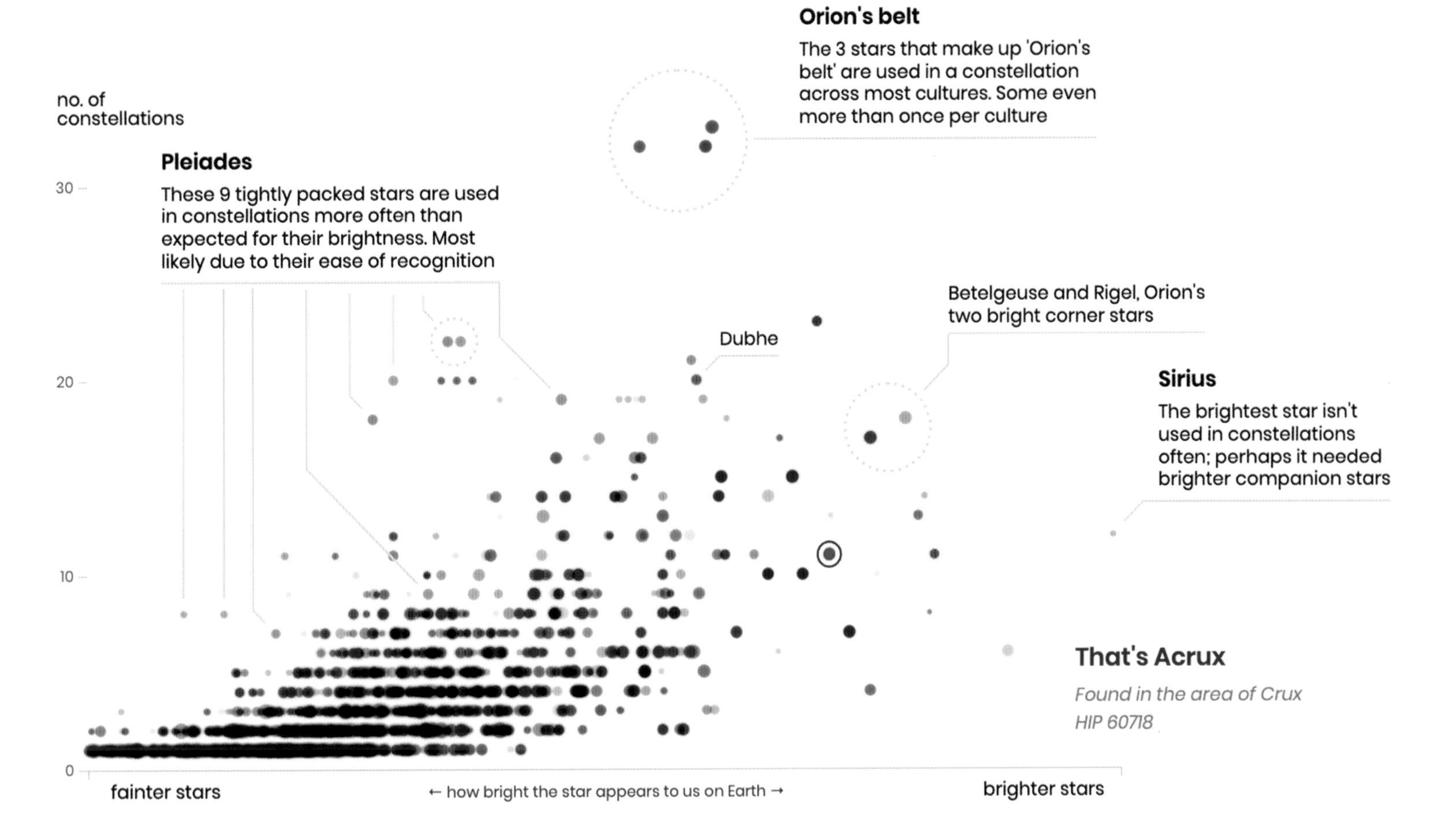

These 9 tightly packed stars are used in constellations more often than expected for their brightness. Most likely due to their ease of recognition

The 3 stars that make up 'Orion's belt' are used in a constellation across most cultures. Some even more than once per culture

Betelgeuse and Rigel, Orion's two bright corner stars

That's Acrux
Found in the area of Crux
HIP 60718
fainter stars

brighter stars

Fig.11.29

A scatter plot showing all ±2,200 stars that are included in at least one constellation.

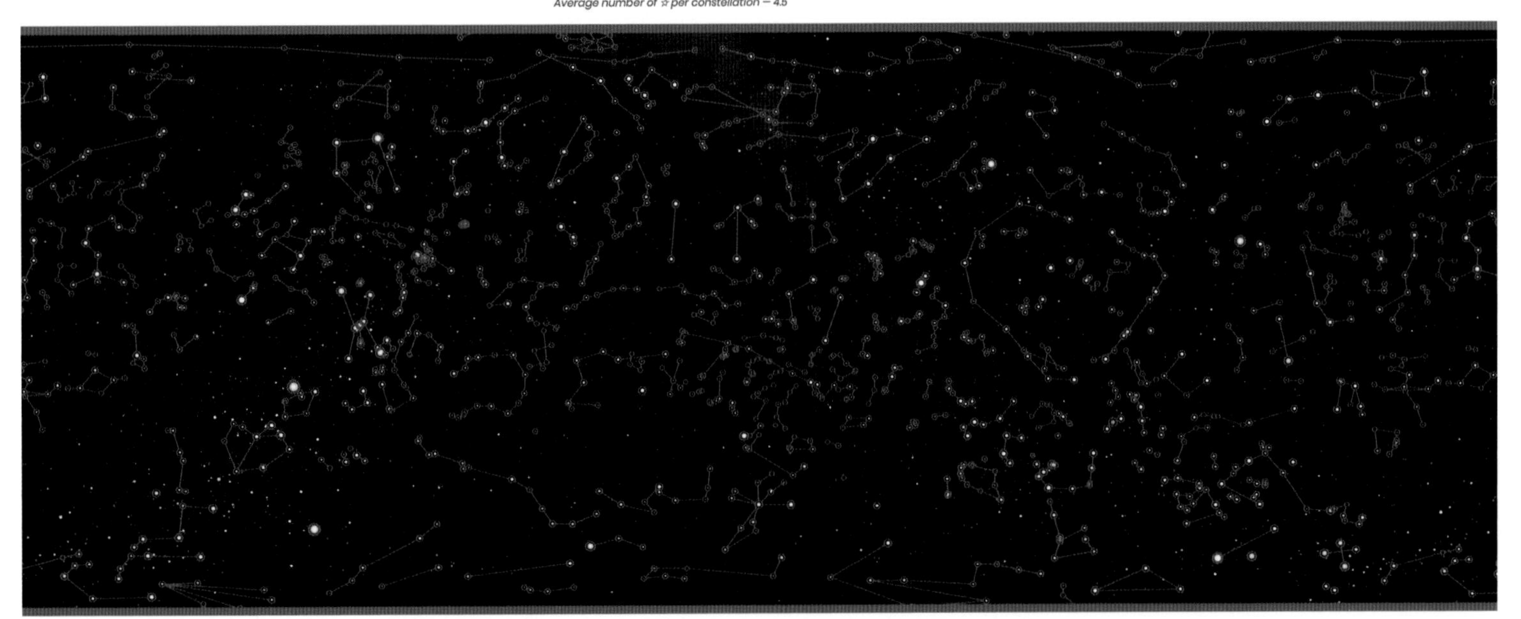

Constellations in the night sky as seen by Chinese culture
Number of Constellations — 318
Average number of ☆ per constellation — 4.5

Fig.11.30

The full sky map of the 318 different Chinese constellations.

Legends

SHIRLEY

Just like Nadieh, it took me *forever* to decide on a good dataset and angle for this project. We chose "Myths & Legends" because it sounded like a great topic with a lot of potential, but the ideas I came up with either didn't excite me much or were difficult from a data gathering perspective. I wanted to do something related to my Chinese background and bounced from Chinese and Asian mythology to classic Chinese literature to *Mythbusters* episodes.

Then, the idea came to me after watching *Crazy Rich Asians*. I loved Michelle Yeoh in the movie, but it wasn't until I read more about her that I learned how accomplished and legendary she was. It made me wonder about all of the legendary women across history that I've never heard of, and the idea took shape from there.

Data

Once I had the idea about doing something with legendary women, the next step was figuring out how to get the data. I decided early on that Wikipedia, with its user-generated content, would be a great resource. I wanted to figure out a way to get the "top" women on the platform, but ran into a tricky problem: I didn't have any idea how to define "top." Would it be page views? Page length? Inbound links? And even though I had seen something similar on *The Pudding*—they were even awesome enough to write about their methodology[1]—I still wasn't sure how to go about adopting their approach for my own project.

The Pudding is the same awesome visual essay collective I published my "Hamilton" project on!

Then, it hit me: instead of trying to define "top" myself, I should just look for a definitive list. After a few Google searches, I ended up on a Wikipedia page with a list of the 51 female Nobel Laureates. They were all incredible women, yet I hadn't heard of most of them. It was the perfect dataset.

I copy-pasted the table—with information about the category (Peace, Literature, Physiology or Medicine, Chemistry, Physics, Economic Sciences) and year of their awards, as well as the achievements that led to the award—into a spreadsheet and did some light formatting and cleaning. I exported the spreadsheet as a CSV and used an online converter to get the data into JSON format.

This is one of the most important data gathering lessons I learned from Nadieh: spreadsheets are great for cleaning data. The extended lesson is to use the right tool for the job, instead of trying to use the same hammer (code) every time.

Because I also wanted to get data from the laureates' individual Wikipedia pages, I researched for ways to access the Wikipedia API. It was a little hard to navigate (I wasn't sure if I should use Wikidata or MediaWiki, which are the two options that came up when I searched "Wikipedia API"), but thankfully *The Pudding* had me covered. A quick dig through their repo led me to their code that used `wiki.js`, and I could interface with that Node.js package instead of the actual Wikipedia API.

After that, I wrote a script using `wiki.js` to programmatically get additional data from each woman's Wikipedia page, including basic biographical information, the number of links into their page ("backlinks"), and the number of sources at the bottom of their page.

Sketch

In early 2018, my friend and I pitched a design to a potential client. We didn't get the contract, but the core idea—to represent the individual data points as multifaceted crystals—really stuck with me. I pinned some gorgeous photos and paintings of crystals and mused how I could programmatically recreate them.

Fast-forward to fall 2018 and I had the idea of legendary women but wasn't sure how I wanted to visualize them. Then, one day as I was going through the "Information Is Beautiful Awards" shortlist and pinning my favorite entries, I came across the crystals again in one of my Pinterest boards. As soon as I saw artist Rebecca Chaperon's gorgeous paintings of crystals,[2] I knew I wanted to represent each of the legendary women as one of those bright, colorful crystals—because how beautiful would that be?

[1] The Pudding, "What Does the Path to Fame Look Like?": https://pudding.cool/2018/10/wiki- breakout/
[2] Rebecca Chaperon, "Crystals": https://www.thechaperon.ca/gallery/crystals

It wasn't long before I had come up with the other details. The size of the crystal would represent the number of articles linking back to a laureate's Wikipedia page (her "influence"). The number of faces on the crystal would map to the number of sources at the bottom of her page (because she's "multi-faceted"; get it? (*≧艸≦), and the colors would represent the award category.

The only thing that evaded me was how to position the crystals. For the longest time, I could only think to lay them out in a two-dimensional grid and have the reader scroll through them. But then I took Matt DesLauriers's "Creative Coding" workshop on Frontend Masters[3], where he taught (among other things) `Three.js` and WebGL. The workshop inspired me to try out a 3D layout, and I knew immediately that I would use the z-axis for the date they received their award: the closer to the foreground, the more recent her award.

For more explanation of `Three.js` and WebGL, see "Technologies & Tools" at the beginning of the book.

All of this came to me so quickly and naturally, that I didn't draw a single sketch of the idea.

Code

For years I wanted to make *something* physical, and in 2018, I made it a goal to create a physical installation. But every time I thought about it, I got stuck thinking about 2D projections (or TVs) on walls; I didn't know how to take advantage of all that *floor space*.

And then one day, it hit me: of course I didn't know how to think in physical spaces, I worked digitally in 2D all day long. So if I could teach myself to work in 3D digitally, then it should (hopefully) follow that I could think in physical spaces also. I put `Three.js` and WebGL at the top of my list of technologies to learn.

I took Matt's Frontend Masters workshop and learned the basics of `Three.js`, fragment shaders, and vertex shaders. I learned the "right-hand rule" to orient myself in WebGL's coordinate system: with my right palm facing me, use the thumb for the x-axis (increases going right), index finger for the y-axis (increases going up, which is the opposite of SVG and canvas), and the middle finger for the z-axis (increases coming out of the screen and towards us). I learned that WebGL's coordinate system doesn't operate in pixels, but rather in "units" of measurement that we can think of as feet or meters or whatever we like, as long as we're consistent.

In mid-November, David Ronai asked me if I was interested in participating in *Christmas Experiments*, an annual WebGL advent calendar. I was hesitant to accept, since I had never worked with WebGL before, but David encouraged me to give it a try and promised to put me later in the month to give me more time. I agreed, knowing that the deadline would give me the motivation I needed to complete the project. I made it a goal to do a little bit each weekday starting December 1st, until I could get to something presentable on the 23rd—the slotted date for my Christmas Experiment.

I started by reading the first two chapters of WebGL Programming Guide,[4] which taught me how WebGL was set up. I then rewatched the `Three.js` section of Matt's workshop so that I could see what heavy lifting it was doing for me. After the workshop, I created an Observable notebook to figure out the minimum amount of setup required to draw with `Three.js`.

I always like understanding the most fundamental building blocks required to make something work—the core foundation of a technology or library.

[3] Matt DesLauriers on Frontend Masters: https://frontendmasters.com/teachers/matt-deslauriers/

After setting up the notebook, I wanted to create a crystal shape. I decided to use `Three.js`'s `PolyhedronGeometry` because I could define a set of points and specify which three points would make up each triangular face. On the first day, I only managed to create a triangle, but on the second day, I managed a 3D (rotating!) crystal as well as the crystal shape I had in mind (Figure 11.1). And even though I later found a better and easier way to create the crystals I wanted, I'm really grateful for the practice `PolyhedronGeometry` gave me to think through WebGL's *x*, *y*, and *z* coordinates.

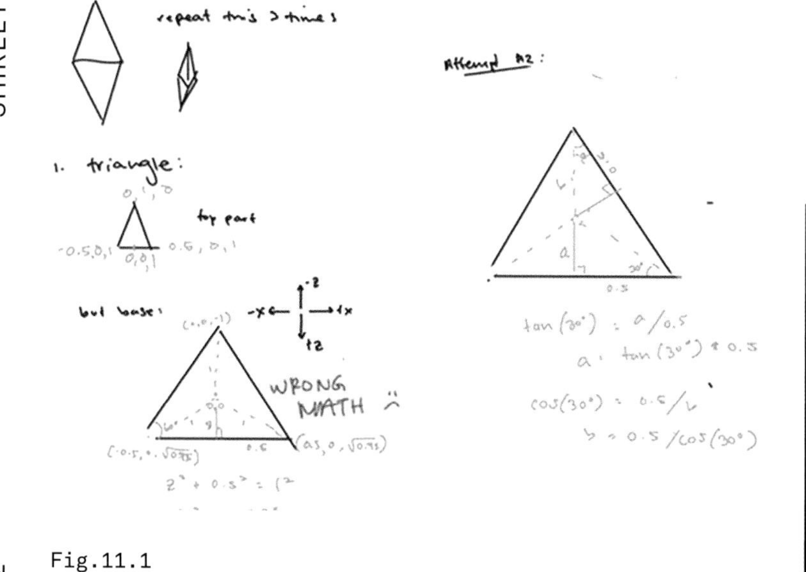

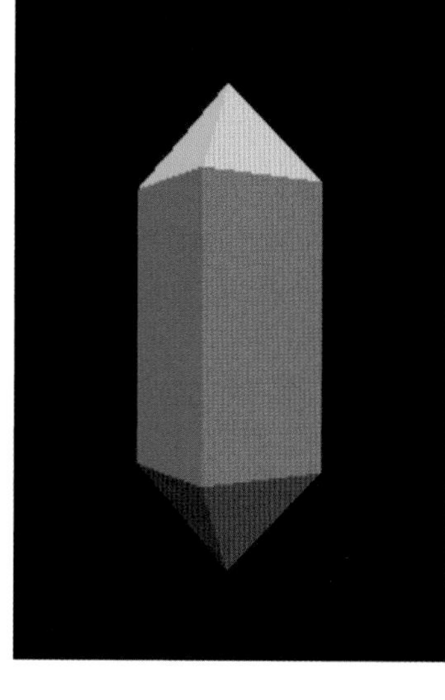

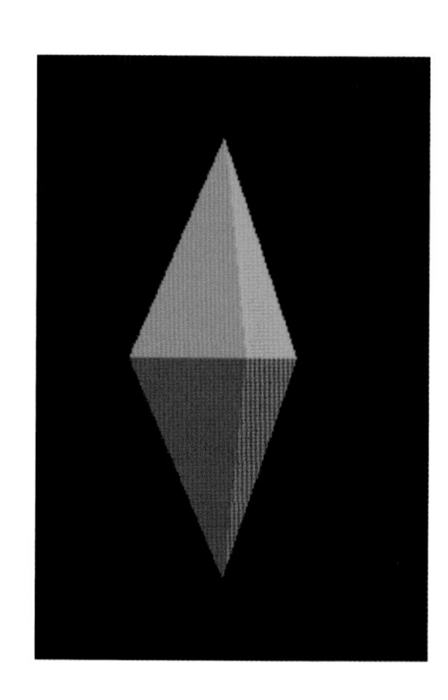

Fig.11.1

Notes trying to figure out the math for where to position the points, and the two resulting successful crystal shapes.

Once I was satisfied with the shapes, I moved on to learning how to add color to them. I went back through the workshop's section on vertex and fragment shaders and played around with Matt's fragment shader code (Figure 11.2), which taught me some commonly-used GLSL (OpenGL Shading Language) functions.

Shaders were originally computer programs used for shading, but are now used in a variety of ways in computer graphics. In WebGL, fragment shaders are used to color a shape, and vertex shaders are used to manipulate the position of each point in a shape.

Fig.11.2

Experimenting with Matt's fragment shader code.

[4] Matsuda & Lea, "WebGL Programming Guide": https://sites.google.com/site/webglbook/

The next goal was to use the fragment shader to color the crystal shape, but I wasn't a fan of how it ended up looking, so I decided to read more into the different ways of outputting colors with shaders. This is when I turned more heavily to Patricio Gonzalez Vivo's *The Book of Shaders*,[5] and in particular, the chapter on "Shaping Functions." I learned about GLSL's built-in trigonometry and math functions, which take in a number (or a set of numbers) and output another. I also learned about the `mix()` function, where I could input two colors and not only get a new color between those two, but also individually mix the Red, Green, and Blue channels to make completely new colors that aren't between the two original colors. Once I combined those two functions, I was able to get beautiful gradients and shapes.

Fig.11.3

Gradients that were created by mixing blue and yellow, and tweaking the Red, Green, and Blue channels at each "pixel" with `power()`, `sine()`, `absolute()`, `step()`, and `smoothstep()`.

When I felt happy with the colors, I switched gears and started plugging in the data: number of faces for number of references at the bottom of the Wikipedia page, size for number of Wikipedia backlinks, and color for the Nobel Prize award category.

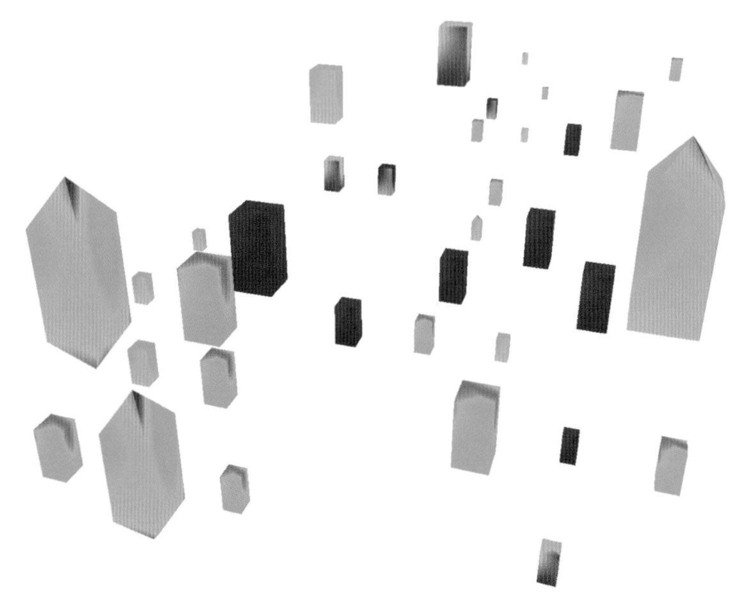

Fig.11.4

First attempt at the crystals with the data plugged in.

[5] Patricio Gonzalez Vivo, The Book of Shaders: https://thebookofshaders.com/

I really didn't like this output for two reasons:

- I was hard-coding each shape, so I only had three distinct shapes: a rectangle, a rectangle with a triangle on top, and a rectangle with a triangle on the top and bottom. This meant most of the "crystals" were just rectangles.
- I also wanted one color per category for a total of six colors; so for each crystal I only passed in one color and programmatically manipulated that color with a shaping function. The resulting colors didn't look good at all.

It was around this time that I came across a demo for Bloom[6] (a post-processing effect that gives objects a glow around them), and in that example there were round, gem-like objects that looked quite similar to the crystals I wanted.

When I looked at the code, I learned that it used `Three.js`'s `SphereGeometry` but with the `flatShading` option on, which shows each distinct face of a shape instead of the default smooth, round surface. This made me realize that in addition to being able to manipulate and mix colors to get new ones, I could also manipulate the settings on an existing `Three.js` geometry to get a new (and completely different looking) shape—very cool! So I swapped out the `PolyhedronGeometry` with `SphereGeometry`, set the width and height segments to programmatically vary the number of faces depending on the data, stretched out the shape by setting the vertical scale to twice the horizontal scale, added jitter to the vertices so that each crystal looked a little different, and ended up with a much more interesting set of shapes:

Fig.11.5

The new crystals where I programmatically set the number of faces, instead of hard-coding every shape.

[6] https://vanruesc.github.io/postprocessing/public/demo/#bloom

Once I had my new shapes, I went back to my fragment shader code to apply what I learned about color mixing and shaping functions (Figure 11.6).

I liked the new colors, but didn't like that I had lost the hard edges. Thankfully, Matt taught me I could get the definition back by calling `computeFlatVertexNormals()` on the geometry and adding it to each "pixel" color in the fragment shader. This not only made the edges really apparent, but also faked a light source, making the bottom and back of the crystals appear darker and in shadow (Figure 11.7).

From there, I played around with two sets of gradients: one for awards in the humanities (Peace, Literature, Economic Sciences), and another for awards in the natural sciences (Physics, Chemistry, Medicine) (Figure 11.8).

Fig.11.6

The crystals with the new fragment shader code applied.

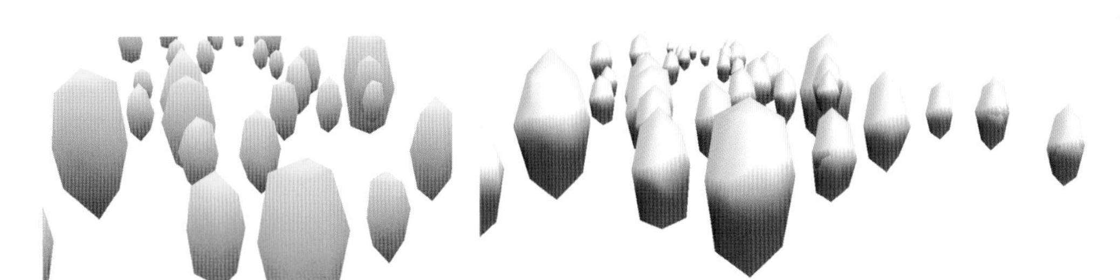

I love how much the first one looks like sweet potatoes, which is seriously my dad's favorite food. (*≧▽≦)

Fig.11.7

The crystals with the edges defined.

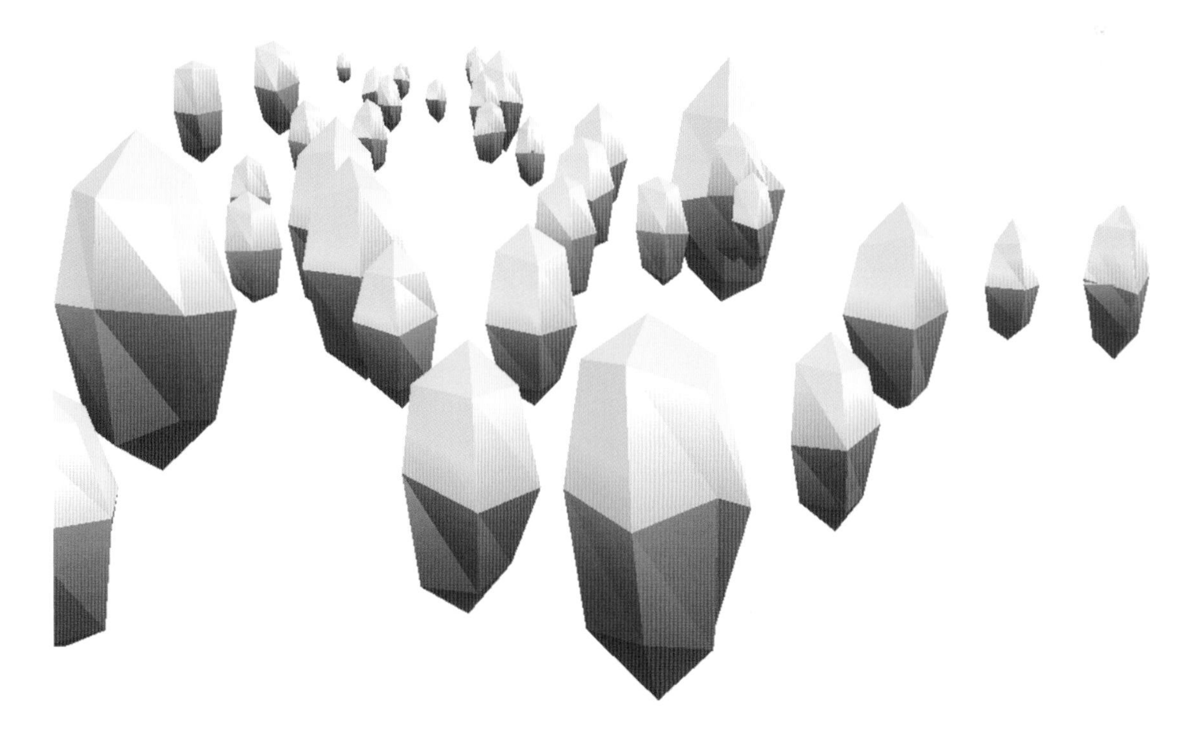

Fig.11.8

Experimenting with two different color schemes to distinguish awards in the humanities and natural sciences.

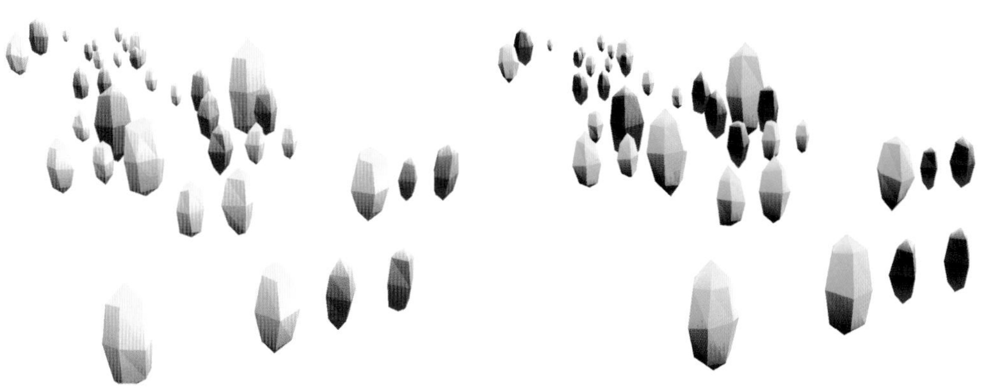

Two because I knew that I couldn't come up with two distinct gradients

Next came the background. I created the "floor" by using a `PlaneGeometry` placed below the crystals, dividing it up into a constant number of segments and jittering the *y*-position of those segments to create a sense of uneven ground. I created the "sky" by placing a huge sphere around the scene with the viewer inside of it. I experimented with three different kinds of lights: hemisphere lights and ambient lights to give the "sky" a nice sunrise glow, and directional lights to cast shadows from the crystals to the "floor" (Figure 11.9).

To finish the piece, I added "stars" to represent all the men who won Nobel Prizes in the same time period, as well as annotations for each crystal. It was a fun challenge trying to add text for the annotations: I tried to create the text using `Three.js`'s `TextGeometry` at first, but the page became completely unresponsive because `Three.js` was rendering each letter as a 3D object. After some Google searches, I found a solution to render the text within canvas, create a `PlaneGeometry`, and use that canvas as an image texture to fill the `PlaneGeometry`—a much more performant solution!

I was quite upset when I learned that there have been 866 male Nobel Laureates but only 53 women Nobel Laureates. There's always a gasp of disbelief when I reveal to an audience that the stars are male Nobel Laureates, because there are so many of them compared to the crystals.

My favorite part of the visualization is how I decided to use the third dimension to represent time. The crystals are placed according to the decade the laureate received her award, so the closer to the front (and the viewer), the more recent the award. But the decades are only revealed when a user "flies up" to view the crystals from above; if they "walk through" the crystals at ground level, they will only see information about each woman. I did this because I really wanted the user to learn more about each laureate first, before they "flew up" for a holistic view.

Unfortunately, because the text is rendered in canvas, it is treated as an image and thus isn't a11y (accessibility) compliant—something I only realized later on.

To finish, I created a landing page with a legend for my legends (hehe), and managed the componentization and interaction with `Vue.js`.

Fig.11.9

Background with a floor and sky added.

Reflections

This project was super fun, and I'm so proud that I was able to finish it in three weeks—something I hadn't been able to do since the "Travel" project. I was also able to teach myself `Three.js` and a little bit of GLSL, which I wanted to do for a very long time. This project was a great opportunity to use the third dimension and also gave me the confidence to experiment with more 3D and physical installation projects in the future. But most importantly, I'm so glad I chose this dataset of women Nobel Laureates; it has since motivated me to work with and highlight datasets featuring underrepresented groups.

Legends

↳ shirleywu.studio/projects/legends

Malala Yousafzai
Peace, 2014
for their struggle against the suppression of children and young people and for the right of all children to education

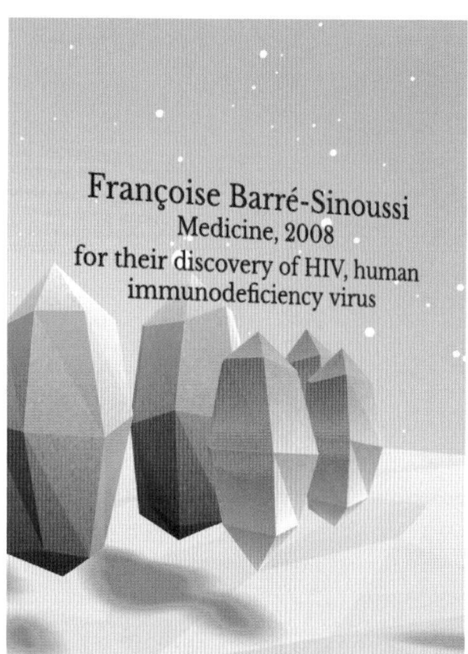

Françoise Barré-Sinoussi
Medicine, 2008
for their discovery of HIV, human immunodeficiency virus

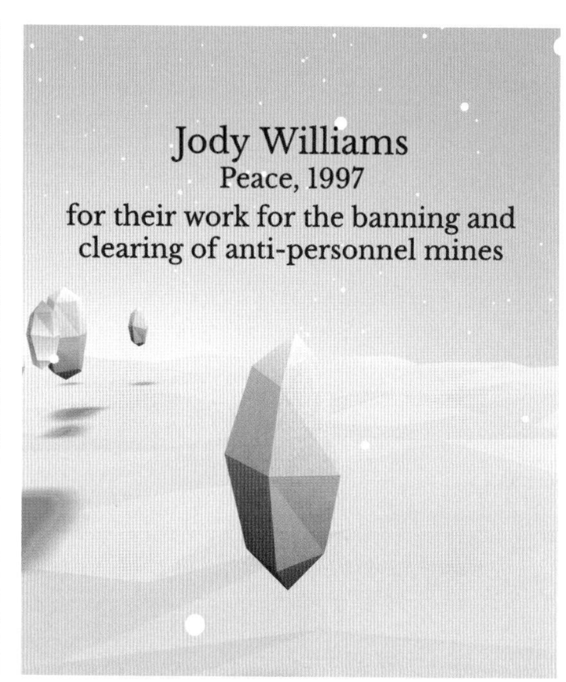

Jody Williams
Peace, 1997
for their work for the banning and clearing of anti-personnel mines

Fig.11.11

Recent women Nobel Laureates I find incredibly inspiring.

Fig.11.12

I also turned this project into washi tape! But because each crystal wasn't distinct enough, I turned them into flowers instead.

FLOWER POWER
by Alice Lee & Shirley Wu
April 2019
Node, D3, SVG, Illustrator, Photoshop, Wood, Paint
Since its inception in 1901, 51 women have been awarded the Nobel Prize. These legendary women are represented as flowers, encoded by data from their Wikipedia pages. They are colored by the category of their award, sized by their "influence" (mentions in other Wikipedia pages), and positioned by the decade they received their prize.
This installation was a collaboration between Alice and Shirley, mixing Shirley's data and software background with Alice's focus on all things illustrative and visual. It is based on Shirley's original 3D data visualization, Legends.
This project wouldn't have been possible without the help of Catherine Madden, Alex Lam, and HJ Chen.
2010

Fig.11.13

My main goal with this project was to create a physical installation, and I got the opportunity when I collaborated with my studio-mate and illustrator-muralist-extraordinaire Alice Lee! It was a fun learning experience as I had to pay attention to details I never had to when working digitally, like physically organizing the pieces by prize category so we painted them the correct colors (top left), and then again by decade so that we hung them up in the right order (middle left).

FEAR

LESS

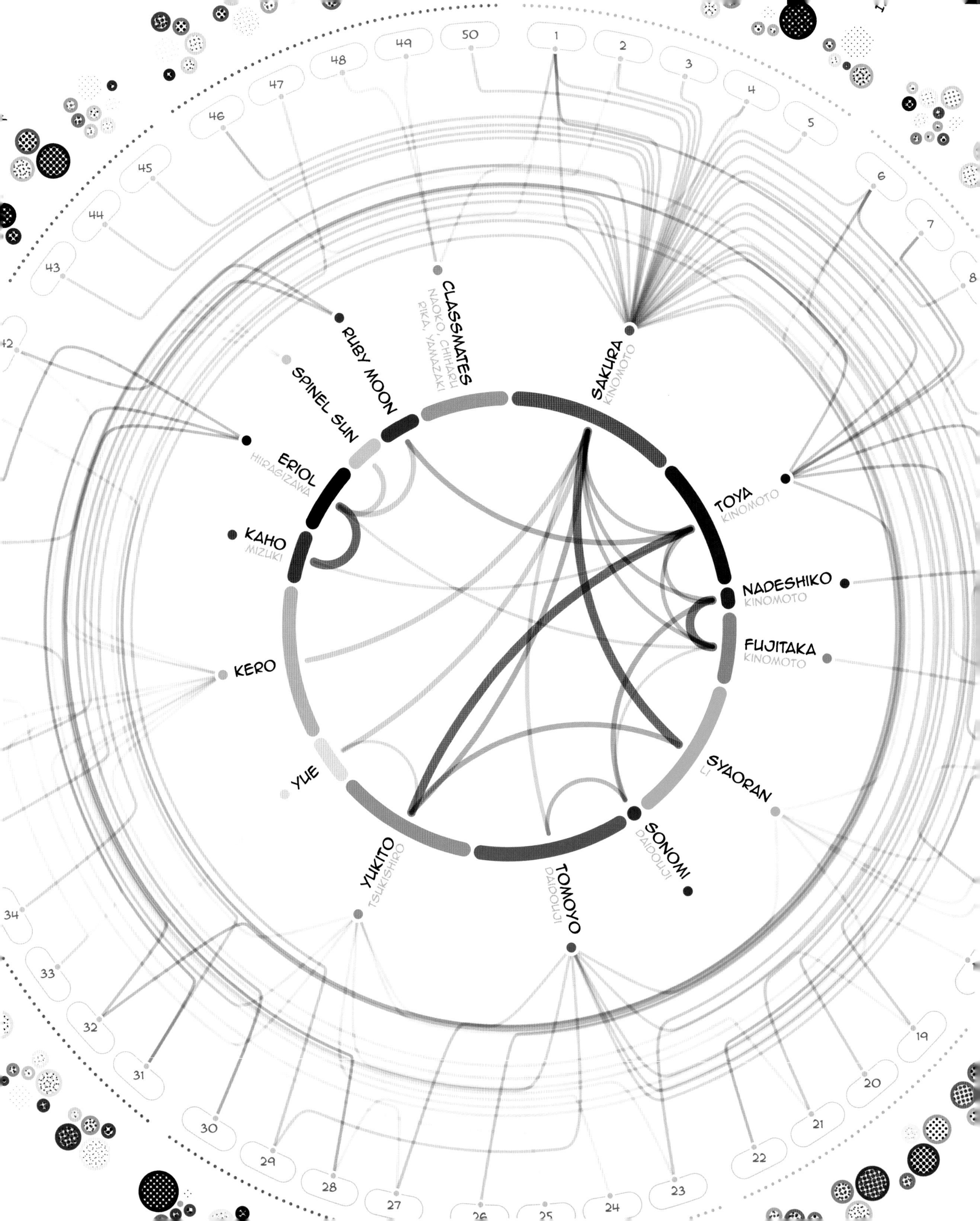

1
2
3
4
5
6
7
8
19
20
21
22
23
24
25
26
27
28
29
30
31
32
33
34
42
43
44
45
46
47
48
49
50
SAKURA
KINOMOTO
CLASSMATES
NAOKO, CHIHARU
RIKA, YAMAZAKI
RUBY MOON
SPINEL SUN
ERIOL
HIIRAGIZAWA
KAHO
MIZUKI
KERO
YUE
YUKITO
TSUKISHIRO
TOMOYO
DAIDOUJI
SONOMI
DAIDOUJI
SYAORAN
LI
FUJITAKA
KINOMOTO
NADESHIKO
KINOMOTO
TOYA
KINOMOTO

An Ode to Cardcaptor Sakura

NADIEH

I'm starting to understand that the topics which make me really enthusiastic are quite niche topics. Not many other people know them. I can only hope this visual will be a joy to explore. I chose to dive into "Cardcaptor Sakura."

Cardcaptor Sakura (or CCS) is a magical-girl manga (a Japanese comic or graphic novel) that was released about 20 years ago and revolves around a kind, brave, and fearless girl called Sakura. It was the very first manga I owned, and I've read it dozens of times. I picked up CCS because the cover looked like the most *beautiful* comic I had ever seen, and basically every panel within was just mind-blowingly perfect and cute.

Even though it's been more than 20 years since the last chapter was published, I'd recently learned about a new "arc" of CCS that was released and thus it was on my mind again. One of my favorite things about CCS is how beautiful each page looks. The covers of each chapter are tiny works of art. I wanted to investigate those covers through data somehow, as I had never done any kind of analysis based on image data before. I loved the idea of creating a visualization that abstracted the colors of each cover into a few colors and thought it would be a fun way to explore and learn. Due to copyright, I'm sadly unable to show any images of CCS in this book (╥_╥).

Data

There are 50 chapters in the original CCS manga, divided into two “arcs.” I flipped through the 12 complete CCS volumes to see what image was on the cover of the chapters. However, these chapter covers were printed in black and white. Thankfully, all of those chapter covers have since been published in full-color in several CCS art books. A quick search revealed that they were all available as a color image from the CCS Wiki page.[1]

A volume combines several chapters.

Using the `imager` package, I loaded the images into R where each pixel was transformed into a multidimensional array of RGBA values. I converted this complex array into something more simple, having three columns (for the RGB values), with the number of rows being equal to the number of pixels. To figure out which algorithm would cluster the data points (pixels) into decent color groups, I tried several things. First, I experimented with different clustering techniques: from the standard K-means, to hierarchical clustering, and even tSNE. Then I also converted the RGB values of each pixel into other color spaces (where colors have different “distances” to each other, which leads to different clustering results per color space), using (among others) the `colorspace` package in R.

RGBA stands for red, green, blue, and alpha. Alpha refers to the opacity/transparency.

Examples of colors spaces are: RGB, HSL, HSV, and LAB.

I converted the results of each test into a bar chart such as the one in Figure 12.1 to see the color groups. Eventually, I found that using K-means together with the colors converted to LAB color-space visually gave the best fitting results, with the most distinct groupings of colors.

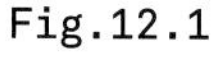

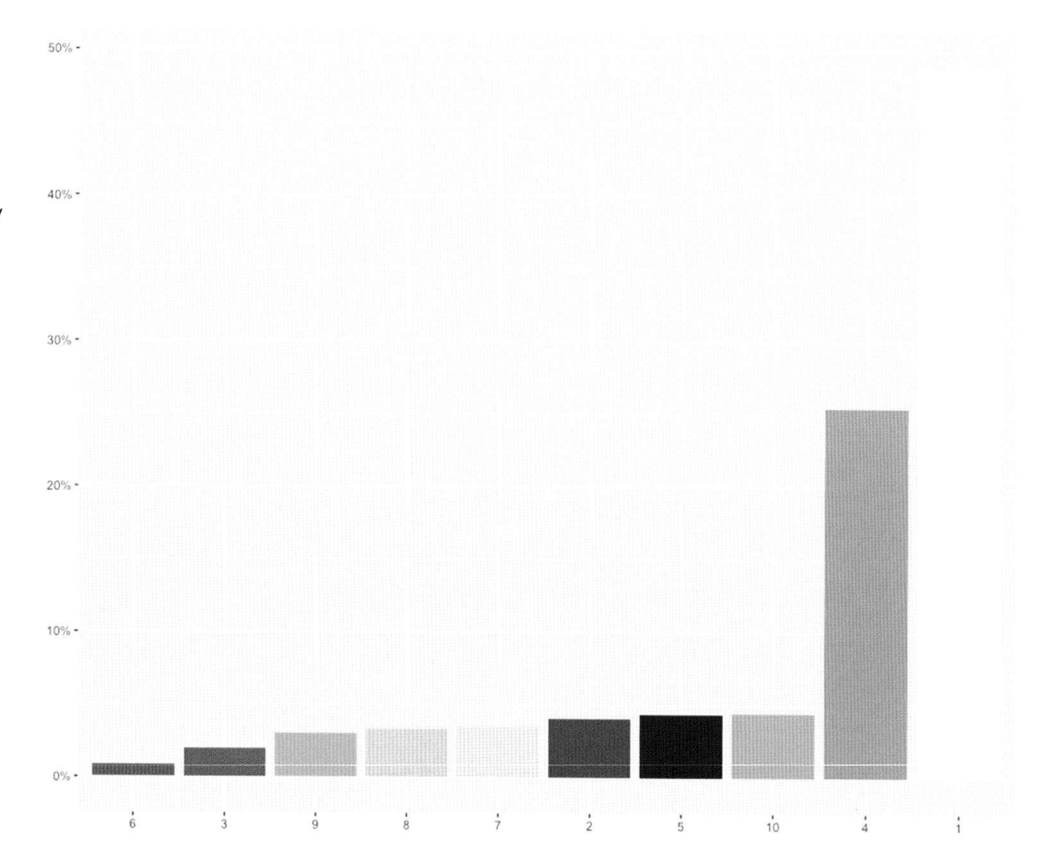

Fig.12.1

Color distribution of the first CCS chapter using a K-means clustering. Imagine a Japanese girl, in front of a mostly blue sky with clouds, wearing a black and white school uniform.

[1] Cardcaptor Sakura Wiki: https://ccsakura.fandom.com/wiki/Artbook

However, one of the tricky things with K-means is figuring out exactly how many clusters are needed to create groups. Eventually, I settled for a method that could gauge this better than any of the techniques I used before: my own eyes! I created a graph such as Figure 12.2 for each chapter which shows the color distribution for three groups, up to 11. I then compared the actual cover to these groups and chose the one that achieved the best balance between capturing all the colors and having a good blend of distinct colors. The color hex codes were then saved into a JSON file, along with the cluster/color percentage. (This is the percentage of the image's pixels that were grouped into the corresponding color.)

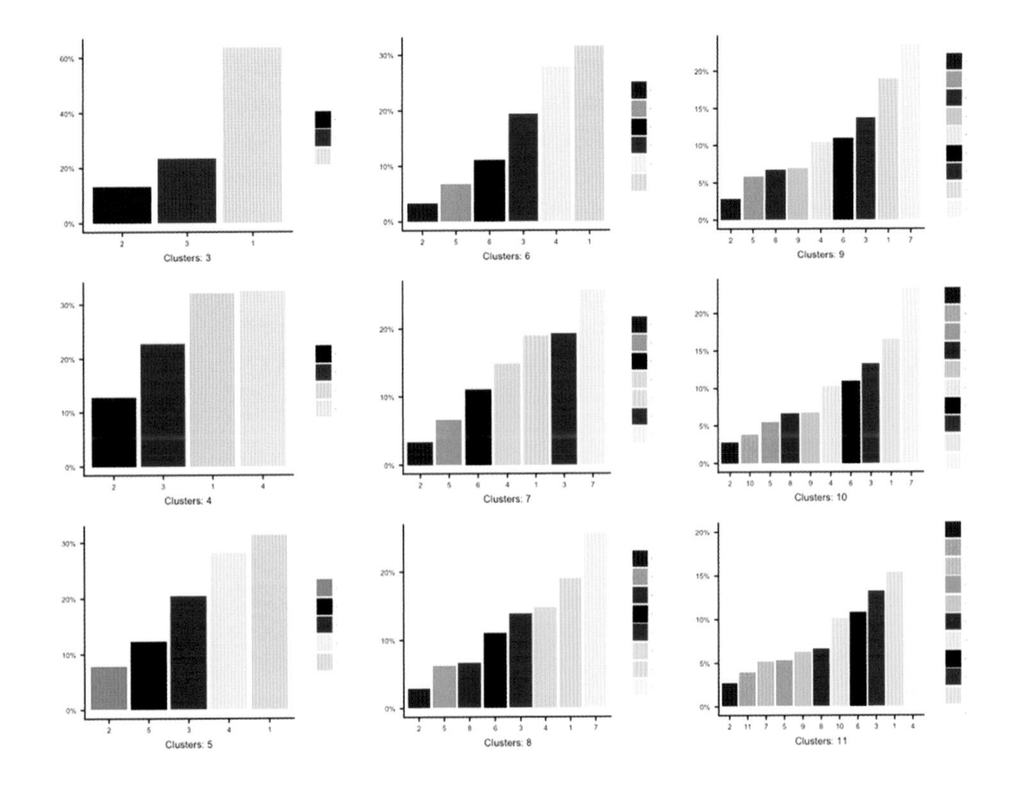

Fig.12.2

Color distributions of several K-means run on the same image, but with a different number of groups predetermined.

To complement this data about the chapter covers, I also wanted to gather information about which characters appeared in each chapter and which "card" was captured in which chapter. The CCS Wiki page on each chapter seemed like *just* the resource I needed, but only the first eight chapters contained this information. There was nothing else to do but re-read every chapter myself and manually populate an Excel file with the information I needed about characters, cards, and covers. ┐(̄ヮ ̄)┌

CCS is about a magical girl, Sakura, who has to collect so-called magical "Clow Cards."

↳ Precalculate "Visual" Variables

Use the tool that is best suited for data preparation and collection work. In this case, I created several different files with R that I needed for each of the information layers in the CCS visual instead of keeping all the data in one large file, such as a separate file about the colors per chapter cover, but also the chapters each character appeared in, the relationships between characters, and more. Doing it in R saved me having to add more data preparation steps and aggregations, filters, and nesting in my JavaScript code, which (for me) wasn't the easiest/best place for those kinds of operations.

Sketch

Figuring out the design for this project came slowly. It was more of a domino effect; a concrete idea for one aspect led to a vague idea for the next part of the data which I then explored. I started with visualizing the results of the color analysis. Having a cluster of small colored circles per chapter in a radial layout seemed like a logical and interesting option. At first I wanted to do a semi-circle with the color clusters to the right and character information to the left (see the top-right page of Figure 12.3). However, with 50 chapters, I needed as much space as I could get. So I went with an "onion-layered approach": a smaller, inner circle related to the characters and a bigger, outer circle related to the color clusters. These two circles would then be connected by lines to show which characters appeared in which chapters (the bottom-right page of Figure 12.3).

I've always been fascinated by the *CMYK dot printing process*; you can see the separate dots when you're looking at it up close, but move farther back and the bigger picture comes into view. The CMYK dot technique for a visualization about a printed manga seemed like a proper style, and a challenge. And a challenge it was! I'll explain more about the technical difficulties in the "Code" section, but the right page of Figure 12.4 shows some of my "how to understand CMYK" scribbles.

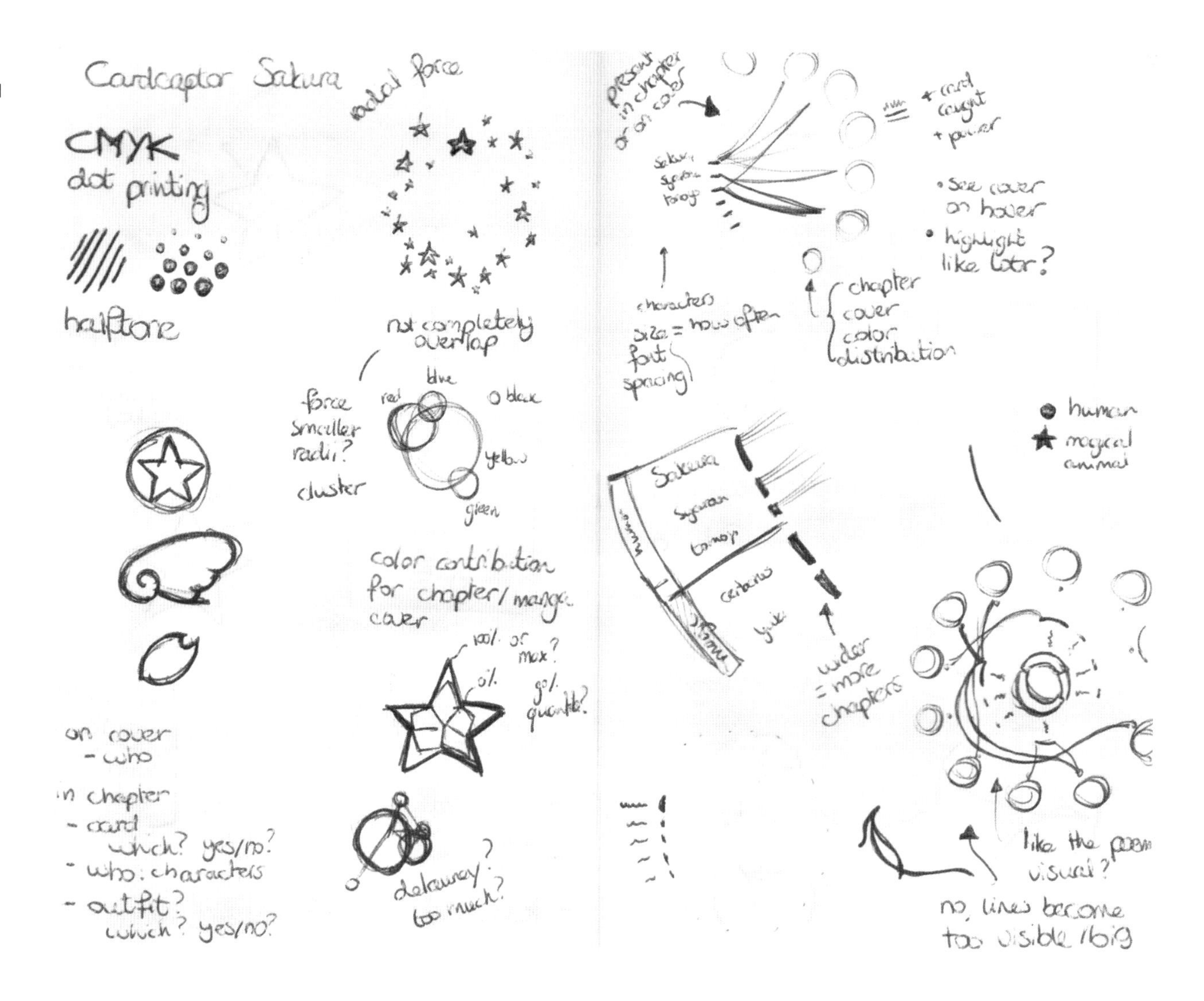

Fig.12.3

Sketch of the overall design for the CCS visualization.

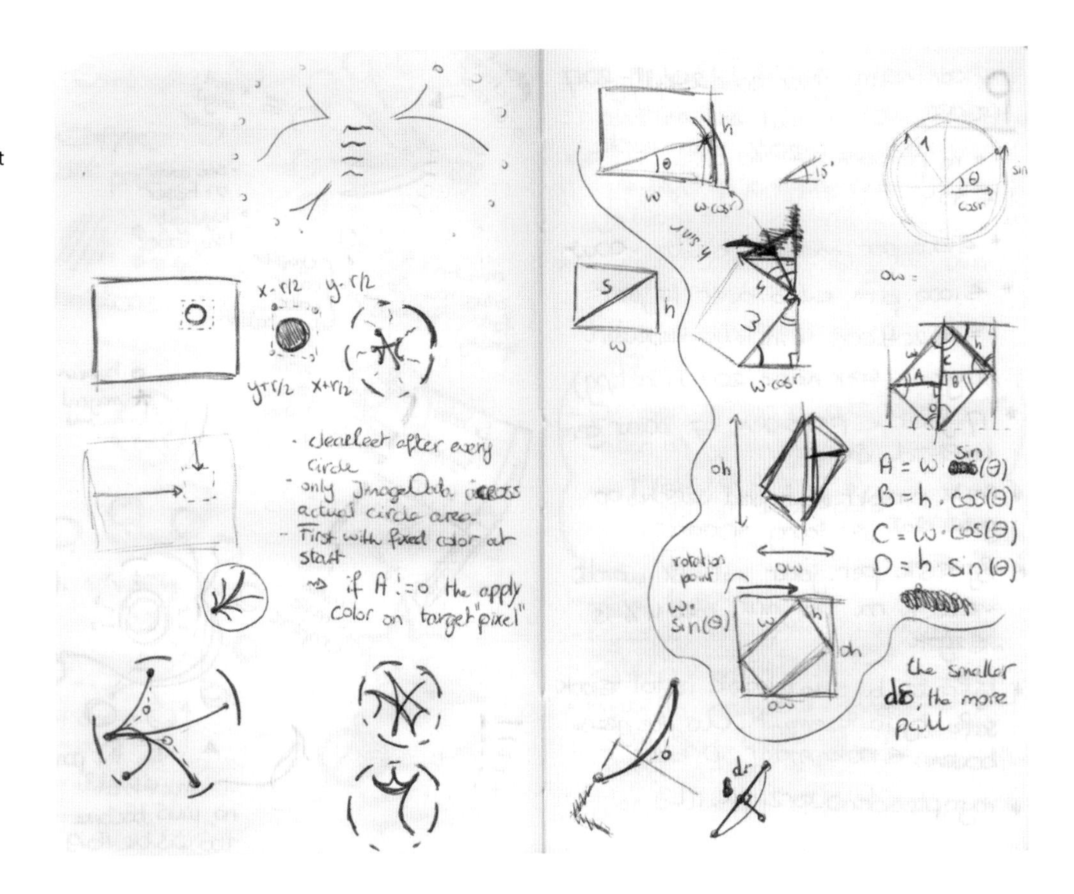

Fig.12.4

Sketch of trigonometric functions to figure out CMYK dots.

Fig.12.5 (a & b)

Sketch of the swirling lines between characters and chapters and figuring out tangent lines.

Figure 12.5 b shows how I handled the calculation of a tangent line to a circle for different circumstances. I needed this information for those swirly lines from Figure 12.5 a. But even though I ended up with different lines in the final result, I was able to use some of these ideas to easily convert the lines to the formulas needed for the final result (which are more circular running lines).

a

b

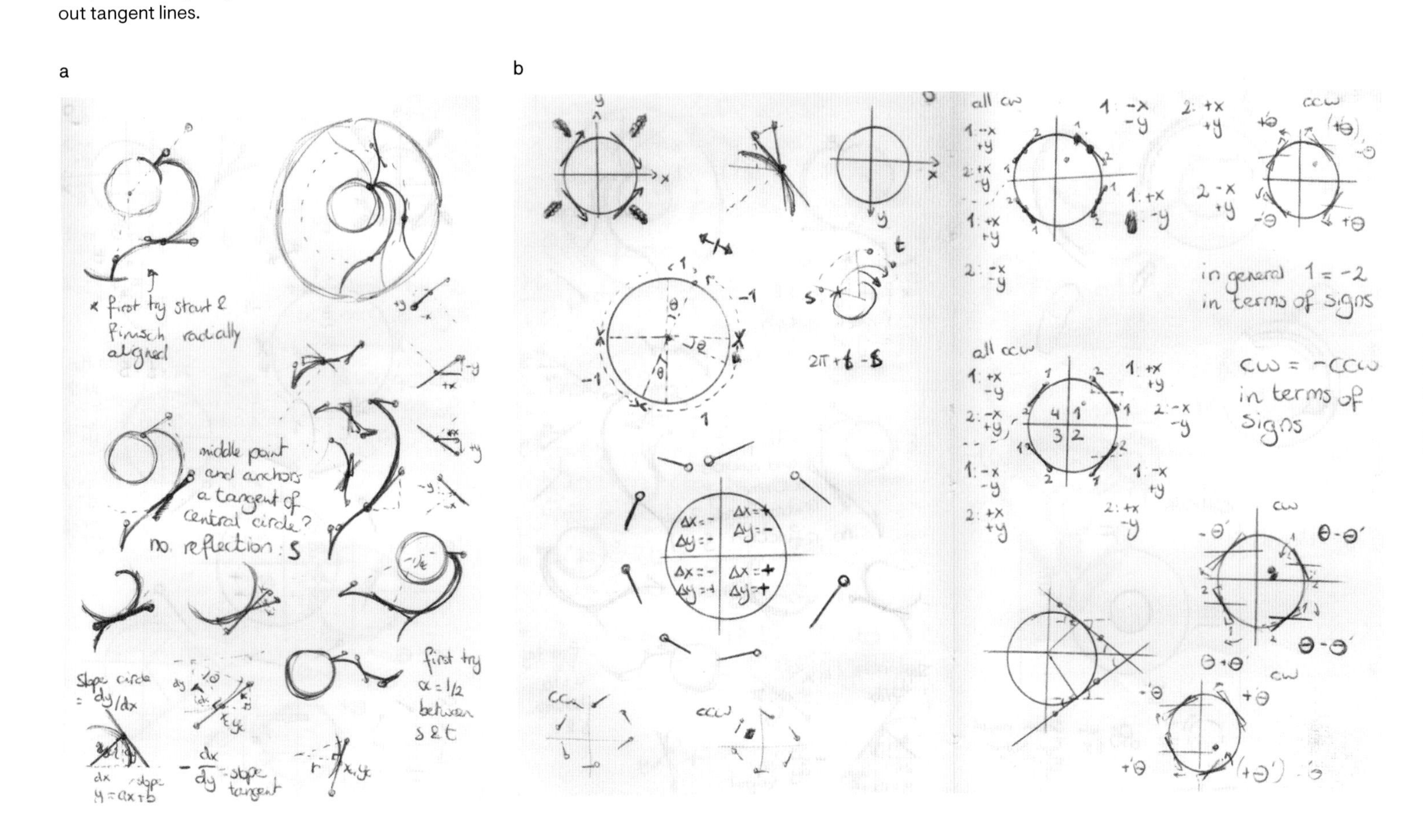

↳ Learn to Love Math

Having an understanding of the SVG Cubic Bézier Curve paths was of vital importance at the start of the project. However, to set up a function that would somewhat correctly make the lines flow around the inner circle, I definitely relied on my knowledge of geometry. This was especially true for thinking in polar coordinates (radius and angle), and how lines should move from start to end point.

As I've mentioned in previous projects, I strongly recommend learning the basics of both geometry (especially trigonometry) and statistics if you want to be able to shape your data visualization to whatever idea you have in mind!

Code

I first focused on getting the ring of color circles on my screen. Mostly because I wanted to see how that CMYK idea would look as soon as possible. Thanks to an excellent test that Shirley made for her "Books" project[2] to distribute clusters of circles in a grid, making this was actually a piece of cake! But darn it, those circles had to become quite small to make room for all 50 chapters (Figure 12.6). I was starting to have my doubts if the CMYK effect would work as well in this particular design as I'd hoped ...

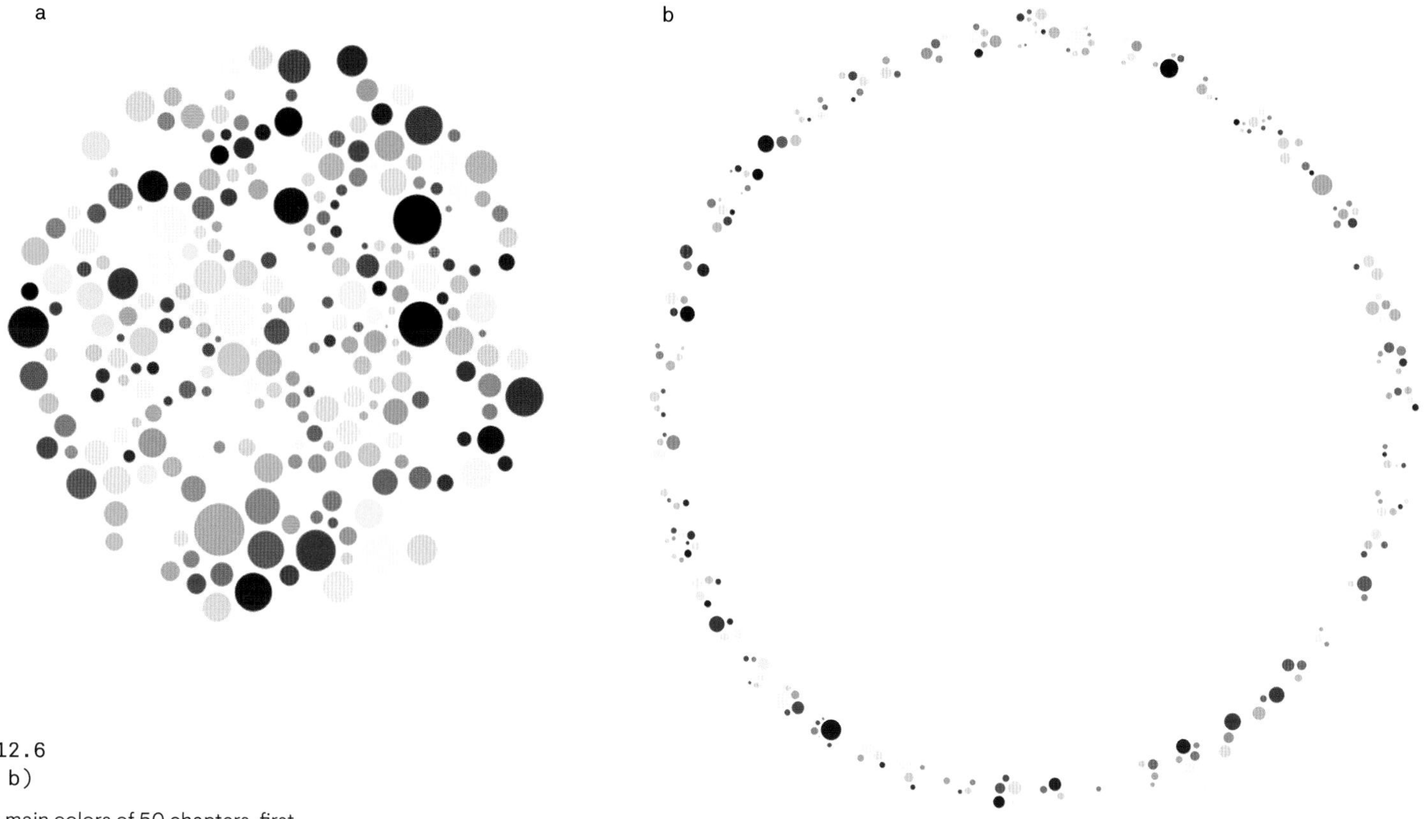

Fig.12.6 (a & b)

All the main colors of 50 chapters, first together and then clustered per chapter around a circle.

[2] Test Shirley created for her "Books" project: https://bl.ocks.org/sxywu/570df88e66e420191d33dc5b5650aaf4

I wasn't going to give up without trying. I found an example[3] that neatly coded up a CMYK dot effect as SVG patterns. I rewrote it to create a separate pattern for each color in my data, which resulted in a ring full of CMYK-based colored circles. But on closer inspection I found something I didn't quite like. Although the circles on the inside looked exactly how I wanted them, they were still SVGs. So they had been perfectly clipped into a circle (Figure 12.7). But I wanted my CMYK dots to *smoothly* fade out, not end abruptly. Furthermore, I also wanted to play around with partially overlapping circles and having the colors mix even further, which wasn't possible with this technique.

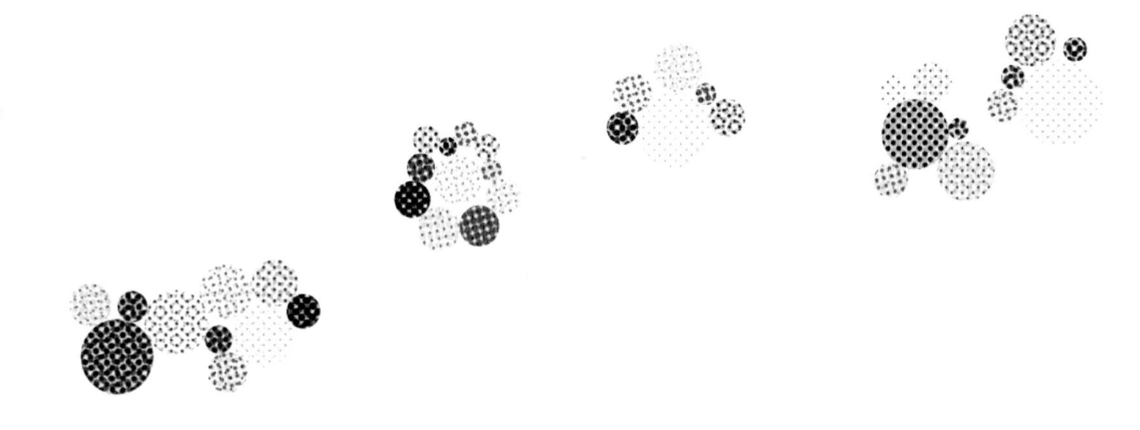

Fig.12.7

Each SVG circle filled with a CMYK dot pattern, but too perfectly cut to a circle at the edges.

I already suspected that using canvas was probably the way to go. I found two interesting code snippets[4] that took me a good 3–4 hours to wrap my head around and combine them mathematically into one. [¬°-°]¬

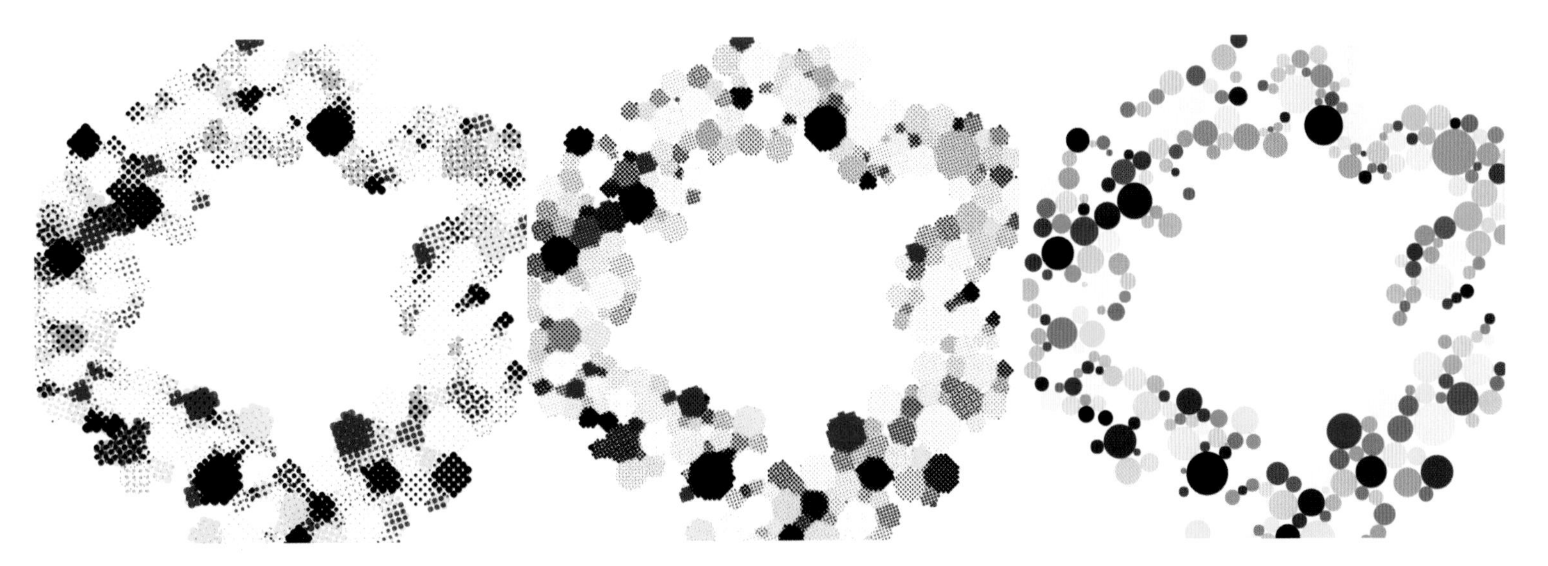

Fig.12.8

Comparing the CMYK transformations to the original group of circles.

[3] CMYK Halftone Printing by Noah Veltman: https://bl.ocks.org/veltman/50a350e86de82278ffb2df248499d3e2
[4] Color Halftone Filter by Patrick Matte: http://www.patrickmatte.com/?p=352 and
Canvas Halftoning by @ucnv: https://gist.github.com/ucnv/249486

Eventually I got the smooth edges I was going for (with the dots getting smaller around the sides instead of being clipped—see the images of Figure 12.9). Plus, I was able to plot the circles on top of each other which combined the CMYK effects of these circles.

I applied it to all the circles from the 50 chapters and ..., the circles were so small that there wasn't "enough CMYK" going on. It was often too difficult to actually get a feeling for the circle's true color because it contained only a few CMYK dots in itself. o(╥﹏╥)o

Well, that's how (dataviz) design works; sometimes you spend hours of work on something that never makes the cut. I took a closer look at the original SVG version again, and wondered what to do about the crisp outer edges. Hmmm ... what about adding a thick stroke? And yup, that fixed it enough for me ┐(￣ヮ￣)┌ (see Figure 12.10).

Fig.12.9 (a,b,c)

Final canvas CMYK circles that have smooth edges (left images) and can overlap (rightmost image).

Fig.12.10

Going back to the SVG version, but now with a stroke to "mask" the sharp edges of the CMYK dots a little.

I thought the swirly lines between the inner and outer circles would probably be the second most difficult thing to tackle. So I first completed the donut-chart-like inner circle where the lines would connect to. Still, I wanted to see if I could get the connections/relations between the characters "visually working" in the inner circle before I moved to the outer lines. If the inner lines didn't work, I would have to think of a different general layout. Time to write some custom SVG paths again! Figure 12.11 shows the progress from the simplest approach, straight lines in Figure 12.11 a, to the final nicely curved lines in Figure 12.11 d.

I colored the lines according to the type of connection (e.g., family, love, friend). With the final result on my screen I could see that there weren't too many lines in there, no visual overload. (๑•̀ㅂ•́)و✧ Finally, it was time to dive into those swirly lines connecting the chapters to the characters.

The paths in the final version are made up of circle segments. I reused code that I initially wrote for the small arcs in my "Books" project about fantasy books.

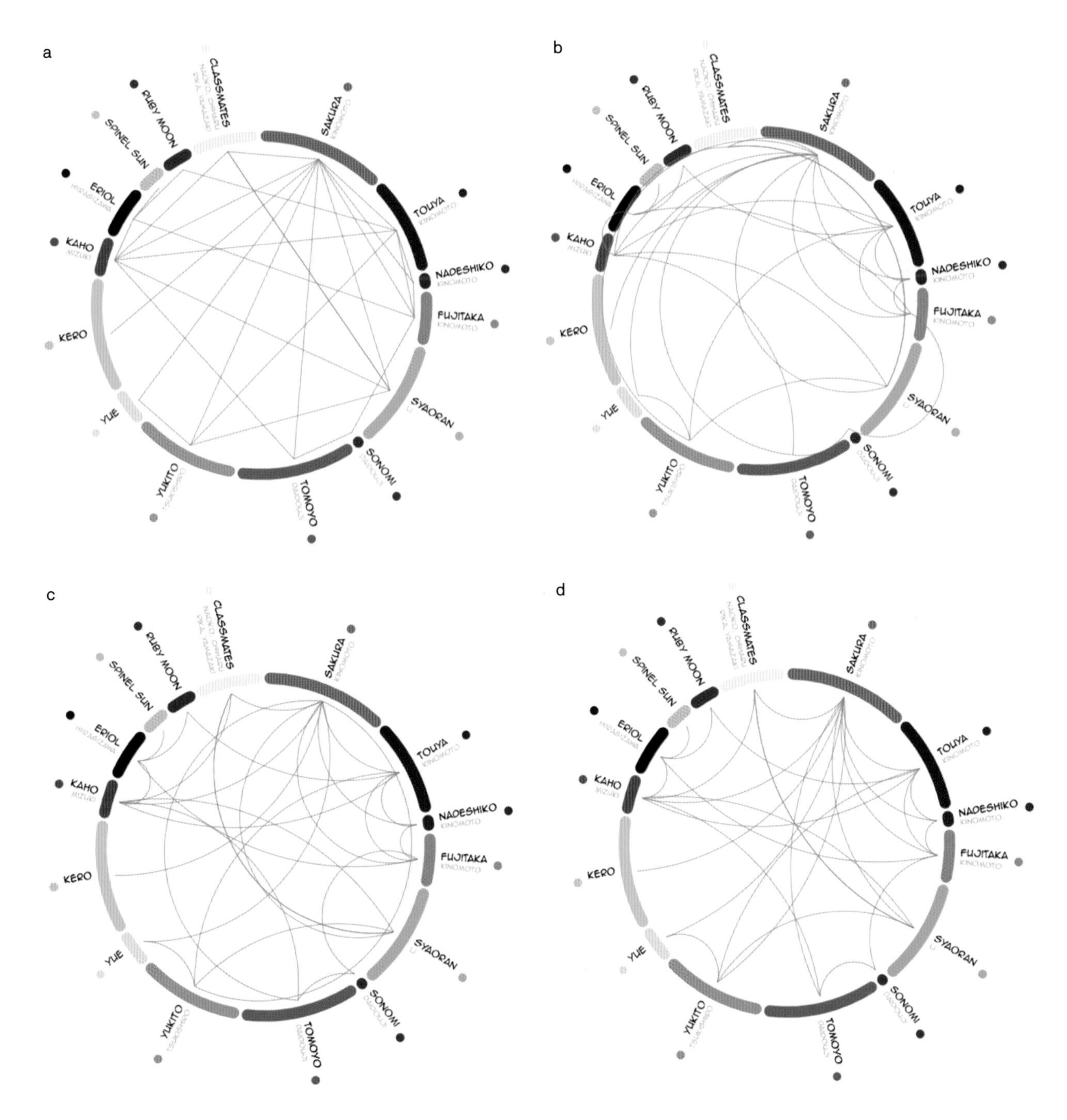

Fig.12.11 (a,b,c,d)

Figuring out the SVG paths of the "inner relationship" lines.

The most extreme and difficult lines that I could create would run from a character to a chapter that's on the other side of the circle. The line would have to swirl around the inner circle, without touching any of the other character's names. I thought I could probably pull that off by combining two Cubic Bézier Curves. But making one of those curves act as you want, depending on the data, can be a hassle. And I found out that two was more than twice the hassle. (◍•﹏•)

With such difficult paths I started out by placing small dots along the line path itself (the red circle in the center of Figure 12.12 a) and the anchor points (the blue, green, and yellow-orange circle [ignore the pink circle]). It gave me an idea of what "handles" I was moving around and how that affected the shape of the visible line.

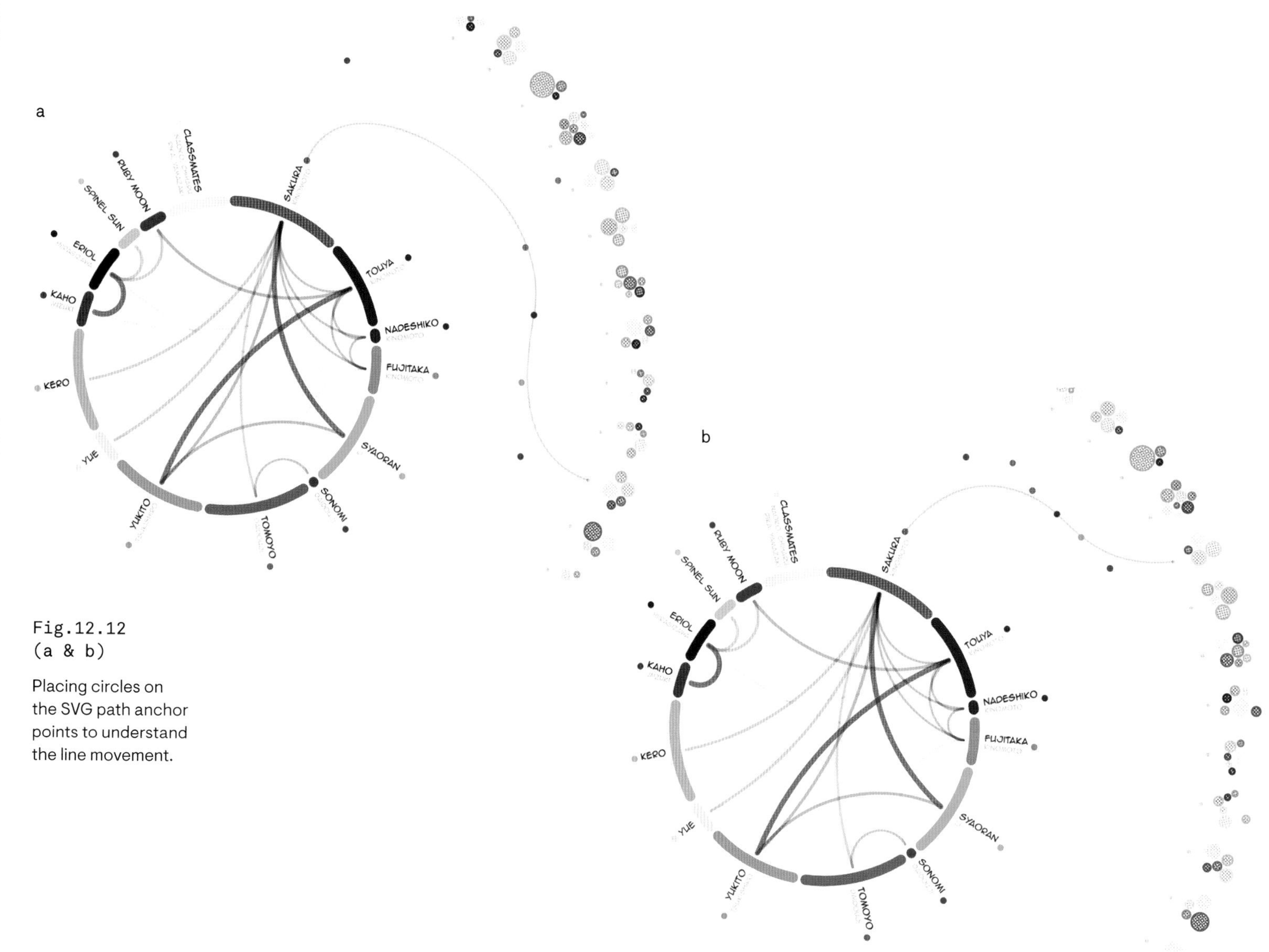

Fig.12.12 (a & b)

Placing circles on the SVG path anchor points to understand the line movement.

After some manual tweaking with fixed numbers I had a shape for the longer line that I liked. I saved those settings and made a shorter line. I then inspected how all of the settings changed between the two options. This gave me hints on how to infer several formulas that would hopefully create nice looking lines, no matter the start and end points. But, like I said, that wasn't as easy as I'd hoped.

I don't really want to think back on the journey that eventually led me to the lines I needed, but for a long, long time I kept messing it all up completely (see Figure 12.13).

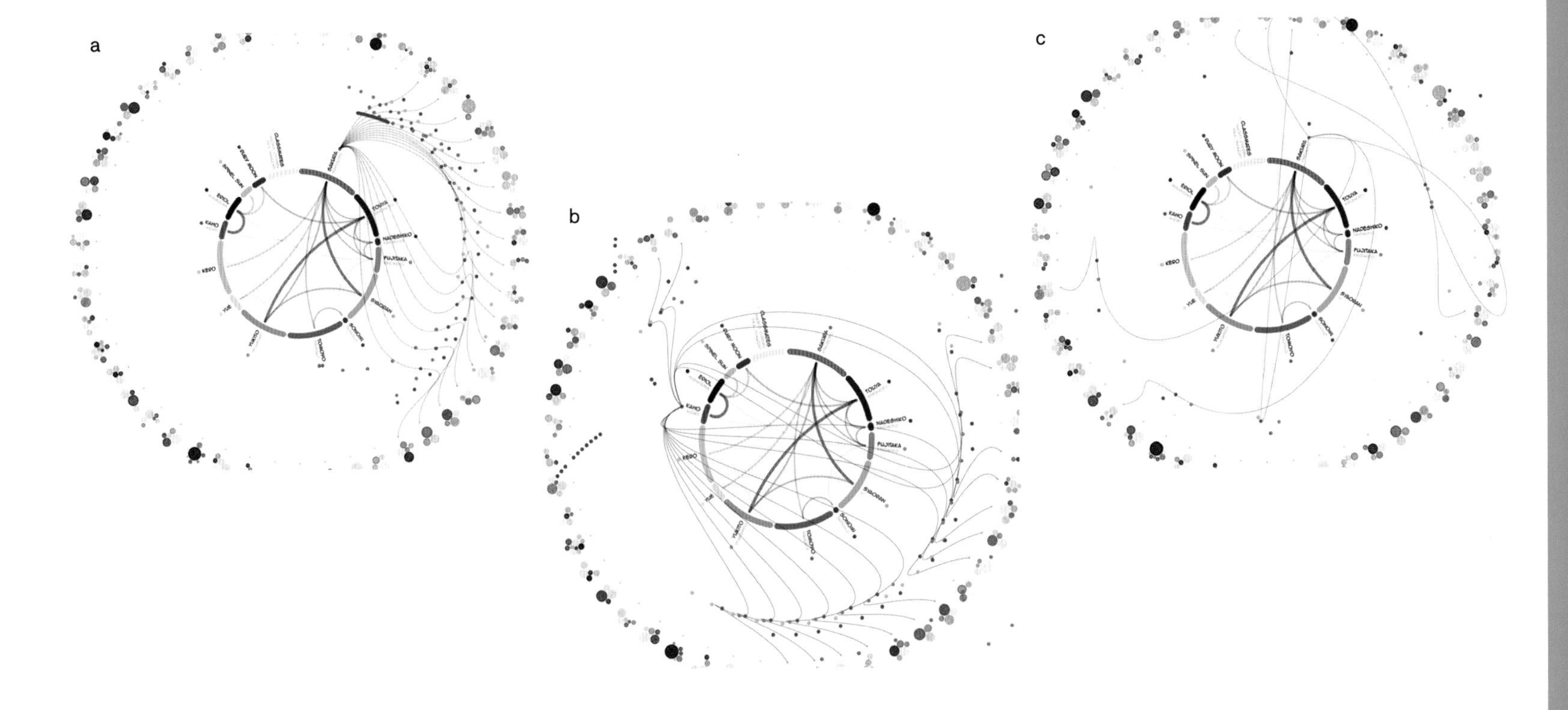

Fig.12.13 (a,b,c)

Some lines might look ok, but for others the angle might not be right anymore and the lines would get very strange loops.

Only through many, many hours of testing, drawing, thinking, and fiddling did I inch closer to having all the lines sort of going around the center, and I was left with the image of Figure 12.14 a where I visualized all of the lines. Awesome, that looked like one big mess! There were too many lines in there to glean any insight. (—_—*;) I also didn't like those thin lines. Perhaps they should have more *body*? So another session of line formula tweaking ensued. I decided to taper the lines, similar to what I did in my Dragon Ball Z visualization. It felt like an improvement over the same-width lines I had before, but I was still not happy (Figure 12.14 b).

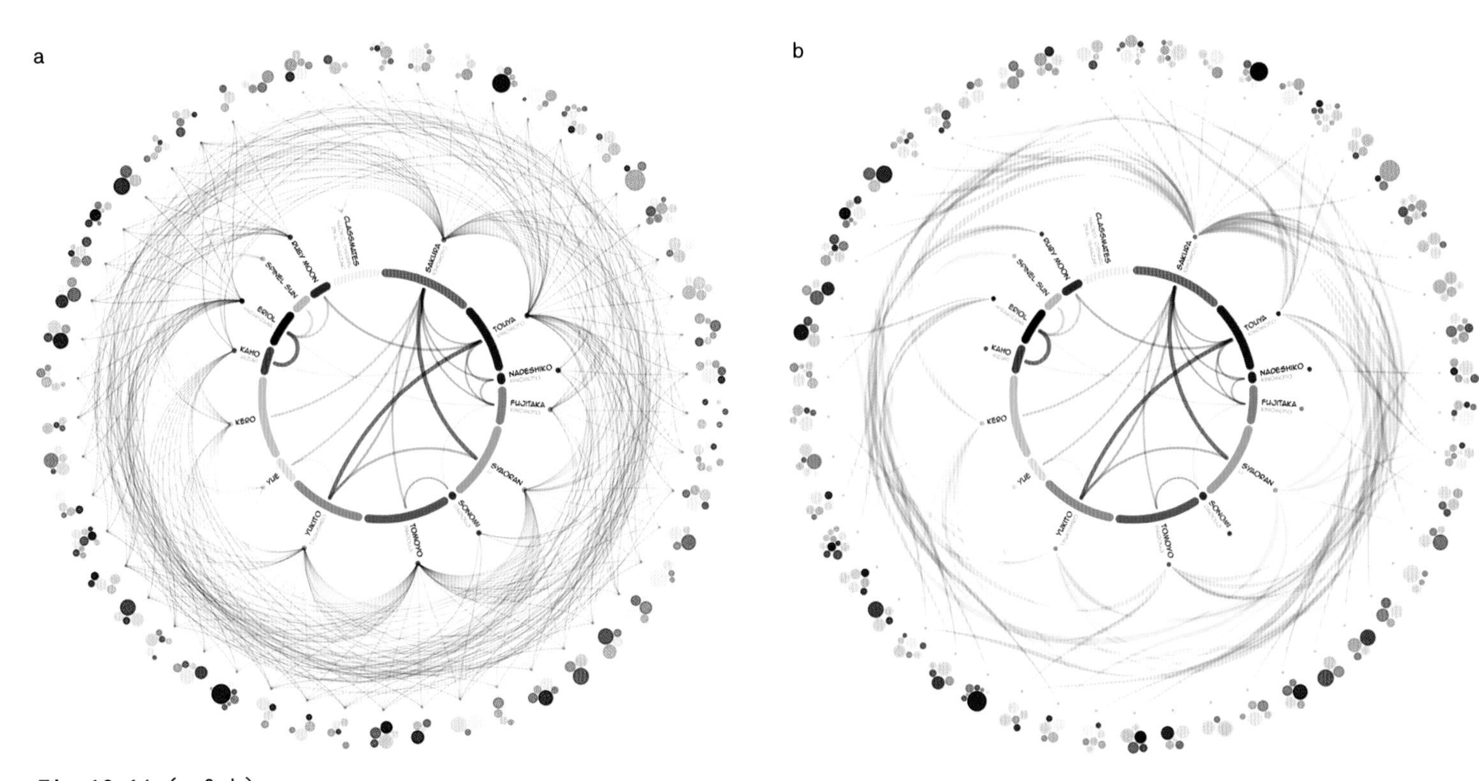

Fig.12.14 (a & b)

Thin lines at first, but trying tapered lines to make it look better visually.

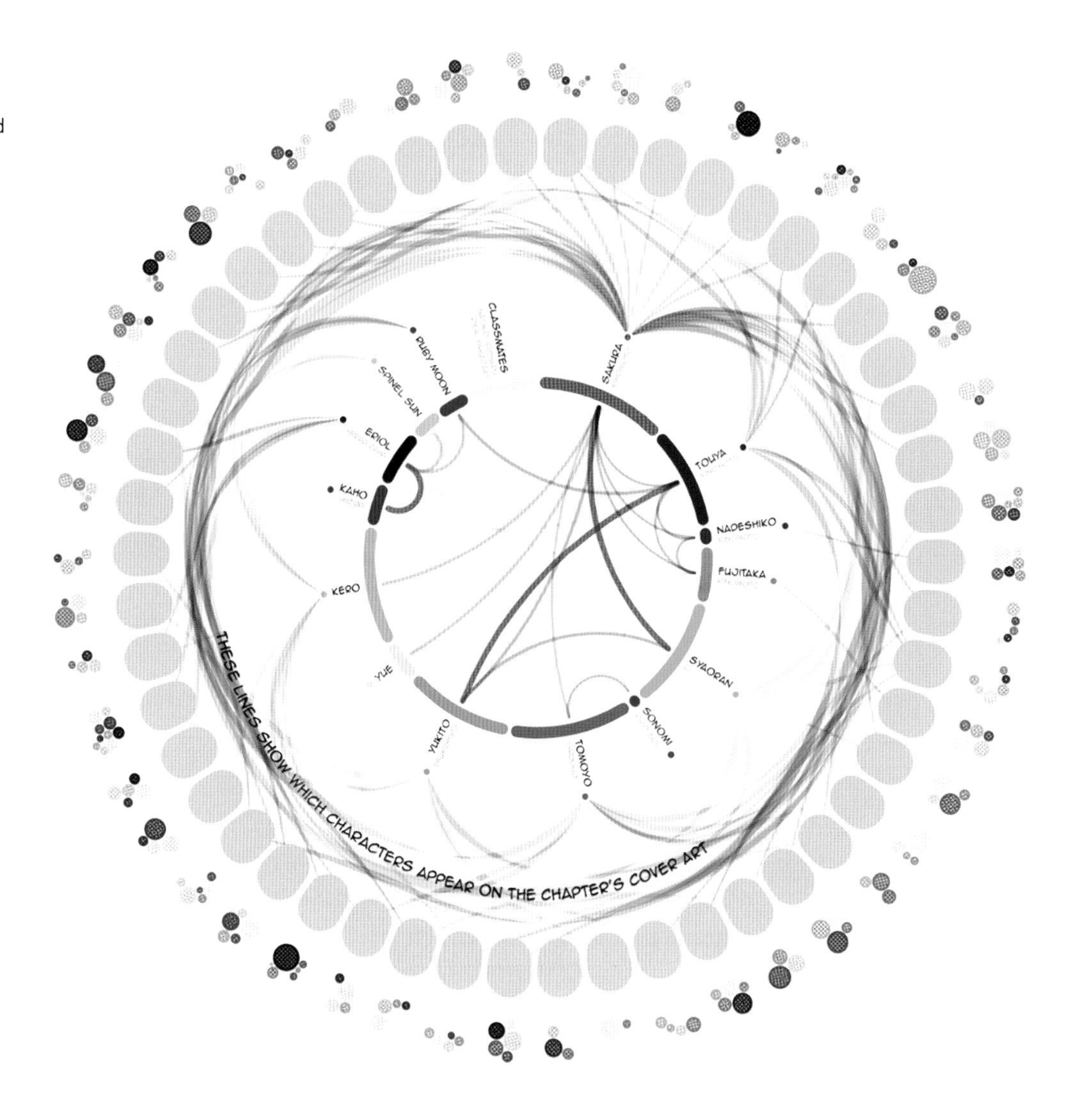

Fig.12.15

Setting up the 50 chapter "pills" around the outside.

I moved on from the lines for a bit to adding the chapter and volume numbers to the visual. The inner donut chart inspired me to try something similar for the chapters as well, so I created a donut chart with 50 equally sized, rounded-off sections (shaped like small "pills") in which I could place a number (See a work-in-progress shot in Figure 12.15). I was happy with the end result and moved on.

Next, I organized the volumes (typically a collection of four chapters) into a thinner donut chart which I placed outside the ring of circles. I wasn't as happy with that as the chapter "pills," but in the meantime something else was bothering me even more. Now that I'd placed more elements on the page and the look and feel of the visualization was more consistent, those inner lines didn't appear to fit the design. ಠ_ಠ

Accepting that the swirly lines weren't working, I decided to try something radically different. Inspired by the way the rest of the visual looked with all of the added elements, my new idea was to make the lines swirl around and have them run along *circular paths*. (Think of subway map lines or pipes in a home, but then transformed to a radial layout.) Calculating the small array of radii and angles that I now needed to draw a line was a walk in the park compared to my previous SVG Cubic Bézier Curve shenanigans! Naturally, all that didn't go right on the first try, but thankfully I could use some of the work I'd done with the swirly lines.

The screenshots from Figure 12.16 were all made within two hours; not bad in terms of progress!

With the change of how the lines looked, I felt that the resulting lines one saw when hovering over a character or chapter fit perfectly with the "straight-roundedness" of the rest. And as an added benefit, no more lines overlapped! (See Figure 12.16 f)

Fig.12.16
(a,b,c,d,e,f)

Different steps in the process of creating the third iteration of lines.

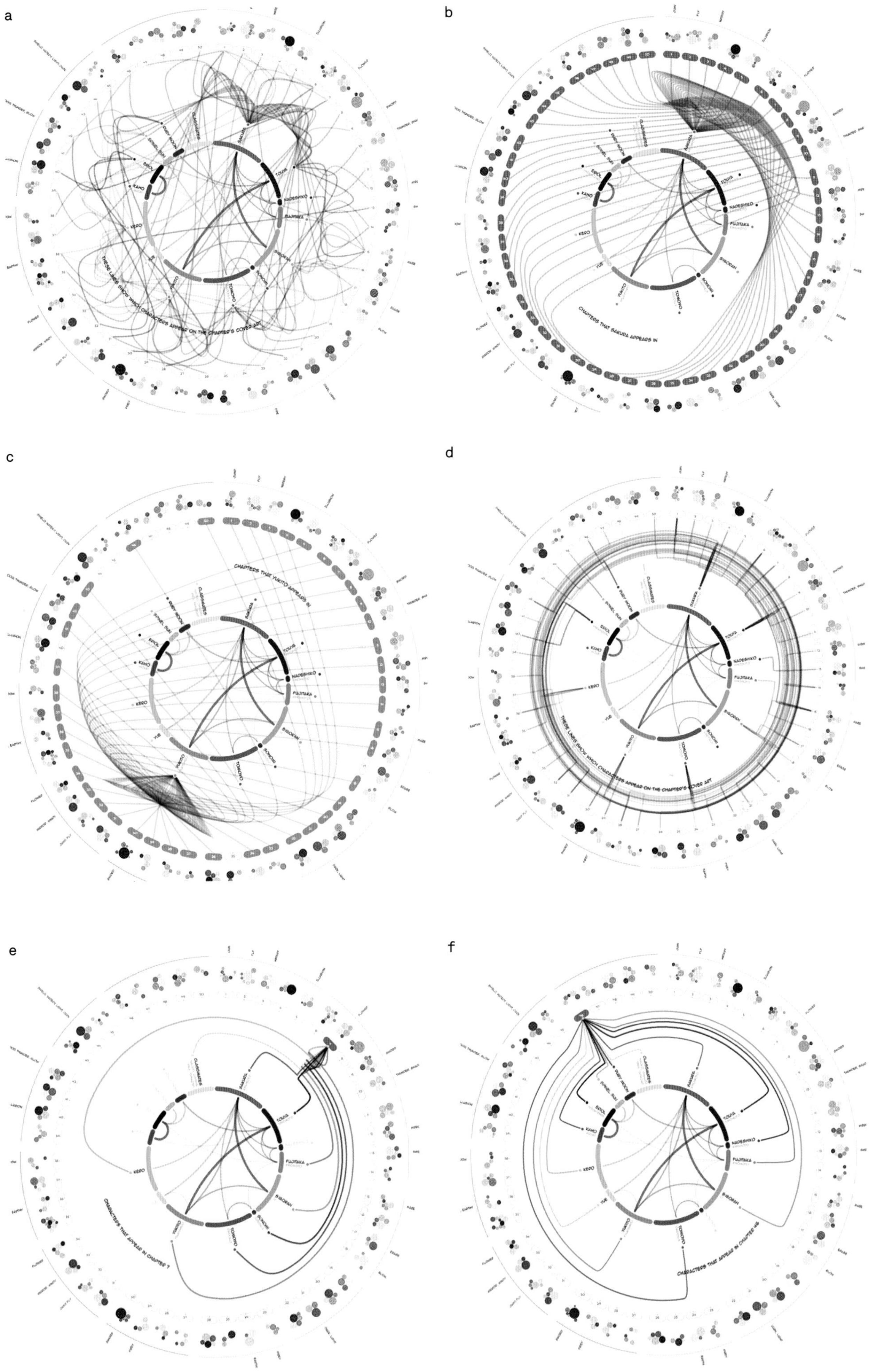

a

b

c

d

e

f

↳ Design to Maximize for Delight

Cutting out square images from the 50 chapters and 15 characters was yet another to-do in this already long month, but I knew that having actual images of CCS in the visual would make the whole visual a lot more fun to investigate for readers. Plus, it was an important part of teaching people about CCS and therefore worth the effort!

Since part of the visualization was about the covers of the chapters, and because I knew most people that landed on the page would probably not know about CCS, I wanted to incorporate some of its imagery. Thankfully, I just so happened to have a nice large circular area in the center! It took a while to manually "cut out" a good looking square image from all 50 chapter covers, but at that time I was in an airport and on a plane anyway. These covers would then be shown in the center when someone hovers over any of the chapters.

I was flying back from a great night at the "Information Is Beautiful Awards" where *Data Sketches* won GOLD!! Woohoo!

While reading through all the chapters, I took some notes of interesting story points that I wanted to highlight. These were easy to add around the outer circle with the help of the excellent `d3-annotation.js` library. The super handy "edit" mode let me drag the annotations around, see where I wanted them, and then hardcode and add those locations back into my code. I did end up creating my own lines connecting the visual with the annotations because I wanted them radiating outward from the center (Figure 12.17 b).

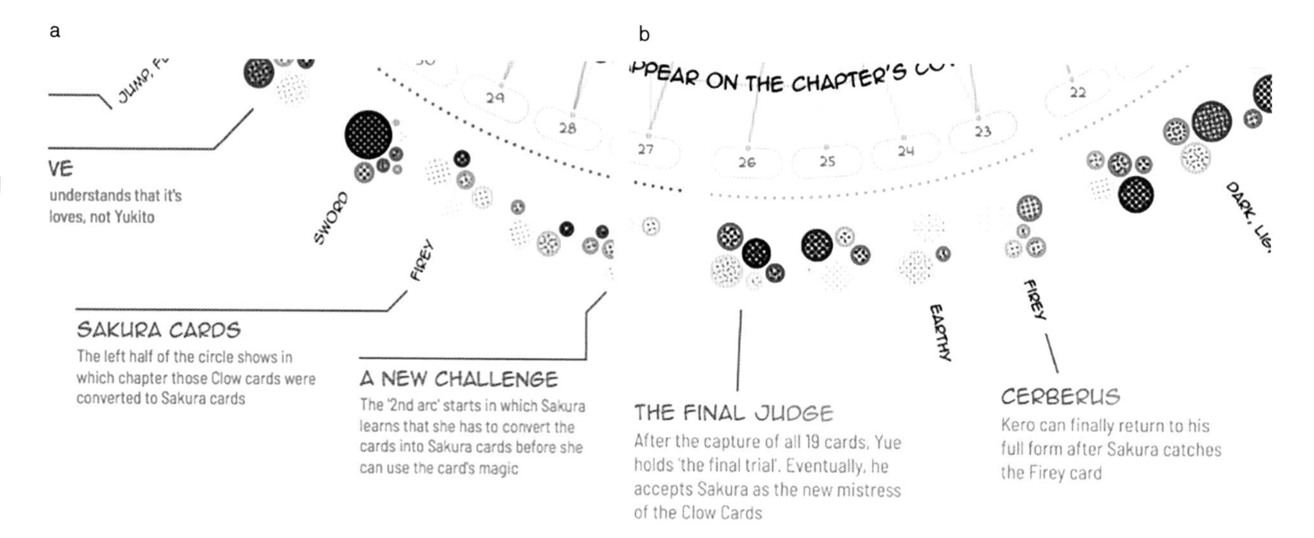

Fig.12.17 (a & b)

Starting out with the original annotation lines, but eventually creating my own radial lines.

With all these elements and layers of information in the visualization, I really needed a legend. I created it using Adobe Illustrator (Figure 12.18), saving me *a lot* of time over creating them through code. However (as I was getting used to for this project), this legend did not turn out to be the final result.

I added an introduction at the top to explain the series and its main characters, the legend at the bottom, and with the full visualization ready to preview I shared it with some friends to ask for feedback. I received great suggestions on how to improve the (interactivity) understanding and a much better idea for the legend that would show the visualization with its rings and explain what each ring truly meant (see Figure 12.26 later in this chapter). And after all that time and effort, I finally had a visualization to share with everybody!

Fig.12.18

The first legend I created in Adobe Illustrator, explaining how to read the visualization.

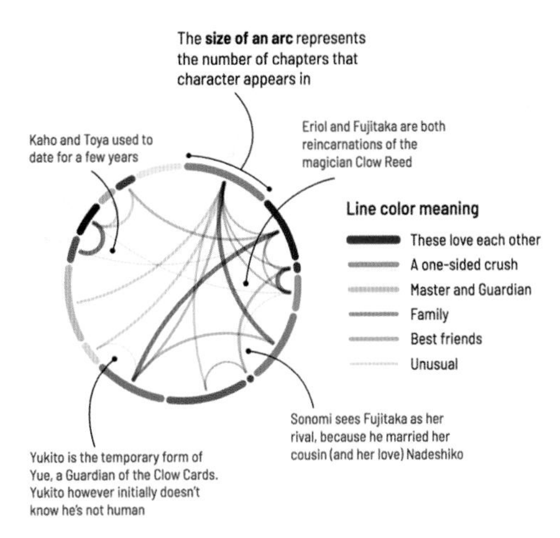
INNER RELATIONSHIPS
The size of an arc represents the number of chapters that character appears in
Kaho and Toya used to date for a few years
Eriol and Fujitaka are both reincarnations of the magician Clow Reed
Line color meaning
These love each other
A one-sided crush
Master and Guardian
Family
Best friends
Unusual
Sonomi sees Fujitaka as her rival, because he married her cousin (and her love) Nadeshiko
Yukito is the temporary form of Yue, a Guardian of the Clow Cards. Yukito however initially doesn't know he's not human

COLOR CLUSTERS

The pixels of each chapter's cover image are grouped into 3 - 10 main colors

The best-matching number of groups has been manually determined for each cover

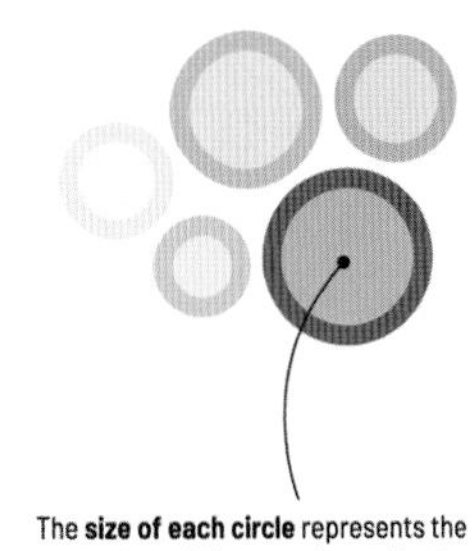

The **size of each circle** represents the percentage of the cover image that is captured in that color

All circles from one chapter cluster together add up to 100%

MAIN CHARACTERS

Sakura Kinomoto The main character and master of the Clow Cards

Toya Kinomoto Sakura's bigger brother, also has magical powers

Fujitaka Kinomoto Sakura's father, an archeology professor

Nadeshiko Kinomoto Sakura's mother, who died when Sakura was only 3

Tomoyo Daidouji Sakura's best friend. Makes Sakura's outfits

Sonomi Daidouji Mother of Tomoyo and cousin of Nadeshiko

Kero A Guardian of the Clow Cards. Sun based

Yukito Tsukishiro Toya's best friend/love. Yue's temporary form

Yue A Guardian of the Clow cards. Moon based

Syaoran Li A distant relative of Clow Reed. Sakura's rival at first

Kaho Mizuki Sakura's substitute Math teacher & friend

Eriol Hiiragizawa Sakura's friend and main "antagonist" of the 2nd arc

Spinel Sun One of Eriol's Guardians. The counterpart of Kero

Ruby Moon One of Eriol's Guardians. The counterpart of Yue

Classmates Naoko, Rika, Chiharu and Yamazaki

Normal human *Human with magical powers* *Magical being* *Reincarnation of Clow Reed*

↳ Combine Tools

For a very long time, I created most of my visuals completely using JavaScript code. However, I noticed that some parts of the visual, such as legends and annotations, took up both a lot of time to create and many lines of code.

I later realized this was a silly process for visuals that ended up as static images anyway. For those, I now only create the "base" visualization while programming. Then I copy the SVG visual from my browser into a vector drawing program and add any surrounding annotations, pointers, titles, and other elements.

I use either Adobe Illustrator or Affinity Designer for my vector work, but Inkscape, GIMP, or Sketch are other options.

I can still make use of the ease of a vector program even when the final visual remains interactive. For example, you can create a legend, either completely as an image, and place this next/within your visual (which I did for this project). Or you can save your legend as an SVG and load it into your visual. There are certain ways in which the (legend) SVG remains changeable by your JavaScript code. Which means that you can adjust the colors of each element separately, resize it, or even add an animation to it!

So keep an open mind when deciding which tool to use; not everything has to be done in the tool that you use for the (interactive) visualization. Tools are typically not meant to do everything, but specialize in certain things.

Reflections

This visual took me 86 hours to create, and a lot of that time went into things that weren't used in the final visualization. I spent ±5 hours on a CMYK canvas-based dot effect, ±15 hours on the swirly lines, and ±6 hours on a page layout that is only really visible when you press "Read More."

Still, I'm happy with how the visualization turned out. I feel I haven't quite seen a similar radial visualization like it before. And *Data Sketches* is all about experimenting with new ideas, so it's always a joy when a project turns out refreshing. I hope you enjoy interacting with the visualization online, even if you've never heard of *Cardcaptor Sakura* before. (It's great, btw!)

An Ode to Cardcaptor Sakura

↳ CardcaptorSakura.VisualCinnamon.com

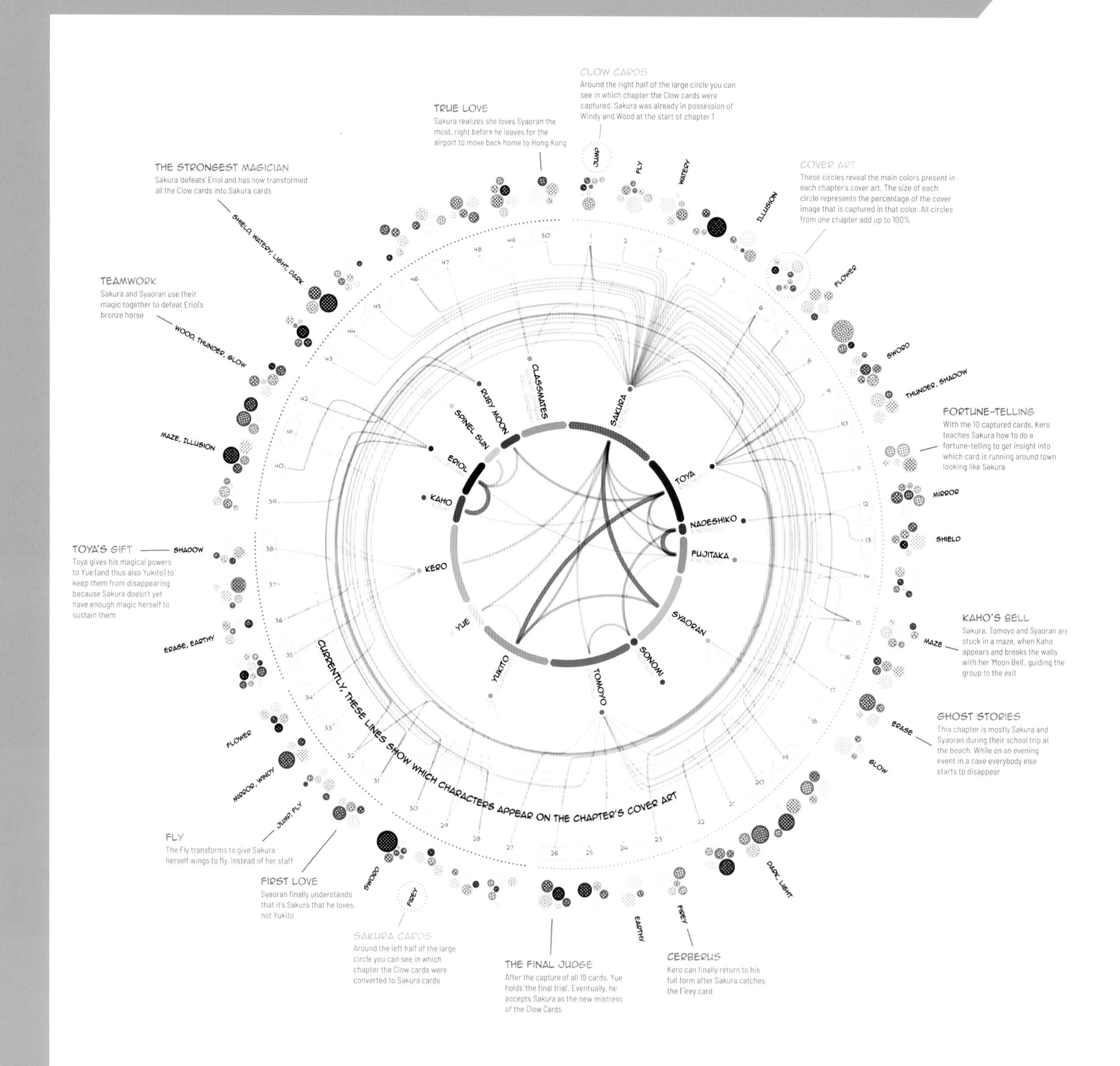

Fig.12.19

The main visual outlining the 50 chapters around the outside and the characters along the inside circle.

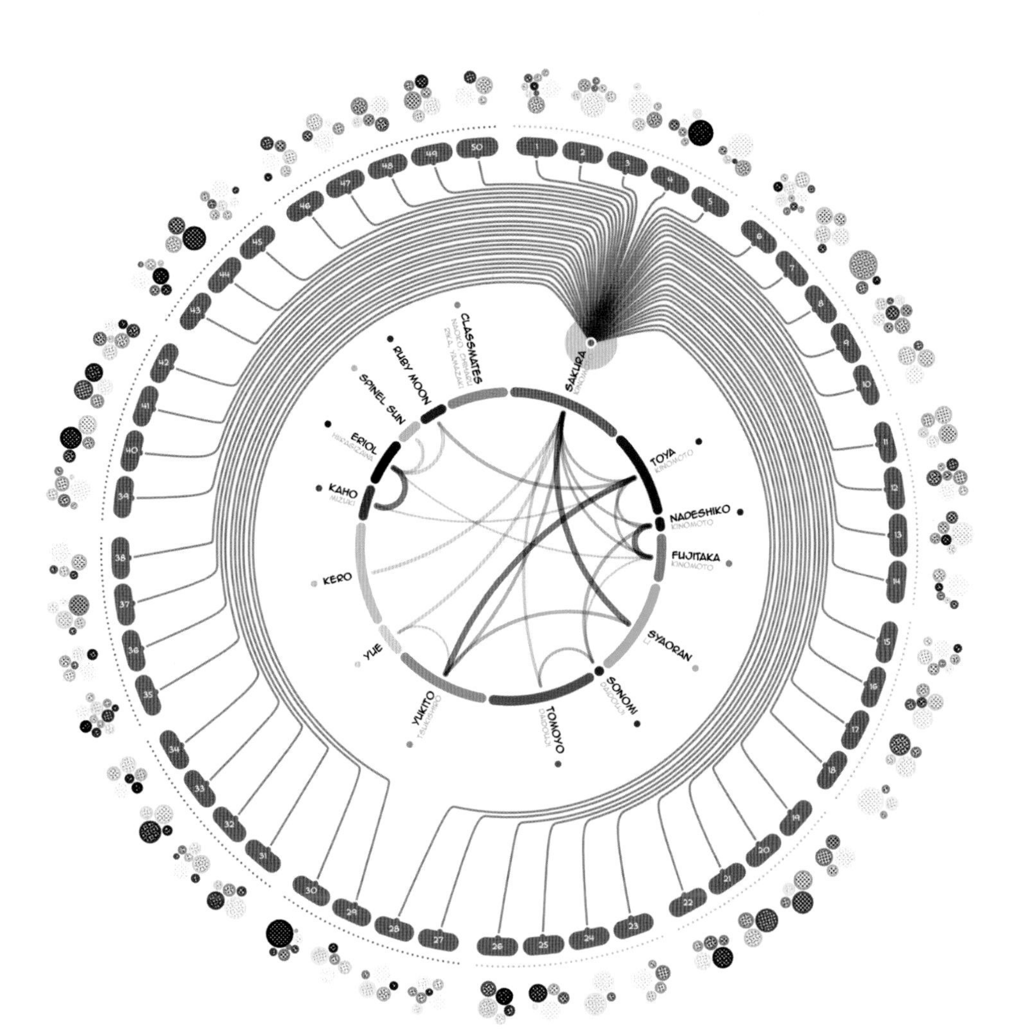

Fig.12.20

When hovering over a character, such as Sakura here, the lines reveal in which chapter they have appeared (while awake). Sakura is the main character of the story and is in all chapters.

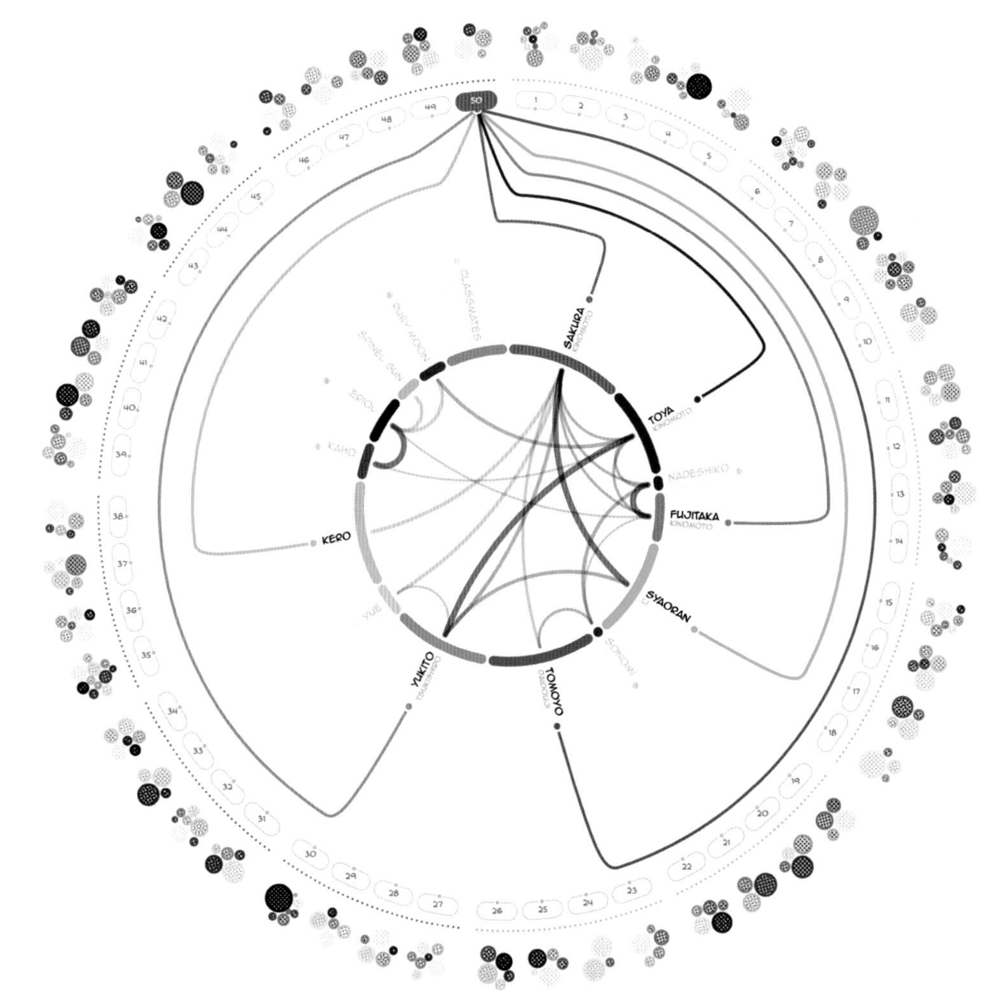

Fig.12.21

Hovering over a chapter "pill" in the outer ring, such as Chapter 50 here, shows all the characters that appeared in that chapter.

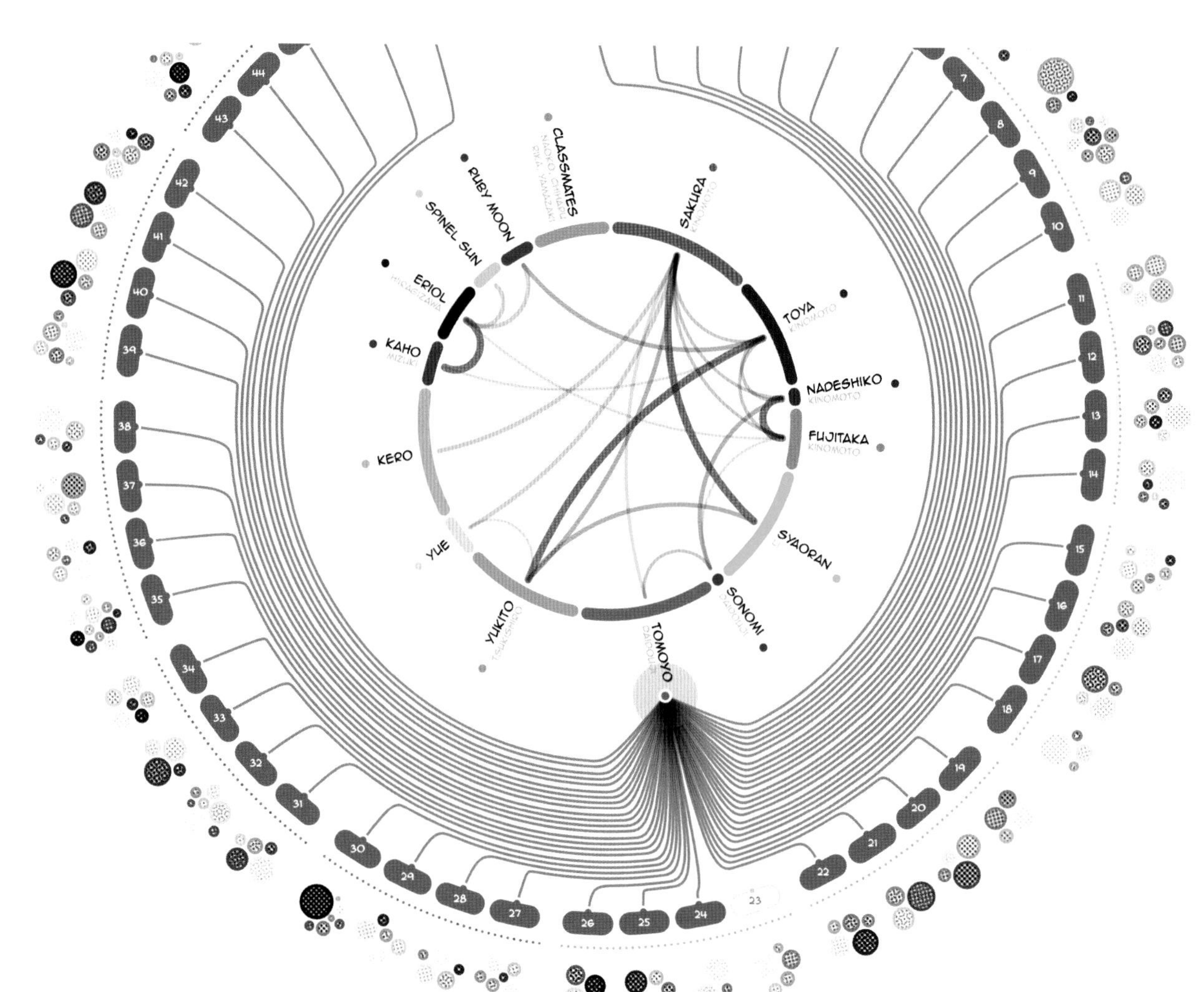

Fig.12.22

Being Sakura's best friend (and classmate and dress creator and maker of videos), Tomoyo also appears in almost every chapter.

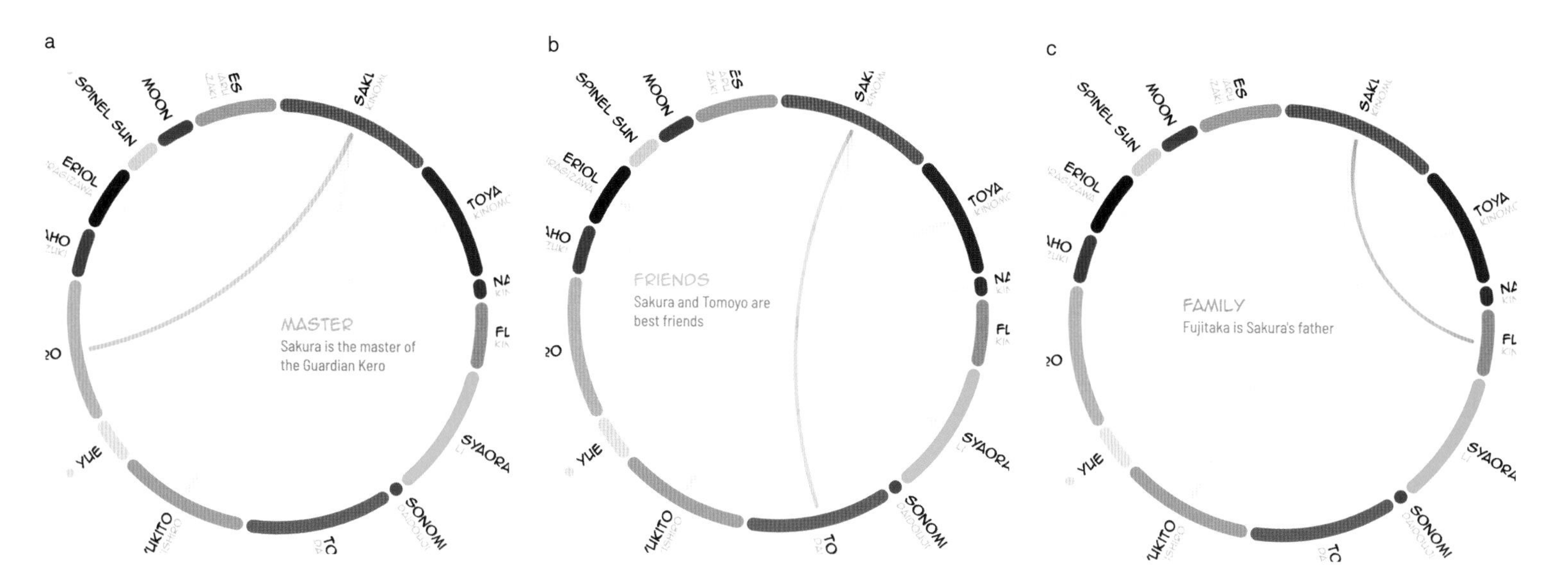

Fig.12.23
(a,b & c)

When hovering over any of the "relationship lines," a small annotation provides more detail about it.

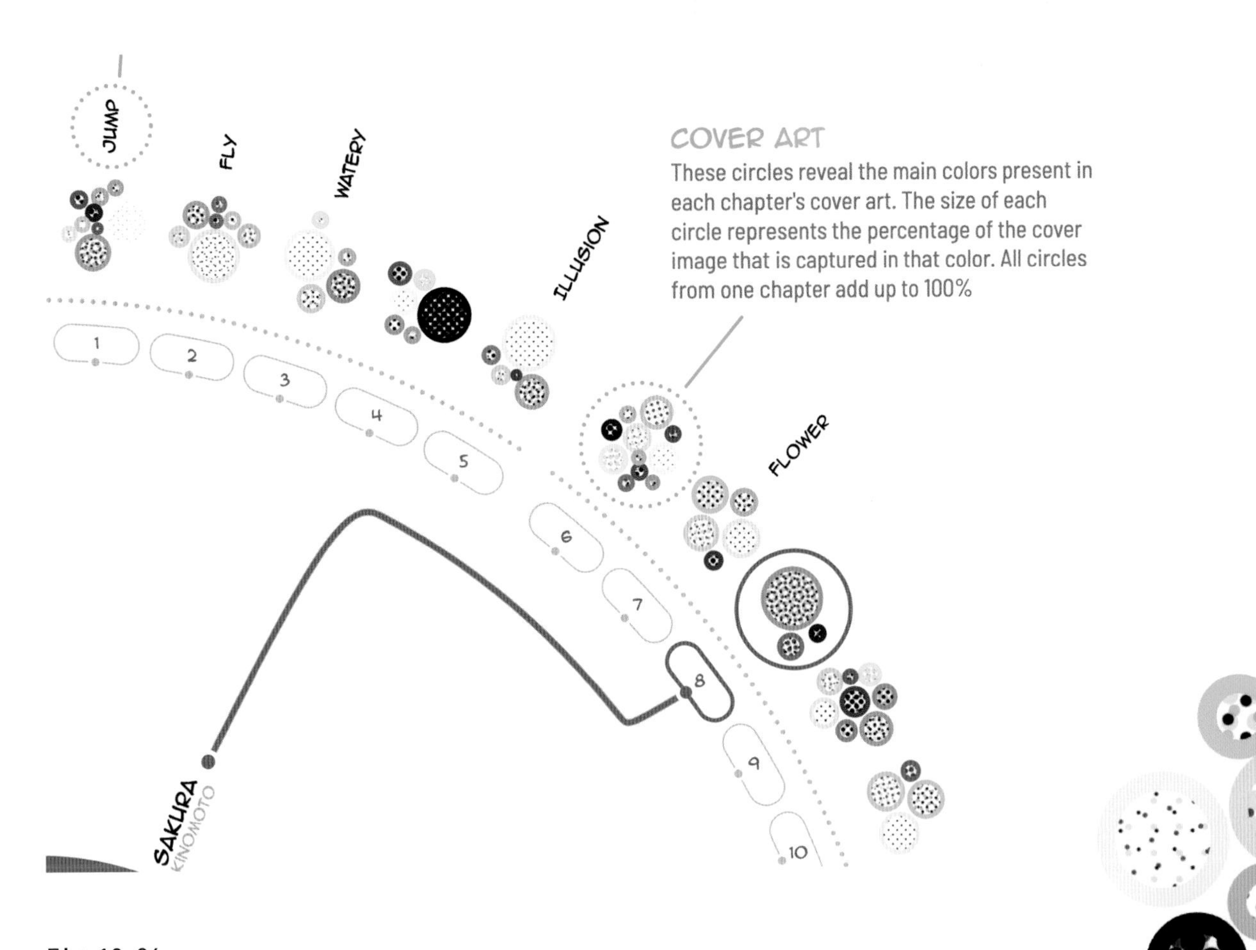
JUMP
FLY
WATERY
ILLUSION
COVER ART
These circles reveal the main colors present in each chapter's cover art. The size of each circle represents the percentage of the cover image that is captured in that color. All circles from one chapter add up to 100%
FLOWER
1
2
3
4
5
6
7
8
9
10
SAKURA
KINOMOTO

Fig.12.24

A close-up of the outer CMYK dotted circles, where each cluster of circles gives an idea of the colors used on the (gorgeous) cover art of each chapter. Hovering over any cluster will reveal a line for the characters appearing on the cover (here just Sakura).

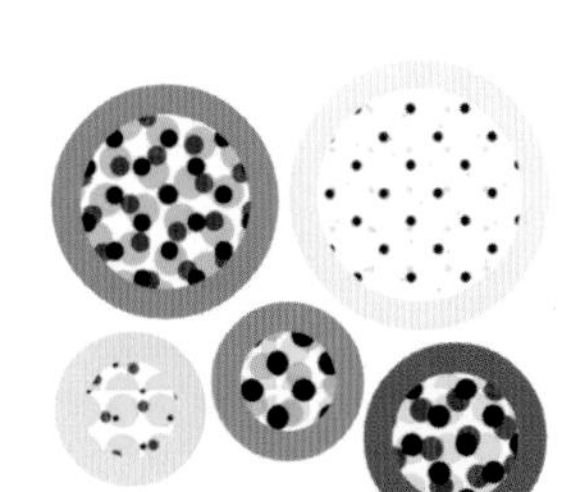

Fig.12.26

A zoom-in on several CMYK dotted circles.

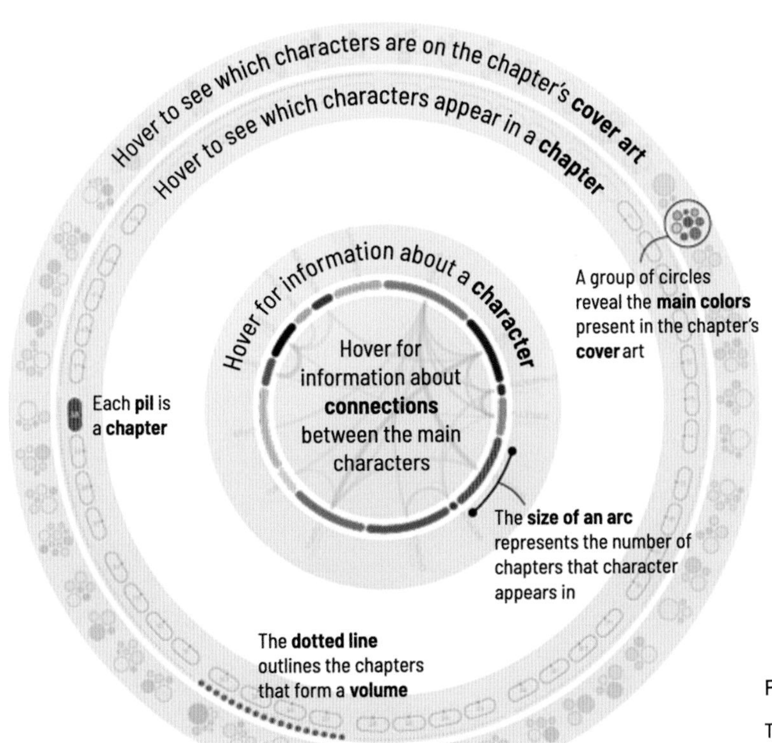
Hover to see which characters are on the chapter's cover art
Hover to see which characters appear in a chapter
Hover for information about a character
Hover for information about connections between the main characters
A group of circles reveal the main colors present in the chapter's cover art
Each pil is a chapter
The size of an arc represents the number of chapters that character appears in
The dotted line outlines the chapters that form a volume

Fig.12.25

The final legend using the visual itself to overlay explanations on what each of the different rings represent.

One Amongst Many

One Amongst Many

SHIRLEY

For the last few years, I've been wanting to bring my data visualizations into the physical world, to express my data-driven stories as immersive installations that people can walk through, explore, touch, and feel. I struggled with this dream for years, until I realized why I felt such a mental block: because I designed and coded primarily in 2D and not 3D, I had no idea how to even think in the third dimension. As soon as I realized that, I created *Legends* with `Three.js` and WebGL, and it really helped me understand how to think about space and lighting in the third dimension.

But being able to code in 3D isn't, by itself, enough to create a physical installation; I also needed to learn hardware. In the fall of 2019, I did an Artist-in-Residency at New York University's Interactive Telecommunications Program (ITP), a master's program that is at the intersection of art and technology. I went to ITP with two major hopes: to learn more about hardware and to collaborate with talented people on a physical data installation.

To learn hardware, I audited "Introduction to Fabrication," where I learned about power tools and machines and played with the laser cutter, and "Introduction to Physical Computation," where I learned about circuits and sensors and lights and Arduinos. Through those classes (and the office hours I was asked to hold as a resident), I was able to meet my two collaborators: Christina Dacanay[1] with her beautiful illustration style and bold uses of color, and Tina Rungsawang[2] with her architectural background and keen awareness of space. When I asked them if they wanted to collaborate with me on a project for ITP's Winter Show, they immediately agreed.

[1] Christina Dacanay: http://www.cdacanay.com
[2] Tina Rungsawang: http://tina.pizza

Data

From the very beginning, we agreed that we should focus the project on women in computing. It's a topic that I've been interested in since I created *Legends* with women Nobel Laureates, and Tina and Christina agreed that it would also be a perfect topic for ITP, with its emphasis on combining code and computation with art.

I started my data gathering with a Wikipedia article—the same way as *Legends*. I was able to find a page titled "Women in Computing," and from there I discovered the Wikipedia category "Women computer scientists." I started by scraping the list of women in the "Women computer scientists" category, but after a day of cleaning and exploring the data, I realized that both the data I pulled and the analysis I did were too simplistic to tell me anything interesting. Instead of trying to devise a more sophisticated analysis (we were quite limited on time), I decided to take a step back and try my hand at another dataset. I had initially avoided scraping the "Women in Computing" article because the women's names were scattered throughout the paragraphs instead of cleanly in a list, and that would be more work to extract. But once I realized that the data in there was much richer—the paragraphs succinctly described each woman's accomplishments and even gave an approximate timeframe for most women's contributions—I knew immediately it would be worth the trouble.

For my data gathering process, please see the data section of "Myths & Legends" where I did almost exactly the same thing. For the analysis, I looked at distributions of birth years and backlinks, and intersection of keywords in the summary—none of which produced anything noteworthy.

For this second attempt at scraping the data, I wrote a script to:

1. Use `wikijs` to access all the links in the "Women in Computing" page, and filter down to only those that link to other Wikipedia pages.
2. Use Google's Knowledge Graph Search to determine whether the Wikipedia page is for a person or thing, and only keep those that the Knowledge Graph returned as "Person."
3. For each person, use `wikijs` to pull their Wikipedia article summaries, and for each summary, see if it includes "she/her" or "he/him." Keep those that have more mentions of "she/her."
4. If there were no mentions of either, mark the person for a manual check.
5. Save name, summary, and certainty of gender into a spreadsheet.

Once I had the spreadsheet, I went through each person to double-check their gender. For each of the remaining 46 women, I wrote a one sentence summary and found an approximate year for their contributions by reading through the "Women in Computing" article and cross-referencing their own Wikipedia pages.

Because of the space constraint for the ITP Winter Show, we could only design an installation with a maximum of 20 women. While gathering and cleaning the data, I highlighted approximately 30 women that I found particularly inspiring and brilliant. For the final pass, Christina, Tina, and I sat down and narrowed the list down to the 16 women that we found undeniably inspiring. (We achieved "undeniably inspiring" if all three of us, upon reading her summary, said "wow, she's so cool, I want to learn more about her!"—a very scientific process.)

Sketch

For our very first brainstorm, we asked ourselves what we'd make if we had no restrictions on time, space, or money. We went through many different ideas, but at the core of all of them was the desire to highlight the invisibility of these women and raise awareness of their incredible contributions through our installation. We also wanted to create an experience that would leave people feeling inspired, stimulated, and wanting to go through it again. Finally, we wanted to create something interactive, where people's interactions with the piece could potentially alter it and leave something of themselves behind.

This was the most unique and interesting sketch section out of all the *Data Sketches* projects I've done, and should more accurately be called "Prototype."

For the second brainstorm, we narrowed our ideas down to the more budget-and time-friendly ones and looked at existing installations for inspiration. We especially liked Judith G Levy's "Memory Cloud,"[3] where nostalgic photographs are placed within small souvenir viewers and people are invited to walk through dozens of these viewers and peek in to see each 35mm slide. We loved the idea of people interacting with each object and decided to do something similar: mapping each object to a woman and placing their information inside (Figure 12.1). We wanted people to take their time reading and learning about each woman, and I especially loved the motion of looking inside, like looking into a secret world and learning something precious.

Fig.12.1

Initial notes on the aesthetic, interaction, and data mapping of the installation

Hanging tubes:
- glass, reflective tubes w/ light at top.
- hang above head level
- pulling it down "triggers" it & pulling it again retracts it?
- while pulled down, the LED is bright, when it goes back it dims to its normal level.
- Color: "cluster", height: backlinks, x-position: year/date. brightness: cumulative # of interactions

Summary board
- corresponding name lights up
- give something to participant to stick next to people they liked.
- at the "end" of the path a person walks down.

From there, the rest of the data mappings came quickly: the women would be organized in the z-axis by the year of accomplishment, from earliest to most recent. The y-axis would correlate to the number of their backlinks, so the more backlinks they had and the more "renowned" they were, the higher up we would suspended them from the ceiling because they were more "out of reach." And for the women with fewer backlinks and less "renown," we placed them lower and easier to reach, so that more people could read and learn about them.

[3] Judith G Levy, "Memory Cloud": https://judithglevy.com/section/421361-Memory-Cloud-1-Indianapolis-Museum-of-Art-an-interactive-installation.html

To encourage people to “leave something behind,” we wanted to include a “summary board” near the exit with a prompt for participants to place a (gold, star-shaped ★☆) sticker by women they felt particularly inspired by because of the installation (Figure 12.2).

When we shared our initial ideas in class, our professor suggested an interaction where when one object is pulled down, it triggers others to drop down. This feedback inspired us to group the women so that when a visitor walks up and reads about one woman, the others in the group would drop down and create a path, guiding them through the rest of the installation.

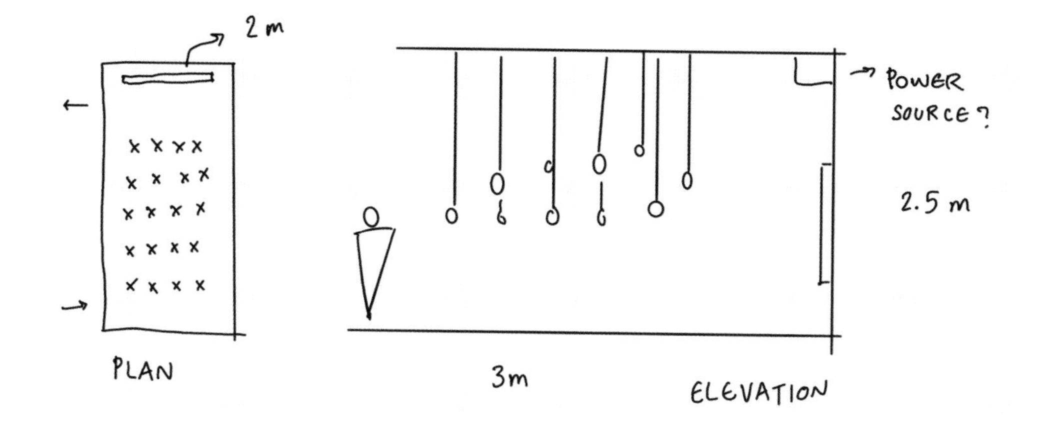

Fig.12.2

Initial floor plan with the “summary board” placement courtesy of Tina. We originally wanted a clear enter and exit, so that people would have a path they could follow, but we weren’t able to do this in the Winter Show with the space we were given.

With the majority of the idea fleshed out, Christina went about sourcing materials, and Tina and I began work on figuring out how to detect a person interacting with our objects in the installation. We brainstormed several ideas: a pulley system that would complete the circuit when pulled, a motor at the top that would unwind the cord when pulled, a capacitive touch sensor on the object, infrared motion sensors placed somewhere nearby ... all the options felt overly engineered, finicky, or sensitive.

When we went to our Fabrication professor’s office hours, he suggested we go simpler. He gave us a vibration sensor (where the circuit completes when we shake the sensor, Figure 12.3), and a tilt sensor (the circuit completes when it’s tilted up, Figure 12.4) to experiment with.

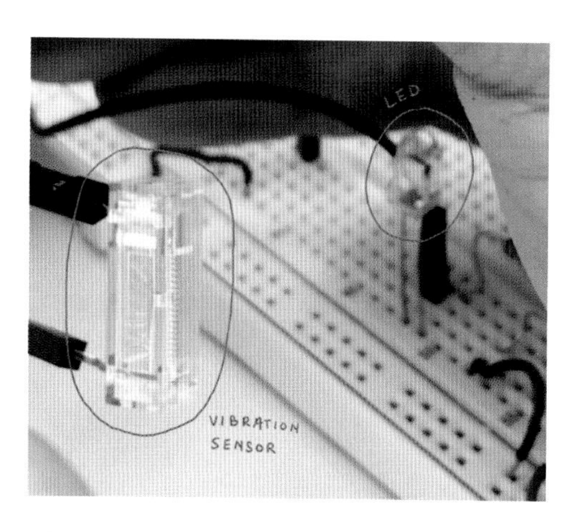

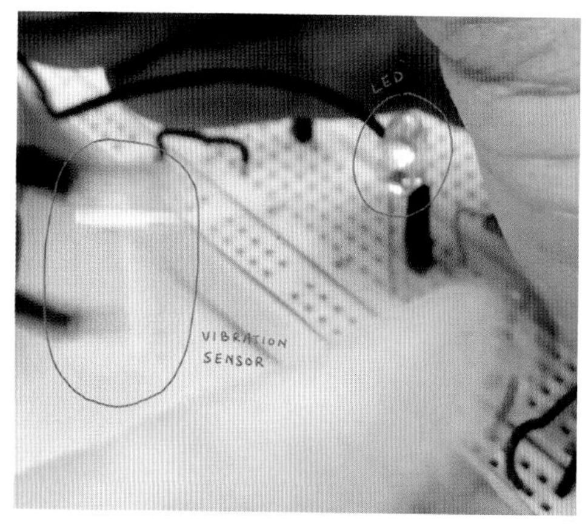

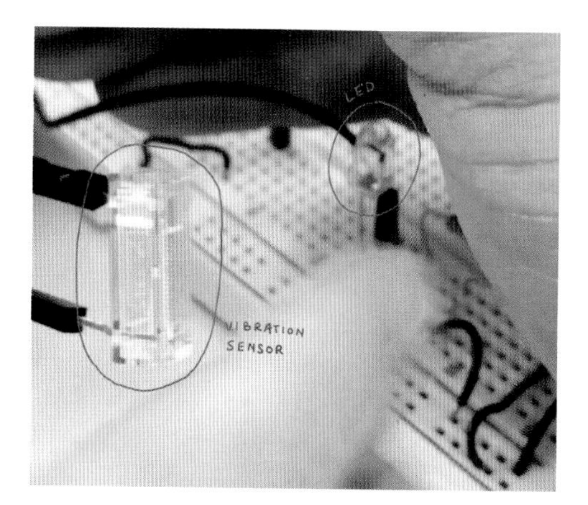

Fig.12.3

Vibration sensor, which unfortunately would only trigger when I flicked it at full force.

Fig.12.4

Tilt sensor, which completes the circuit when tilted above a certain angle. Slightly finicky but fairly reliable.

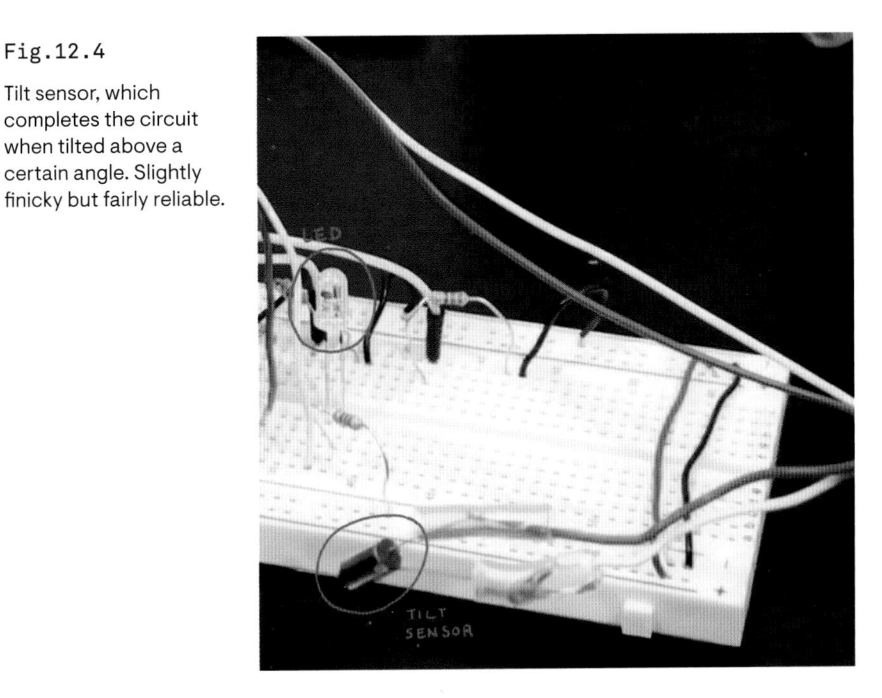

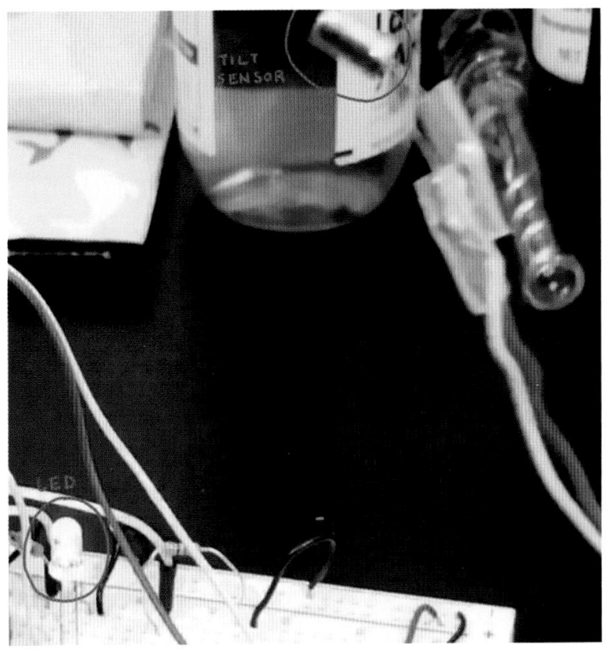

Unfortunately, the vibration sensor took quite a bit of force to trigger, which meant that if we used it in our object a person would have to really shake it (instead of gently picking it up) for us to know that they'd interacted with it. On the other hand, though the tilt sensor wasn't perfect—there didn't seem to be an exact angle that it started sending "on" signals, and even then those signals were quite noisy—it was a much better option.

We also considered an accelerometer, which would have more sophisticated readings than a tilt sensor but was 4x more expensive. We quickly decided against it.

Through this process, we realized how finicky and fragile a motor would be, especially with the thousands of people (many with kids!) coming through the Winter Show and pulling on the cords. Even though we really liked the idea—motion is always great for capturing attention—we decided to nix the idea. Instead, we agreed to put lights in the objects since changes in light are quite noticeable and even more importantly, much easier to control.

As soon as we settled on using lights, we knew the perfect way to take advantage of it: we would start the installation with each of the objects dimmed—a metaphor that despite their important contributions to computing, these women are still largely unknown—and as people went through and read about each woman, their lights would grow brighter and brighter, until by the end of the show, all of their lights would be brightly lit. This not only gave us a great overarching metaphor, it also accomplished our goal of having people's interactions alter the piece itself.

For the lights themselves, we went with Adafruit's brand of individually controllable RGB color lights called the Neopixel. We originally went with them because we wanted to map data to the color of the lights, but quickly decided against the idea when we realized how hard it was to distinguish most colors from each other (unless they were *very* different, like between red, blue, and green). We did end up keeping the Neopixels though, because they gave us better programming control than most LEDs (and also we'd already spent money on them (〃￣ω￣〃ゞ).

The hardest part about prototyping and working with these Neopixels was having to solder them to wires. Because the conductive areas on the Neopixel strips were so small and close to each other, we ran the risk of the solder touching and causing short-circuits if we weren't being careful. Fortunately, a kind ITP student taught me about soldering paste (Figure 12.5), which makes the solder stick only to the conductive part and nowhere else and it was such a magical time-saver. I also learned to use a hot glue gun after soldering to make sure that the wires wouldn't fall off easily. These are precisely the sorts of lessons I went to ITP for. (๑•̀ㅂ•́)و✧

On the bright side, I eventually learned how to use a multimeter to test for shorts—but only after accidentally short-circuiting my (and Tina's) Arduino boards...

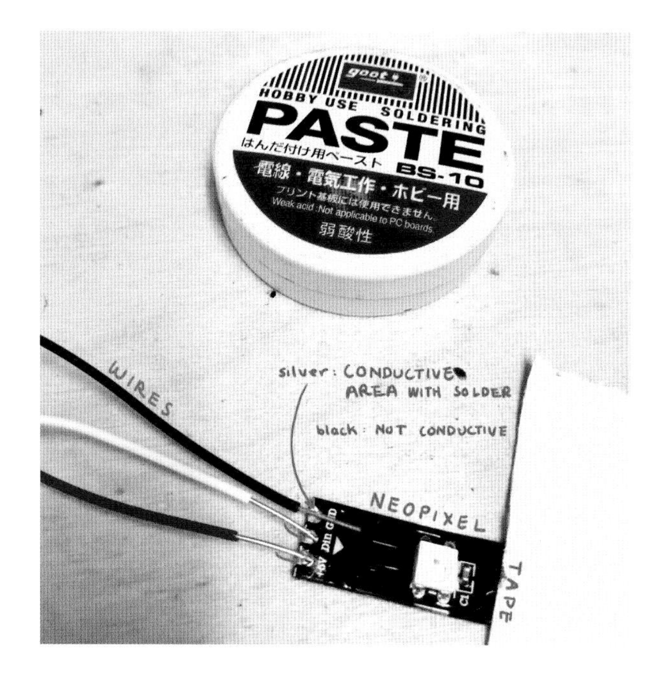

Fig.12.5

Soldering a Neopixel strip with the help of soldering paste, which was life changing.

Fig.12.6

The spray painted orbs, which unfortunately did not have a mirror-like finish.

At the same time we were figuring out the electrical components for each object, Christina was sourcing and experimenting with materials. She quickly landed on using fillable Christmas ornaments, and we went about acquiring one from a nearby Michael's to prototype with, while also ordering three dozen from Amazon at the same time. Christina then went about sanding the nobs off of each ornament and carefully drilling holes into the center for running the wires through. To hide the tilt sensor and Neopixel within the orb, we first tried to spray paint the interior of the orbs. We had bought reflective spray paint that was supposed to give a mirror-like finish, but it didn't turn out well. That was the night we learned reflective spray paint only works on glass and not plastic (Figure 12.6). (◠_◠;)

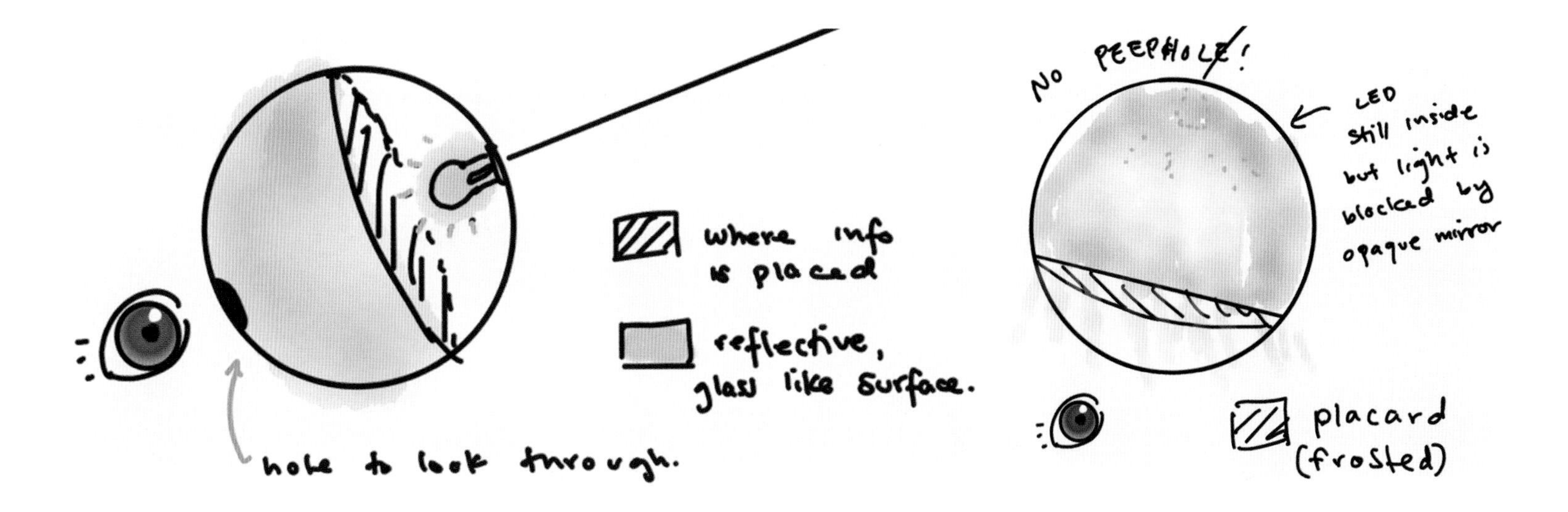

Fig.12.7

The two options we considered for how the orb should look.

We also took the opportunity to test out two competing ideas we had: we couldn't decide whether we wanted people to peek in through a small hole to read the information, or whether we should leave the orb as is and have a larger area to read the information from (Figure 12.7). So it was really great to have the spray painted prototypes, because right away we could tell that the peephole would be way too small and too much of a struggle for people to read large amounts of information through.

We also realized that we should place the information placard at a certain angle so that people would: 1) know that there was a placard inside they needed to read, and 2) be incentivized to pick it up in the correct direction to trigger the tilt sensor:

Fig.12.8

Prototype of the second option with no peephole and a large transparent area to read the information through.

As the reflective spray paint didn't work out, Christina suggested cellophane (Christina was an absolute pro when it came to materials). This not only hid the tilt sensor and neopixel, it also had a beautiful effect when lit up in the dark:

Fig.12.9

A quick prototype with cellophane and the information printed on vellum, a translucent paper to help mimic frosted acrylic and diffuse the light, that can also be printed on.

It was really fun to prototype the materials. The whole process was so different from working digitally, where we never have to worry about gathering materials, that it felt like a fun new challenge. Overall, it took about two weeks for us to settle on all the parts.

> We had to be strategic about buying just enough material to test with and ordering the rest early enough to get them by the time we had to start production.

I also learned a lot about prototyping with cheap, readily available materials (like masking tape, cardboard, and tin foil) to mimic the effect and quickly try out an idea.

Code

Once we had all the physical components figured out for the orb, I started on the coding. To control the tilt sensor and Neopixel we settled on, we used the Arduino:

Fig.12.10

An Arduino Nano on a breadboard.

Arduino is a brand of microcontrollers, and the way we use it is by wiring the pins on the Arduino (circled in orange in the photo) to our lights, sensors, motors, etc. Each of those pins have a corresponding number it is identified by, and in our code, we can tell Arduino to read a certain pin as input (usually a sensor) and another as output to write to and control.

(You never thought you were gonna get a run-down of Arduino from *Data Sketches*, did you? Well, we aim to go above and beyond! (๑•̀ㅂ•́)و✧)

Because this was my first full project with Arduino, I started by keeping my code simple. I used a button as a stand-in for the tilt sensor and an LED in place of the Neopixel. I used an array to keep track of which pins the buttons were wired to and another for which pins the LEDs were wired to. I also had an array for keeping track of which buttons were pressed and another for each LED's brightness. Every time the button was pressed it was the equivalent of a "tilt" and I'd increase the brightness of the LED and save it to the brightness array. I wanted to use the button and LED (which I was already familiar with how to code) first so that I could get the functionality working first, before having to worry about incorporating the tilt sensor and Neopixel—both things I hadn't worked with before.

Once I had the basic functionality, I went about replacing the LED and push button with the Neopixel and tilt sensor in my code. The Neopixel actually comes with its own library of functions, but because it assumes that I'm programming a strip of many Neopixels, it took me a long time and a lot of guesswork to setup each orb's Neopixel individually.

We needed to set up the Neopixels individually because the wiring between each Neopixel was too long, and the electronic signals that carry the instructions would decay at those lengths—another physical constraint I'd have never expected.

Reading the tilt sensor was a little more straightforward, though I did have to adapt an Adafruit example to accomodate for noisy signals. One of the most interesting lessons I learned working with hardware is that nothing in the physical world is clear cut like it is in digital and all sensor readings are noisy.

With the Neopixel and tilt sensor code incorporated, the rest of the code from my first iteration worked perfectly.

With one orb working, I had to make sure it could work both scaled up and with groups. As a team, we'd decided to group our list of 16 women into four categories: mathematicians, computer scientists, creatives, and other (executives, inventors, etc.). If any one orb was interacted with and lit up, the rest of the women in the same group would also light up in a delayed stagger.

To implement the delayed stagger, I needed to incorporate time into my code. The only way to do this in Arduino is with `mills()`, which returns the number of milliseconds since the Arduino began running. I refactored my code to store the following information: whether the whole group should be lit up or not, which orb in the group triggered that change, and at what millisecond that change was triggered (which I called `startTime`). I could then use that information to loop through each orb in the group, and given the index of the orb that triggered the change (`activeOrb`), calculate each orb's `startTime` based on the distance of that orb from the `activeOrb`. I would then set the orb's brightness if the time was past that orb's `startTime`.

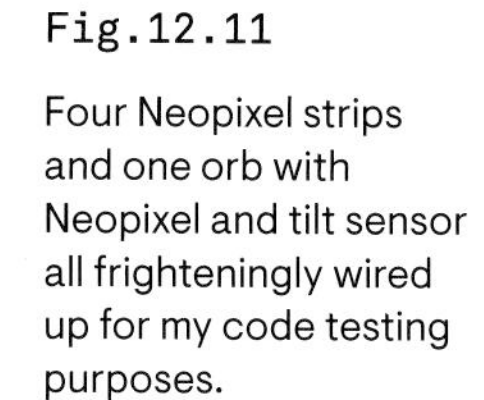

Fig.12.11

Four Neopixel strips and one orb with Neopixel and tilt sensor all frighteningly wired up for my code testing purposes.

I also included a fade animation by storing a starting and ending brightness for each orb, which made the change in lighting much gentler. Finally, I added in code for a pulsating introductory orb to catch visitor's attention and explain the installation.

Coding in Arduino was an interesting challenge in trying to make sure I was being memory efficient, especially when I had to add in groups and remember milliseconds (which were stored as unsigned longs). It definitely made my code more complex, but I had a really fun time trying to figure out the challenge.

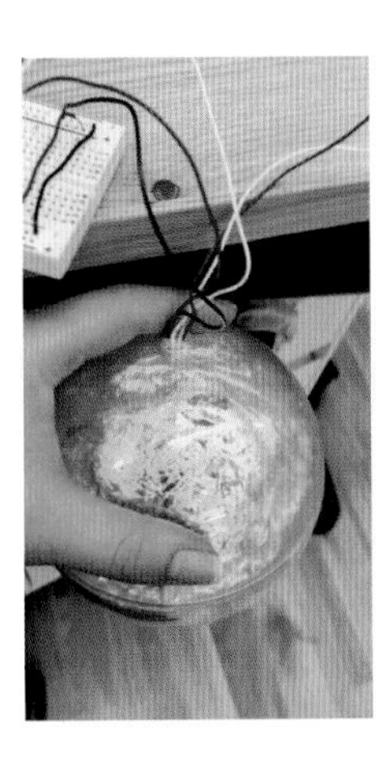
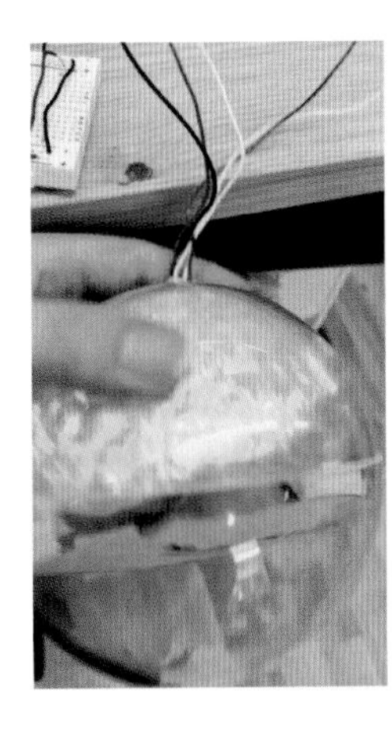
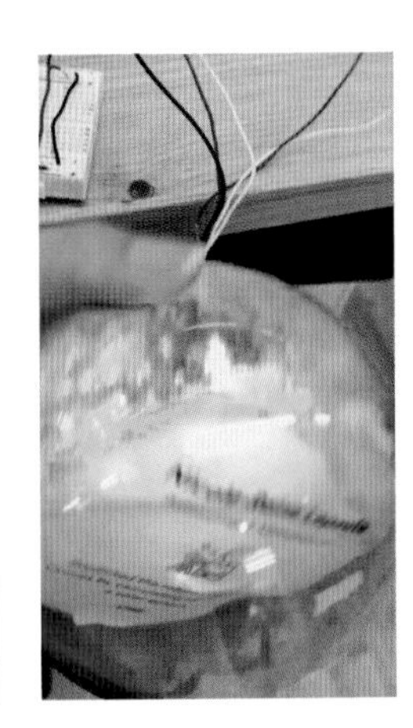

Fig.12.12

Lifting up the orb makes it shine at full brightness, and putting it back down leaves it slightly brighter than before. I got this interaction working around 1 AM after a few late nights and I was so deliriously happy.

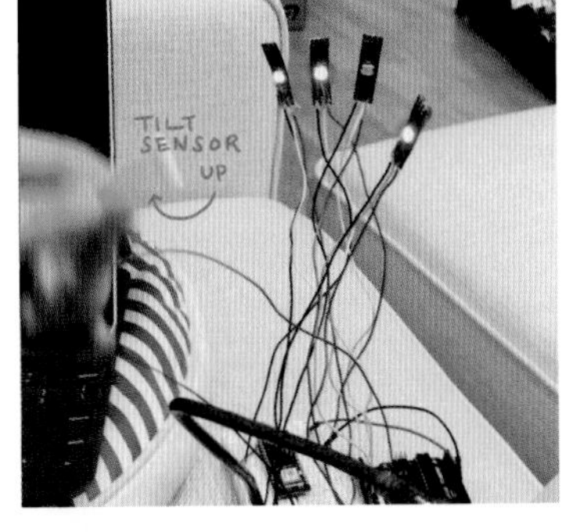

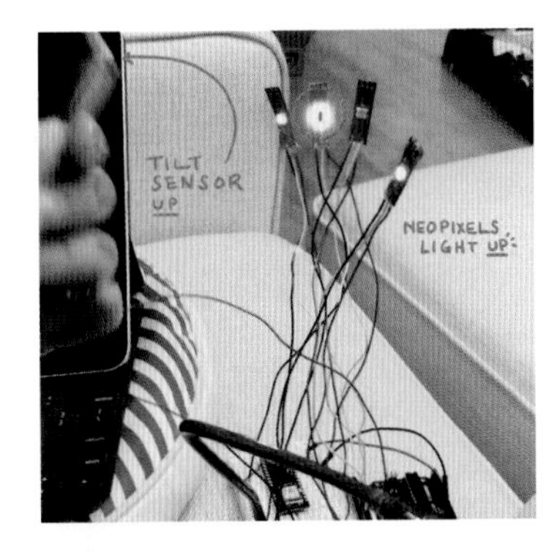

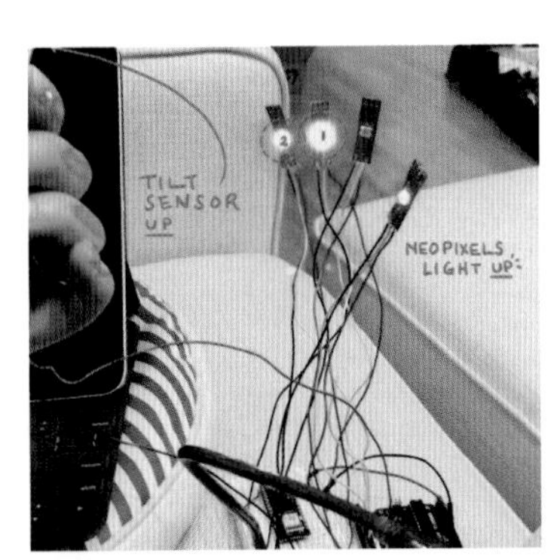

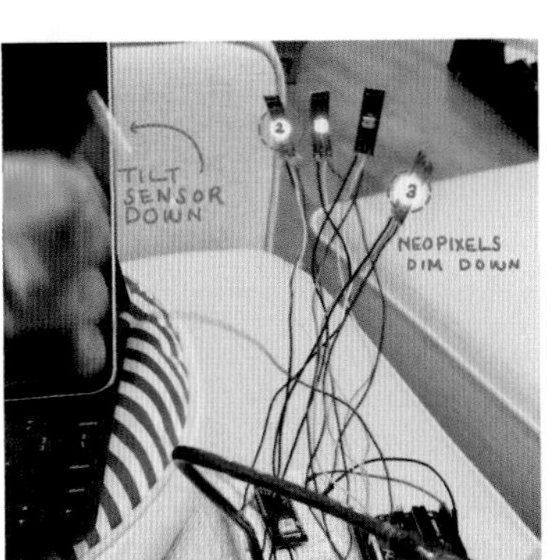

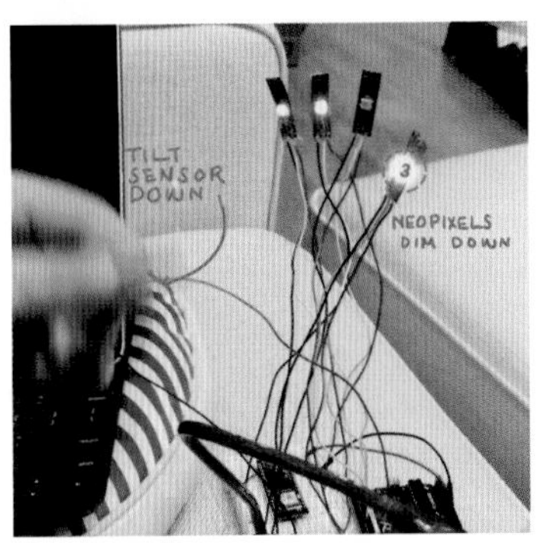

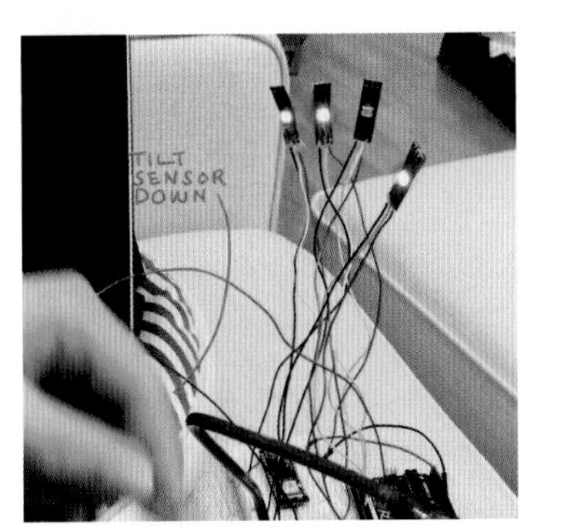

Fig.12.13

The Neopixels lighting up with a staggered delay when the tilt sensor in my hand is lifted, and dimming down when the tilt sensor is down.

Production

One of the first things that Tina did to kick-start the production phase was to 3D model an enclosure to hold both the tilt sensor (which is very sensitive to the angle of its placement) and Neopixel in place (Figure 12.15, top left). We also received a brilliant suggestion to use a phone jack to connect our orbs to our wires, so that we could make the orbs easily detachable (Figure 12.15, bottom left). That way if an orb malfunctioned, we could easily separate it and debug it away from the rest of the installation. Tina incorporated the phone jack into her enclosure design, and together with Christina, 3D printed seventeen of them and spent roughly 30 hours soldering them all.

At this point, I had to return home (and then to a conference in Tokyo), so I wasn't able to help through production. I highly recommend reading both Tina's[4] and Christina's[5] write-ups for more details on this section.

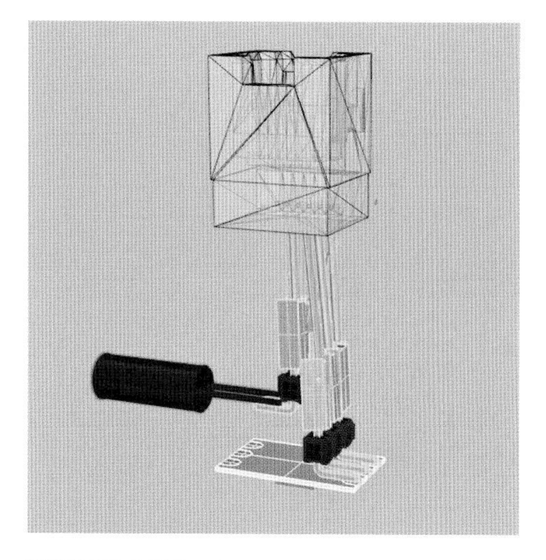

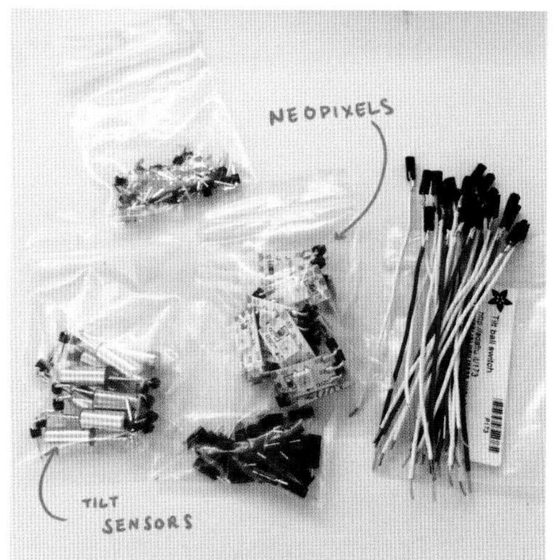

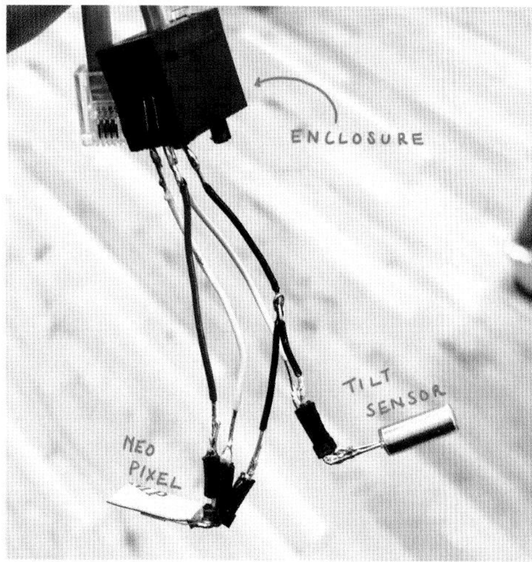

Fig.12.14

3D model of the enclosure (top left), the soldered parts (top right), tilt sensor and Neopixel soldered to a phone jack within the 3D printed enclosure (bottom left), and the 3D printed enclosure inside an orb (bottom right).

[4] Tina Rungsawang, One Amongst Many: https://tina.pizza/one-amongst-many
[5] Christina Dacanay, One Amongst Many—Connecting Women in Computing: http://www.cdacanay.com/itp-blog/2019/12/23/one-amongst-many-connecting-womxn-in-computing

To hang the orbs from the ceiling, they mapped each woman's position data to long planks of wood and drilled holes to run the wires through (Figure 12.18). The wires were then plugged into an Arduino Mega, which was powered by an iPhone USB power adapter connected to a wall plug.

They pre-labeled the wires to make sure the correct orbs were plugged into the correct Arduino pins.

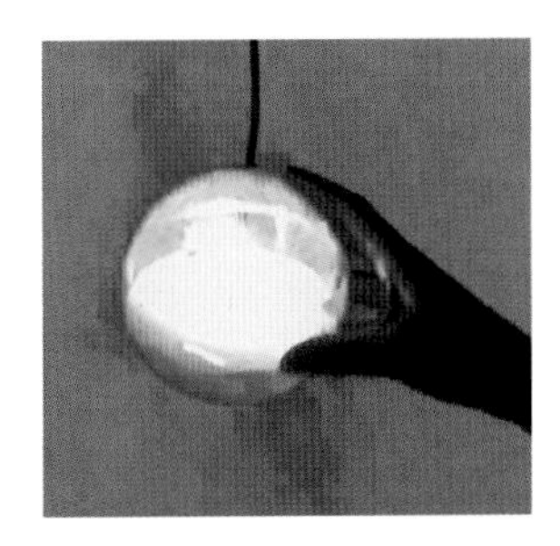

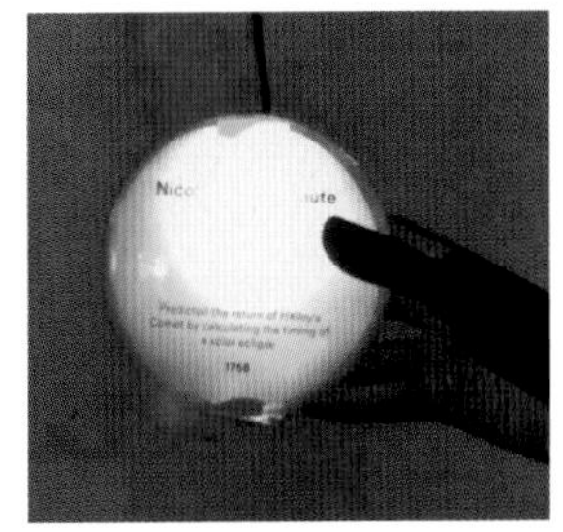

Fig.12.15

A prototype of the orb wired up and working as a stand-alone piece!

Fig.12.16

Each woman's information laser-etched into acrylic.

Fig.12.17

Position data mapped to each plank of wood, the wires taped to the wood and Christina plugging the wires into the Arduino Mega.

Reflections

It was absolutely incredible to see this whole project come together. We received great feedback at the Winter Show, where more than 200 people experienced our installation. Christina and Tina told me stories of young women looking inspired, of little girls giddily sticking gems on our summary board, but my favorite story is that of an older woman who had spent her entire life working in computing. As she went through our installation, she became emotional with the thought that someday, future generations might come to highlight and recognize her contributions as well.

I'm so happy that we created such meaningful work that has resonated with so many people, and I'm really grateful for Tina and Christina, the best teammates I could have pulled off this ambitious project with. I'm also grateful to the teachers, residents, and students who graciously offered their time and wisdom; I don't think we could have done so much anywhere else except at ITP.

I'm also really happy and grateful that I get to wrap-up *Data Sketches* with this project. It encapsulates all of my personal and professional growth from the last three years; I've grown from someone who used to only care about the code, to someone who also values design and the end experience. I started freelancing with the hope of learning the kind of challenges I want to work on and the problems I want to solve. I've found them. I want to tell human-centric, data-driven, and personally meaningful stories. I've been able to expand the formats in which I do this, from 2D to 3D, from digital to physical. As I finish *Data Sketches* and enter a new decade, I'm excited to move onto the next phase of my career and to boldly, unapologetically, courageously pursue my biggest dreams.

Thank you, *Data Sketches*, for all of that courage.

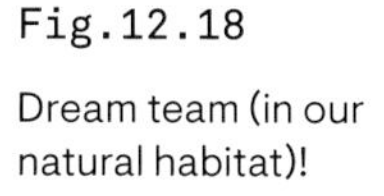

Fig.12.18

Dream team (in our natural habitat)!

Fig.12.19

My friend Alice gave me this card before I left for ITP and it perfectly encapsulates my feels as I wrap up *Data Sketches* and move onto the next phase of my career.

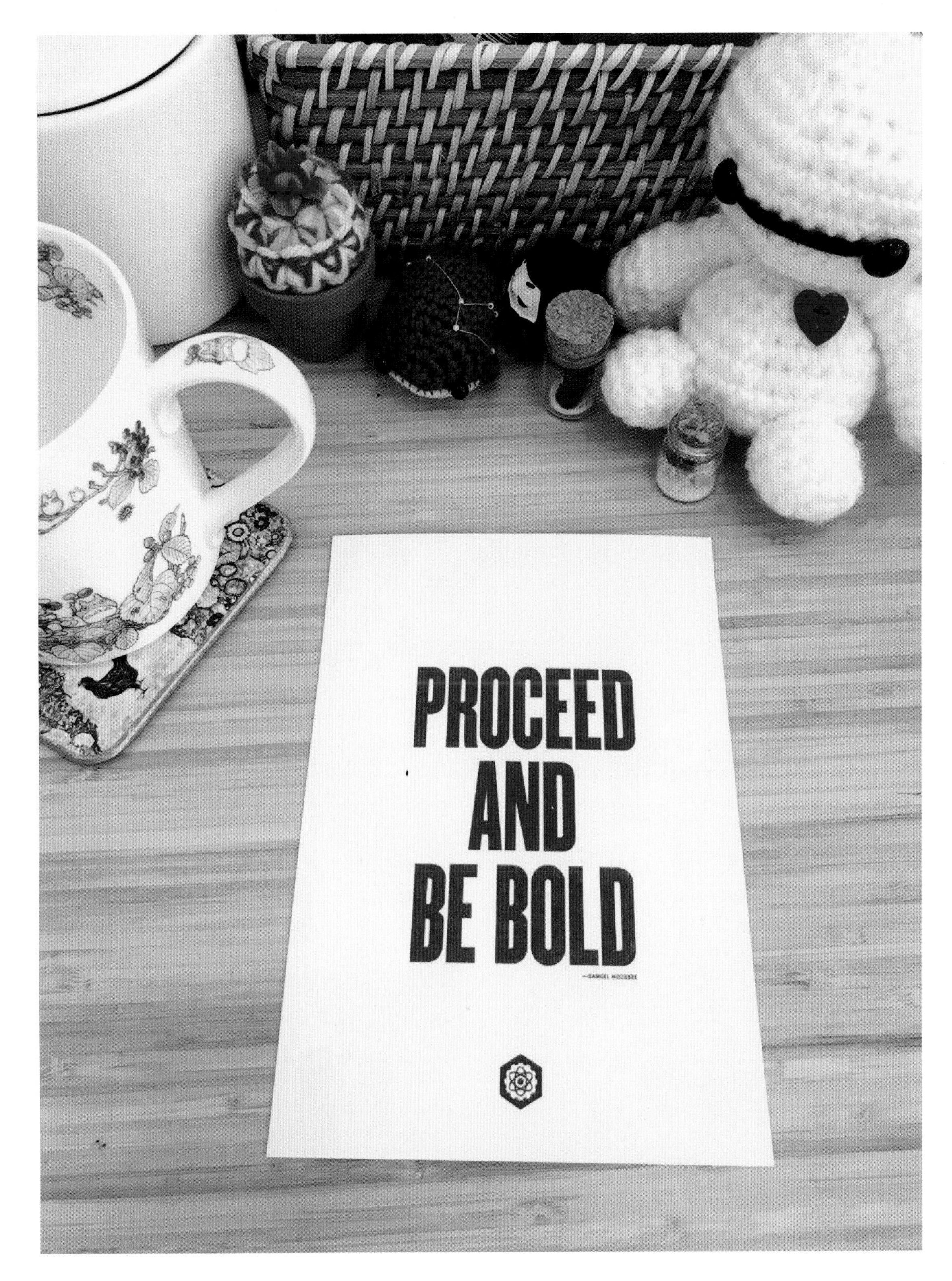

One Amongst Many

↳ oneamongstmany.com

`Fig.12.23`

Summary board where people could place a gem in the categories (mathematicians, computer scientists, creatives, and other) they identify most with.

`Fig.12.21`

A timelapse of the Winter Show. More than 200 people came through our installation!

`Fig.12.24`

The major downside of a physical installation is that people outside of that area can't experience it. Because we wanted people around the world to learn about these women in computing, I created a digital counterpart with `D3.js`, `Three.js`, `Greensock`, and `Vue.js`.

The lessons are grouped into the sections in which they appear in the chapters. However, a few lessons appear in multiple chapters. These have been placed in the section most appropriate.

Data

Sketch

Code

THE STARTING AND THE FINISHING

Interview with Series Editor Tamara Munzner

In this interview, we sit down with Tamara Munzner, professor of computer science at the University of British Columbia and editor for this series of visualization books, to discuss how we came up with the idea for *Data Sketches*, the "bookification" process, and what we plan to do when it's finally printed.

This interview has been edited for brevity and clarity.

Tamara Munzner: This has been a long process, but you're almost done. What do you know now that you didn't know before?

Nadieh Bremer: I learned a lot of new techniques, such as working with WebGL, getting a better grip on canvas, and figuring out how to visually render bigger datasets on the screen. I also learned how to go from making singular and smaller visuals to making an entire dataviz-driven story. This taught me that it's more than just the visuals and the layout; you must be able to design for multiple visuals and make them flow together. That's what I took away from *Data Sketches* in terms of the technical part. I also learned how extremely valuable it is to have a teammate that you can go through this process with. I didn't have to do everything on my own, and knowing that Shirley was there for me helped immensely.

Shirley Wu: I don't know if I can say that any better than Nadieh did. It was so valuable to have someone to bounce my ideas off of and to discuss visualizations with. We also went through this whole freelance journey together, learning together. From a personal perspective, for *Data Sketches*, the biggest thing for me is that I've really expanded my skills beyond just the technical. When I went into *Data Sketches*, I had about four years working as a software engineer on the front end. I knew I was strong with `D3.js` and would be able to create visualizations in SVG or canvas. But when I first started, I didn't know what I didn't know. I didn't fully understand how important the design component would be. I didn't know how much there was in terms of data analysis. I also didn't know what it took to tell a compelling and easily navigable story. All of this is so important to me now, but when I first started with *Data Sketches*, I often approached it with a focus on fun and less of a focus on the end user. But now I recognize that the end user is extremely important, that their ability to read and understand the whole visualization is critical and I should not do something just for the technical fun of it. I think that's my biggest learning lesson, coming in as an engineer but then refocusing to place more emphasis on design.

TM: What fantastic answers. You learn so much when writing a book, and what, if anything, did you learn from the process of bookification?

SW: Writing is hard! I don't know about Nadieh, but I know I really struggled with the bookification part. *Data Sketches* started out as something we were just doing for fun, so I never felt the obligation to have the write-ups make 100% sense. In fact, sometimes I would just dump all the screenshots, add some words in between and that was my write-up. I think that was still helpful for people because they got to see all of those notebook sketches and all of the different iterations of my code. But for the bookification part itself, the biggest struggle I had was having to really think through and explain my process; not just show it, but also explain it. I had to help people understand my thought process and explain why I did certain things—which is often, honestly, for no apparent reason other than simply trying it out. So, I was really challenged by turning all my write-ups into applicable learning lessons.

NB: For me in general, the bookification and the writing of the chapters themselves, wasn't too difficult. I don't mind writing. However, I have a tendency to want to explain everything in lists: "And then I did A and then I did B and then I did C," and my write-ups were a lot like that. So as the projects became bigger, the write-ups became longer and longer and longer. For example, my first write-up was about nine pages in the book manuscript, and my final one started out at 25! This process taught me how to downsize what information I give to a reader and to present it more succinctly. This, in turn, helped me get a better grip on what is interesting and, conversely, what is just rambling. I learned that it's not easy and it was a little bit painful to have to downsize but I'm thrilled with the result and I know people will enjoy the writing better. I didn't want to be responsible for making this a 600-page book, so making it more concise was critical for me. I wanted to focus on the actual story without going on too many tangents and potentially losing the reader. The most painful part for me, overall, was mostly the fact that this process has been a long one. This book was always in the back of my mind and that can get kind of stressful, so I really wanted to have it finished.

TM: Fair enough. I think the process of creating a book is always a mix of the logistic and technical, and then just the emotional part of doing the thing and making it be done. It is not at all easy. Can you say a little more about the mindset you had when you started the project? Why did you decide this would be a good thing to do?

NB: Well, it seemed like a fun way to make new projects. We started *Data Sketches* because we felt as if we hadn't created enough actual projects in the previous months. We felt like having this collaboration and forcing each other to work on it would be a great way to consistently create new things—and with a high level of fun woven into it. I did not expect it to become the thing that it became.

SW: Yeah, definitely. When we started this project, we were chatting back and forth on Slack. It's not even that we were like, "Okay, we're going to do this super serious, big project. It's going to be a big deal." We were actually saying "Hey, do you want to do something together?" We were very casual about the format, too, and we figured we could just pass code back and forth and maybe it could be a 12-month thing with 12 projects. I didn't even think about how big that would be in terms of time commitment. We came up with the format in an hour, just shooting ideas back and forth, and we spent much more time on the name. We literally took two days just brainstorming the name.

TM: Do you have any advice for people who might want to do something similar?

NB: My advice is to thoroughly understand the project before you start and consider placing constraints on how you approach your work. For example, if we'd placed one, saying that whatever we made after 20 hours, we'd use, it may have been a little easier. You don't want to constrain your creativity, of course, but sometimes constraints can actually make you more creative. I also recommend doing it with somebody else. Working with Shirley helped with this process a lot and I think the partnership works really well.

SW: I appreciated that a lot too. Basically, between the two of us, I prefer to look more into the future and plan for the future and Nadieh is so good at making sure that what needs to be done in the present is done in the present. It's a very good balance.

TM: I've got to say, I think it was a match made in heaven. You were both able to push each other towards doing things you might not have done individually.

SW: Yeah, definitely.

TM: Both for the starting and for the finishing, which are two excellent verbs. What are you going to do the day you physically pick up a book that is your book?

SW: Cry in joy. Cry with joy. It's done!

NB: Yeah, happy dance!

SW: Just hug it, cry, tears, dance around, and then go on with life.

Even though this is the end of our *Data Sketches* "Awesome Collaboration Marathon," we intend to keep working on projects that excite us, together or individually. Follow our latest progress on Twitter and Instagram at @sxywu and @NadiehBremer, and on our websites at shirleywu.studio and VisualCinnamon.com!

Acknowledgments

No book (or large-scale project) happens in a vacuum. We want to thank our loved ones, our friends, and all of the friendly strangers on the Internet that have supported us through this journey and gotten us to this point.

To Alberto Cairo, for giving us the idea of creating a book and all of the encouragement, feedback, and mentorship he's given us; Tamara Munzner, for never hesitating to champion us, including us in her data visualization book series, and encouraging us every time we needed that extra push or piece of advice to keep going.

To our editors Stephanie Morillo and Tianna Mañón, for making sure our rambles and tangents were turned into beautiful, coherent sentences; Kirsten Barr, Elliott Morsia, Sunil Nair, and the staff of CRC Press who have guided us throughout the publishing process.

To the team at Praline who turned our written manuscript and images into this wonderfully designed book weaving together the images and text in such a way we were once afraid was impossible.

To Alice Lee, the creator of our gorgeous cover, who had endless patience and listened to every one of our exceptionally specific requests.

To the many people who have looked at our works in progress to provide thoughtful feedback, gave tips to solve particularly gnarly coding problems, or replied to our Twitter calls for help. In particular, we wouldn't have reached the same level of sophistication in our projects without the help of RJ Andrews, Taylor Baldwin, Tony Chu, Erik Cunningham, Matt Daniels, Matt DesLauriers, Sarah Drasner, Russell Goldenberg, Ian Johnson, Isaac Kelly, Susie Lu, Elijah Meeks, Philippe Rivière, and Sam Saccone.

For giving feedback on our final manuscript to polish it as much as we could: Zan Armstrong, Peter Beshai, Mike Brondbjerg, Chen Hui Jing, Jen Christiansen, Michael K Freeman, Andy Kirk, Maarten Lambrechts, John Burn-Murdoch, Mollie Pettit, Visnu Pitiyanuvath, Jane Pong, Lisa Charlotte Rost, Sara Sprinkhuizen, Alli Torban, Jan Willem Tulp, Jim Vallandingham, Jonni Walker, and Amelia Wattenberger.

To Sonja Kuijpers and Nicholas Rougeux for joining us as guests for two of the topics in the online version of *Data Sketches*, and experiencing the stress of creating an elaborate data visualization from scratch with us.

From Nadieh

My biggest hug to Ralph, my highschool sweetheart for 16 years and counting now, for understanding that my visuals needed many evenings of sitting behind my laptop and still making sure that I remembered to drink, eat, and go to bed on time. To Snookie, my weird laser-pointer-obsessed cat, who has been my cuddle support throughout it all, sleeping on her pillow next to me on my desk while I was drawing, working, and writing on each project.

And thank you so, so very much, Shirley, for taking me along on this amazing journey which I'd never dared to have done on my own.

Liefs, Nadieh ʕつ•ᴥ•ʔつ

From Shirley

To Alex, who's been my boyfriend, fiancé, and now husband through Data Sketches and the creation of this book, who's stayed with me and cooked delicious food for me as motivation—even through all of my whining about this book; my (Asian) parents, who really don't understand my career choices but are (silently) proud of me nonetheless; and my (Asian-American) sister, who frequently tells me how proud of me she is to make up for how much our parents rarely voice it.

And, finally, a huge thank you to Nadieh without whom I don't think I could have ever finished this gigantic, monumental, ambitious project.

(ノ^ワ^)ノ*:･ﾟ✧! ♡ Shirley

LU

The different ways Nadieh/Shirley/Bremer/Wu would combine into one person, as imagined by our friend Susie Lu.